Lecture Notes in Computer Science 5917

Commenced Publication in 1973
Founding and Former Series Editors:
Gerhard Goos, Juris Hartmanis, and Jan van Leeuwen

Jianer Chen   Fedor V. Fomin (Eds.)

Parameterized and Exact Computation

4th International Workshop, IWPEC 2009
Copenhagen, Denmark, September 10-11, 2009
Revised Selected Papers

 Springer

Volume Editors

Jianer Chen
Texas A&M University
Department of Computer Science and Engineering
College Station, Texas 77843, USA
E-mail: chen@cse.tamu.edu

Fedor V. Fomin
Universitetet i Bergen
Institutt for informatikk
Postboks 7803
5020 Bergen, Norway
E-mail: Fomin@ii.uib.no

Library of Congress Control Number: 2009941300

CR Subject Classification (1998): B.2.4, F.2, G.1, G.2, G.4, I.1, E.1, I.2.8

LNCS Sublibrary: SL 1 – Theoretical Computer Science and General Issues

ISSN   0302-9743
ISBN-10   3-642-11268-4 Springer Berlin Heidelberg New York
ISBN-13   978-3-642-11268-3 Springer Berlin Heidelberg New York

springer.com

Typesetting: Camera-ready by author, data conversion by Scientific Publishing Services, Chennai, India
Printed on acid-free paper   SPIN: 12823598   06/3180   5 4 3 2 1 0

Preface

The Workshop on Parameterized and Exact Computation (IWPEC) is an international workshop series that covers research in all aspects of parameterized and exact algorithms and complexity, and especially encourages the study of parameterized and exact computations for real-world applications and algorithmic engineering. The goal of the workshop is to present recent research results, including significant work-in-progress, and to identify and explore directions for future research.

IWPEC 2009 was the fourth workshop in the series, held in Copenhagen, Denmark, during September 10-11, 2009. The workshop was part of ALGO 2009, which also hosted the 17th European Symposium on Algorithms (ESA 2009), the 9th Workshop on Algorithmic Approaches for Transportation Modelling, Optimization, and Systems (ATMOS 2009), and the 7th Workshop on Approximation and Online Algorithms (WAOA 2009). Three previous meetings of the IWPEC series were held in Bergen, Norway, 2004, Zürich, Switzerland, 2006, and Victoria, Canada, 2008.

At IWPEC 2009, we had two plenary speakers, Noga Alon (Tel Aviv University, Israel) and Hans Bodlaender (Utrecht University, The Netherlands), giving 50-minute talks each. Professor Alon spoke on "Color Coding, Balanced Hashing and Approximate Counting, " and Professor Bodlaender on "Kernelization: New Upper and Lower Bound Techniques." Their respective abstracts accompanying the talks are included in these proceedings.

In response to the Call for Papers, 52 papers were submitted. Each submission was reviewed by at least three reviewers (most by at least four). The reviewers were either Program Committee members or invited external reviewers. The Program Committee held electronic meetings using the EasyChair system, went through thorough discussions, and selected 25 of the submissions for presentation at the workshop and inclusion in this LNCS volume.

We are very grateful to the Program Committee, and the external reviewers they called on, for the hard work and expertise which they brought to the difficult selection process. We also wish to thank all the authors who submitted their work for our consideration. Special thanks go to Thore Husfeldt for the local organization of the ALGO 2009 conference in Copenhagen.

Finally, we would like to thank the members of the Editorial Board of *Lecture Notes in Computer Science* and the Editors at Springer for their encouragement and cooperation throughout the preparation of these proceedings.

November 2009

Jianer Chen
Fedor Fomin

Organization

IWPEC 2009 Program Committee

Faisal Abu-Khzam	Beirut, Lebanon
Hans Bodlaender	Utrecht, The Netherlands
Jonathan Buss	Waterloo, Canada
Leizhen Cai	Hong Kong, China
Liming Cai	Georgia, USA
Jianer Chen	College Station, USA (Co-chair)
Michael Fellows	Newcastle, Australia
Henning Fernau	Trier, Germany
Jörg Flum	Freiburg, Germany
Fedor Fomin	Bergen, Norway, (Co-chair)
Jiong Guo	Jena, Germany
Edward A. Hirsch	St. Petersburg, Russia
Thore Husfeldt	Copenhagen, Denmark
Iyad Kanj	Chicago, USA
Dániel Marx	Budapest, Hungary
Catherine McCartin	Wellington, New Zealand
Igor Razgon	Cork, Ireland
Saket Saurabh	Bergen, Norway
Uwe Schöning	Ulm, Germany
Stefan Szeider	Durham, UK
Dimitrios Thilikos	Athens, Greece
Mark Weyer	Berlin, Germany

IWPEC Steering Committee

Jianer Chen	College Station, USA
Frank Dehne	Ottawa, Canada
Rod Downey	Wellington, New Zealand
Michael Fellows	Newcastle, Australia
Mike Langston	Knoxville, USA
Rolf Niedermeier	Jena, Germany
Venkatesh Raman	Chennai, India

External Reviewers

Adler, Isolde	Lampis, Michael
Antipov, Dmitry	Liedloff, Mathieu
Betzler, Nadja	Liers, Frauke
Chandrasekaran, Venkat	Liu, Yang
Chen, Xi	Lokshtanov, Daniel
Chen, Yijia	Mathieson, Luke
Daligault, Jean	Meister, Daniel
Dantchev, Stefan	Mnich, Matthias
Dom, Michael	Moser, Hannes
Dorn, Frederic	Mouawad, Amer
Gao, Yong	Müller, Moritz
Gaspers, Serge	Naeher, Stefan
Giannopoulou, Archontia	Nederlof, Jesper
Gravin, Nikolai	Ordyniak, Sebastian
Gutner, Shai	Paturi, Ramamohan
Hoang, Thanh Minh	Pelsmajer, Michael
Hoogeveen, Han	Penninkx, Eelko
Huang, Xiuzhen	Philip, Geevarghese
Itsykson, Dmitry	Raible, Daniel
Jansen, Bart	Rossmanith, Peter
Kaminski, Marcin	Schaefer, Marcus
Kanté, Mamadou Moustapha	Serna, Maria
Kim, Eun Jung	Shareghi, Pooya
Knauer, Christian	Todinca, Ioan
Kojevnikov, Arist	Villanger, Yngve
Kratochvil, Jan	Wagner, Fabian
Kratsch, Dieter	Woltran, Stefan
Kreutzer, Stephan	Xia, Ge
Krokhin, Andrei	Yang, Lin
Kuegel, Adrian	Zhang, Fenghui
Kulikov, Alexander	van Rooij, Johan M.M.

Table of Contents

Balanced Hashing, Color Coding and Approximate Counting

Noga Alon[1,*] and Shai Gutner[2,**]

[1] Schools of Mathematics and Computer Science, Tel-Aviv University,
Tel-Aviv, 69978, Israel and IAS, Princeton, NJ, 08540, USA
nogaa@tau.ac.il
[2] School of Computer Science, Tel-Aviv University, Tel-Aviv, 69978, Israel
gutner@tau.ac.il

Abstract. Color Coding is an algorithmic technique for deciding efficiently if a given input graph contains a path of a given length (or another small subgraph of constant tree-width). Applications of the method in computational biology motivate the study of similar algorithms for counting the number of copies of a given subgraph. While it is unlikely that exact counting of this type can be performed efficiently, as the problem is #$W[1]$-complete even for paths, approximate counting is possible, and leads to the investigation of an intriguing variant of families of perfect hash functions. A family of functions from $[n]$ to $[k]$ is an (ϵ, k)-balanced family of hash functions, if there exists a positive T so that for every $K \subset [n]$ of size $|K| = k$, the number of functions in the family that are one-to-one on K is between $(1 - \epsilon)T$ and $(1+\epsilon)T$. The family is perfectly k-balanced if it is $(0, k)$-balanced.

We show that every such perfectly k-balanced family is of size at least $c(k)n^{\lfloor k/2 \rfloor}$, and that for every $\epsilon > \frac{1}{poly(k)}$ there are explicit constructions of (ϵ, k)-balanced families of hash functions from $[n]$ to $[k]$ of size $e^{(1+o(1))k} \log n$. This is tight up to the $o(1)$-term in the exponent, and supplies deterministic polynomial time algorithms for approximately counting the number of paths or cycles of a specified length k (or copies of any graph H with k vertices and bounded tree-width) in a given input graph of size n, up to relative error ϵ, for all $k \leq O(\log n)$.

Keywords: Approximate counting of subgraphs, color-coding, derandomization, expanders, perfect hashing, k-wise. independence.

1 Introduction

1.1 Motivation and Background

Color Coding is an algorithmic technique for deciding efficiently if a given input graph contains a path or a cycle of a given length, or any other prescribed

---

* Research supported in part by an ERC Advanced grant, by a USA-Israel BSF grant, by NSF grant CCF 0832797 and by the Ambrose Monell Foundation.
** This paper forms part of a Ph.D. thesis written by the author under the supervision of Prof. N. Alon and Prof. Y. Azar in Tel Aviv University.

subgraph of bounded tree-width. Focusing, for simplicity, on paths, the method supplies a deterministic algorithm for deciding, in time $2^{O(k)}|E|\log|V|$, whether or not a given input (directed or undirected) graph $G = (V, E)$ contains a (simple) path on k vertices. The basic approach, introduced in [8], is very simple. One first gives a randomized algorithm, and then converts it into a deterministic one. The randomized algorithm works by first coloring the vertices of G randomly by k colors. Call a path on k vertices (a k-path, for short) colorful if its vertices get all the distinct k colors. It is not difficult to check in time $O(k2^k|E|)$, using dynamic programming, if there is a colorful path. As the probability of a k-path to become colorful in a random coloring is $k!/k^k > e^{-k}$, repeating the above procedure some Ce^k times provides a randomized algorithm in which the probability not to find a path in case one exists is smaller than e^{-C}. The crucial point in the derandomization of this algorithm is the observation that known constructions of families of hash functions given by [22] following [14], supply an explicit family of $2^{O(k)}\log|V|$ colorings of the vertices of G by k colors, so that the members of every set of k vertices get distinct colors in at least one of the colorings. Thus one can simply run the dynamic programming algorithm for each of these colorings, getting a deterministic algorithm for the problem.

The above technique has found several recent applications in computational biology (see [23], [24], [25], [17]), where it has been applied for detecting signaling pathways in protein interaction networks. These applications suggest the problem of counting, or approximating the number of k-paths (or other graphs of bounded tree-width) in a given graph. As using dynamic programming it is easy to count precisely the number of colorful k-paths in a given graph with colored vertices, the existence of efficient randomized approximation algorithms for counting follows quite easily by following the same approach; this is done in [2].

In order to derandomize the randomized counting (or approximate counting) procedures, one needs a strengthening of the usual notion of hash functions. This is given in the following definition.

A family of functions from $[n]$ to $[\ell]$ is an (ϵ, k)-balanced family of hash functions, if for every $S \subset [n]$, $|S| = k$, the number of functions that are one-to-one on S is between $(1 - \epsilon)T$ and $(1 + \epsilon)T$ for some constant $T > 0$. The family is perfectly k-balanced if it is $(0, k)$-balanced, that is, it is (ϵ, k)-balanced for $\epsilon = 0$.

Note that with a perfectly k-balanced family one can count precisely the number of k-paths in a graph on n vertices: we simply count, by dynamic programming, the number of colorful k-paths for each of the functions (considered as a coloring of the vertices), sum the results and divide by T. Similarly, an (ϵ, k)-balanced family will enable us to approximate the number of paths up to a relative error of ϵ. This suggests the study of the smallest possible size of such families, and the problem of constructing explicitly such families.

1.2   Related Work

The problem of counting paths and cycles in graphs has been considered by various researchers. In [9] the authors describe an $O(|V|^\omega)$ algorithm for counting the number of cycles of size at most 7, where $\omega < 2.38$ is the exponent in fast matrix multiplication. The method of this paper does not extend to longer paths, and indeed Flum and Grohe [13] proved that the problem of counting *exactly* the number of paths and cycles of length k in both directed and undirected graphs, considered as a problem parameterized by k, is $\#W[1]$-complete. This implies that it is unlikely that there is an $f(k) \cdot n^c$-time algorithm for counting the precise number of paths or cycles of length k in a graph of size n for any computable function $f : \mathbb{N} \to \mathbb{N}$ and constant c. The best known algorithms for computing exactly the number of k-paths in an n vertex graph run in time $n^{k/2+O(1)}$, see [11], [26].

However, the problem of approximating these numbers is more tractable. Arvind and Raman (see [10]) obtained a *randomized* fixed-parameter tractable algorithm to approximately count the number of copies of k-paths (or any fixed subgraph with bounded tree-width) within a large graph. A similar approximation appears in [2].

In an earlier paper [4] we considered *deterministic* approximation counting algorithms for this problem. To this end, we introduced the notion of (ϵ, k)-balanced families of hash functions and used them to exhibit a deterministic polynomial time algorithm for approximating the number of paths of length k up to any $k \leq O(\frac{\log n}{\log \log \log n})$ in a graph with n vertices. This was done by constructing explicitly (ϵ, k)-balanced families from $[n]$ to $[k]$, where the size of the family is $2^{O(k \log \log k)} \log n$ and the time for construction is $2^{O(k \log \log k)} n \log n$. The main open problem raised in [4] is to find such a construction of size $2^{O(k)} \log n$ (in time $2^{O(k)} n^{O(1)}$), which is optimal, even for standard (non-balanced) families of hash functions, and will supply polynomial time deterministic approximation algorithms for counting the number of paths of length k in a given graph of size n, for all $k \leq O(\log n)$. This problem is settled in the present paper.

1.3   The New Results

The results of Flum and Grohe mentioned above suggest that there is no perfectly k-balanced family of hash functions from $[n]$ to $[k]$ of size $f(k)n^{O(1)}$. We prove a stronger result, showing that every perfectly k-balanced family of hash functions from $[n]$ to $[\ell]$ is of size at least $c(k, \ell)n^{\lfloor k/2 \rfloor}$, where $c(k, \ell)$ is a positive constant depending only on k and ℓ. We also observe that this is not far from being tight, as for every $n > k$ there is a perfectly k-balanced family of functions from $[n]$ to $[k]$ of size $\binom{n}{k-1}$. This shows that the Color Coding approach cannot supply an algorithm for counting k-paths in an n vertex graph in time $o(n^{\lfloor k/2 \rfloor})$.

Our main positive result is an explicit construction, for every $\frac{1}{poly(k)} < \epsilon \leq 1$, of an (ϵ, k)-balanced family of hash functions from $[n]$ to $[k]$ of size $e^{k+O(\log^3 k)} \log n$. The running time of the procedure that provides the construction is $e^{k+O(\log^3 k)} n \log n$. Note that the size of the family is optimal up to the error term $O(\log^3 k)$ in the exponent, as there is a known lower bound of $\Omega(e^k \log n/\sqrt{k})$ for the size of any family of hash functions from $[n]$ to $[k]$, (even if it is not balanced and the only requirement is that every set of size $[k]$ is mapped in a one-to-one fashion at least once).

This supplies deterministic approximation algorithms for counting the number of simple k-paths in a graph $G = (V, E)$ up to a relative error of $\epsilon = \frac{1}{poly(k)}$ in time $2^{O(k)}|E| \log |V|$. Similar results hold for counting approximately the number of copies of any graph of size k with constant tree-width. Note that this is polynomial for all $k \leq O(\log n)$, and it is unlikely that one can do better, as this would imply the existence of a $2^{o(n)}$-time algorithm for the Hamilton path problem, contradicting the Exponential Time Hypothesis of [18,19].

1.4   Methods and Organization

Our lower bound for the size of perfectly balanced families are proved by Linear Algebra tools, combining the basic approach of [1] in the proof of the lower bound for the size of sample spaces supporting k-wise independent random variables with two additional ideas.

The construction of (ϵ, k)-balanced families combines several ingredients. Two of them are rather standard and are based on nearly pairwise independent random variables and on the method of conditional expectations. The third one is more challenging, and combines the approach of [21] with an iterative construction based on properties of expanders. It is convenient to apply here (some version of) the expanders of [5], though other expanders could have been used as well.

Since our main motivation is the application for the subgraph approximate counting problem using Color Coding, there is no reason to provide explicit constructions of (ϵ, k)-balanced families of functions which are more efficient than the time of writing these functions down, as anyway our counting algorithm will have to go through these functions. We thus describe the constructions in this way, without trying to describe separately which parts of them admit more efficient descriptions. It is worth noting, however, that the part of our construction which applies the method of conditional expectations indeed requires the time stated in its description.

The rest of this paper is organized as follows. In section 2 we describe the main ingredients of the construction: balanced families of hash functions and balanced splitters, a modified version of a notion introduced in [21]. Section 3 contains the results concerning perfectly balanced families of hash functions. The explicit construction of expanders presented in section 4 is used in section 5 for constructing small sample spaces supporting a certain relaxed version of nearly k-wise independent random variables. This is used to obtain a construction of what we call balanced (n, k, ℓ)-splitters, which is later applied in section 6 as a crucial ingredient in the construction of balanced families of hash functions. The constructions, together with the color coding technique, are used for designing algorithms for approximately counting the number of copies of subgraphs of bounded tree-width in given graphs. We conclude with some remarks and open problems.

2   The Ingredients of the Construction

In this section we formally define the notions of balanced families of hash functions and balanced splitters. For a positive integer n, denote by $[n]$ the set

$\{1, \ldots, n\}$. For any k, $1 \leq k \leq n$, the family of k-sized subsets of $[n]$ is denoted by $\binom{[n]}{k}$. As usual, $k \bmod \ell$ denotes the unique integer $0 \leq r < \ell$ so that $k = q\ell + r$, for some integer q.

Definition 1. *Suppose that $1 \leq k \leq \ell \leq n$ and $\epsilon \geq 0$. A family of functions from $[n]$ to $[\ell]$ is an (ϵ, k)-balanced family of hash functions if there exists a constant $T > 0$, such that for every $S \in \binom{[n]}{k}$, the number of functions that are one-to-one on S is between $(1 - \epsilon)T$ and $(1 + \epsilon)T$. The family is perfectly k-balanced if it is $(0, k)$-balanced.*

The following definition is motivated by a related notion defined and used in [21].

Definition 2. *Suppose that $1 \leq \ell < k \leq n$ and $\epsilon \geq 0$, and let H be a family of functions from $[n]$ to $[\ell]$. For a set $S \in \binom{[n]}{k}$, let $split_H(S)$ denote the number of functions $h \in H$ so that for every j, $1 \leq j \leq k \bmod \ell$, $|h^{-1}(j) \cap S| = \lceil k/\ell \rceil$, and for all $k \bmod \ell < j \leq \ell$, $|h^{-1}(j) \cap S| = \lfloor k/\ell \rfloor$. The family H is an ϵ-balanced (n, k, ℓ)-splitter if there exists a constant $T > 0$, such that for every $S \in \binom{[n]}{k}$, $(1 - \epsilon)T \leq split_H(S) \leq (1 + \epsilon)T$.*

Note that if ℓ divides k, then in the above definition $split_H(S)$ is the number of functions that split S into equal parts. The splitters of [21] differ from the ones defined here, just as usual families of hash functions differ from balanced families; in [21] it is only required that for every set S there will be some function in H splitting it evenly, while in our splitters each S should be divided evenly by roughly the same number of functions. The construction of balanced splitters is thus much harder than the one of splitters in [21], and is in fact the most challenging part in the explicit description of balanced families of hash functions.

Each function f in our explicit construction of balanced families of hash functions is the composition of members from three families. The first one is an (ϵ_1, k)-balanced family of hash functions from $[n]$ to $[q]$, where $q = \Theta(\frac{k^2}{\epsilon})$. The second one is an ϵ_2-balanced (q, k, ℓ)-splitter from $[q]$ to $[\ell]$, where $\ell = \Theta(\log k)$, and the last one is an $(\epsilon_3, k/\ell)$-balanced family of hash functions from $[q]$ to $[k/\ell]$ (for simplicity assume for now that ℓ divides k). In order to define f we actually need ℓ members of the third family, with each of them being applied to the elements mapped by the members of the second family to a single $j \in [\ell]$.

3   Perfectly Balanced Families

Let $n > \ell \geq k > 0$ be positive integers. Recall that a family $\mathcal{F}$ of functions from $[n]$ to $[\ell]$ is perfectly k-balanced, if there exists a number $T > 0$ so that for every set $K \subset [n]$ of size $|K| = k$, $|\{f \in \mathcal{F} : |f(K)| = k\}| = T$. In this section we show that the size of each such family must be at least $c(k, \ell)n^{\lfloor k/2 \rfloor}$, where $c(k, \ell)$ is a positive constant depending only on k and ℓ.

Theorem 1. *Let $\mathcal{F}$ be a perfectly k-balanced family of functions from $[n]$ to $[\ell]$, where*
$$n > \ell \geq k.$$

(i) If $k = 2r$ is even then

$$|\mathcal{F}| \geq \frac{\binom{n}{r}}{\binom{\ell}{r}\binom{\ell-r}{r}}.$$

(ii) If $k = 2r + 1$ is odd then

$$|\mathcal{F}| \geq \frac{\binom{n-1}{r}}{\binom{\ell-1}{r}\binom{\ell-r-1}{r}}.$$

(iii) If $\ell = k = 2$ then $|\mathcal{F}| \geq n - 1$, and equality can hold if and only if there is a Hadamard matrix of order n. Otherwise, the smallest possible size of $\mathcal{F}$ is precisely n.

Proof. (i) Let $\mathcal{F}$ be a perfectly $2r$-balanced family of functions from $[n]$ to $[\ell]$.

For each $R \subset [n]$ of size $|R| = r$, define two vectors u_R and w_R, each of length $|\mathcal{F}|\binom{\ell}{r}\binom{\ell-r}{r}$, whose coordinates are indexed by the set of all ordered triples (f, S_1, S_2), with

$$f \in \mathcal{F}, \quad S_1, S_2 \subset [\ell], \quad |S_1| = |S_2| = r, \quad \text{and} \quad S_1 \cap S_2 = \emptyset.$$

These vectors are defined as follows:

$$u_R(f, S_1, S_2) = 1 \quad \text{if} \quad f(R) = S_1, \quad \text{and} \quad u_R(f, S_1, S_2) = 0 \quad \text{otherwise.}$$

$$w_R(f, S_1, S_2) = 1 \quad \text{if} \quad f(R) = S_2, \quad \text{and} \quad w_R(f, S_1, S_2) = 0 \quad \text{otherwise.}$$

Note that the inner product of two such vectors u_{R_1} and w_{R_2} is zero if $R_1 \cap R_2 \neq \emptyset$. Indeed, in this case $f(R_1)$ must have a nonempty intersection with $f(R_2)$ for all $f \in \mathcal{F}$, and thus there is no coordinate (f, S_1, S_2) as above in which both v_{R_1} and w_{R_2} do not vanish. Similarly, if $R_1 \cap R_2 = \emptyset$, the inner product of u_{R_1} and w_{R_2} is precisely the number of functions $f \in \mathcal{F}$ which are one-to-one on $R_1 \cup R_2$. Indeed, for each such f there is a unique pair of disjoint sets S_1, S_2, each of size r, so that $f(R_1) = S_1$ and $f(R_2) = S_2$, while if f maps two elements of $R_1 \cup R_2$ to the same image, there is no such pair. Since $\mathcal{F}$ is a perfectly balanced $2r$-family, there exists a positive T so that for every disjoint R_1, R_2 as above, the inner product of u_{R_1} with w_{R_2} is T.

Let U be the $\binom{n}{r}$ by $|\mathcal{F}|\binom{\ell}{r}\binom{\ell-r}{r}$ matrix whose rows are all vectors u_R with $R \subset [n], |R| = r$, and let W be the matrix whose rows are all vectors w_R. By the above discussion, the product $U \cdot W^t = T \cdot DIS_{n,r}$, where $DIS_{n,r}$ is the disjointness matrix whose rows and columns are indexed by the r-subsets of $[n]$, defined by $DIS_{n,r}(R_1, R_2) = 1$ if $R_1 \cap R_2 = \emptyset$ and $DIS_{n,r}(R_1, R_2) = 0$ otherwise. It is well known (see, e.g., [20]) that the matrix $DIS_{n,r}$ is nonsingular (over the reals) for all $n \geq 2r$, and as this is the case here and T is nonzero, it follows that the rank of U is at least that of $U \cdot W^t$ which is $\binom{n}{r}$. As this rank is at most the number of columns of U, we conclude that

$$|\mathcal{F}|\binom{\ell}{r}\binom{\ell-r}{r} \geq \binom{n}{r},$$

completing the proof of part (i).

(ii) The proof is similar to that of part (i), with a few modifications. Here are the details. Let $\mathcal{F}$ be a perfectly $2r + 1$-balanced family of functions from $[n]$ to $[\ell]$.

For each $R \subset [n-1]$ of size $|R| = r$ define two vectors u_R and w_R, each of length $|\mathcal{F}|\binom{\ell-1}{r}\binom{\ell-r-1}{r}$, whose coordinates are indexed by the set of all ordered triples (f, S_1, S_2), satisfying

$$f \in \mathcal{F}, \; S_1, S_2 \subset [\ell] - \{f(n)\}, \; |S_1| = |S_2| = r, \; \text{and} \; S_1 \cap S_2 = \emptyset.$$

These vectors are defined as before:

$$u_R(f, S_1, S_2) = 1 \; \text{if} \; f(R) = S_1, \; \text{and} \; u_R(f, S_1, S_2) = 0 \; \text{otherwise.}$$

$$w_R(f, S_1, S_2) = 1 \; \text{if} \; f(R) = S_2, \; \text{and} \; w_R(f, S_1, S_2) = 0 \; \text{otherwise.}$$

It is clear that just as before, the inner product of two such vectors u_{R_1} and w_{R_2} is zero if $R_1 \cap R_2 \neq \emptyset$. Similarly, if $R_1 \cap R_2 = \emptyset$, the inner product of u_{R_1} and w_{R_2} is precisely the number of functions $f \in \mathcal{F}$ which are one-to-one on $R_1 \cup R_2 \cup \{n\}$. Indeed, for each such f there is a unique pair of disjoint subsets S_1, S_2 of $[\ell] - \{f(n)\}$, each of size r, so that $f(R_1) = S_1$ and $f(R_2) = S_2$, while if f does not map $R_1 \cup R_2 \cup \{n\}$ in a one-to-one manner, there is no such pair. As before, since $\mathcal{F}$ is a perfectly balanced $2r + 1$-family, there exists a positive T so that for the matrices U and W whose rows are all vectors u_R and w_R, respectively, with $R \subset [n-1], |R| = r$, the product $U \cdot W^t = T \cdot DIS_{n-1,r}$. The desired result follows as before, since $DIS_{n-1,r}$ is nonsingular and yet its rank cannot exceed the number of columns of U. This completes the proof of part (ii).

(iii) Let $\mathcal{F}$ be a perfectly 2-balanced family of functions from $[n]$ to $[2]$. Note that by part (i), $|\mathcal{F}| \geq n/2$, but here one can improve the constant factor and obtain a tight bound. To do so, define, for each $i \in [n]$, a vector u_i of length $|\mathcal{F}|$, whose coordinates are indexed by the elements of $\mathcal{F}$, where here $u_i(f) = (-1)^{f(i)-1}$. It is easy to check that the inner product of u_i and u_j is $|\mathcal{F}|$ if $i = j$, and is $|\mathcal{F}| - 2T$ if $i \neq j$, where here $T > 0$ is the number of functions $f \in \mathcal{F}$ that map i and j to distinct elements. (This number is the same for all $i \neq j$, as $\mathcal{F}$ is perfectly 2-balanced.) We conclude that all diagonal elements of the gram matrix of the n vectors u_i are $|\mathcal{F}|$, while all other elements are $|\mathcal{F}| - 2T$. It is easy to check that this matrix is nonsingular unless the sum of its elements in each row is zero, in which case it has rank $n - 1$. In fact, all eigenvalues of this matrix are $2T$, with multiplicity $n - 1$, and the sum of all entries in a row, with multiplicity 1. (In case this sum is also $2T$, then the matrix is $2T$ times the identity matrix, and all eigenvalues are equal). We conclude that the length of the vectors, $|\mathcal{F}|$ is always at least $n - 1$. Equality can hold only if the sum of elements in a row of the gram matrix is 0. In this case, $|\mathcal{F}| = n - 1$ and $n - 1 - 2T = -1$, that is, the inner product of each two of our n vectors is -1. For each $1 \leq i \leq n$, let $\overline{u_i}$ denote the vector obtained from u_i by adding to it a coordinate in which its value is 1. Then the vectors $\overline{u_i}$ are n pairwise orthogonal vectors of length n with $\{-1, 1\}$ entries, that is, they form the rows of a Hadamard matrix of order n. Thus, if there is no Hadamard matrix of order n then any family of perfectly 2-balanced functions from $[n]$ to $[2]$ has at least n functions. The family $\mathcal{F} = \{f_1, f_2, \ldots, f_n\}$ in which $f_i(i) = 1$ and $f_i(j) = 2$ for all $j \neq i$ shows that this is tight, completing the proof of the theorem.   □

Remarks:
(i) A well known conjecture (c.f., e.g., [15]) asserts that for $n > 2$ there is a Hadamard matrix of order n iff n is divisible by 4. It is easy to see that if there is such a matrix then n is indeed divisible by 4. The converse is not known, but there are many infinite families of known Hadamard matrices, showing that the $(n-1)$-bound in part (iii) of the theorem is tight in many cases.
(ii) It is easy to see that for every $n > k$ there is a perfectly k-balanced family $\mathcal{F}$ of functions from $[n]$ to $[k]$ of size $|\mathcal{F}| = \binom{n}{k-1}$. Indeed, for each subset $R = \{r_1, r_2, \ldots, r_{k-1}\}$ of $[n]$, with $r_1 < r_2 < \ldots < r_{k-1}$ let f_R denote the function defined by $f_R(r_i) = i$ for all $1 \leq i \leq k-1$, and $f_R(j) = k$ for all $j \in [n] - R$. It is not difficult to check that the family of all these functions f_R is perfectly k-balanced (with $T = k$).
(iii) The lower bounds in Theorem 1 hold for weighted families as well, even if the weight $weight(f)$ of some of the functions f is negative, as long as there is a real $T \neq 0$ so that for every $K \subset [n]$, $|K| = k$, the total weight of functions which are one-to-one on K is exactly T. To see this, repeat the proof above, modifying the definition of the vectors u_R to be

$$u_R(f, S_1, S_2) = weight(f) \text{ if } f(R) = S_1, \text{ and } u_R(f, S_1, S_2) = 0 \text{ otherwise,}$$

keeping the definition of the vectors w_R as before.

4   Expanders

In this section we describe a special case of the Cayley expanders of [5] that we use later. Note that these are not bounded-degree graphs, and their degrees grow with the number of vertices, but they suffice for our purpose. This is a special case of a construction suggested in [5], which is based on one of the codes described in [3].

The following are standard definitions and observations concerning eigenvalues and expanders (c.f., e.g., [7],[16]).

Let $G = (V, E)$ be a d-regular graph and let $A = A_G = (a_{uv})_{u,v \in V}$ be its adjacency matrix. Since G is d-regular, the largest eigenvalue of A is d, corresponding to the all 1 eigenvector. Let $\lambda = \lambda(G)$ denote the largest absolute value of an eigenvalue other than the first one. For two (not necessarily disjoint) subsets B and C of V, let $e(B, C)$ denote the number of ordered pairs (u, v), where $u \in B$, $v \in C$ and uv is an edge of G. The following useful bound is the Expander Mixing Lemma (c.f., e.g., [7], page 146).

Proposition 1. *Let G be a d-regular graph with n vertices and set $\lambda = \lambda(G)$. For every two sets of vertices B and C of G, where $|B| = bn$ and $|C| = cn$, we have*

$$|e(B, C) - bcdn| \leq \lambda \sqrt{bc}\, n.$$

We need the following explicit expanders, described, for example, in [6], following [5]. Let $bin : GF(2^k) \mapsto \{0,1\}^k$ be a one-to-one mapping satisfying $bin(0) = 0^k$ and $bin(x+y) = bin(x) \oplus bin(y)$, where $\alpha \oplus \beta$ means the bit-by-bit xor of the binary strings α and β. (The standard representation of $GF(2^k)$ as a vector space satisfies the above conditions.) Given $x, y \in GF(2^k)$, let $< x, y >$ denote the bit $(bin(x), bin(y))_2$, where $(\alpha, \beta)_2$ is the inner-product mod 2 of the binary vectors α and β. For a fixed d and $x, y \in GF(2^k)$, the binary vector u_{xy} is defined as $< x, y >< x^2, y > \cdots < x^d, y >$. For every $d, k \geq 1$, we define a 4^k-regular graph $G_{d,k}$ with 2^d vertices, as follows. The vertex set is $\{0,1\}^d$ and every vertex v is adjacent to $v \oplus u_{xy}$ for all $x, y \in GF(2^k)$.

Theorem 2. *For every two positive integers d and k satisfying $4^k < 2^d$ there is an explicit construction of a 4^k-regular graph $G_{d,k}$ on 2^d vertices so that $\lambda(G_{d,k}) \leq d \cdot 2^k$.*

Proof. Denote $F = GF(2^k)$, $D = \{0,1\}^d$, and let A be the $2^d \times 2^d$ adjacency matrix of $G_{d,k}$. For every $a = a_1 a_2 \cdots a_d \in D$, let v_a be the vector whose bth entry, where $b \in D$, satisfies $v_a(b) = (-1)^{(a,b)_2}$. Let $p_a(x)$ be the polynomial $\sum_{i=1}^d a_i x^i$ and denote $\lambda_a = \sum_{x,y \in F}(-1)^{<p_a(x),y>}$. We now prove that v_a is an eigenvector of A over $\mathbb{R}$ with eigenvalue λ_a.

$$(Av_a)(b) = \sum_{c \in D} A_{bc} v_a(c) = \sum_{x,y \in F} v_a(b \oplus u_{xy}) = v_a(b) \sum_{x,y \in F} v_a(u_{xy})$$

$$= v_a(b) \sum_{x,y \in F}(-1)^{(a,u_{xy})_2} = \lambda_a v_a(b).$$

It is easy to verify that the 2^d vectors $\{v_a\}_{a \in D}$ are orthogonal, and therefore we found all the eigenvalues of A. It remains to bound the absolute value of λ_a. For a fixed $x \in F$, the term $\sum_{y \in F}(-1)^{<p_a(x),y>}$ is equal to 2^k if $p_a(x) = 0$, and to zero in case $p_a(x) \neq 0$. If $a \neq 0^d$, then $p_a(x)$ is a non-zero polynomial of degree at most d, and therefore has at most d roots. Thus, $|\lambda_a| \leq d \cdot 2^k$, as needed. $\square$

Note that this construction is applicable for a wide range of parameters, that is, the number of vertices of the expander can be any power of 2, whereas the degree can be any power of 4.

5 Partially Independent Variables

In this section we introduce a certain relaxation of almost k-wise independence and describe an appropriate explicit construction, which will give the main building block required in the construction of balanced families of hash functions of optimal size. For notational convenience, we give the following definitions related to the probabilities implied by a multinomial distribution.

Definition 3. *Suppose that* $1 \leq \ell \leq k$ *and* $k = k_1 + k_2 + \cdots + k_\ell$, *where* $k_i \geq 0$ *for every* i. *Define* $m(k_1, \ldots, k_\ell)$ *to be the following probability:*

$$\frac{\binom{k}{k_1, k_2, \ldots, k_\ell}}{\ell^k} = \frac{k!}{k_1! k_2! \cdots k_\ell! \ell^k}.$$

For random variables $X_1, \ldots, X_k$, *let* Y_i *denote the number of variables* X_j *that are equal to* i. *Define* $M(X_1, \ldots X_k; k_1, \ldots k_\ell)$ *to be the event that* $Y_i = k_i$ *for every* i, $1 \leq i \leq \ell$.

We now construct probability distributions which are uniform over a set of strings of length q in the alphabet $[\ell]$. In the standard notion of almost k-wise independence, it is required that in any k positions, each substring of length k appears with probability close to ℓ^{-k}. Here we are interested in a weaker condition. Our objective is to construct small probability spaces of the following type.

Definition 4. *A sequence* $X_1, \ldots X_q$ *of random variables that take values from* $[\ell]$ *is* (ϵ, k)-*partially-independent if for any* $p \leq k$ *positions* $i_1 < \cdots < i_p$ *and any* ℓ *values* $k_1, \ldots, k_\ell$ *such that* $k_1 + \cdots + k_\ell = p$, *we have*

$$|Pr[M(X_{i_1}, \ldots X_{i_p}; k_1, \ldots k_\ell)] - m(k_1, \ldots, k_\ell)| < \epsilon.$$

Observe that we require the property to hold for any p variables, where $1 \leq p \leq k$. This is needed since the fact that the property is satisfied for a value p does not imply that it holds for $p' < p$. Furthermore, requiring that it applies for every value $p \leq k$ is crucial for the correctness of our recursive construction. To demonstrate the definition, here is what it means for $\ell = 2$. A sequence $X_1, \ldots X_q$ of random Boolean variables (taking values from $\{0,1\}$) is (ϵ, k)-partially-independent if for any $p \leq k$ positions $i_1 < \cdots < i_p$ and any r, $0 \leq r \leq p$, we have $\left| Pr[X_{i_1} + \cdots + X_{i_p} = r] - \binom{p}{r} 2^{-p} \right| < \epsilon$.

Theorem 3. *For any* $\ell \leq k \leq q$ *and* $0 < \epsilon \leq 1$, *a sample space of size* $\left(\frac{qk^\ell}{\epsilon} \right)^{O(\log q)}$ *that supports* q *variables that take values from* $[\ell]$ *and are* (ϵ, k)-*partially-independent can be constructed in time* $\left(\frac{qk^\ell}{\epsilon} \right)^{O(\log q)}$.

Proof. Assume, without loss of generality, that q is a power 2. Otherwise, q can be simply rounded to the next power of 2. Assume also that $\epsilon \leq \frac{1}{k^\ell}$. If this is not the case, then ϵ can be replaced by $\frac{\epsilon}{k^\ell}$. We recursively construct sample spaces that support an increasing number of variables. For every $t = 0, 1, \ldots, \log_2 q$, we shall construct a sample space C_t that supports 2^t variables that take values from $[\ell]$ and are $\left(\frac{4^t \epsilon}{q^2}, k \right)$-partially-independent. The sample space C_t will consists of strings of length 2^t over the alphabet $[\ell]$.

We start with $t = 0$. To support one variable, it is possible to simply define a sample space that consists of the ℓ strings of length 1, and there will be no error at all in this case. For our purpose, the size of each sample space should be a power of 2, so let N_0 be the result of rounding the value $4(\frac{20q^2k^\ell}{\epsilon})^4$ to the next higher power of 2. The sample space consists of N_0 strings, where each string of length 1 appears either $\lfloor\frac{N_0}{\ell}\rfloor$ or $\lceil\frac{N_0}{\ell}\rceil$ times. Obviously $N_0 \le 8(\frac{20q^2k^\ell}{\epsilon})^4$ and we have one variable which is certainly $\left(1, \frac{\epsilon}{q^2}\right)$-partially-independent.

Let D be the result of rounding the value $\left(\frac{20q^2k^\ell}{\epsilon}\right)^4$ to the next higher power of 4. Suppose that in step t, a sample space of size $N_t \le 8D^{t+1}$ that supports 2^t variables that are $\left(\frac{4^t\epsilon}{q^2}, k\right)$-partially-independent has been constructed. We now describe step $t+1$. Let G be the D-regular expander with N_t vertices described in section 4 (note that $D < N_t$). It follows from Theorem 2 that

$$\frac{\lambda(G)}{D} \le \frac{\log_2 N_t}{\sqrt{D}} \le \frac{3+(t+1)\log_2 D}{\sqrt{D}} \le \frac{(\log_2 D)^2}{\sqrt{D}} \le \frac{20}{D^{1/4}} \le \frac{\epsilon}{q^2k^\ell}.$$

To every vertex of the graph G we assign one of the N_t strings of length 2^t from C_t that were constructed in step t. For every ordered pair (u,v) such that uv is an edge of G, the concatenation of the string assigned to u followed by the string assigned to v is added to the sample space C_{t+1}. The resulting sample space is of size $N_{t+1} = DN_t$.

Suppose that in step t, a sample space C_t of size N_t that supports 2^t variables that are (γ, k)-partially-independent has been constructed, where $\gamma = \frac{4^t\epsilon}{q^2}$. We now prove that the approximation error is increased in step $t+1$ by a multiplicative factor of at most 4, that is, the sample space C_{t+1} supports 2^{t+1} variables that are $(4\gamma, k)$-partially-independent. Suppose that $p \le k$ and take any p positions $1 \le i_1 < \cdots < i_p \le 2^{t+1}$ and any ℓ values $k_1, \ldots, k_\ell$ such that $k_1 + \cdots + k_\ell = p$. We further assume that among the p positions selected, exactly p' positions are in the first half of the string. Therefore $Pr[M(X_{i_1}, \ldots X_{i_p}; k_1, \ldots k_\ell)]$ is equal to

$$\sum_{k_1'+\cdots+k_\ell'=p'} Pr[M(X_{i_1}, \ldots X_{i_{p'}}; k_1', \ldots, k_\ell') \cap M(X_{i_{p'+1}}, \ldots X_{i_p}; k_1-k_1', \ldots, k_\ell-k_\ell')].$$

We would like $Pr[M(X_{i_1}, \ldots X_{i_p}; k_1, \ldots k_\ell)]$ to be close to:

$$m(k_1, \ldots k_\ell) = \sum_{k_1'+\cdots+k_\ell'=p'} m(k_1', \ldots, k_\ell')m(k_1-k_1', \ldots, k_\ell-k_\ell').$$

Note that the number of terms in the two summations above is at most k^ℓ and that obviously $\sum_{k_1'+\cdots+k_\ell'=p'} m(k_1', \ldots, k_\ell') \le 1$. Since C_t is (γ, k)-partially-independent, it follows from Proposition 1 that the estimation error is as follows:

$$\left| Pr[M(X_{i_1}, \ldots X_{i_p}; k_1, \ldots k_\ell)] - m(k_1, \ldots k_\ell) \right| \leq$$

$$\sum_{k_1' + \cdots + k_\ell' = p'} \left[(m(k_1', \ldots, k_\ell') + \gamma)(m(k_1 - k_1', \ldots, k_\ell - k_\ell') + \gamma) + \frac{\lambda(G)}{D} \right] -$$

$$\sum_{k_1' + \cdots + k_\ell' = p'} m(k_1', \ldots, k_\ell') m(k_1 - k_1', \ldots, k_\ell - k_\ell') =$$

$$\sum_{k_1' + \cdots + k_\ell' = p'} \gamma[m(k_1', \ldots, k_\ell') + m(k_1 - k_1', \ldots, k_\ell - k_\ell')] + \gamma^2 + \frac{\lambda(G)}{D} \leq$$

$$2\gamma + k^\ell \left(\gamma^2 + \frac{\lambda(G)}{D} \right) \leq 4\gamma,$$

where the last inequality follows from the inequalities $\gamma \leq \epsilon \leq \frac{1}{k^\ell}$ and $\frac{\lambda(G)}{D} \leq \frac{\epsilon}{q^2 k^\ell} \leq \frac{\gamma}{k^\ell}$. After step $\log_2 q$, the sample space constructed is (ϵ, k)-partially-independent, as needed. $\qquad \square$

6   Balanced Families and Approximate Counting

The following inequality is Robbins' formula [12] (a tight version of Stirling's formula).

Claim. For every integer $n \geq 1$,

$$\sqrt{2\pi} n^{n+1/2} e^{-n+1/(12n+1)} < n! < \sqrt{2\pi} n^{n+1/2} e^{-n+1/(12n)}.$$

This supplies the following simple lower bound for the multinomial distribution (recall Definition 3).

Lemma 1. *If $k \geq \ell > 0$, then*

$$m(\underbrace{\lceil k/\ell \rceil, \ldots, \lceil k/\ell \rceil}_{k \bmod \ell}, \underbrace{\lfloor k/\ell \rfloor, \ldots, \lfloor k/\ell \rfloor}_{\ell - (k \bmod \ell)}) > (15k/\ell)^{-\ell/2}.$$

Proof. (sketch) Assume first that ℓ divides k. Using Robbins' formula, we get:

$$m(\underbrace{k/\ell, \ldots, k/\ell}_{\ell}) = \frac{k!}{(k/\ell)!^\ell \ell^k} > (2\pi k/\ell)^{-\ell/2} e^{-\ell^2/12k}$$

$$\geq (2\pi e^{1/6} k/\ell)^{-\ell/2} > (7.5k/\ell)^{-\ell/2}.$$

The result for general k and ℓ follows similarly. $\qquad \square$

The previous Lemma shows that the events we would like to estimate have a relatively high probability, enabling us to give the following construction.

Theorem 4. *For any $k \geq \ell$ and $0 < \epsilon \leq 1$, an ϵ-balanced (q, k, ℓ)-splitter of size $\left(\frac{qk^\ell}{\epsilon} \right)^{O(\log q)}$ can be constructed in time $\left(\frac{qk^\ell}{\epsilon} \right)^{O(\log q)}$.*

Proof. As implied by Theorem 3, we use an explicit probability space of size $\left(\frac{qk^{\ell}}{\gamma}\right)^{O(\log q)}$ that supports q random variables that take values from $[\ell]$ and are (γ, k)-partially-independent, where $\gamma = (15k/\ell)^{-\ell/2}\epsilon$. We attach one of the random variables to each element of $[q]$. If follows from Lemma 1 that the splitter achieves the required approximation. $\square$

We can now describe our main construction of balanced families of hash functions, using the ingredients mentioned at the end of section 2. Recall that there are three ingredients in this construction. Two of them are relatively simple, and are given in the next two propositions.

Proposition 2. *For any $0 < \epsilon \le 1$, an (ϵ, k)-balanced family of hash functions from $[n]$ to $[q]$, where $q = \lceil \frac{2k^2}{\epsilon} \rceil$, of size $\frac{k^{O(1)} \log n}{\epsilon^{O(1)}}$ can be constructed in time $\frac{k^{O(1)} n \log n}{\epsilon^{O(1)}}$.*

Proposition 3. *For any $0 < \epsilon \le 1$, an (ϵ, g)-balanced family of hash functions from $[m]$ to $[g]$ of size $O(\frac{e^g \sqrt{g} \log m}{\epsilon^2})$ can be constructed in time $\binom{m}{g} \frac{e^g g^{O(1)} m \log m}{\epsilon^2}$.*

The first proposition is proved using a standard construction of nearly pairwise independent random variables. Here n is the number of variables, they attain values in $[q]$, and the number of functions is the size of the sample space. Since every two variables are equal with probability close to $1/q$, for every fixed set S of k variables, the values of the random variables in S are pairwise distinct in at least a fraction of $(1 - \epsilon)$ of the functions.

The second proposition is proved using the method of conditional expectations. The details appear in [4].

The main part of the construction is the balanced (q, k, ℓ)-splitter described in Theorem 4. The three ingredients are combined as follows. Each function f of our final family is described by a member f_1 of an $(\epsilon/6, k)$-balanced family of Proposition 2, a member f_2 of the ϵ_2-balanced splitter of Theorem 4 with $\epsilon_2 = \frac{\epsilon}{6}, q = \lceil \frac{2k^2}{\epsilon_2} \rceil$ and $\ell = \lceil \log k \rceil$, and ℓ members $\phi_1, \ldots, \phi_\ell$ of the $(\frac{\epsilon}{6\ell}, g)$-balanced family of Proposition 3 with $m = q$ and $g = k/\ell$. (For simplicity we assume here that ℓ divides k.) To compute the value of f on some $x \in [n]$, we first apply f_1 to x, getting a value y in $[q]$, then we apply f_2 to y, getting as a result some $i \in [\ell]$, and finally we apply ϕ_i to y, where the final result is $(i - 1)k/\ell + \phi_i(y)$. A k-set $S \subset [n]$ can be mapped in a one-to-one manner by such an f only if it is mapped in a one-to-one manner by f_1, and then only if it is split evenly into ℓ parts by f_2, and then only if its elements mapped to each of the ℓ parts are mapped in a one-to-one manner by each of the functions ϕ_i. Since all the ingredients in the construction are sufficiently balanced, this gives the required balanced family. The detailed computation, which yields the following theorem, is postponed to the full version of the paper.

Theorem 5. *For $\frac{1}{poly(k)} < \epsilon \le 1$, an (ϵ, k)-balanced family of hash functions from $[n]$ to $[k]$ of size $e^{k + O(\log^3 k)} \log n$ can be constructed deterministically in time $e^{k + O(\log^3 k)} n \log n$.*

Using Color-Coding we can now approximate the number of paths and cycles (or other fixed graphs of bounded tree-width) in a given input graph. Let $G = (V, E)$ be a directed or undirected graph. The algorithms use the construction of (ϵ, k)-balanced families of hash functions from V to $[k]$. Each such function defines a coloring of the vertices of the graph. Recall that a path is colorful if each vertex on it is colored by a distinct color. Using dynamic programming one can count efficiently the exact number of colorful paths in each of these colorings. The properties of the balanced family of hash functions then provide the following deterministic polynomial time algorithms for approximately counting the number of paths or cycles of size k in a given input graph of size n for all $k \leq \log n$. Similar results apply for approximate counting of prescribed subgraphs of size k and bounded tree-width.

Theorem 6. *For any $\frac{1}{poly(k)} < \epsilon \leq 1$, the number of simple (directed or undirected) paths of k vertices in a (directed or undirected) graph $G = (V, E)$ can be approximated deterministically up to relative error ϵ in time $2^{O(k)}|E|\log|V|$.*

Theorem 7. *For any $\frac{1}{poly(k)} < \epsilon \leq 1$, the number of simple (directed or undirected) cycles of size k in a (directed or undirected) graph $G = (V, E)$ can be approximated deterministically up to relative error ϵ in time $2^{O(k)}|E||V|\log|V|$.*

7 Concluding Remarks

- The notion of balanced families of hash functions seems natural and useful, and it will be interesting to find additional applications of it.
- An easy combination of Proposition 2 and Theorem 5 supplies, for any $\epsilon \geq \frac{1}{k^\ell}$, explicit ϵ-balanced (n, k, ℓ)-splitters of size at most $e^{O(\ell \log^2 k)} \log n$. In particular, for $\ell = 2$ the size is $e^{O(\log^2 k)} \log n$. A simple probabilistic argument shows, however, that for any fixed $\epsilon > 0$ there are ϵ-balanced $(n, k, 2)$-splitters of size $O(k\sqrt{k} \log n)$, and although this is not crucial for our application here, it will be interesting to find an explicit construction of such splitters of size polynomial in k and $\log n$.
- Our results settle the problem of approximately counting the number of paths and cycles of length $k = \Theta(\log n)$ in an n-vertex graph in deterministic polynomial time. As mentioned in the introduction, it is probably impossible to extend the result for larger values of k, since even a polynomial time algorithm for deciding whether there exists one simple path of length k where $\log n = o(k)$ would imply a sub-exponential time algorithm for the Hamiltonian cycle problem. This follows easily by padding a graph on k vertices by $n - k = 2^{o(k)}$ isolated ones, thus converting the above decision algorithm to one that decides in time $2^{o(k)}$ whether a graph on k vertices is Hamiltonian, contradicting the Exponential Time Hypothesis (ETH) [18,19].
- Our method here, combined with the Color Coding technique, easily yields results for additional approximate counting problems for graphs. In particular, given a weighted graph G on n vertices, we can approximate deterministically, in polynomial time, the number of minimum (or maximum)

weight paths or cycles (or copies of any prescribed subgraph of bounded tree width) on k vertices in G up to any fixed desired relative accuracy, for all $k \leq O(\log n)$.

References

1. Alon, N., Babai, L., Itai, A.: A fast and simple randomized parallel algorithm for the maximal independent set problem. Journal of Algorithms 7(4), 567–583 (1986)
2. Alon, N., et al.: Biomolecular network motif counting and discovery by color coding. In: ISMB (Supplement of Bioinformatics), pp. 241–249 (2008)
3. Alon, N., Goldreich, O., Håstad, J., Peralta, R.: Simple construction of almost k-wise independent random variables. Random Struct. Algorithms 3(3), 289–304 (1992)
4. Alon, N., Gutner, S.: Balanced families of perfect hash functions and their applications. In: Arge, L., Cachin, C., Jurdziński, T., Tarlecki, A. (eds.) ICALP 2007. LNCS, vol. 4596, pp. 435–446. Springer, Heidelberg (2007)
5. Alon, N., Roichman, Y.: Random Cayley graphs and expanders. Random Struct. Algorithms 5(2), 271–285 (1994)
6. Alon, N., Schwartz, O., Shapira, A.: An elementary construction of constant-degree expanders. Combin. Probab. Comput. 17(3), 319–327 (2008)
7. Alon, N., Spencer, J.H.: The probabilistic method. Wiley-Interscience Series in Discrete Mathematics and Optimization. John Wiley & Sons Inc., Hoboken (2008)
8. Alon, N., Yuster, R., Zwick, U.: Color-coding. Journal of the ACM 42(4), 844–856 (1995)
9. Alon, N., Yuster, R., Zwick, U.: Finding and counting given length cycles. Algorithmica 17(3), 209–223 (1997)
10. Arvind, V., Raman, V.: Approximation algorithms for some parameterized counting problems. In: Bose, P., Morin, P. (eds.) ISAAC 2002. LNCS, vol. 2518, pp. 453–464. Springer, Heidelberg (2002)
11. Björklund, A., Husfeldt, T., Kaski, P., Koivisto, M.: Counting paths and packings in halves. In: Fiat, A., Sanders, P. (eds.) ESA 2009. LNCS, vol. 5757, pp. 578–586. Springer, Heidelberg (2009)
12. Feller, W.: An introduction to probability theory and its applications, 3rd edn., vol. I. Wiley, New York (1968)
13. Flum, J., Grohe, M.: The parameterized complexity of counting problems. SIAM Journal on Computing 33(4), 892–922 (2004)
14. Fredman, M.L., Komlós, J., Szemerédi, E.: Storing a sparse table with O(1) worst case access time. Journal of the ACM 31(3), 538–544 (1984)
15. Hall Jr., M.: Combinatorial theory, 2nd edn. Wiley-Interscience Series in Discrete Mathematics. John Wiley & Sons Inc., A Wiley-Interscience Publication, New York (1986)
16. Hoory, S., Linial, N., Wigderson, A.: Expander graphs and their applications. Bull. Amer. Math. Soc (N.S.) 43(4), 439–561 (2006) (electronic)
17. Hüffner, F., Wernicke, S., Zichner, T.: Algorithm engineering for color-coding to facilitate signaling pathway detection. In: Sankoff, D., Wang, L., Chin, F. (eds.) APBC. Advances in Bioinformatics and Computational Biology, vol. 5, pp. 277–286. Imperial College Press (2007)
18. Impagliazzo, R., Paturi, R.: On the complexity of k-SAT. J. Comput. Syst. Sci 62(2), 367–375 (2001)

19. Impagliazzo, R., Paturi, R., Zane, F.: Which problems have strongly exponential complexity? J. Comput. Syst. Sci. 63(4), 512–530 (2001)
20. Jukna, S.: Extremal combinatorics. Texts in Theoretical Computer Science. An EATCS Series. Springer, Berlin (2001); With applications in computer science
21. Naor, M., Schulman, L.J., Srinivasan, A.: Splitters and near-optimal derandomization. In: 36th Annual Symposium on Foundations of Computer Science, pp. 182–191 (1995)
22. Schmidt, J.P., Siegel, A.: The spatial complexity of oblivious k-probe hash functions. SIAM Journal on Computing 19(5), 775–786 (1990)
23. Scott, J., Ideker, T., Karp, R.M., Sharan, R.: Efficient algorithms for detecting signaling pathways in protein interaction networks. Journal of Computational Biology 13(2), 133–144 (2006)
24. Sharan, R., Ideker, T.: Modeling cellular machinery through biological network comparison. Nature Biotechnology 24(4), 427–433 (2006)
25. Shlomi, T., Segal, D., Ruppin, E., Sharan, R.: QPath: a method for querying pathways in a protein-protein interaction network. BMC Bioinformatics 7, 199 (2006)
26. Vassilevska, V., Williams, R.: Finding, minimizing, and counting weighted subgraphs. In: Mitzenmacher, M. (ed.) Proceedings of the 41st Annual ACM Symposium on Theory of Computing, pp. 455–464. ACM, New York (2009)

Kernelization: New Upper and Lower Bound Techniques

Hans L. Bodlaender

Department of Information and Computing Sciences, Utrecht University, P.O. Box
80.089, 3508 TB Utrecht, the Netherlands
hansb@cs.uu.nl

Abstract. In this survey, we look at *kernelization*: algorithms that
transform in polynomial time an input to a problem to an equivalent
input, whose size is bounded by a function of a parameter. Several results of recent research on kernelization are mentioned. This survey looks
at some recent results where a general technique shows the existence of
kernelization algorithms for large classes of problems, in particular for
planar graphs and generalizations of planar graphs, and recent lower
bound techniques that give evidence that certain types of kernelization
algorithms do not exist.

Keywords: fixed parameter tractability, kernel, kernelization, preprocessing, data reduction, combinatorial problems, algorithms.

1 Introduction

In many cases, combinatorial problems that arise in practical situations are
NP-hard. As we teach our students in algorithms class, there are a number of
approaches: we can give up optimality and design approximation algorithms
or heuristics; we can look at special cases or make assumptions about the input
that one or more variables are small; or we can design algorithms that sometimes
take exponential time, but are as fast as possible. In the latter case, a common
approach is to start the algorithm with preprocessing.

So, consider some hard (say, NP-hard) combinatorial problem. We start our
algorithm with a *preprocessing* or *data reduction* phase, in which we transform
the input I to an equivalent input I' that is (hopefully) smaller (but never
larger). Then, we solve the smaller input I' optimally, with some (exponential
time) algorithm. E.g., in practical settings, we can use an ILP-solver, branch
and bound or branch and reduce algorithm, or a SATISFIABILITY-solver. After
we obtained an optimal solution S' for S, we transform this solution back to an
optimal solution for I.

In this overview paper, we want to focus on the following question for given combinatorial problems: suppose the preprocessing phase takes polynomial time; what
can we say about the size of the reduced instance, as a function of some parameter
of the input? This question is nowadays phrased as: does the problem we consider
have a kernel, and if so, how large is the kernel? So, kernelization gives us quanti-
tative insights in what can be achieved by polynomial time preprocessing. In this

J. Chen and F.V. Fomin (Eds.): IWPEC 2009, LNCS 5917, pp. 17–37, 2009.

paper, we aim to first give a general introduction to the field of kernelization, and then survey a number of very recent general techniques from the field.

As a simple example, let us look at the VERTEX COVER problem. Here, we are given a graph $G = (V, E)$ and some integer k, and ask if there is a set $W \subseteq V$ of at most k vertices, such that for each edge $\{v, w\} \in E$ at least one endpoint belongs to W ($v \in W$ or $w \in W$). We can use the following 'kernelization' algorithm, due to Buss, see [26]: while there is at least one vertex $v \in V$ with degree at least $k + 1$, remove v and its incident edges, and set k to $k - 1$. This gives equivalent instances: v must belong to an optimal solution, because if we do not take v, we must take all neighbors of v, which are more than k vertices. Also, remove all vertices of degree 0, without changing k. If at some point, $k < 0$, we can decide **no**: there clearly is no solution. Now, if we have more than k^2 edges, we decide **no**. Each remaining vertex has degree at most k, so with k vertices, we cannot cover more than k^2 edges. If we did not return **no**, we end with an equivalent instance with at most k^2 edges (and less than k^2 vertices).

The simple algorithm given above is not the best (in terms of 'kernel sizes') kernelization algorithm for VERTEX COVER: a clever algorithm by Nemhauser and Trotter [72] gives reduced instances with $2k$ vertices. The algorithm above, however, does give a nice example of a methodology that is used in many kernelization algorithms: we have a set of 'safe reduction rules', i.e., rules that give a smaller, equivalent instance, and we have a mathematical analysis on the size of yes-instances when no rule applies. When our input is larger than this size, we return **no**; otherwise, we have a small reduced instance.

For the analysis of kernels, we can fortunately make use of a large toolbox from the field of fixed parameter algorithms, which was pioneered by Downey and Fellows. For background in this field, we refer to [44,52,73]. We use a number of definitions from this field in a form that is useful for this exposition.

A *parameterized* problem is a subset of $\Sigma^* \times \mathbf{N}$ for some fixed alphabet Σ. I.e., we look at decision problems where some specific part of the input, called the *parameter*, is an integer.

The theory of fixed parameter complexity is used to distinguish between the running time for parameterized problem, where we pay attention how this time depends on the parameter and on the input size. Three important types of behavior can be seen:

- **NP-complete**: the problem is NP-complete for some fixed values of k. E.g., GRAPH COLORING is NP-complete, even when the number of colors is 3.
- XP: for every fixed k, there is a polynomial time algorithm, but the exponent of the running time grows with k, i.e., the running time is $\Theta(n^{f(k)})$ for some function f with $\lim_{k \to \infty} f(k) = \infty$.
- FPT: there is an algorithm, that solves the problem in time $O(f(k)n^c)$ for some function f on inputs of size n with parameter k, with c a constant. FPT is defined as the class of all parameterized problems that have such a kernel. FPT is short for *fixed parameter tractable*.

A *kernelization* algorithm for a parameterized problem P is an algorithm A, that transforms inputs (I, k) of P to inputs (I', k') of P, such that

1. the algorithm uses time polynomial in $|I| + k$;
2. the algorithm transforms inputs to equivalent inputs: $(I, k) \in P \Leftrightarrow A(I, k) \in P$;
3. $k' \leq k$;
4. $|I'| \leq f(k)$ for some function f: the value of the new parameter and the size of the new input are bounded by a function of the value of the old parameter.

We say that P has a *kernel of size* f. Throughout this paper, we focus on the question for given problems P: does it have a kernel, and if so, of what size. Of course, we prefer kernels of small size, and asks ourselves for problems: do they have polynomial kernels. A variant of the definition, giving a slightly different notion of kernelization, has the condition $k' \leq f(k)$ instead of $k' \leq k$.

The topic of kernelization has become a very active of research. An excellent survey on the field was made by Guo and Niedermeier in 2007 [59]. In the past two years, more work has been done, and this paper aims at focussing at some recent developments, where general methods were obtained.

A little technical remark. In several cases, we can decide the problem directly. Applying the notion of kernelization would require that we instead transform the problem to an equivalent instance. This can be easily resolved as follows: we take a yes-instance and a no-instance, both of small (constant bounded) size, and instead of deciding, we transform the input to the yes- or no-instance of constant bounded size. This little trick also shows that when a problem (seen as decision problem, we take the parameter in unary) belongs to P, then it has a kernel of $O(1)$ size. Also, when a problem has an $O(1)$ kernel, say of size at most c, it belongs to P: we first make a kernel, and then check if this belongs to the set P_c of yes-instances with size at most c. This latter set does not depend on the input, and thus can be hardwired in our algorithm. Thus, when the problem is NP-hard, then it does not have an $O(1)$ kernel, unless $P = NP$.

The theory of fixed parameter complexity gives us an excellent tool to see if a problem has a kernel (i.e., without considering its size.) First, consider the following result, which is nowadays folklore. The main idea of the result is due to Cai et al. [28], while its first statement is due to Neal Koblitz [47] and appeared in [45] in 1997. In a few cases in the literature, the result or its proof are slightly incorrectly stated, see the discussion below. The following proof is given by Flum and Grohe [52] (for the variant of strongly uniformly FPT), see also [73].

Theorem 1. *Let P be a parameterized problem. Then P belongs to the class FPT, if and only if P is decidable and P has a kernel.*

Proof. If P is decidable and has a kernel, then we can use the following algorithm for P. If we have an input (I, k) of size n, then we can build in $p(n + k)$ time for some polynomial p an equivalent instance (I', k'), with $\max(|I'|, k') \leq f(k)$ for some function f, by using the kernelization algorithm. Then, decide on (I', k') by any algorithm for P; thus, for some function g, this costs in total $O(p(n + k) + g(f(k)))$ time.

Suppose P belongs to FPT. For some function f and constant c, we have an algorithm A that solves instances of P in $f(k)(n + k)^c$ time. Now, run algorithm

A for $(n+k)^{c+1}$ steps. If the algorithm decides the problem in this time, we are done: report an $O(1)$ yes- or no-instance accordingly. Otherwise, we know that $n+k \leq f(k)$: we return the original input, which is of the desired size. □

The result and its proof are interesting for two reasons. First, we see how kernels can be applied to obtain FPT-algorithms. Secondly, while the algorithm in the other direction does not give an interesting kernelization algorithm, it does give us a method to obtain negative evidence: if we have evidence that a problem does not belong to the class FPT, then we also have evidence that there does not exist a kernel. Such evidence is available. Downey and Fellows defined a number of complexity classes of parameterized problems (see [42,43]), for which for our exposition, $W[1]$ is the most relevant. For the precise definition of $W[1]$, see for example [42,43,44]. It is widely believed that $FPT \neq W[1]$, while $FPT \subseteq W[1]$. So, decidable problems that are hard for $W[1]$ are believed to not belong to FPT, and thus are believed not to have a kernel. Moreover, if $FPT = W[1]$, i.e., if a $W[1]$-hard decidable problem has a kernel, then the *Exponential Time Hypothesis* does not hold, see [1,29]. There are many problems that are known to be $W[1]$-hard. For example, INDEPENDENT SET is $W[1]$-complete [42,43] and DOMINATING SET is $W[2]$-complete and hence $W[1]$-hard [42]. (When not specified otherwise, the parameter of a problem is assumed to be the upper or lower bound of the size of the set to be found.) Hence, these problems have no kernel, unless the Exponential Time Hypothesis does not hold.

In the literature, the condition that the problem is decidable is sometimes forgotten; however, the condition is necessary. Consider the following parameterized problem. Let X be some undecidable set of integers. Now, consider the language $\{(I, k) \mid k \in X\}$, i.e., the first part of the input is ignored, and we just ask if the parameter belongs to X. It has a trivial linear kernel: map each (I, k) to (ϵ, k), ϵ the empty string. But, $\{(I, k) \mid k \in X\}$ is also undecidable and hence cannot belong to FPT.

FPT, as defined here, is also known as *uniformly FPT*. Different versions exist: *strongly uniformly FPT* requires in addition that f is computable. A parameterized problem L is *nonuniformly FPT*, if there is a constant c, such that for each fixed k, there is an $O(n^c)$ algorithm that solves all instances of L with parameter k. A typical example of proofs of membership in nonuniformly FPT comes from the graph minor theorem of Robertson and Seymour [12,76,77]: for a graph parameter f that does not increase when taking minors, this theory tells us that the problem: '*given a graph G and integer k, is $f(G) \leq k$?*' is non-uniformly FPT. The three classes are different, see the discussion in [44] Theorem 1 holds for *uniformly FPT*. A variant of the proof of Theorem 1 shows that a problem with a kernel belongs to nonuniformly FPT (build the kernel, and then we have to check, for fixed k, a constant number of possibilities), and that a problem belongs to strongly uniformly FPT, if and only if it is decidable and has a kernel of size f for some recursive function f [52]. There are computable problems that are nonuniformly FPT but not uniformly FPT [44]; by Theorem 1, these are nonuniformly FPT problems that do not have a kernel.

Besides an argument for the non-existence of kernels, $W[1]$-hardness proofs give evidence for the non-existence of FPT-algorithms for the parameterized problem at hand, i.e., assuming $FPT \neq W[1]$, a $W[1]$-hard problem does not have an algorithm with running time $O(f(k)n^c)$ for some function f and constant c. Recently, stronger results have been obtained for several parameterized problems. For a number of problems, it is shown that — under certain complexity theoretic assumptions — the problems do not have algorithms that solve them in $O(n^{o(k)})$ time, see [30,32,33,53].

It is desirable to have kernels of small size. Fortunately, many problems have small kernels, and we tabulate a number of examples below in Tables 1 and 2. In Table 2, sizes of kernels for several problems on planar graphs are given, but also some negative results are mentioned. **W** means: No kernel unless the Exponential Time Hypothesis fails. These problems are $W[1]$-hard. For problems, marked **X**, there is evidence that these problems do not have a kernel of polynomial size. Each of these belongs to FPT, so a kernel (usually of exponential size) exists. More precisely, for each of the problems marked **X** in Table 2, we have that they do not have a polynomial sized kernel, unless $NP \subseteq coNP/poly$, which in its turn implies that the polynomial time hierarchy collapses to the third level. For more details, see Section 3. The entry marked **?** is open to my knowledge. It is also open if EDGE CLIQUE COVER has a kernel of polynomial size. The tables are incomplete, and often only list the smallest kernel known to me.

In case of the positive results for planar graphs, there is a common underlying methodology that allows to obtain kernels for many problems on planar graphs.

In the remainder of the paper, we will give an introduction to two developments in the theory of kernelization as discussed above: meta theorems that allow us to obtain kernels for collections of problems, and techniques to show that certain problems do not have a polynomial kernel. Several other important topics on kernelization will not be covered here; more information can for example be obtained from [59].

Table 1. Kernel sizes for various problems. For graph problems, the bounds express the number of vertices.

Problem	Kernel	Reference
CLUSTER EDITING	$4k$	[58]
CONVEX RECOLORING OF TREES	$O(k^2)$	[16]
FEEDBACK ARC SET IN TOURNAMENTS	$O(k)$	[10]
EDGE CLIQUE COVER	2^k	[57]
KEMENY SCORE	$2k$	[11]
MAXEXACT-q-SAT	$O(k)$	[66]
MAX NON-LEAF OUT-BRANCHING	$O(k^2)$	[63]
MULTICUT IN TREES	$O(k^6)$	[25]
NONBLOCKER	$5/3k + 3$	[38]
ROOTED k-LEAF-OUT-BRANCHING	$O(k^2)$	[36]

Table 2. Kernels for problems on general graphs and on planar graphs. Sizes are expressed in number of vertices. X = kernel, but unless $NP \subseteq coNP/poly$ no polynomial kernel; W = $W[1]$-hard: 'no' kernel (see text); ? = open.

Problem	Kernel (all graphs)	Kernel (planar)	References
CONNECTED VERTEX COVER	W	$O(k)$	[41,60]
CAPACITATED VERTEX COVER	X	?	[41,62]
CAPACITATED DOMINATING SET	W	W	[42,18]
CONNECTED DOMINATING SET	W	$O(k)$	[42,69]
DISJOINT CYCLES	X	$O(k)$	[21,20]
DOMINATING SET	W	$O(k)$	[42,6,31]
EDGE DOMINATING SET	$8k^2$	$O(k)$	[49,60]
FEEDBACK VERTEX SET	$4k^2$	$O(k)$	[78,19]
INDEPENDENT SET	W	$4k$	[43,7]
LONG CYCLE	X	X	[15]
LONG PATH	X	X	[15]
MAX LEAF SPANNING TREE	$3.75k$	$3.75k$	[46]
WEIGHTED MAX LEAF SPANNING TREE	W	$78k$	[65]
TRIANGLE PACKING	$O(k^2)$	$O(k)$	[70,60]

2   Upper Bounds: Meta Theorems

For many concrete problems, polynomial kernels have been found. Very recently, results have been obtained that 'go one step further': they show that for certain classes of problems, each problem in such a class has a polynomial kernel.

2.1   Meta Theorems for Approximation Classes

The first such result was obtained by Kratsch [67]. The classes MIN $F^+\Pi_1$ and MAX NP are known from the field of approximation: each problem in these classes contains a constant factor polynomial time approximation algorithm. The subclass MAX SNP is well known. The result by Cai and Chen [27] that all problems in these classes are in FPT is strengthened by Kratsch [67] as follows.

Theorem 2 (Kratsch [67]). *For each problem in MIN $F^+\Pi_1$ and MAX NP, its version where the parameter is the value to optimize has a polynomial kernel.*

2.2   Meta Theorems for Graphs on Surfaces

Bodlaender et al. [17] consider problems on fixed surfaces with certain properties. There is a large number of parameterized problems on planar graphs that have a polynomial kernel. The first of these was the seminal result by Alber, Fellows, and Niedermeier [6], who gave a linear kernel for DOMINATING SET on planar graphs. More linear size kernels were obtained for a large number of other problems, including CONNECTED VERTEX COVER, CYCLE PACKING, EFFICIENT EDGE DOMINATING SET, FEEDBACK VERTEX SET, FULL-DEGREE SPANNING TREE, INDUCED MATCHING, MAXIMUM TRIANGLE PACKING, and MINIMUM EDGE

DOMINATING SET [19,20,31,60,61,69,71]. See also for example [55]. Guo and Niedermeier [60] gave a general method to obtain such algorithms, based on decompositions of the planar input graph in 'regions' and rules that decrease the size
of such regions. Then, in [17] it is shown that general conditions on the problem
statement yield rules that always reduce regions to bounded size, and thus result
in kernels of either linear, quadratic, or cubic size on planar graphs for a large class
of problems. Also, these results are generalized to problems on other surfaces.

Here, we will give an example of a very simplified version of the proof method
for a concrete problem, namely the RED-BLUE DOMINATING SET problem on
planar graphs, and state the general theorems shown in [17].

In the RED-BLUE DOMINATING SET problem, we are given a bipartite graph
$G = (R \cup B, E)$ and an integer k, and ask for a subset $S \subseteq R$ of at most k 'red'
vertices from R, such that each 'blue' vertex from B is adjacent to a vertex in
S. We consider this problem, restricted to planar graphs. See for example [51]
for an FPT algorithm.

A set of vertices S in a graph G is d-dominating, if each vertex in G is at
distance at most d from a vertex in S. We may assume there are no isolated
vertices. Now, each solution S is 2-dominating, as each red vertex is adjacent to
a blue vertex and each blue vertex is adjacent to a red vertex.

A t-boundaried graph is a graph $G = (V, E)$ with t distinguished vertices
(called *terminals*, uniquely labeled from 1 to t, together called the *boundary* of
G) and a boundaried graph is a t-boundaried graph for some t. The following
result is a variant of a result, shown by Guo and Niedermeier [60].

Lemma 1. *Let d be a positive integer. For all planar graphs G, given with a d-
dominating set S, there is a collection of $O(|S|)$ boundaried graphs $G_1, \ldots, G_r$,
such that*

- *Each vertex belongs to at least one graph G_i, $1 \leq i \leq r$.*
- *If a vertex belongs to more than one graph G_i, it belongs to the boundary of
  all G_i's it belongs to.*
- *Each G_i has diameter at most $2d$ and a boundary of size at most $4d$.*

Given S, this collection can be found in polynomial time.

This decomposition is called a *region decomposition* [60]; each region has a small
boundary and diameter.

Suppose we are given a planar graph $G = (V, E)$. First, with a Baker-style
polynomial time approximation scheme for RED-BLUE DOMINATING SET [9],
we either find a solution S of size at most $2k$, or determine that G has no red-
blue dominating set of size at most k. (For other problems, we can approximate
the MINIMUM d-DOMINATING SET PROBLEM with a Baker-style PTAS.) In the
latter case, we are done. In the former case, we then find the collection of $O(k)$
boundaried graphs with diameter and boundary $O(d)$.

What remains to be done to get a kernel of linear size is to have rules, that replace each boundaried graph by a new boundaried graph, with the same boundary,
such that the answer does not change. Doing so, we possibly update k. To describe
safety of such rules, we use the following definitions.

If we have two t-boundaried graphs G and H, $G \oplus H$ is the t-boundaried graph, obtained by taking the disjoint union of G and H while identifying, for $i = 1 \cdots t$, the ith terminal of G with the ith terminal of H, and then dropping parallel edges. For a graph property P and integer t, we can define the equivalence relation $\sim_P^t$ on t-boundaried graphs as follows: for t-boundaried graphs G and H, $G \sim_P^t H$, if and only if for each t-boundaried graph K, $P(G \oplus K) \Leftrightarrow P(H \oplus K)$. $G \sim_P^t H$ gives a 'reduction rule' for algorithms that want to test if P holds for a given input graph. Suppose H is smaller than G. If we have a graph of the form $G \oplus K$ for some K, we can replace the input by $H \oplus K$. It was shown by Arnborg et al. [8], that for each integer k and each graph property P that is formulated in *Monadic Second Order Logic*, there is a finite set of such 'safe' reduction rules, such that each graph with treewidth at most k and with property P can be reduced to a graph of size $O(1)$. Moreover, the total time of the reduction algorithm is linear (for fixed k). This gives a linear time algorithm for testing P on graphs of bounded treewidth, based solely on reduction rules, i.e., no tree decomposition of the graph is needed. See also [2,24,35,48]. In particular, for each MSOL-expressible property P and t, the relation $\sim_P^t$ has a finite number of equivalence classes [8]. We use this here for P the property of being planar.

This idea was generalized to some optimization problems by Bodlaender and van Antwerpen-de Fluiter [22,37]. Let f be a function, mapping graphs to integers. For t-boundaried graphs G and H, and integer $i \in \mathbf{Z}$, we write $G \rightarrow_{f,i} H$, if for all t-boundaried graphs K, $f(G \oplus H) = f(G \oplus K) + i$.

Generalizing this in the natural way to colored graphs (for example, blue terminals remain blue) and letting f be the minimum size of a red-blue dominating set, we see two examples of $\rightarrow_{f,i}$ in Figure 1. Terminal vertices are drawn with a square. E.g., if we have a path of length 5 with only its two red endpoints adjacent to other vertices, then, if we replace this by a path of length 2, the size of a minimum red-blue dominating set in the graph drops by exactly one.

Let $\sim_f^t$ be the equivalence relation on t-boundaried graphs, defined by $G \sim_f^t H$ iff there is an i with $G \rightarrow_{f,i} H$. We say that f is *finite integer index*, if for each fixed t, $\sim_f^t$ has a finite number of equivalence classes. Similar to [22], one can show that RED-BLUE DOMINATING SET is finite integer index. Let for t-boundaried graphs G and H hold $G \sim_{planar,rbds}^t H$, if there exists an integer i, such that for all t-boundaried graphs K: the size of the minimum red-blue dominating set in $G \oplus K$ is exactly the size of the minimum red-blue dominating set in $H \oplus K$ plus i, and $G \oplus K$ is planar, if and only if $H \oplus K$ is planar.

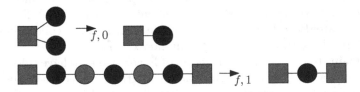

Fig. 1. Example reductions for RED-BLUE DOMINATING SET

The discussion above shows that for each i, the relation $\sim^t_{planar,rbds}$ has a finite number of equivalence classes. For each $t \leq 4d$ and each equivalence class, we select a *representative*. We can do this such that, whenever $G \rightarrow_{planar,rbds,i} H$ for a representative H, $i \geq 0$. This ensures that the parameter does not increase when we carry out a reduction. As these representatives and d are only problem dependent, we can assume that the largest representative has size $O(1)$.

The main step of the kernelization algorithm is the following: we replace each of the $O(k)$ boundaried graphs G_i in the decomposition, implied by Lemma 1 by its representative for the relation $\sim^t_{planar,rbds}$, and update the parameter k accordingly. That is, if we replace t-boundaried subgraph H by K, and $H \rightarrow_{rbds,i} K$, then we subtract i from k. Now, each of these reductions keep the graph planar. They are also 'safe' with respect to the answer to the RED-BLUE DOMINATING SET problem.

As each of the graphs G_i has bounded diameter, it also has bounded treewidth (see for example [14, Theorem 83] or [75]), and thus, we can compute its equivalence class for $\sim^t_{planar,rbds}$. After transforming each G_i in the decomposition to the representative of its equivalence class, we have a partition of the input graph in $O(k)$ parts, each of size $O(1)$, and thus obtained an equivalent input of size $O(k)$.

The sketch above does neither give the most efficient, nor the simplest kernelization algorithm for RED-BLUE DOMINATING SET on planar graphs, but it illustrates an approach that works for a large collection of problems, as was shown in [17]. Several other techniques and generalizations are used to obtain the following results.

For a fixed g, we consider parameterized problems on graphs that can be embedded into a surface of Euler-genus at most g. For a graph $G = (V, E)$ given with an embedding, the *radial distance* for two vertices $v, w \in V$ is the minimum length of a sequence of vertices, starting with v and ending with w, where each two successive vertices share a face. A parameterized problem is *compact*, if for each yes-instance (G, k), we can embed G on a surface of Euler-genus g and select a set S of $O(k)$ vertices, such that each vertex in G is at radial distance at most r, for some fixed r that only depends on the problem. An additional technical (and in all relevant cases trivially fulfilled) condition is that $k \leq |V|^r$. For example, the FEEDBACK VERTEX SET problem is compact: if S is a set of vertices such that each cycle contains a vertex in S, then each vertex in G shares a face with a vertex in S so is at radial distance at most 1.

A generalization of compactness is *quasi-compactness*: now, we are allowed to split the vertices in two sets, one inducing a subgraph of bounded treewidth, and one that are at bounded radial distance to a set of size $O(k)$; and again, $k \leq |V|^r$.

Theorem 3 (Bodlaender et al. [17]). *Let g be a fixed integer. Every parameterized problem on graphs of Euler-genus g that is finite integer index and that is quasi-compact or whose complement is quasi-compact has a linear kernel.*

So, e.g., FEEDBACK VERTEX SET restricted to graphs of Euler-genus g has a linear kernel for all g. (This generalizes the result of [19].)

A weaker result was obtained in [17] for compact optimization problems that can be formulated with Counting Monadic Second Order Logic (CMSO). Consider a predicate expressed in CMSO that formulates a property of graphs and vertex sets $P(G, S)$. E.g., S is a dominating set in G can be expressed as:

$$\forall v \in V : \exists w \in V : w \in S \wedge (\{v, w\} \in E \vee v = w)$$

We formulate Theorem 1 from [17] in a slightly weaker but easier to understand form.

Theorem 4 (Bodlaender et al. [17]). *Let g be a fixed integer. Let P be a CMSO-expressible property of graphs and vertex sets. Consider a problem Q, whose input consists of a graph $G = (V, E)$ of Euler-genus at most g, a set of vertices $Y \subseteq V$, and an integer k. Suppose Q is compact or the complement of G is compact.*

1. *If Q is of the form: $\exists S \subseteq Y : |S| \leq k \wedge P(G, S)$, then Q has a kernel of size $O(k^2)$.*
2. *If Q is of the form: $\exists S \subseteq Y : |S| = k \wedge P(G, S)$, then Q has a kernel of size $O(k^3)$.*
3. *If Q is of the form: $\exists S \subseteq V : |S \cap Y| \geq k \wedge P(G, S)$, then Q has a kernel of size $O(k^2)$.*

The set Y plays the role of *annotations*, e.g., in parts 1 and 2 of Theorem 4, vertices in $V - Y$ are 'annotated' in the sense that they cannot belong to the solution. The theorem leaves room for improvement: can we get rid of these annotations, and can we obtain linear kernels for these problems? These are important, but probably not easy, open problems. Several applications of Theorems 3 and 4 can be found in [17].

2.3   Meta Theorems for Graphs Avoiding a Minor

Fomin et al. [54] obtained a characterization of a large collection of problems that have a small kernel on graphs that avoid a minor. A central tool in their results is the notion of *bidimensionality*: a notion that has played an important role in several important meta-results for problems on graphs avoiding a minor, both with respect to parameterized algorithms and with respect to approximation algorithms. See the overview paper by Demaine and Hajiaghayi [40]. We sketch a few notions used for stating the meta-theorem from [54]. A graph H is a *minor* of a graph $G = (V, E)$ if H can be obtained from G by a series of zero or more vertex deletions, edge deletions and edge contractions. Consider graph parameter f that maps each graph to an integer, and the corresponding parameterized problem P_f to determine for a given graph G and parameter value k, if $f(G) \leq k$. We say that P_f is *minor-bidimensional*, if for any minor H of G, $f(H) \leq f(G)$ (i.e., f cannot increase when taking minors), and there is some $\delta > 0$, such that for the r by r grid GR_r, $f(GR_r) \geq \delta r^2$. The notion of *contraction-bidimensional* is defined similarly. Now f does not increase when contracting edges, and instead of a grid, a grid with additional triangulation edges is used. For precise definitions, see for example [53]. The *separation property* is a technical condition, that holds for several problems, and is often easy to verify; for the precise definition, we refer again to [53]. A graph $G = (V, E)$ is an *apex* graph, if there is a vertex $v \in V$, such that the graph, obtained by removing v from G and its incident edges is planar. Theorem 5 also uses the notion of finite integer index, which was explained above.

Theorem 5 (Fomin et al [54]). *(i) Let H be a graph, and P be a parameterized problem, that is minor-bidimensional, has the separation property, and is finite integer index. Then P, restricted to graphs that do not have H as minor, has a quadratic kernel.*
(ii) Let H be an apex graph, and P be a parameterized problem, that is contraction-bidimensional, has the separation property, and is finite integer index. Then P, restricted to graphs that do not have H as minor, has a quadratic kernel.

Theorem 5 proves the existence of quadratic kernels for several problems, e.g., a quadratic kernel for DISJOINT CYCLES on H-minor free graphs for any fixed graph H, and a quadratic kernel for DOMINATING SET on H-minor free graphs for any fixed apex graph H. See also [74,64] for related results.

It would be very interesting to try to obtain more general meta-kernelization results, with simpler or less conditions on the problem, and with linear kernels.

3   Lower Bounds: No Polynomial Kernels

In this section, we discuss *lower bounds* techniques for kernels. A number of linear lower bounds for kernel sizes were found by Chen et al. [31]; recent techniques allow to show larger lower bounds, building upon complexity theoretic assumptions.

In this section, we will sometimes view a problem as a *parameterized* problem, and sometimes as a 'classic' decision problem. To a parameterized problem P, we can associate the decision problem P^c, where we assume the parameter to be given in unary, and which has the same set of yes-instances. So, to the parameterized VERTEX COVER problem, we can associate the 'classic' VERTEX COVER problem with the output size k given in unary.

Showing for parameterized problems whose classic variant is NP-complete that they do not have a kernel of polynomial size is very hard, as such a proof would imply that $P \neq NP$. However, we have proofs for several concrete problems that they do not have a kernel of polynomial size, unless $NP \subseteq coNP/poly$, or, in a few cases, a weaker condition. As a first example, consider the LONG PATH problem:

LONG PATH
Instance: undirected graph $G = (V, E)$, integer k
Parameter: k
Question: Does G have a simple path with at least k edges?

The classic variant of this problem is NP-complete (containing HAMILTONIAN PATH as a special case), and the parameterized variant belongs to FPT. Much study has been done on parameterized algorithms for this problem; recently, an algorithm with $O(4^{k+o(k)}m)$ time algorithm for the problem was found by Chen et al. [34]. While using Theorem 1 gives us a kernel whose size is exponential in k, it is unlikely that there exists a kernel whose size is polynomial in k.

One can have the following intuition. Suppose there would be a kernelization algorithm, giving kernels to LONG PATH with at most k^c vertices and edges, for some constant c. Now, take a graph G with, say, k^{2c} connected components. There is a long path with k edges in G if and only if at least one of the connected components of G has a path with k edges. A solution in one connected component in G does not seem to have impact on a solution for another connected component. Thus, as we have much more connected components than the kernel size, it seems that we must solve some connected components to get this small kernel. But solving the LONG PATH problem for a connected component cannot be done in polynomial time, unless $P = NP$.

With the present state of theory, we need an assumption, different from $P \neq NP$, namely $NP \nsubseteq coNP/poly$. Still, for many problems we can show, under this assumption, that they do not have a kernel of polynomial size. Central in the theory is the notion of *compositionality*, which states thet, given several instances with the same parameter, we can build one instance of polynomial size with bounded parameter. More formally, we have the following definition.

Definition 1. *An* or-composition *algorithm for a parameterized problem $Q \subseteq \Sigma^* \times \mathbf{N}$ is an algorithm, that gets as input a sequence $((x_1, k), \ldots, (x_r, k))$, with each $(x_i, k_i) \in \Sigma^* \times \mathbf{N}$, and outputs a pair (x', k'), such that*

- *the algorithm uses time polynomial in $\sum_{1 \leq i \leq r} |x_i| + k$;*
- *k' is bounded by a polynomial in k*
- *$(x', k') \in Q$, if and only if there exists an i, $1 \leq i \leq r$, with $(x_i, k) \in Q$.*

We have a similar definition for *and-compositionality*; the last condition is replaced by

- *$(x', k') \in Q$, if and only if $(x_i, k) \in Q$ for all i, $1 \leq i \leq r$.*

LONG PATH is or-compositional: a series of inputs to LONG PATH with the same parameter $(G_1, k), \ldots, (G_r, k)$ can be mapped to $(G_1 \cup \cdots G_r, k)$, i.e., we just take the disjoint union of the graphs. It is easy to see that the conditions of or-compositionality are fulfilled. Actually, the same proof can be used for all problems where we want to maximize a graph parameter for which the value of a graph is the maximum value of its connected components. If we want to minimize a variable where the value of a graph is the maximum value of its connected components, like for TREEWIDTH, then we have and-compositionality. One further ingredient are two conjectures, by Bodlaender et al. [15].

Conjecture 1 (Or-distillation conjecture [15]). Let R be an NP-complete problem. There is no algorithm D, that gets as input a series of m instances of R, and outputs one instance of R, such that

- If D has as input m instances, each of size at most n, then D uses time polynomial in m and n, and its output size is bounded by a function that is polynomial in n.
- If D has as input instances $x_1, \ldots, x_m$, then $D(x_1, \ldots, x_m) \in R$, if and only if $\exists_{1 \le i \le m} x_i \in R$.

Conjecture 2 (And-distillation conjecture [15]). Let R be an NP-complete problem. There is no algorithm D, that gets as input a series of m instances of R, and outputs one instance of R, such that

- If D has as input m instances, each of size at most n, then D uses time polynomial in m and n, and its output size is bounded by a function that is polynomial in n.
- If D has as input instances $x_1, \ldots, x_m$, then $D(x_1, \ldots, x_m) \in R$, if and only if $\forall_{1 \le i \le m} x_i \in R$.

The relation between the existence of polynomial kernels, compositionality, and these conjectures is given by the following theorem.

Theorem 6 (Bodlaender et al. [15]). *Let P be a parameterized problem with P^c its corresponding classic decision variant.*

1. *If P is or-compositional and P^c is NP-complete, then P has no kernel of polynomial size, unless the or-distillation conjecture does not hold.*
2. *If P is and-compositional and P^c is NP-complete, then P has no kernel of polynomial size, unless the and-distillation conjecture does not hold.*

We can sharpen Theorem 6(i), by using a result by Fortnow and Santhaman [56].

Theorem 7 (Fortnow and Santhaman [56]). *If the or-distillation conjecture does not hold, then $NP \subseteq coNP/poly$.*

Corollary 1 (Bodlaender et al. [15], Fortnow and Santhaman [56]). *Let P be a parameterized problem with P^c its corresponding classic decision variant. If P is or-compositional and P^c is NP-complete, then P has no kernel of polynomial size, unless $NP \subseteq coNP/poly$.*

Corollary 1 can frequently be used to obtain evidence that problems have no polynomial kernel: we need a proof that the problem is or-compositional (which, in several cases, is not hard to establish), and a proof of NP-completeness. Note that we do not have a variant of the Theorem 7 for and-compositionality; this is an important open problem. Thus, for and-compositional problems, the evidence of non-existence of polynomial kernels is weaker.

There also exist several problems that cannot be easily seen to be compositional, but for which we can still derive evidence for non-existence of polynomial kernels, using transformations. The arguments are a variant of the theory of NP-completeness.

Definition 2. *Let P and Q be parameterized problems. We say that P is polynomial time and parameter reducible to Q, written $P \leq_{Ptp} Q$, if there exists a polynomial time computable function $f : \{0,1\}^* \times \mathbf{N} \rightarrow \{0,1\}^* \times \mathbf{N}$, and a polynomial $p : \mathbf{N} \rightarrow \mathbf{N}$, and for all $x \in \{0,1\}^*$ and $k \in \mathbf{N}$, if $f((x,k)) = (x',k')$, then the following hold:*

- *$(x,k) \in P$, if and only if $(x',k') \in Q$, and*
- *$k' \leq p(k)$.*

We call f a polynomial time and parameter transformation from P to Q.

The main difference between the 'usual' polynomial time transformations from the theory of NP-completeness is that now, in addition, we demand that the parameter is mapped to a parameter whose value is bounded by a polynomial of the old parameter. Also, note that the fixed parameter transformations as introduced by Downey and Fellows (see [42,43,44]) are similar, except that these allow non-polynomial growth of the parameter. Also, fixed parameter transformations are used in general to show hardness for $W[1]$ or a related class, and thus are used for problems of which we expect that there exist no kernel at all; while polynomial time and parameter transformations are used to for problems to show that we do not expect the existence of a polynomial kernel. The following result is a 'folklore' theorem.

Theorem 8. *Let P and Q be parameterized problems, and suppose that P^c and Q^c are the derived classical problems. Suppose that P^c is NP-complete, and $Q^c \in NP$. Suppose P is polynomial time and parameter reducible to Q. If Q has a polynomial kernel, then P has a polynomial kernel.*

Proof. We sketch the proof. Suppose Q has a polynomial kernel. We build a polynomial kernel for P as follows. Take an input (I,k) to P. Apply the polynomial time and parameter reduction to this input, and obtain (I',k') as equivalent input to Q. Apply the kernelization algorithm for Q to this input, and we obtain an input (I'',k'') to Q. $|I''|$ and k'' are polynomially bounded in k', and k' is polynomially bounded in k. Now, NP-completeness of P^c shows that we can transform (I'',k'') to an equivalent input (I''',k''') to P, whose size is polynomially bounded in $|I''| + k''$, and hence polynomially bounded in k. One easily sees that this input is also equivalent. Here we use that the parameter is encoded in unary in the derived classical problems. $\square$

The technique is used in several recent papers to obtain non-trivial proofs of the non-existence of polynomial kernels, under the usual assumption $NP \not\subseteq coNP/poly$. Fernau et al. [50] apply this technique to the k-Leaf-Out-Branching problem: given a digraph, give a rooted oriented spanning tree with at least k leaves. Curiously, the variant Rooted k-Leaf-Out-Branching does have a kernel of size $O(k^2)$, as was shown very recently by Daligault and Thomassé [36] improving upon an $O(k^3)$ kernel by Fernau et al. [50]. Thus, k-Leaf-Out-Branching has something what is called a "cheat kernelization" in

[50]: we can transform the input to $O(n)$ inputs, each of size at most $O(k^2)$, by building a kernel for each of the n choices of a root.

An interesting open problem is to find techniques to give evidence of the non-existence of such "cheat kernels", i.e., transformations to a polynomial (in the input size n) number of inputs of size polynomial in the parameter k, for problems in FPT.

Dom, Lokshtanov and Saurahb [41] obtain results for the non-existence of polynomial kernels (unless $NP \subseteq coNP/poly$) for a large number of problems, where involved reduction techniques, based on colored versions of problems and identifications of vertices are used. These include natural parameterized versions of CONNECTED VERTEX COVER, CAPACITATED VERTEX COVER, STEINER TREE, RED-BLUE DOMINATING SET, DOMINATING SET, UNIQUE COVERAGE, and SMALL SUBSET SUM. Bodlaender, Thomassé and Yeo [21] apply the techniques to get non-polynomial kernel-results for DISJOINT CYCLES and DISJOINT PATHS. Kratsch and Wahlstrom [68], answering an open problem by Cai at IW-PEC 2006, show that there exists a graph H on seven vertices such that the H-FREE EDGE DELETION and H-FREE EDGE EDITING problems do not have polynomial kernels, unless $NP \subseteq coNP/poly$.

A very interesting and very recent development is work by Dell and van Melkebeek [39], who obtained lower bounds for compressibility of instances for vertex cover problems, satisfiability problems, and subgraph-deletion type problems.

Theorem 9 (Dell and van Melkebeek [39]). *Let $d \geq 2$ be an integer and $\epsilon > 0$ a real number. If $NP \nsubseteq coNP/poly$, then there is no polynomial-time mapping reduction from* VERTEX COVER *for d-uniform hypergraphs to any language such that instances with n vertices are mapped to instances of bitlength at most $O(n^{d-\epsilon})$.*

Note that the result is more general than kernel lower bounds in two ways: lower bounds are given also for reductions to other problems and not only reductions to the problem itself, and the bound is expressed as function of the input size n. The bound is essentially tight, as an input for VERTEX COVER for d-uniform hypergraphs has size at most $O(n^d)$.

For appreciation of the result, let us briefly look at what it implies for kernelization for VERTEX COVER on undirected graphs. The Nemhauser-Trotter kernel [72] gives a kernel with at most $2k$ vertices, but it can have $\Theta(n^2)$ edges. Theorem 9 with $d = 2$ shows that we should also expect this many edges for such a kernel: we cannot expect a kernel with $O(n^{2-\epsilon})$ edges for any $\epsilon > 0$.

Dell and van Melkebeek use Theorem 9 to obtain lower bounds for compressibility of several other problems, including SATISFIABILITY for d-CNF formulas and a large class of subgraph deletion problems. For example, they show:

Theorem 10 (Dell and van Melkebeek [39]). *If $NP \nsubseteq coNP/poly$, then there is no polynomial-time mapping reduction from* FEEDBACK VERTEX SET *to any language instances with parameter k are mapped to instances of bitlength at most $O(k^{2-\epsilon})$.*

The kernelization algorithm for FEEDBACK VERTEX SET given by Thomassé [78] gives reduced instances with $O(k^2)$ vertices and $O(k^2)$ edges. Thus, by Theorem 10, Thomassé's algorithm is asymptotically optimal with respect to the number of edges.

4 Conclusions

Kernelization is a very interesting modern topic of algorithm design and analysis, giving new insights to the ancient techniques of preprocessing, simplification and data reduction. In this overview paper, a few of the new theoretical methods were discussed; in particular, we looked at meta-theorems that imply the existence of small kernels for various problems on planar graphs and generalizations of planar graphs, and at lower bound techniques, i.e., methods that give evidence for various problems that they do not have kernels of polynomial size.

Besides a theoretical analysis of kernelization algorithms, it is also very interesting to experimentally evaluate kernelization algorithms. Experiments on kernelization have been carried out for several important problems, e.g., for CLIQUE COVER [57], CLUSTER EDITING [13], DOMINATING SET [5], FEEDBACK VERTEX SET [23], and VERTEX COVER [3,4].

The topic of kernelization is a relatively new area, with a lot of new developments and techniques. Fellows calls the area in his invited talk *the lost continent of polynomial time*. Indeed, there remains a lot to explore. Let me end with another metaphor. In the family of algorithmics, kernelization is a new member. As a child of fixed parameter tractability theory, she is getting a life of her own. While still dependent on her parent, she also often is of help to her parent. At what life stage would she be? Rapid growth and still sometimes with problems to understand herself, the life stage would be infancy or adolescence. However, the prospects for her future are great, and I hope that the reader will contribute to this future with new results, insights, techniques, and applications.

Acknowledgments

This paper would not have been possible for me to write without the help of several colleagues, in the form of comments, answers to questions, discussions, and cooperation. In particular, I thank Holger Dell, Thomas van Dijk, Rod Downey, Mike Fellows, Danny Hermelin, Bart Jansen, Eelko Penninkx, Johan van Rooij, Stephan Thomassé, and Anders Yeo. I apologize to authors whose work should have been mentioned here, but was missed.

References

1. Abrahamson, K.A., Downey, R.G., Fellows, M.R.: Fixed-parameter tractability and completeness IV: On completeness for W[P] and PSPACE analogues. Annals of Pure and Applied Logic 73, 235–276 (1995)

2. Abrahamson, K.R., Fellows, M.R.: Finite automata, bounded treewidth and well-quasiordering. In: Robertson, N., Seymour, P. (eds.) Proceedings of the AMS Summer Workshop on Graph Minors, Graph Structure Theory. Contemporary Mathematics, vol. 147, pp. 539–564. American Mathematical Society (1993)

3. Abu-Khzam, F.N., Collins, R.L., Fellows, M.R., Langston, M.A., Suters, W.H., Symons, C.T.: Kernelization algorithms for the vertex cover problem: Theory and experiments. In: Proceedings of the 6th Workshop on Algorithm Engineering and Experimentation and the 1st Workshop on Analytic Algorithmics and Combinatorics, ALENEX/ANALCO 2004, pp. 62–69. ACM-SIAM (2004)

4. Abu-Khzam, F.N., Fellows, M.R., Langston, M.A., Suters, W.H.: Crown structures for vertex cover kernelization. Theory of Computing Systems 41, 411–430 (2007)

5. Alber, J., Betzler, N., Niedermeier, R.: Experiments in data reduction for optimal domination in networks. Annals of Operations Research 146, 105–117 (2006)

6. Alber, J., Fellows, M.R., Niedermeier, R.: Polynomial-time data reduction for dominating sets. J. ACM 51, 363–384 (2004)

7. Appel, K., Haken, W.: Every planar map is 4-colorable. Illinois J. Math. 21, 429–567 (1977)

8. Arnborg, S., Courcelle, B., Proskurowski, A., Seese, D.: An algebraic theory of graph reduction. J. ACM 40, 1134–1164 (1993)

9. Baker, B.S.: Approximation algorithms for NP-complete problems on planar graphs. J. ACM 41, 153–180 (1994)

10. Bessy, S., Fomin, F.V., Gaspers, S., Paul, C., Perez, A., Saurabh, S., Thomassé, S.: Kernels for feedback arc set in tournaments. The Computing Research Repository, abs/0907.2165. To appear in proceedings FSTTCS 2009 (2009)

11. Betzler, N., Fellows, M.R., Guo, J., Niedermeier, R., Rosamond, F.A.: Fixed-parameter algorithms for Kemeny rankings. Theor. Comp. Sc. 410, 4554–4570 (2009)

12. Bienstock, D., Langston, M.A.: Algorithmic implications of the graph minor theorem. In: Ball, M.O., Magnanti, T.L., Monma, C.L., Nemhauser, G.L. (eds.) Handbook of Operations Research and Management Science: Network Models, pp. 481–502. North-Holland, Amsterdam (1995)

13. Böcker, S., Briesemeister, S., Klau, G.W.: Exact algorithms for cluster editing: Evaluation and experiments. To appear in Algorithmica (2009) doi 10.1007/s00453-009-9339-7

14. Bodlaender, H.L.: A partial k-arboretum of graphs with bounded treewidth. Theor. Comp. Sc. 209, 1–45 (1998)

15. Bodlaender, H.L., Downey, R.G., Fellows, M.R., Hermelin, D.: On problems without polynomial kernels (Extended abstract). In: Aceto, L., Damgård, I., Goldberg, L.A., Halldórsson, M.M., Ingólfsdóttir, A., Walukiewicz, I. (eds.) ICALP 2008, Part I. LNCS, vol. 5125, pp. 563–574. Springer, Heidelberg (2008)

16. Bodlaender, H.L., Fellows, M.R., Langston, M., Ragan, M., Rosamond, F., Weyer, M.: Quadratic kernelization for convex recoloring of trees. In: Lin, G. (ed.) COCOON 2007. LNCS, vol. 4598, pp. 86–96. Springer, Heidelberg (2007)

17. Bodlaender, H.L., Fomin, F.V., Lokshtanov, D., Penninkx, E., Saurabh, S., Thilikos, D.M. (Meta) kernelization. To appear in Proceedings FOCS 2009 (2009)

18. Bodlaender, H.L., Lokshtanov, D., Penninkx, E.: Planar capacitated dominating set is W[1]-hard. In: Proceedings IWPEC 2009 (2009)

19. Bodlaender, H.L., Penninkx, E.: A linear kernel for planar feedback vertex set. In: Grohe, M., Niedermeier, R. (eds.) IWPEC 2008. LNCS, vol. 5018, pp. 160–171. Springer, Heidelberg (2008)

20. Bodlaender, H.L., Penninkx, E., Tan, R.B.: A linear kernel for the k-disjoint cycle problem on planar graphs. In: Hong, S.-H., Nagamochi, H., Fukunaga, T. (eds.) ISAAC 2008. LNCS, vol. 5369, pp. 294–305. Springer, Heidelberg (2008)

21. Bodlaender, H.L., Thomassé, S., Yeo, A.: Kernel bounds for disjoint cycles and disjoint paths. In: Fiat, A., Sanders, P. (eds.) ESA 2009. LNCS, vol. 5757, pp. 635–646. Springer, Heidelberg (2009)

22. Bodlaender, H.L., van Antwerpen-de Fluiter, B.: Reduction algorithms for graphs of small treewidth. Information and Computation 167, 86–119 (2001)

23. Bodlaender, H.L., van Dijk, T.C.: A cubic kernel for feedback vertex set and loop cutset. To appear in Theory of Computing Systems (2009) doi: 10.1007/s00224-009-9234-2

24. Borie, R.B., Parker, R.G., Tovey, C.A.: Automatic generation of linear-time algorithms from predicate calculus descriptions of problems on recursively constructed graph families. Algorithmica 7, 555–581 (1992)

25. Bousquet, N., Daligault, J., Thomassé, S., Yeo, A.: A polynomial kernel for multicut in trees. In: Albers, S., Marion, J.-Y. (eds.) Proceedings 26th International Symposium on Theoretical Aspects of Computer Science, STACS 2009, Schloss Dagstuhl, Germany. Dagstuhl Seminar Proceedings, vol. 09001, pp. 183–194. Leibniz-Zentrum für Informatik (2009)

26. Buss, J.F., Goldsmith, J.: Nondeterminism within P. SIAM J. Comput. 22, 560–572 (1993)

27. Cai, L., Chen, J.: On fixed-parameter tractability and approximability of NP optimization problems. Journal of Computer and System Sciences 54, 465–474 (1997)

28. Cai, L., Chen, J., Downey, R.G., Fellows, M.R.: Advice classes of parameterized tractability. Annals of Pure and Applied Logic 84, 119–138 (1997)

29. Cai, L., Juedes, D.: On the existence of subexponential parameterized algorithms. Journal of Computer and System Sciences 67, 789–807 (2003)

30. Chen, J., Chor, B., Fellows, M., Huang, X., Juedes, D.W., Kanj, I.A., Xia, G.: Tight lower bounds for certain parameterized NP-hard problems. Information and Computation 201, 216–231 (2005)

31. Chen, J., Fernau, H., Kanj, I.A., Xia, G.: Parametric duality and kernelization: Lower bounds and upper bounds on kernel size. SIAM J. Comput. 37, 1077–1106 (2007)

32. Chen, J., Huang, X., Kanj, I.A., Xia, G.: On the computational hardness based on linear FPT-reductions. Journal of Combinatorial Optimization 11, 231–247 (2006)

33. Chen, J., Huang, X., Kanj, I.A., Xia, G.: Strong computational lower bounds via parameterized complexity. Journal of Computer and System Sciences 72, 1346–1367 (2006)

34. Chen, J., Lu, S., Sze, S.-H., Zhang, F.: Improved algorithms for path, matching, and packing problems. In: Bansal, N., Pruhs, K., Stein, C. (eds.) Proceedings of the 17th Annual ACM-SIAM Symposium on Discrete Algorithms, SODA 2007, pp. 298–307 (2007)

35. Courcelle, B.: The monadic second-order logic of graphs I: Recognizable sets of finite graphs. Information and Computation 85, 12–75 (1990)

36. Daligault, J., Thomassé, S.: On finding directed trees with many leaves. In: Proceedings IWPEC 2009 (2009)

37. de Fluiter, B.: Algorithms for Graphs of Small Treewidth. PhD thesis, Utrecht University (1997)

38. Dehne, F., Fellows, M., Fernau, H., Prieto, E., Rosamond, F.: Nonblocker: Parameterized algorithms for minimum dominating set. In: Wiedermann, J., Tel, G., Pokorný, J., Bieliková, M., Štuller, J. (eds.) SOFSEM 2006. LNCS, vol. 3831, pp. 237–245. Springer, Heidelberg (2006)
39. Dell, H., van Melkebeek, D.: Satisfiability allows no nontrivial sparsification unless the polynomial-time hierarchy collapses. In: To appear in: Electronic Colloquium on Computational Complexity (ECCC), vol. 16 (2009)
40. Demaine, E.D., Hajiaghayi, M.: The bidimensionality theory and its algorithmic applications. The Computer Journal 51, 292–302 (2008)
41. Dom, M., Lokshtanov, D., Saurabh, S.: Incompressibility through colors and IDs. In: Albers, S., Marchetti-Spaccamela, A., Matias, Y., Nikoletseas, S.E., Thomas, W. (eds.) ICALP 2009, Part I. LNCS, vol. 5555, pp. 378–389. Springer, Heidelberg (2009)
42. Downey, R.G., Fellows, M.R.: Fixed-parameter tractability and completeness I: Basic results. SIAM J. Comput. 24, 873–921 (1995)
43. Downey, R.G., Fellows, M.R.: Fixed-parameter tractability and completeness II: On completeness for W[1]. Theor. Comp. Sc. 141, 109–131 (1995)
44. Downey, R.G., Fellows, M.R.: Parameterized Complexity. Springer, Heidelberg (1999)
45. Downey, R.G., Fellows, M.R., Stege, U.: Parameterized complexity: A framework for systematically confronting computational intractability. In: DIMACS Series in Discrete Mathematics and Theoretical Computer Science, pp. 49–99. American Mathematical Society (1997)
46. Estivill-Castro, V., Fellows, M.R., Langston, M.A., Rosamond, F.A.: Fpt is P-time extremal structure I. In: Broersma, H., Johnson, M., Szeider, S. (eds.) Proceedings of the 1st Workshop on Algorithms and Complexity in Durham, ACiD 2005. Text in Algorithms, vol. 4, pp. 1–41. King's College, London (2005)
47. Fellows, M.R.: Personal communication
48. Fellows, M.R., Langston, M.A.: An analogue of the Myhill-Nerode theorem and its use in computing finite-basis characterizations. In: Proceedings of the 30th Annual Symposium on Foundations of Computer Science, FOCS 1989, pp. 520–525 (1989)
49. Fernau, H.: Edge dominating set: Efficient enumeration-based exact algorithms. In: Bodlaender, H.L., Langston, M.A. (eds.) IWPEC 2006. LNCS, vol. 4169, pp. 140–151. Springer, Heidelberg (2006)
50. Fernau, H., Fomin, F.V., Lokshtanov, D., Raible, D., Saurabh, S., Villanger, Y.: Kernel(s) for problems with no kernel: On out-trees with many leaves (extended abstract). In: Albers, S., Marion, J.-Y. (eds.) Proceedings 26th International Symposium on Theoretical Aspects of Computer Science, STACS 2009, Schloss Dagstuhl, Germany. Dagstuhl Seminar Proceedings, vol. 09001, pp. 421–432. Leibniz-Zentrum für Informatik (2009)
51. Fernau, H., Juedes, D.W.: A geometric approach to parameterized algorithms for domination problems on planar graphs. In: Fiala, J., Koubek, V., Kratochvíl, J. (eds.) MFCS 2004. LNCS, vol. 3153, pp. 488–499. Springer, Heidelberg (2004)
52. Flum, J., Grohe, M.: Parameterized Complexity Theory. Springer, Heidelberg (2006)
53. Fomin, F.V., Golovach, P.A., Lokshtanov, D., Saurabh, S.: Algorithmic lower bounds for problems parameterized by clique-width. To appear in Proceedings SODA 2010 (2009)
54. Fomin, F.V., Lokshtanov, D., Saurabh, S., Thilikos, D.M.: Bidimensionality and kernels. To appear in Proceedings SODA 2010 (2009)

55. Fomin, F.V., Thilikos, D.M.: Fast parameterized algorithms for graphs on surfaces: Linear kernel and exponential speedup. In: Díaz, J., Karhumäki, J., Lepistö, A., Sannella, D. (eds.) ICALP 2004. LNCS, vol. 3142, pp. 581–592. Springer, Heidelberg (2004)
56. Fortnow, L., Santhanam, R.: Infeasibility of instance compression and succinct PCPs for NP. In: Proceedings of the 40th Annual Symposium on Theory of Computing, STOC 2008, pp. 133–142. ACM Press, New York (2008)
57. Gramm, J., Guo, J., Hüffner, F., Niedermeier, R.: Data reduction and exact algorithms for clique cover. ACM Journal of Experimental Algorithms13(2.2) (2008)
58. Guo, J.: A more effective linear kernelization for cluster editing. Theor. Comp. Sc. 410, 718–726 (2009)
59. Guo, J., Niedermeier, R.: Invitation to data reduction and problem kernelization. ACM SIGACT News 38, 31–45 (2007)
60. Guo, J., Niedermeier, R.: Linear problem kernels for NP-hard problems on planar graphs. In: Arge, L., Cachin, C., Jurdziński, T., Tarlecki, A. (eds.) ICALP 2007. LNCS, vol. 4596, pp. 375–386. Springer, Heidelberg (2007)
61. Guo, J., Niedermeier, R., Wernicke, S.: Fixed-parameter tractability results for full-degree spanning tree and its dual. In: Bodlaender, H.L., Langston, M.A. (eds.) IWPEC 2006. LNCS, vol. 4169, pp. 203–214. Springer, Heidelberg (2006)
62. Guo, J., Niedermeier, R., Wernicke, S.: Parameterized complexity of vertex cover variants. Theory of Computing Systems 41, 501–520 (2007)
63. Gutin, G., Razgon, I., Kim, E.J.: Minimum leaf out-branching and related problems. Theor. Comp. Sc. 410, 4571–4579 (2009)
64. Gutner, S.: Polynomial kernels and faster algorithms for the dominating set problem on graphs with an excluded minor. In: Chen, J., Fomin, F.V. (eds.) IWPEC 2009. LNCS, vol. 5917, pp. 246–257. Springer, Heidelberg (2009)
65. Jansen, B.: Fixed parameter complexity of the weighted max leaf problem. Master's thesis, Department of Computer Science, Utrecht University (2009)
66. Kneis, J., Mölle, D., Richter, S., Rossmanith, P.: On the parameterized complexity of exact satisfiability problems. In: Jedrzejowicz, J., Szepietowski, A. (eds.) MFCS 2005. LNCS, vol. 3618, pp. 568–579. Springer, Heidelberg (2005)
67. Kratsch, S.: Polynomial kernelizations for MIN $F^+\Pi_1$ and MAX NP. In: Albers, S., Marion, J.-Y. (eds.) Proceedings 26th International Symposium on Theoretical Aspects of Computer Science, STACS 2009, Schloss Dagstuhl, Germany. Dagstuhl Seminar Proceedings, vol. 09001, pp. 601–612. Leibniz-Zentrum für Informatik (2009)
68. Kratsch, S., Wahlström, M.: Two edge modification problems without polynomial kernels. In: Proceedings IWPEC 2009 (2009)
69. Lokshtanov, D., Mnich, M., Saurabh, S.: Linear kernel for planar connected dominating set. In: Chen, J., Cooper, S.B. (eds.) TAMC 2009. LNCS, vol. 5532, pp. 281–290. Springer, Heidelberg (2009)
70. Moser, H.: A problem kernelization for graph packing. In: Nielsen, M., Kucera, A., Miltersen, P.B., Palamidessi, C., Tuma, P., Valencia, F.D. (eds.) SOFSEM 2009. LNCS, vol. 5404, pp. 401–412. Springer, Heidelberg (2009)
71. Moser, H., Sikdar, S.: The parameterized complexity of the induced matching problem. Disc. Appl. Math. 157, 715–727 (2009)
72. Nemhauser, G.L., Trotter, L.E.: Vertex packing: Structural properties and algorithms. Mathematical Programming 8, 232–248 (1975)

73. Niedermeier, R.: Invitation to fixed-parameter algorithms. Oxford Lecture Series in Mathematics and Its Applications. Oxford University Press, Oxford (2006)
74. Philip, G., Raman, V., Sikdar, S.: Solving dominating set in larger classes of graphs: FPT algorithms and polynomial kernels. In: Fiat, A., Sanders, P. (eds.) ESA 2009. LNCS, vol. 5757, pp. 694–705. Springer, Heidelberg (2009)
75. Robertson, N., Seymour, P.D.: Graph minors. III. Planar tree-width. J. Comb. Theory Series B 36, 49–64 (1984)
76. Robertson, N., Seymour, P.D.: Graph minors. XIII. The disjoint paths problem. J. Comb. Theory Series B 63, 65–110 (1995)
77. Robertson, N., Seymour, P.D.: Graph minors. XX. Wagner's conjecture. J. Comb. Theory Series B 92, 325–357 (2004)
78. Thomassé, S.: A quadratic kernel for feedback vertex set. In: Proceedings of the 19th Annual ACM-SIAM Symposium on Discrete Algorithms, SODA 2009, pp. 115–119 (2009)

A Faster Fixed-Parameter Approach to Drawing Binary Tanglegrams

Sebastian Böcker[1], Falk Hüffner[2], Anke Truss[1], and Magnus Wahlström[3]

[1] Lehrstuhl für Bioinformatik, Friedrich-Schiller-Universität Jena,
Ernst-Abbe-Platz 2, 07743 Jena, Germany
{sebastian.boecker,anke.truss}@uni-jena.de
[2] Algorithms in Computational Genomics group, School of Computer Science,
Tel Aviv University, Tel Aviv 69978, Israel
hueffner@tau.ac.il
[3] Max-Planck-Institut für Informatik, Department 1: Algorithms and Complexity,
Building 46.1, Stuhlsatzenhausweg 85, 66123 Saarbrücken, Germany
wahl@mpi-inf.mpg.de

Abstract. Given two binary phylogenetic trees covering the same n species, it is useful to compare them by drawing them with leaves arranged side-by-side. To facilitate comparison, we would like to arrange the trees to minimize the number of crossings k induced by connecting pairs of identical species. This is the NP-hard TANGLEGRAM LAYOUT problem. By providing a fast transformation to the BALANCED SUBGRAPH problem, we show that the problem admits an $O(2^k n^4)$ algorithm, improving upon a previous fixed-parameter approach with running time $O(c^k n^{O(1)})$ where $c \approx 1000$. We enhance a BALANCED SUBGRAPH implementation based on data reduction and iterative compression with improvements tailored towards these instances, and run experiments with real-world data to show the practical applicability of this approach. All practically relevant ($k \leq 1000$) TANGLEGRAM LAYOUT instances can be solved exactly within seconds. Additionally, we provide a kernel-like bound by showing how to reduce the BALANCED SUBGRAPH instances for TANGLEGRAM LAYOUT on complete binary trees to a size of $O(k \log k)$.

1 Introduction

In phylogenetics, researchers often wish to compare different phylogenetic trees with the same set of leaves: This can be two trees that resulted from applying different tree-building methods to the same dataset, a gene tree vs. species tree comparison, or a host-parasite comparison. The TANGLEGRAM LAYOUT problem (TL) deals with visually comparing a pair of binary rooted trees with identical leaf sets [4, 9]: The trees are drawn such that the leaves of both trees face each other, and each leaf is connected to the corresponding leaf in the opposing tree by an edge, see Fig. 1. A layout with many crossings of connecting edges can be hard or even impossible to analyze. Hence, our goal is to find a layout of the two trees such that we can draw connecting edges with as few crossings as possible.

J. Chen and F.V. Fomin (Eds.): IWPEC 2009, LNCS 5917, pp. 38–49, 2009.
© Springer-Verlag Berlin Heidelberg 2009

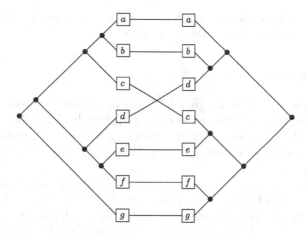

Fig. 1. A *tanglegram*

The TANGLEGRAM LAYOUT problem is NP-hard, even if both trees are complete [3]. In the same publication, Buchin et al. provide a $O(n^3)$ 2-approximation and a $O(4^k n^3)$ fixed-parameter algorithm for this special case, where k is the number of crossings in an optimal tanglegram. They also show that under the Unique Games Conjecture, there is no constant-factor approximation for the problem on general binary trees, that is, trees that are not necessarily complete. For general binary trees, the fastest fixed-parameter algorithm is due to Fernau et al. [4] and has running time $O(c^k n^{O(1)})$, where c was estimated by the authors to be about 1024.

On the application side, Nöllenburg et al. [8] compare different heuristics and exact algorithms for TANGLEGRAM LAYOUT. Baumann et al. [2] and Bansal et al. [1] study the generalization where the perfect matching between leaves is replaced by an arbitrary bipartite graph, and present heuristics and Integer Linear Programs for its solution.

In this paper, we transform TANGLEGRAM LAYOUT instances to the BALANCED SUBGRAPH problem. The fastest fixed-parameter algorithm for the latter problem is due to Hüffner et al. [7] and has running time $O(2^k m^2)$ time, where k is the number of edges violated in an optimal solution and m is the number of graph edges. As an algorithm engineering technique, the authors also provide a set of efficient data reduction rules for BALANCED SUBGRAPH.

Our contributions. We show in Sec. 3 that we can transform a TANGLEGRAM LAYOUT instance into a BALANCED SUBGRAPH instance in polynomial time, so that TANGLEGRAM LAYOUT is solvable in $O(2^k n^4)$ time, where k is the minimum number of crossings in a tanglegram of the two input trees and n the number of leaves in each input tree. In Sec. 4, we present a $O(k \log k)$ kernel-like bound on the size of BALANCED SUBGRAPH instances derived from tanglegrams with complete binary trees. In experiments described in Sec. 5, we give some improvements to the BALANCED SUBGRAPH solver tailored towards our application. We then apply the algorithm to synthetic and real-world tanglegram datasets and thus show that we can compute exact solutions for all practically relevant tanglegram instances within seconds.

2   Preliminaries

For an inner node v of a binary tree, let $T(v)$ be the subtree rooted at v, and $l(v)$ and $r(v)$ the left and right child of v, respectively. Then $L(v) := T(l(v))$ and $R(v) := T(r(v))$ are the subtrees rooted at $l(v)$ and $r(v)$, respectively. Let $\mathcal{L}(v)$ denote the set of leaf labels in $T(v)$. We identify leaves with their labels for the sake of readability.

The *last common ancestor* $\mathrm{lca}_T(l_1, l_2)$ of two leaves l_1, l_2 is the inner node v of T where $\{l_1, l_2\} \subseteq \mathcal{L}(v)$ but $\{l_1, l_2\} \not\subseteq \mathcal{L}(l(v))$ and $\{l_1, l_2\} \not\subseteq \mathcal{L}(r(v))$. The last common ancestor of two nodes is defined accordingly. To *switch* v means that we interchange the order of the children $l(v)$ and $r(v)$ such that the former $L(v)$ becomes $R(v)$, and vice versa, without changing node and leaf orders in $L(v)$ or $R(v)$.

Given two not necessarily complete binary trees S, T with identical leaf sets, a *tanglegram* is a planar embedding of S and T, where the two trees are contrasted in such a way that the leaves of each tree are arranged on one of two parallel straight lines, and identical leaves are connected by additional edges, see Figure 1. The task of the TANGLEGRAM LAYOUT problem is to find a tanglegram which minimizes the number of crossings between leaf label edges.

The BALANCED SUBGRAPH problem is defined as follows: Given an undirected multigraph with m edges, each of which is either labeled with $=$ or with $\neq$, the task is to find a two-coloring of the vertices that *violates* as few edges as possible. We say that a two-coloring *violates* an $=$-edge (an $\neq$-edge) if the incident vertices have different colors (the same color).

For ease of presentation, in the following we will assume all graphs to be multigraphs.

3   Transformation

In this section, we present the transformation from a TANGLEGRAM LAYOUT instance to a BALANCED SUBGRAPH instance. This, in turn, can be solved in time $O(2^k n^4)$ for input trees with n leaves that admit a tanglegram with at most k crossings [7]. A similar construction was used by Nöllenburg et al. [8] and by Buchin et al. [3] in the context of approximation and integer linear programming modelling, respectively.

Given two trees S, T, our starting point is an arbitrary tanglegram of the trees. We now transform this tanglegram into an instance of the BALANCED SUBGRAPH problem, that is, an undirected graph G where all edges are labeled '$=$' or '$\neq$'. This graph may contain multiple edges between two vertices.

Let G be a bipartite graph with vertex set $V_S \cup V_T$, where the vertices in V_S (or V_T) correspond to all inner vertices of S (or T, respectively). For each pair of leaf labels l_1, l_2, we draw an edge between the vertices corresponding to the last common ancestors $\mathrm{lca}_S(l_1, l_2)$ and $\mathrm{lca}_T(l_1, l_2)$ of the leaves labelled l_1, l_2 in

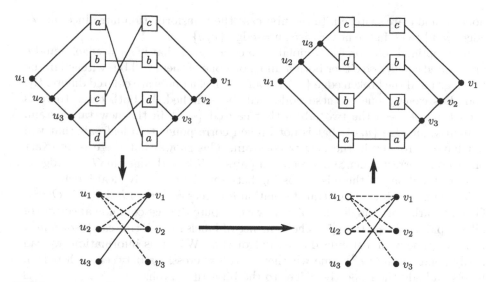

Fig. 2. An example of the transformation to BALANCED SUBGRAPH. An arbitrary tanglegram of the input trees (upper left) is transformed into a bipartite graph (lower left). Continuous lines denote =-edges, dashed lines ≠-edges. This instance can be solved by violating one edge, e.g. $\{u_2, v_2\}$, which leads to a valid two-coloring of the vertices (lower right). The vertices of one color, here u_1 and u_2, are switched to obtain an optimal tanglegram (upper right).

S and T. For each such edge we test whether there is a crossing between edges $l_1 - l_1$ and $l_2 - l_2$ in the current tanglegram. If so, we label the edge with '≠', else with '='.

We use G as input for a BALANCED SUBGRAPH algorithm which returns an optimal two-coloring of the vertices of G, such that the vertices incident to =-edges have the same color, whereas those incident to ≠-edges have different colors, while at the same time, violating as few edge labels as possible. This corresponds to leaving as few crossings as possible in the tanglegram: The optimal tanglegram is obtained by picking one of the colors, say white, and switching all inner vertices in S and T which correspond to white vertices in G. See Fig. 2 for an example.

Lemma 1. *The balanced subgraph instance generated by the transformation above is solvable with violating at most k edges if and only if the original trees admit a tanglegram with at most k crossings. The transformation can be done in quadratic time.*

Proof. Let S, T be two trees of a tanglegram instance and $G = (V_S \cup V_T, E)$ be the graph after the transformation.

It is easy to see that a crossing between two labels l_1, l_2 in the initial tanglegram is resolved if and only if exactly one of the last common ancestors $u = \text{lca}_S(l_1, l_2)$ and $v = \text{lca}_T(l_1, l_2)$ is switched. Analogously, if the edges between these labels are not crossed initially, they stay uncrossed if and only if both or none of the inner nodes u and v are switched. In the first case the transformation introduces an $\neq$-edge $\{u, v\}$ to G, in the latter case an $=$-edge $\{u, v\}$.

If an optimal solution for G violates an edge $\{u, v\}$ that is an $\neq$-edge, u and v are colored equally, so either both u and v or none of them will be switched in the tanglegram. If the broken edge $\{u, v\}$ is an $=$-edge, u and v are colored differently, and thus exactly one of these nodes will be switched. In both cases there is a crossing between the two labels that caused $\{u, v\}$ in the new tanglegram. Similarly, an edge $\{u, v\}$ that is not broken corresponds to a leaf pair that will not have a crossing in the new tanglegram. This shows that there is an exact one-to-one correspondence between leaf pairs in S, T and edges in G: An edge is broken if and only if there is a crossing between the respective leaf labels.

The running time of our transformation for trees S, T with n leaves is $O(n^2)$: Given an arbitrary tanglegram of S, T, we compute the last common ancestors of all leaf pairs in linear time [5]. The sequence of labels in each tree can be obtained in linear time with a standard search algorithm. With this information, we can easily compute in $O(n^2)$ time whether there are crossings between each pair of labels and add the respective edges to the bipartite graph. □

The BALANCED SUBGRAPH instance we generate has $2n - 2$ vertices and $\binom{n}{2}$ edges, as a binary tree has $n - 1$ inner nodes. The running time of the BALANCED SUBGRAPH algorithm from [7] is $O(2^k m^2)$ for a graph with m edges, so the total running time of our algorithm is $O(2^k n^4)$. Note, however, that the n^4 factor can be reduced in practice; see Section 5.

Baumann et al. [2] present an ILP for the generalized problem where leaves of the two trees are not necessarily connected by a perfect matching but instead by an arbitrary bipartite graph. We note that the above transformation can be applied to instances of this more general problem, too. This possibility has been also noted in [1].

Our transformation allows us to prove that BALANCED SUBGRAPH is NP-complete on bipartite graphs, but this can also be seen directly: For an arbitrary instance of BALANCED SUBGRAPH, insert dummy vertices into every edge, and replace an $=$-edge by two $=$-edges, and an $\neq$-edge by an $=$-edge and an $\neq$-edge. The resulting instance is bipartite, and has a solution if and only if the original instance has a solution.

Grötschel and Pulleyblank [6] showed that EDGE BIPARTIZATION can be solved in polynomial time for weakly bipartite graphs, a class of graphs that includes both bipartite and planar graphs. Hüffner et al. [7] wrongly claimed that using this result, BALANCED SUBGRAPH can be solved in polynomial time for (weakly) bipartite input graphs. The reason is that starting from a weakly bipartite instance of BALANCED SUBGRAPH, the resulting EDGE BIPARTIZATION instance is no longer weakly bipartite.

4   A Kernel-Like Result for Complete Binary Trees

In parameterized complexity, a *kernel* is a polynomial-time self-reduction of a parameterized problem after which the problem size can be bounded by a function only depending on the parameter. Here, we give a bound of $O(k \log k)$ on the size of the BALANCED SUBGRAPH instance which is the result of the transformation from Sect. 3, for the case that the input binary trees are complete. To show this, we begin with some definitions.

For a pair of leaves a, b with $u = \mathrm{lca}_S(a, b)$ and $v = \mathrm{lca}_T(a, b)$, there is a corresponding edge $\{u, v\}$ in G; call this the *edge associated with a, b*. Likewise, a, b is a leaf pair associated with the edge $\{u, v\}$. Obviously, there can be many leaf pairs associated with a single edge.

A pair of *contradictory edges* are two edges between the same pair of nodes in G, with different edge labels. A node u is involved in contradictory edges if there is a pair of contradictory edges with one end in u.

Let a *mirror node* of a node u be a node with identical leaf set. If u has a mirror node v, then the nodes of $T(u)$ and $T(v)$ generate a subgraph of G which is separate from the rest of the graph (the subgraph is not necessarily connected, but every edge of G that is incident to a node of $T(u)$ is incident to a node of $T(v)$ and vice versa). If furthermore $l(u)$ and $r(u)$ have mirror nodes (necessarily among $l(v)$ and $r(v)$), then v is an *identical mirror* of u, and u and v will end up in a two-vertex component in G. Such a component is a *trivial component* and can be solved immediately. Note that one of $l(u)$ and $r(u)$ has a mirror node if and only if both have mirror nodes.

We need the following reduction rules. The first three are generic, i.e. applicable not only in the case of complete trees; their correctness is immediate.

Rule 1 (Remove contradictory edges). *If there are multiple edges $\{u, v\}$, of which $n_1 > 0$ are $=$-edges and $n_2 > 0$ are $\neq$-edges, let $t = \min\{n_1, n_2\}$. Lower k by t, and delete t edges of each kind; reject if $t > k$.*

Assume from now on that Rule 1 has been applied exhaustively.

Rule 2 (High-multiplicity edges). *If there is any edge $\{u, v\}$ of multiplicity more than k, then set it to permanent.*

The following rule describes how edges set to "permanent" can be contracted.

Rule 3 (Vertex merging). *If an $=$-edge between two vertices u, v is set to permanent, then replace each edge $\{v, w\}$, $w \neq u$ with an equally labeled edge $\{u, w\}$ and delete v.*

If an $\neq$-edge between two vertices u, v is set to permanent, then replace each edge $\{v, w\}$, $w \neq u$ with a contrarily labeled edge $\{u, w\}$ ($=$ becomes $\neq$ and vice versa) and delete v.

Finally, we have two rules which are specific for the complete binary case. The essence is that given the restriction on the tree shapes, a node which does not have a mirror node will have leaves involved in edge crossings in any drawing.

Rule 4. *Let S, T be complete binary trees defining a TL instance. Let two nodes be incomparable if they are not in an ancestor-offspring relationship. If there are more than $2k$ pairwise incomparable nodes in S without mirror nodes, then reject the instance.*

Rule 5. *Let S, T be complete binary trees defining a TL instance. If there is any node u in S such that for every node v in T of the same depth as u, more than k leaves of $T(u)$ are missing from $T(v)$, then reject the instance.*

Lemma 2. *Rules 4 and 5 are correct.*

Proof. Since the trees S and T are both complete, the grouping of leaves to nodes with respect to any ordering of the trees follows the same structure: every node u in S will be placed directly opposite a node v in T (on the same level as u) sharing the same section of the leaf ordering (e.g. if the leftmost eight leaves of S meet in u, then the leftmost eight leaves of T meet in v). Now, consider an imaginary line drawn on the left side of $T(u)$ and $T(v)$. The number of edges that cross this line coming from the left side of it in S is the same as the number of crossing edges from the right side in S, and every edge from the left side crosses every edge from the right side; likewise for a line drawn to the right of $T(u)$ and $T(v)$. For every leaf in $T(u)$ not present in $T(v)$, the corresponding matching edge must cross one of these lines, and thus must be involved in a crossing.

Rule 5 follows immediately. Rule 4 follows since incomparable nodes have disjoint leaf sets; if the rule applies, then there are more than $2k$ separate edges involved in crossings.   □

Let U be the set of lowest nodes without mirror nodes, i.e. all nodes u without mirror nodes such that any other node in $T(u)$ does have a mirror node. These nodes are pairwise incomparable, so $|U| \leq k$. Furthermore, any internal node of S which is beneath a node in U, or incomparable to all nodes in U, belongs to a trivial component in the BALANCED SUBGRAPH instance. Repeating the argument from the root of S (which does have a mirror), we find that what remains of S after Rules 4 and 5 have been checked and trivial components removed is a set of binary trees whose leaves are the nodes of U. In principle, this already gives us a kernel of size $O(k^2)$: at most $O(k \log n)$ nodes remain, and if $\log n > k$, then solving the problem exactly in time $O(2^k n^{O(1)})$ counts as polynomial processing in n. We next show that our reduction rules take care of this in a different way, leaving at most $O(k \log k)$ nodes in the BALANCED SUBGRAPH instance.

Theorem 1. *Let S, T be complete binary trees. Applying rules 4 and 5, processing trivial components, and repeatedly merging heavy edges and removing contradictory edges either leads to a rejection of the instance or leaves a BALANCED SUBGRAPH instance with at most $O(k \log k)$ remaining nodes.*

Proof. Call a node *fat* if both children have at least $4k$ leaves. We will essentially show that fat nodes contribute nothing to the size of the final graph (because all but a bounded number of them will be merged into other nodes). We make three claims to show the result.

1. Every fat node has an identifiable *partner* on the same level in the opposite tree, which shares the same leaf set with at most k exceptions. If not, then

Rule 5 would apply. The same rule holds for the children of a fat node; by a counting argument, the partner matching must map the children of a fat node u to different children of its partner v. In particular, for any fat partner nodes u, v, there is an edge $\{u, v\}$ with multiplicity more than k.

2. Let u be a fat node, with partner v. Let v' be an ancestor of v. We claim that if there is an edge $\{u, v'\}$, then there is such an edge with multiplicity more than k. Let a, b be a leaf pair associated with an edge $\{u, v'\}$, and assume w.l.o.g. that $a \in L(v')$, $b \in R(v')$, and $v \in R(v')$.

   Let $\hat{L}$ be the set of leaves that $l(u)$ shares with its partner, and $\hat{R}$ the same for $r(u)$. By the above, $|\hat{L}|, |\hat{R}| \geq 3k$, and $\hat{L} \cup \hat{R} \subseteq \mathcal{L}(v)$. If $a \in L(u)$, then every pair of leaves a, c for $c \in \hat{R}$ has an associated edge $\{u, v'\}$; if $a \in R(u)$, then the same holds for a, c for $c \in \hat{L}$. Thus the claim is shown.

3. Let u be a fat node with partner v. If u and v are not mirror nodes, then we claim that there is an edge $\{u, v'\}$ where v' is an ancestor of v. Note that by the previous claim, there must then exist such an edge that is heavy. Also recall that if u and v are mirror nodes, then for any node $u' \in \mathcal{T}(u)$ and any edge $\{u', v'\}$, $v' \in \mathcal{T}(v)$.

   Assume $a \in \mathcal{L}(u)$, $a \notin \mathcal{L}(v)$. Then for any pair of leaves a, b with $\mathrm{lca}_S(a, b) = u$ there is an associated edge $\{u, v'\}$ where $v' \notin \mathcal{T}(v)$. If v' is not itself an ancestor of v, then there is another edge $\{u, \mathrm{lca}_T(v, v')\}$, which can be found by combining leaves associated with edges $\{u, v\}$ and $\{u, v'\}$.

The last claim has strong implications about the structure of the BALANCED SUBGRAPH instance. In particular, for fat partner nodes u and v, if there is an edge from u to an ancestor v' of v, then for any node v'' between v and v', there is in turn an edge from v'' to an ancestor of its partner (perhaps to the partner of v'). Thus the fat nodes of each connected component in the BALANCED SUBGRAPH instance are merged into one. To finalize the proof, we need to bound the number of connected non-trivial components.

Consider a node u with fat children $l(u)$, $r(u)$. If both children have mirror nodes (which are then their partners) then u has an identical mirror node and ends up in a trivial component. Otherwise, u shares a component with at least one child. In either case, we see that no connected non-trivial component contains u but no child of u. Thus the number of nodes that remain after the merging process is bounded by the number of non-fat ancestor nodes of the nodes U previously defined, which is in turn bounded by $O(k \log k)$.   □

5   Implementation and Experiments

For our experiments, we used Falk Hüffner's implementation of the BALANCED SUBGRAPH algorithm [7]. It is based on a combination of data reduction and iterative compression for solving the unreducible parts. The program consists of about 1900 lines of Objective Caml code and about 300 lines of C code that implements the time-critical compression routine of the iterative compression method. All experiments were run on a dual AMD Opteron 275 machine with 2.2 GHz, 1024 KB cache, and 6 GB main memory running under the Solaris 10 8/07 operating system (only one core was used). The program was compiled with Objective Caml 3.11.1 and the GNU gcc 3.4.3 compiler using the options "-O3 -march=athlon".

Two properties of the instances obtained by the reduction from TANGLEGRAM LAYOUT are notable here. First, they have a particular degree distribution (at least for well-balanced trees): there are vertices with both very low and very high degrees, and the distribution follows a power law, thus the networks are scale-free. Second, there are edges with very high multiplicity (up to several hundred). This works to our advantage. The data reduction rules of the algorithm depend on the existence of small separators, that is, vertex sets whose deletion disconnects the graph. The existence of many small-degree vertices in our instances makes finding such sets likely. Moreover, the exponential part of the running time of the iterative compression algorithm $(O(2^k))$ can be more precisely bounded by $O(2^c)$, where c is the maximum size of a vertex cover needed to cover an (intermediary) balancing set of edges (see [7] for details). Because of the high multiplicity and the existence of "hubs" (vertices with high degree), these vertex covers are much smaller than k.

Another notable property is that BALANCED SUBGRAPH instances resulting from our transformation are bipartite. However, since we noted above that arbitrary instances can be made bipartite, it seems unlikely that this can be exploited.

The special structure also motivated us to add two modifications to the solver, both of which are correct for general BALANCED SUBGRAPH instances but tailored towards such instances.

First, we added a data reduction rule that can get rid of edges with high multiplicity, without needing to know the value of k. The correctness is easy to see.

Rule 6 (Cut with heavy edge). *Let G be a* BALANCED SUBGRAPH *instance, where all pairs of contradictory edges have been removed. If there is an edge cut of G separating two vertices u and v in which at least half the edges of the cut are edges $\{u, v\}$, then the edges $\{u, v\}$ can be made permanent. In particular, this rule applies if there are vertices u and v such that at least half the edges of u are edges $\{u, v\}$.*

After we have decided that an edge is permanent, we can simplify the instance using Rule 3. This rule applies in particular when two nodes in the two trees are similar (that is, they have similar leaf sets, split roughly the same way).

In fact, we implemented only the special case of Rule 6, since our experiments showed that almost always a cut between u and v when an edge $\{u, v\}$ is present either isolates u or isolates v by deleting all adjacent edges. We also did not implement rules that depend on knowing k in advance, such as Rules 2, 4, and 5. The reason is that we either would have to try increasing values of k, which

Table 1. Running times with 60 s time limit

		k		time [s]	
Set	solved [%]	median	maximum	median	maximum
A	69.0	630	58697	0.04	49.30
B	100	83	4639	0.04	2.41
C	39.9	844	8815	1.00	59.79
D	100	172	975	0.06	0.20
E	100	0	10085	< 0.01	2.56
F	99.7	28	34811	< 0.01	31.45
G	100	0	555	< 0.01	2.49

would add a large polynomial factor, or use a heuristic upper bound on k, which is less likely to yield effective reduction.

The second modification concerns the iterative compression process (we assume familiarity with the approach). When building up the instance, we need to add edges one-by-one. Since some instances have extremely many edges (up to 148 240 before data reduction), it is desirable to avoid this factor of n^2. For this, we start with a heuristic solution and compress it repeatedly until no more compression is possible. This typically requires only up to 20 rounds of compression. The initial solution is found using a simple Kernighan–Lin style algorithm: Starting from a random coloring, repeatedly change the color of a vertex as long as this decreases the number of nonsatisfied edges. The disadvantage is that we forfeit the worst-case bound on the running time, and instances can be constructed for which this would give a slowdown. However, for the dense instances we encountered, this is not a problem. To make for a more robust implementation, we could try both methods in parallel.

Data. The seven datasets we used stem from Nöllenburg et al. [8]: Sets A–D are artificial datasets. Set A contains 600 pairs of random complete binary trees of sizes 16–512, set B consists of pairs of mutated complete binary trees, and sets C and D contain 2900 more naturally generated general binary trees with 20–300 leaves and additional mutations in set D. Sets E–F comprise 1303 tree pairs generated with real-world data of animal families. Set E compares Maximum Likelihood and Neighbor joining trees, set F and G Neighbor Joining trees that used different distances. See [8] for details.

The results of the computations are listed in Table 1. We observe that as expected, the algorithm struggles most with sets A and C, which are synthetic random instances that are not expected to have a low number of crossings. From the real-world instances, only 4 instances from set F remain unsolved within a minute. These have $k \geq 10000$.

Instances with $k > 1000$ are unlikely to be of practical interest, since with more than 1000 crossings, the visualization will not be helpful. If we restrict ourselves to the real-world instances with $k \leq 1000$, we can solve all instances with a median of < 0.01 s and a maximum of 2.55 s. This means we can get optimal solutions for all practically relevant instances within seconds.

In general, performance is similar to the ILP approach of Nöllenburg et al. [8], which also can solve most of the instance with k not too high. The advantage of our approach is that it has useful worst-case running time bounds and does not require the proprietary CPLEX software.

6   Conclusion

With improving the previously best-known fixed-parameter running time for the TANGLEGRAM LAYOUT problem from $O(c^k n^{O(1)})$ with $c \approx 1024$ [4] to $O(2^k n^4)$, where k is the minimum number of crossings in a drawing, we managed to make fixed-parameter algorithms applicable for sizes that are interesting for visualization of phylogenetic trees. Experiments showed that we can usually solve instances with $k \leq 1000$ in well below one second.

Consequential challenges are working towards a problem kernel for general binary trees and extending the algorithm to nonbinary phylogenetic trees. We plan to do further algorithm engineering and to integrate the algorithm into the EPoS[1] framework, a modular framework for phylogenetic analysis and visualization, to make it easily available to biologists.

Acknowledgments. We thank Martin Nöllenburg for providing us with the tanglegram datasets from [8].

References

1. Bansal, M.S., Chang, W.-C., Eulenstein, O., Fernández-Baca, D.: Generalized binary tanglegrams: Algorithms and applications. In: Rajasekaran, S. (ed.) BICoB 2009. LNCS (LNBI), vol. 5462, pp. 114–125. Springer, Heidelberg (2009)
2. Baumann, F., Buchheim, C., Liers, F.: Exact crossing minimization in general tanglegrams. Technical Report zaik2009-581, Zentrum für Angewandte Informatik Köln (Mar 2009)
3. Buchin, K., Buchin, M., Byrka, J., Nöllenburg, M., Okamoto, Y., Silveira, R.I., Wolff, A.: Drawing (complete) binary tanglegrams. In: Tollis, I.G., Patrignani, M. (eds.) GD 2008. LNCS, vol. 5417, pp. 324–335. Springer, Heidelberg (2009)
4. Fernau, H., Kaufmann, M., Poths, M.: Comparing trees via crossing minimization. In: Sarukkai, S., Sen, S. (eds.) FSTTCS 2005. LNCS, vol. 3821, pp. 457–469. Springer, Heidelberg (2005)
5. Gabow, H.N., Tarjan, R.E.: A linear-time algorithm for a special case of disjoint set union. In: Proc. of ACM Symposium on Theory of Computing (STOC 1983), pp. 246–251. ACM Press, New York (1983)
6. Grötschel, M., Pulleyblank, W.R.: Weakly bipartite graphs and the max-cut problem. Oper. Res. Lett. 1(1), 23–27 (1981)

---

[1] http://bio.informatik.uni-jena.de/epos/

7. Hüffner, F., Betzler, N., Niedermeier, R.: Optimal edge deletions for signed graph balancing. In: Demetrescu, C. (ed.) WEA 2007. LNCS, vol. 4525, pp. 297–310. Springer, Heidelberg (2007)
8. Nöllenburg, M., Holten, D., Völker, M., Wolff, A.: Drawing binary tanglegrams: An experimental evaluation. In: Proc. of Workshop on Algorithm Engineering and Experiments (ALENEX 2009), pp. 106–119. SIAM, Philadelphia (2009)
9. Page, R.D.M. (ed.): Tangled Trees: Phylogeny, Cospeciation, and Coevolution. University of Chicago Press, Chicago (2002)

Planar Capacitated Dominating Set Is $W[1]$-Hard

Hans L. Bodlaender[1], Daniel Lokshtanov[2], and Eelko Penninkx[1]

[1] Department of Information and Computing Sciences, Universiteit Utrecht, PO Box
80.089, 3508TB Utrecht, The Netherlands
{hansb,penninkx}@cs.uu.nl
[2] Department of Informatics, University of Bergen, N-5020 Bergen, Norway
daniello@ii.uib.no

Abstract. Given a graph G together with a capacity function $c : V(G) \to$
$\mathbb{N}$, we call $S \subseteq V(G)$ a capacitated dominating set if there exists a
mapping $f : (V(G) \setminus S) \to S$ which maps every vertex in $(V(G) \setminus S)$ to
one of its neighbors such that the total number of vertices mapped by f
to any vertex $v \in S$ does not exceed $c(v)$. In the PLANAR CAPACITATED
DOMINATING SET problem we are given a planar graph G, a capacity
function c and a positive integer k and asked whether G has a capacitated
dominating set of size at most k. In this paper we show that PLANAR
CAPACITATED DOMINATING SET is $W[1]$-hard, resolving an open problem
of Dom et al. [*IWPEC, 2008*]. This is the first bidimensional problem
to be shown $W[1]$-hard. Thus PLANAR CAPACITATED DOMINATING SET
can become a useful starting point for reductions showing parameterized
intractablility of planar graph problems.

1   Introduction

In the DOMINATING SET problem we are given a graph G and asked for the
smallest set of vertices such that every vertex in the graph either belongs to this
set or has a neighbor which does. This basic problem in algorithms and complexity has been studied extensively, and finds applications in various domains.
DOMINATING SET has a special place in parameterized complexity [5,8,13]. It is
the most well-known $W[2]$-complete problem and is a standard starting point for
reductions that show intractability of parameterized problems [5]. Even though
the DOMINATING SET problem is a fundamentally hard problem in the parameterized W-hierarchy, it has been used as a benchmark problem for developing
sub-exponential time FPT algorithms [1,3,11], and also for obtaining *linear kernels* on planar graphs [2,8,12,13], and more generally, graphs that exclude a fixed
graph H as a minor.

Different applications of DOMINATING SET have initiated studies of different
generalizations and variations of the problem. These include CONNECTED DOMI-
NATING SET, PARTIAL DOMINATING SET, and CAPACITATED DOMINATING SET
to name a few. In this paper we focus on one such generalization, namely CAPAC-
ITATED DOMINATING SET. Given a graph G together with a capacity function

J. Chen and F.V. Fomin (Eds.): IWPEC 2009, LNCS 5917, pp. 50–60, 2009.

$c : V(G) \to \mathbb{N}$, we call $S \subseteq V(G)$ a capacitated dominating set if there exists a mapping $f : (V(G) \setminus S) \to S$ which maps every vertex in $(V(G) \setminus S)$ to one of its neighbors such that the total number of vertices mapped by f to any vertex $v \in S$ does not exceed $c(v)$. The CAPACITATED DOMINATING SET problem is defined as follows.

CAPACITATED DOMINATING SET (CDS): Given a graph G, a capacity function c and a positive integer k, determine whether there exists a capacitated dominating set S of G containing at most k vertices.

Dom et al. initiated the study of CDS from the perspective of Parameterized Complexity, and showed that CDS is $W[1]$-hard parameterized by solution size and the treewidth of the input graph [4]. Like DOMINATING SET, CDS has become a useful source for showing W-hardness, especially when the parameter is the structure of the input graph [7,10]. It has been recently used to show the first W-hardness results for problems parameterized by the cliquewidth of the input graph [10].

Many graph problems that are W-hard in general turn out to be FPT when restricted to planar graphs. This is true for DOMINATING SET and many of its variants, and hence it is very natural to consider the parameterized complexity of PLANAR CAPACITATED DOMINATING SET, the restriction of CDS to planar graphs. For most planar graph problems, an FPT algorithm can be obtained by combining a combinatorial bound on the treewidth of non-trivial instances with a dynamic programming algorithm for graphs of bounded treewidth. In fact for most problems restricted to planar graphs we have subexponential time parameterized algorithms using *bidimensionality* theory [3]. PCDS, however, is an exception to this rule. In particular, it can easily be shown by using bidimensionality that any planar graph that has a capacitated dominating set of size at most k has treewidth $O(\sqrt{k})$. On the other hand, Dom et al. showed that CDS is $W[1]$-hard when parameterized by solution size and the treewidth of the input graph [4]. Thus, bidimensionality alone was not enough to tackle this problem and it was an intriguing question whether PCDS could still turn out to be FPT by a non-trivial use of planarity. We show that these hopes were futile by giving a $W[1]$-hardness reduction for PCDS. PLANAR CAPACITATED DOMINATING SET is the first bidimensional problem to be shown $W[1]$-hard. We believe that PLANAR CAPACITATED DOMINATING SET can become a useful starting point for reductions showing parameterized intractablility of planar graph problems.

2   Preliminaries

We will work with both undirected and directed graphs. Given a graph G, the vertex set of G is $V(G)$ and the edge set of G is $E(G)$. . For a graph G, $n = |V(G)|$ and $m = |E(G)|$. With $N_G(u)$ we denote all vertices that are adjacent to u and the *degree* of u is $d_G(u) = |N_G(u)|$. Let f be the function associated with a capacitated dominating set S. Given $u \in S$ and $v \in V \setminus S$, we say that u *dominates* v if $f(v) = u$; moreover, every vertex $u \in S$ dominates itself. Note that the capacity of a vertex v only limits the number of neighbors that v can dominate, that is, a vertex $v \in S$ can dominate $c(v)$ of its neighbors plus v itself.

For a directed graph D the node set of D is $N(D)$ and the arc set of D is $A(D)$. For a node u, $N_D^+(u) = \{v : uv \in A\}$ is the set of *outneighbours* of u, $N_D^-(u) = \{v : vu \in A\}$ is the set of *inneighbours* of u and $N_D(u) = N_D^+(u) \cup N_D^-(u)$ is the set of *neighbours* of u. We define $d_D^+(u) = |N_D^+(u)|$, $d_D^-(u) = |N_D^-(u)|$ and $d_D(u) = |N_D(u)|$ to be the *outdegree, indegree* and *degree* of u respectively.

We use the notions of a parameterized problem, Fixed Parameter Tractability, hardness for the complexity class $W[1]$ and our hardness proofs involve *FPT*-reductions. For an introduction to these notions, the reader is referred to the textbooks [5,8,13]. For ease of reference we provide the definition of *FPT*-reductions here.

Definition 1. *[5,8,13] Let A, B be parameterized problems. We say that A is FPT-reducible to B if there is an algorithm Φ which transforms (x, k) into $(x', g(k))$ in time $f(k) \cdot |x|^\alpha$, where $f, g : \mathbb{N} \to \mathbb{N}$ are arbitrary functions and α is a constant independent of $|x|$ and k, so that $(x, k) \in A$ if and only if $(x', g(k)) \in B$.*

It is well known that if A is hard for $W[1]$ and A is FPT-reducible to B, then B is also $W[1]$-hard [5,8,13].

3   PCDS Is W[1]-Hard Parameterized by Solution Size

In this section we show that PCDS is $W[1]$-hard when parameterized by solution size. We reduce from MULTI-COLOR CLIQUE, a restriction of the k-CLIQUE problem.

> MULTI-COLOR CLIQUE (MCC) Given an integer k and a connected undirected graph $G = (V[1] \cup V[2] \cdots \cup V[k], E)$ such that for every i the vertices of $V[i]$ induce an independent set, is there a k-clique C in G?

For $i \leq k$ the sets $V[i]$ are called *color classes* of G. Since each color class forms an independent set, a k-clique in G must contain exactly one vertex from each color class. For two distinct integers i, j between 1 and k the set $E[i, j]$ is the set of edges of G with one endpoint in $V[i]$ and the other in $V[j]$. The MULTI-COLOR CLIQUE problem is known to be $W[1]$-hard [6,9] and is used as a starting point for many hardness reductions.

We will reduce to a slightly modified version of PLANAR CAPACITATED DOMINATING SET, PLANAR MARKED CAPACITATED DOMINATING SET (PMCDS) where we *mark* some vertices and demand that all marked vertices must be in the dominating set. We can then reduce from PMCDS to PLANAR CAPACITATED DOMINATING SET by attaching $k + 1$ leaves to each marked vertex and increasing the capacity of each marked vertex by $k + 1$. It is easy to see that the new instance has a k-capacitated dominating set if and only if the original one had a k-capacitated dominating set that contained all marked vertices, and that this operation preserves planarity of the graph. Thus, to prove that

PLANAR CAPACITATED DOMINATING SET is $W[1]$-hard when parameterized by solution size, it is sufficient to prove that PMCDS is. We will show how given an instance (G, k) of MULTICOLOR CLIQUE, we can build an instance (H, c, k^*) of PMCDS such that $k^* = O(k^3)$ and G has a clique of size k if and only if H has a capacitated dominating set of size k^*.

The reduction to PMCDS goes via an intermediate problem as well. We name this problem the PLANAR ARC SUPPLY problem, (PAS). In PAS we are given a planar digraph $D = (N, A)$ with $|N(D)| + |A(D)| = k$, no loops and no double arcs. Every node $u \in N$ has a demand $\zeta(u)$, which is a natural number, and every arc $uv \in A$ has a list $L(uv)$ of ordered integer pairs, called the *supply pairs* of uv. The task is to decide whether there is a function $f_a : A \to \mathbb{N}$ and a function $f_b : A \to \mathbb{N}$ such that for every arc $uv \in A$ we have $(f_a(uv), f_b(uv)) \in L(uv)$ and for every node $u \in N$ we have that $\zeta(u) \leq \sum_{v \in N^+(u)} f_a(uv) + \sum_{v \in N^-(u)} f_b(vu)$. In essense we are asked to pick a supply pair from the list of every arc such that for every vertex the arcs incident to it are able to cover the demand of the vertex. Therefore the pair of fuctions f_a and f_b are called a *supply selection* and a supply selection is called *satisfying* if the demand of all vertices is met.

3.1 Identification Numbers

Given an instance $(G = (V[1] \cup \ldots \cup V[k], E), k)$ to MCC every vertex and every edge of G gets two pairs of identification numbers. Every vertex gets one pair of *small* and one pair of *medium* identification numbers and every edge gets one pair of *small* and one pair of *large* identification numbers. For the vertices these identification numbers are defined as follows.

- Every vertex v gets assigned a unique number $ID_S^U(v)$ between 1 and n as its *small up-ID*.
- The *small down-ID* of v is $ID_S^D(v) = n^2 - ID_S^U(v)$.
- The *medium up-ID* of v is $ID_M^U(v) = n^3 \cdot ID_S^U(v)$.
- The *medium down-ID* of v is $ID_M^D(v) = n^3 \cdot ID_S^D(v)$.

For edges the identification numbers are defined similarly. In particular:

- Every edge uv of G gets assigned a unique number $ID_S^U(uv)$ between 1 and m as its *small up-ID*.
- The *small down-ID* of uv is $ID_S^D(uv) = n^2 - ID_S^U(uv)$.
- The *large up-ID* of uv is $ID_L^U(uv) = n^6 \cdot ID_S^U(uv)$.
- The *large down-ID* of uv is $ID_L^D(uv) = n^6 \cdot ID_S^D(uv)$.

Observe that $2m < n^2$ and that therefore all small down-ID's are larger than all small up-ID's. We also define the *huge* numbers $\mathcal{U} = n^{20} - n^{19}$, $\mathcal{D} = n^{20} + n^{19}$ and $\mathcal{M} = n^{20}$.

3.2 Reduction to Planar Arc Supply

Given an instance $(G = (V[1] \cup \ldots \cup V[k], E), k)$ to MCC we now construct an instance of PLANAR ARC SUPPLY. For every color class i between 1 and k we make

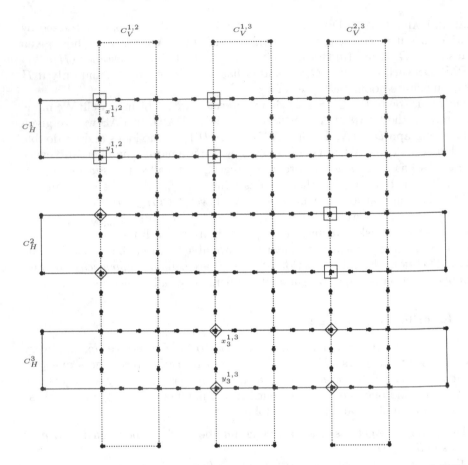

Fig. 1. The construction of D for $k = 3$. Horizontal cycles are shown in full lines, vertical in dotted lines. The left-matching vertices are surrounded by a square, the right-matching vertices are surrounded by a diamond.

a *horizontal cycle* C_H^i. This is a directed cycle of length $12\binom{k}{2}$, oriented clockwise. For every pair of color classes i, j between 1 and k such that $i < j$ we make a *vertical cycle* $C_V^{i,j}$. This is a directed cycle of length $12k$, oriented clockwise. We arrange the cycles in a grid, with the vertical cycles corresponding to vertical lines on the grid and the horizontal cycles corresponding to the horizontal lines on the grid. Every horizontal cycle intersects with every vertical cycle in exactly four points. Following a cycle, there are exactly three edges from one intersection point to the next. This concludes the construction of the directed graph D. Observe that the graph D itself depends only on k. For a construction of D for $k = 3$ see Figure 1.

The number of arcs in D is $k \cdot 12\binom{k}{2} + \binom{k}{2} \cdot 12k$. The number of nodes in D is $k \cdot 12\binom{k}{2} + \binom{k}{2} \cdot 12k - 4k \cdot \binom{k}{2}$. Hence $k' = |N(D)| + |A(D)| = 44k\binom{k}{2} = O(k^3)$.

The horizontal cycle C_H^t and the vertical cycle $C_V^{i,j}$ intersect in exactly four nodes. The top-left intersection node is called $x_t^{i,j}$ and the bottom left intersection node is called $y_t^{i,j}$. The nodes $x_t^{i,j}$ and $y_t^{i,j}$ such that $i = t$ are called *left-matching* nodes and the nodes $x_t^{i,j}$ and $y_t^{i,j}$ such that $j = t$ are called *right-matching* nodes. An arc in a horizontal cycle is called a *horizontal* arc and an arc in a vertical cycle is called a *vertical* arc. An arc whose endpoint is a left-matching node is also called *left-matching* and similarly an arc whose endpoint is a right-matching node is called *left-matching*. If an arc or node is left-matching or right-matching then it is also a *matching* arc (node).

We now describe the demands of all the nodes and the supply lists of all the arcs of D. Observe that every node v of G has degree either 2 or 4. The demand of every vertex of degree 2 in a horizontal cycle is $2\mathcal{M} + n^5$. The demand of every vertex of degree 2 in a vertical cycle is $2\mathcal{M} + n^8$ and the demand of every degree 4 vertex is $4\mathcal{M} + n^8 + n^5$. The description of the supply-lists conclude the description of the PAS instance. Every pair in a supply list of a horizontal arc will correspond to a vertex of G while every pair in a supply list of a vertical arc will correspond to an edge of G.

The intuition is that each horizontal cycle C_H^i encodes the choice which vertex in $V[i]$ will be in the clique. Each vertical cycle $C_V^{i,j}$ encodes the choice of which edge of $E[i, j]$ will be in the clique. The large identification numbers are used in all arcs of the vertical cycles to encode this choice of edges. The medium identification numbers are used in all arcs of the horizontal cycles to encode the choice of vertices. In the intersection of the horizontal cycle C_H^i and the vertical cycle $C_V^{i,j}$ we use the left-matching vertices $x_i^{i,j}$ and $y_i^{i,j}$ to make sure that the vertex selected in the horizontal cycle C_H^i and the edge selected in the vertical cycle $C_V^{i,j}$ are incident. Similarly, in the intersection of the horizontal cycle C_H^j and the vertical cycle $C_V^{i,j}$ we use the right-matching vertices $x_j^{i,j}$ and $y_j^{i,j}$ to make sure that the vertex selected in the horizontal cycle C_H^j and the edge selected in the vertical cycle $C_V^{i,j}$ are incident. The incidence check is performed using small identification numbers. The huge numbers U and D are always present in all supply pairs. These numbers are used in second stage of the reduction (from PAS to PMCDS) and do not play a role in the first part. We now formally describe the supply-lists.

- For every non-matching horizontal arc uv of the cycle C_H^i, for every $a \in V[i]$ there is a pair $(\mathcal{U} + ID_M^U(a), \mathcal{D} + ID_M^D(a))$ in $L(uv)$
- For every matching horizontal arc uv pointing at a node $x_i^{i,j}$, for every $a \in V[i]$ there is a pair $(\mathcal{U} + ID_M^U(a), \mathcal{D} + ID_M^D(a) + ID_S^D(a))$ in $L(uv)$.
- For every matching horizontal arc uv pointing at a node $y_i^{i,j}$, for every $a \in V[i]$ there is a pair $(\mathcal{U} + ID_M^U(a), \mathcal{D} + ID_M^D(a) - ID_S^D(a))$ in $L(uv)$
- For every non-matching vertical arc uv of the cycle $C_V^{i,j}$, for every $ab \in E[i, j]$ there is a pair $(\mathcal{U} + ID_L^U(ab), \mathcal{D} + ID_L^D(ab))$ in $L(uv)$.
- For every left-matching vertical arc uv pointing at a node $x_i^{i,j}$, for every $a \in V[i]$ and $b \in V[j]$ such that $ab \in E$ there is a pair $(\mathcal{U} + ID_L^U(ab), \mathcal{D} + ID_L^D(ab) - ID_S^D(a))$ in $L(uv)$.
- For every left-matching vertical arc uv pointing at a node $y_i^{i,j}$, for every $a \in V[i]$ and $b \in V[j]$ such that $ab \in E$ there is a pair $(\mathcal{U} + ID_L^U(ab), \mathcal{D} + ID_L^D(ab) + ID_S^D(a))$ in $L(uv)$.

- For every right-matching vertical arc uv pointing at a node $x_j^{i,j}$, for every $a \in V[i]$ and $b \in V[j]$ such that $ab \in E$ there is a pair $(\mathcal{U} + ID_L^U(ab), \mathcal{D} + ID_L^D(ab) - ID_S^D(b))$ in $L(uv)$.
- For every right-matching vertical arc uv pointing at a node $y_j^{i,j}$, for every $a \in V[i]$ and $b \in V[j]$ such that $ab \in E$ there is a pair $(\mathcal{U} + ID_L^U(ab), \mathcal{D} + ID_L^D(ab) + ID_S^D(b))$ in $L(uv)$.

Lemma 1. *If G contains a clique C of size k then D has a satisfying supply selection.*

Proof. Any k-clique in G must contain exactly one vertex from each set $V[i]$ and exactly one edge from each set $E[i,j]$. Let c_i be the vertex in $C \cap C_i$. For every i between 1 and k and every horizontal arc in C_H^i select the supply pair corresponding to c_i. For every pair i,j such that $i < j$ and every vertical arc in $C_V^{i,j}$ select the supply pair corresponding to the edge $c_i c_j$. The case analysis below shows that the demand of all vertices is met.

- For every vertex with degree 2 on the horizontal cycle C_H^i, demand is $2M + n^5$, and total supply is $\mathcal{D} + ID_M^D(c_i) + \mathcal{U} + ID_M^U(c_i) = 2\mathcal{M} + n^5$.
- For every vertex with degree 2 on the vertical cycle $C_H^{i,j}$, demand is $2M + n^8$, and total supply is $\mathcal{D} + ID_L^D(c_i c_j) + \mathcal{U} + ID_L^U(c_i c_j) = 2\mathcal{M} + n^8$.
- For every non-matching vertex with degree 4 lying on the horizontal cycle C_H^t and vertical cycle $C_V^{i,j}$, demand is $4M + n^8 + n^5$, and total supply is $\mathcal{D} + ID_M^D(c_t) + \mathcal{U} + ID_M^U(c_t) + \mathcal{D} + ID_L^D(c_i c_j) + \mathcal{U} + ID_L^U(c_i c_j) = 4\mathcal{M} + n^8 + n^5$.
- For every left-matching vertex $x_i^{i,j}$, demand is $4M + n^8 + n^5$, and total supply is $\mathcal{D} + ID_M^D(c_i) + ID_S^D(c_i) + \mathcal{U} + ID_M^U(c_i) + \mathcal{D} + ID_L^D(c_i c_j) - ID_S^D(c_i) + \mathcal{U} + ID_L^U(c_i c_j) = 4\mathcal{M} + n^8 + n^5$.
- For every left-matching vertex $y_i^{i,j}$, demand is $4M + n^8 + n^5$, and total supply is $\mathcal{D} + ID_M^D(c_i) - ID_S^D(c_i) + \mathcal{U} + ID_M^U(c_i) + \mathcal{D} + ID_L^D(c_i c_j) + ID_S^D(c_i) + \mathcal{U} + ID_L^U(c_i c_j) = 4\mathcal{M} + n^8 + n^5$.
- For every right-matching vertex $x_j^{i,j}$, demand is $4M + n^8 + n^5$, and total supply is $\mathcal{D} + ID_M^D(c_j) + ID_S^D(c_j) + \mathcal{U} + ID_M^U(c_j) + \mathcal{D} + ID_L^D(c_i c_j) - ID_S^D(c_j) + \mathcal{U} + ID_L^U(c_i c_j) = 4\mathcal{M} + n^8 + n^5$.
- For every right-matching vertex $y_j^{i,j}$, demand is $4M + n^8 + n^5$, and total supply is $\mathcal{D} + ID_M^D(c_j) - ID_S^D(c_j) + \mathcal{U} + ID_M^U(c_j) + \mathcal{D} + ID_L^D(c_i c_j) + ID_S^D(c_j) + \mathcal{U} + ID_L^U(c_i c_j) = 4\mathcal{M} + n^8 + n^5$.

$\square$

Lemma 2. *If D has a satisfying supply selection then G contains a clique C of size k.*

Proof. Every pair in a supply list of a horizontal arc corresponds to a vertex of G while every pair in a supply list of a vertical arc corresponds to an edge of G. Hence the satisfying supply selection of D represents a choice of an edge of G for every arc in a vertical cycle, and a choice of a vertex for every arc in a horizontal cycle. Consider two consecutive arcs uv and vw on a vertical cycle $C_V^{i,j}$

and let a be the edge of G selected at uv and b be the edge selected at vw. We prove that $ID_L^U(b) \geq ID_L^U(a)$. Suppose for contradiction that $ID_L^U(b) < ID_L^U(a)$. If v has degree 2 then demand is $2M + n^8$ and supply is $\mathcal{D} + ID_L^D(a) + \mathcal{U} + ID_L^U(b) < 2M + n^8$. Observe that since $ID_L^U(a) = n^6 \cdot ID_S^U(a)$ we have that if $ID_L^U(b) < ID_L^U(a)$ then $ID_L^U(b) + n^6 \leq ID_L^U(a)$. Thus, if v has degree 4 then v's demand is $2M + n^8 + n^5$ and the total supply at v is at most $\mathcal{D} + ID_L^D(a) + n^2 + \mathcal{U} + ID_L^U(b) + \mathcal{U} + n^5 + \mathcal{D} + n^5 \leq 4M + n^8 + 2n^5 + n^2 - n^6 < 4M + n^8 + n^5$. Hence $ID_L^U(b) \geq ID_L^U(a)$. Since this holds for every pair of consecutive arcs on the vertical cycle $C_V^{i,j}$, all arcs on the cycle $C_V^{i,j}$ select the same edge of G.

We now prove that a similar observation holds for the horizontal cycles, that is, that all arcs on the horizontal cycle C_H^i select the same vertex of G. Consider two consecutive arcs uv and vw on a horizontal cycle C_H^i and let a be the vertex of G selected at uv and b be the vertex selected at vw. We prove that $ID_M^U(b) \geq ID_M^U(a)$. Suppose for contradiction that $ID_M^U(b) < ID_M^U(a)$. If v has degree 2 then v's demand is $2M + n^5$ and the supply at v is $U + ID_M^D(a) + \mathcal{D} + ID_M^U(b) < 2M + n^5$. Now, suppose v has degree 4, then the demand of v is $4M + n^8 + n^5$. Observe that if $ID_M^U(b) < ID_M^U(a)$ then $ID_M^U(b) + n^3 \leq ID_M^U(a)$. Also, since all arcs on the vertical cycle containing v select the same edge of G, the vertical arcs incident to v supply v with at most $2M + n^8 + n^2$. Hence the total supply at v is at most $2M + n^8 + n^2 + \mathcal{D} + ID_M^D(a) + n^2 + \mathcal{U} + ID_M^U(b) \leq 4M + n^8 + n^5 + 2n^2 - n^3 < 4M + n^8 + n^5$. Hence $ID_L^U(b) \geq ID_L^U(a)$. Since this holds for every pair of consecutive arcs on the horizontal cycle C_H^i, all arcs on the cycle C_H^i select the same vertex of G.

Thus every horizontal cycle C_H^i selects a vertex $c_i \in V[i]$ and every vertical cycle $C_V^{i,j}$ selects an edge $e_{i,j} \in E[i,j]$. It remains to prove that for every i, j, $e_{i,j}$ is incident to both c_i and to c_j. We prove that $e_{i,j}$ is incident to c_i. In particular, let c_i' be the vertex in $V[i]$ incident to $e_{i,j}$ in $V[i]$. We prove that $ID_S^D(c_i) = ID_S^D(c_i')$. Suppose that $ID_S^D(c_i) < ID_S^D(c_i')$. Then the supply at $x_i^{i,j}$ is at most $\mathcal{D} + ID_M^D(c_i) + ID_S^D(c_i) + \mathcal{U} + ID_M^U(c_i) + \mathcal{D} + ID_L^D(e_{i,j}) - ID_S^D(c_i') + \mathcal{U} + ID_L^U(e_{i,j}) < 4M + n^8 + n^5$, a contradiction. Similarly if $ID_S^D(c_i) > ID_S^D(c_i')$ then the supply at $y_i^{i,j}$ is at most $\mathcal{D} + ID_M^D(c_i) - ID_S^D(c_i) + \mathcal{U} + ID_M^U(c_i) + \mathcal{D} + ID_L^D(e_{i,j}) + ID_S^D(c_i') + \mathcal{U} + ID_L^U(e_{i,j}) < 4M + n^8 + n^5$. Hence $ID_S^D(c_i) = ID_S^D(c_i')$ and $e_{i,j}$ is incident to c_i. The proof that $e_{i,j}$ is incident to c_j is similar. This proves that $\{c_1, \ldots, c_k\}$ is a clique in G. $\square$

3.3   Reduction to Planar Marked Capacitated Dominating Set

We now show how to transform an instance D, k' of PLANAR ARC SUPPLY constructed from a MULTI-COLOR CLIQUE instance G, k as in Section 3.2 to an instance H, k^* of PLANAR MARKED CAPACITATED DOMINATING SET. To build H we start with the node set $N(D)$ and make every vertex of $N(D)$ marked. For every arc uv of D we make a gadget between u and v in H. In particular, for an arc $uv \in A(D)$, for every pair of integers $(p,q) \in L(u,v)$ we add a vertex w to H, make w adjacent to u, add p vertices of degree 2 adjacent to u and w and add q vertices of degree 2 adjacent to w and v. We call the vertex w is a *list* vertex. This concludes the construction of the graph H. Since D is planar and the gadget we add to H for every arc of D is planar, H is planar as well. Every marked vertex v of H is also a vertex in D. The capacity of v in H is set to $d_H(v) - \zeta(v) - d_D^+(v)$, that is, the degree of v in H, minus v's demand in D and minus v's outdegree in D. For all unmarked vertices, their capacity in H is equal to their degree in H. Finally, $k^* = |N(D)| + |A(D)|$. This concludes the construction of the PLANAR MARKED CAPACITATED DOMINATING SET instance (H, k^*).

Lemma 3. *If D has a satisfying supply selection then H has a capacitated dominating set of size k^**

Proof. We build a capacitated dominating set S of H. First we insert all the marked vertices of H in S. For every arc uv of D we add a list vertex w to S, namely the list vertex that corresponds to the supply pair in $L(uv)$ that was selected by the satisfying supply selection of D. The size of S is $|N(D)| + |A(D)| = k^*$. We now prove that S is a capacitated dominating set of H.

First, observe that the marked vertices of H form a dominating set of H, so S is a dominating set of H. Now, every unmarked vertex in S has capacity equal to its degree, so all unmarked vertices in S dominate all their neighbours. We now prove that for every marked vertex u, the number of yet undominated neighbours of u is at most the capacity of u. The number of neighbours of u that already have been dominated is at least $\zeta(u)$. The number of neighbours of u that are in S is $d_D^+(u)$. Hence, the total number of yet undominated neighbours of u is at most $d_H(u) - \zeta(u) - d_D^+(u)$ which is the capacity of u. Hence S is a capacitated dominating set of H.   □

Lemma 4. *If H has a capacitated dominating set S of size k^* then D has a satisfying supply selection.*

Proof. There are two kinds of unmarked vertices in H, list vertices and vertices of degree 2. Every degree 2 vertex u has exactly one neighbour that is marked, and one neighbour v that is a list vertex. Since the capacity of v is equal to its degree and all marked vertices must be in S, if $u \in S$ then $S \cup \{v\} \setminus \{u\}$ is a capacitated dominating set of H of size at most k^*. Thus, without loss of generality, all unmarked vertices in S are list vertices.

For an arc uv of D, let $s(uv)$ be the number of vertices in S in the gadget corresponding to the arc uv. For a vertex u of D let $s^+(u) = \sum_{uv \in A(D)} s(uv)$, $s^-(u) = \sum_{vu \in A(D)} s(uv)$ and $s(u) = s^+(u) + s^-(u)$. Since S contains at most $|A(D)|$ unmarked vertices we have that $\sum_{u \in V(D)} s(u) \leq 2|A(D)|$. If $s(u) < d_D(u)$ for a vertex u then the number of vertices in $N_H(u)$ dominated by vertices other than u is at most $s(u) \cdot (\mathcal{D} + n^{10}) < d_D(u)M$. However the capacity of u is at most $d_H(u) - d_D(u)M$, contradicting that S is a capacitated dominating set. Hence, for every node $u \in N(D)$, $s(u) \geq d_D(u)$. If for some node $s(u) > d_D(u)$ then $\sum_{u \in N(D)} s(u) > \sum_{u \in N(D)} d_D(u) = 2|A(D)|$, contradicting that $\sum_{u \in N(D)} s(u) \leq 2|A(D)|$. Thus, for every node $u \in N(D)$, $s(u) = d_D(u)$.

Consider now three consecutive arcs pq, qr and rs in $A(D)$ such that both q and r have degree 2 in D. There are three cases, either $s(pq) = s(qr) = s(rs) = 1$

or $s(pq) = s(rs) = 2$ and $s(qr) = 0$ or finally $s(pq) = s(rs) = 0$ and $s(qr) = 2$. We show that the last two cases lead to a contradiction. If $s(pq) = s(rs) = 2$ and $s(qr) = 0$ then the number of neighbours of r dominated by vertices other than r is at most $2(\mathcal{U} + n^{10}) < 2M$. However the capacity of r is at most $d_H(r) - 2M$, contradicting that S is a capacitated dominating set. Similarly, if $s(pq) = s(rs) = 0$ and $s(qr) = 2$ then the number of neighbours of q dominated by vertices other than q is at most $2(\mathcal{U} + n^{10}) < 2M$. However the capacity of q is at most $d_H(q) - 2M$, contradicting that S is a capacitated dominating set. It follows that $s(pq) = s(qr) = s(rs) = 1$. Because the distance in H between any pair of vertices with degree 4 is at least 3 it follows that $s(pq) = 1$ for every arc $pq \in A(D)$.

We now make a supply selection (f_a, f_b) for D as follows. For every arc uv there is exactly one unmarked vertex x in S in the gadget in H corresponding to the arc uv. This vertex x corresponds to a pair $(p, q) \in L(uv)$ and we make uv select the pair (p, q). Every arc selects a pair from its list in this manner. We now show that this supply selection is satisfying. Suppose for contradiction that this is not the case, then there is some vertex $u \in N(D)$ whose demand is not met. Then u is a marked vertex in H, and the demand of u is $d_H(u) - \zeta(u) - d_D^+(u)$. The number of neighbours of u that are dominated by vertices other than u is at most $\sum_{v \in N^+(u)} f_a(uv) + \sum_{v \in N^-(u)} f_b(vu) < \zeta(u)$. Since $s(pq) = 1$ for every arc $pq \in A(D)$, u is adjacent to exactly $d^+(u)$ vertices in S. Thus u must dominate more than $d_H(u) - \zeta(u) - d_D^+(u)$ vertices, a contradiction. This concludes the proof. $\qquad\square$

The constructions together with Lemmata 1, 2, 3 and 4 yield the main result of this paper.

Theorem 1. PLANAR CAPACITATED DOMINATING SET *is* $W[1]$-*hard*.

References

1. Alber, J., Bodlaender, H.L., Fernau, H., Kloks, T., Niedermeier, R.: Fixed Parameter Algorithms for DOMINATING SET and Related Problems on Planar Graphs. Algorithmica 33(4), 46–493 (2002)
2. Alber, J., Fellows, M.R., Niedermeier, R.: Polynomial-time data reduction for dominating set. J. ACM 51(3), 363–384 (2004)
3. Demaine, E.D., Fomin, F.V., Hajiaghayi, M.T., Thilikos, D.M.: Subexponential parameterized algorithms on bounded-genus graphs and H-minor-free graphs. J. ACM 52(6), 866–893 (2005)
4. Dom, M., Lokshtanov, D., Saurabh, S., Villanger, Y.: Capacitated Domination and Covering: A Parameterized Perspective. In: Grohe, M., Niedermeier, R. (eds.) IWPEC 2008. LNCS, vol. 5018, pp. 78–90. Springer, Heidelberg (2008)
5. Downey, R.G., Fellows, M.R.: Parameterized Complexity. Springer, Heidelberg (1999)
6. Fellows, M.R., Hermelin, D., Rosamond, F.A., Vialette, S.: On the parameterized complexity of multiple-interval graph problems. Theoretical Computer Science 410(1), 53–61 (2009)

7. Fiala, J., Golovach, P.A., Kratochvíl, J.: Parameterized Complexity of Coloring Problems: Treewidth versus Vertex Cover. In: TAMC 2009. LNCS, vol. 5532, pp. 221–230. Springer, Heidelberg (2009)
8. Flum, J., Grohe, M.: Parameterized Complexity Theory. Springer, Heidelberg (2006)
9. Flum, J., Grohe, M., Weyer, M.: Bounded fixed-parameter tractability and $\log^2 n$ nondeterministic bits. J. Comput. Syst. Sci. 72(1), 34–71 (2006)
10. Fomin, F.V., Golovach, P.A., Lokshtanov, D., Saurabh, S.: Clique-width: on the price of generality. In: The Proceedings of SODA, pp. 825–834 (2009)
11. Fomin, F.V., Thilikos, D.M.: Dominating Sets in Planar Graphs: Branch-Width and Exponential Speed-Up. SIAM Journal on Computing 36(2), 281–309 (2006)
12. Guo, J., Niedermeier, R.: Linear Problem Kernels for NP-Hard Problems on Planar Graphs. In: Arge, L., Cachin, C., Jurdziński, T., Tarlecki, A. (eds.) ICALP 2007. LNCS, vol. 4596, pp. 375–386. Springer, Heidelberg (2007)
13. Niedermeier, R.: Invitation to Fixed-Parameter Algorithms. Oxford University Press, Oxford (2006)

Boolean-Width of Graphs*

B.-M. Bui-Xuan, J.A. Telle, and M. Vatshelle

Department of Informatics, University of Bergen, Norway
{buixuan,telle,vatshelle}@ii.uib.no

Abstract. We introduce the graph parameter boolean-width, related to the number of different unions of neighborhoods across a cut of a graph. For many graph problems this number is the runtime bottleneck when using a divide-and-conquer approach. Boolean-width is similar to rank-width, which is related to the number of $GF(2)$-sums ($1+1=0$) of neighborhoods instead of the Boolean-sums ($1+1=1$) used for boolean-width. For an n-vertex graph G given with a decomposition tree of boolean-width k we show how to solve Minimum Dominating Set, Maximum Independent Set and Minimum or Maximum Independent Dominating Set in time $O(n(n+2^{3k}k))$. We show for any graph that its boolean-width is never more than the square of its rank-width. We also exhibit a class of graphs, the Hsu-grids, having the property that a Hsu-grid on $\Theta(n^2)$ vertices has boolean-width $\Theta(\log n)$ and tree-width, branch-width, clique-width and rank-width $\Theta(n)$. Moreover, any optimal rank-decomposition of such a graph will have boolean-width $\Theta(n)$, *i.e.* exponential in the optimal boolean-width.

1 Introduction

Width parameters of graphs, like tree-width, branch-width, clique-width and rank-width, are important in the theory of graph algorithms. Many NP-hard graph optimization problems have fixed-parameter tractable (FPT) algorithms when parameterized by these graph width parameters, see e.g. [12] for an overview. Such algorithms usually have two stages, a first stage computing the right decomposition of the input graph and a second stage solving the problem by a divide-and-conquer approach, or dynamic programming, along the decomposition. For practical applications we must look carefully at the runtimes as a function of the parameter. We may then have to concentrate on heuristic algorithms for the first stage, for example in the way done for tree-width as part of the TreewidthLIB project at University of Utrecht, see e.g. [2]. For the second stage we should carefully design algorithms for each separate problem. When comparing the usefulness of these width parameters, we first need to compare the values of the parameters on various graph classes, we secondly need good algorithms or fast heuristics for the first stage, and we thirdly need to compare the best runtimes for the second stage. In this paper we introduce a graph width parameter called boolean-width, and compare it to other parameters.

---

* Supported by the Norwegian Research Council, project PARALGO.

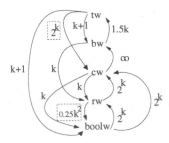

Fig. 1. Upper bounds tying parameters tw=tree-width, bw=branch-width, cw=clique-width, rw=rank-width and boolw=boolean-width. An arrow from P to Q labelled $f(k)$ means that any class of graphs having parameter P bounded by k will have parameter Q bounded by $O(f(k))$, and ∞ means that no such upper bound can be shown. Except for the labels in a box the bounds are known to be tight, meaning that there is a class of graphs for which the bound is $\Omega(f(k))$. For the box containing label 2^k a $\Omega(2^{k/2})$ bound is known [6].

Firstly, we show that the boolean-width of a graph is never more than quadratic in its rank-width, which also constitutes a comparison with other parameters since the rank-width of a graph is known to never be larger than its clique-width, nor its branch-width (resp. tree-width) plus one [19,20,22]. We also know that the boolean-width of a graph is never larger than its tree-width plus one [1]. On the other hand we show a class of graphs, the Hsu-grids, that have boolean-width bounded by k while they have rank-width (and thus also clique-width, branch-width and tree-width) exponential in k. See Figure 1 for a sketch of how the various parameters compare. Note that for any class of graphs we have only three possibilities: either all five parameters are bounded (e.g. for trees) or none of them are bounded (e.g. for grids) or only clique-width, rank-width and boolean-width are bounded (e.g. for cliques).

Secondly, regarding first stage algorithms, since boolean-width is tied to rank-width, when parameterizing by the boolean-width of an input graph we get an FPT algorithm that computes an approximation of an optimal boolean-width decomposition, by applying either the algorithm of Hliněný and Oum [11] computing an optimal rank-decomposition or the approximation algorithm of Oum and Seymour [20]. We have also initiated research into heuristic algorithms for the first stage.

Thirdly, for the second stage, we concentrate in this paper on the Minimum Dominating Set (MDS) problem and show that given a decomposition of boolean-width k of an n-vertex graph we can solve MDS in time $O(n(n + 2^{3k}k))$. See Figure 2 for a comparison of the best runtimes for MDS when parameterized by other width parameters. Combining the information in Figures 1 and 2 we see that the runtime for MDS compares well to the other parameters. For example, clique-width is bounded for a class of graphs exactly when boolean-width is, but as we show in Section 3 boolean-width is never larger than clique-width, and therefore $O^*(2^{3boolw})$ is always better than $O^*(2^{4cw})$. We also show there exists graphs where cliquewidth is exponential in boolean-width and for these graphs $O^*(2^{3boolw})$ is exponentially better than $O^*(2^{4cw})$. In [4] we similarly show that

	tree-width	branch-width	clique-width	rank-width	boolean-width
MDS	$O^*(2^{1.58tw})$[23]	$O^*(2^{2bw})$ [8]	$O^*(2^{4cw})$ [16]	$O^*(2^{0.75rw^2+O(rw)})$ [5,9]	$O^*(2^{3boolw})$ [here]

Fig. 2. Runtimes achievable for Minimum Dominating Set using various parameters

in time $O^*(2^{d \times q \times boolw^2})$, for problem-specific constants d and q, we can solve a large class of vertex subset and vertex partitioning problems.

A main open problem is to approximate the boolean-width of a graph better than what we get by using the algorithm for rank-width [11]. Nevertheless, for many problems it could be advantageous to use boolean-width for the second stage regardless of which decomposition is given. In Figure 3 we illustrate the runtimes achievable by the best second-stage algorithm for the MDS problem using either the algorithm given in Section 4 of this paper or the best runtime when parameterized by rankwidth, which are $O^*(2^{0.75rw^2+O(rw)})$ algorithms in both [5] and [9]. The values in Figure 3 assume we are given a decomposition tree (T, δ) of rank-width rw and is based on the result from Section 3 that the boolean-width of (T, δ) lies between $\log rw$ and $\frac{1}{4}rw^2 + \frac{5}{4}rw + \log rw$.

boolean-width of $(T,\delta)=$	using rank-width	using boolean-width
$0.25rw^2 + O(rw)$	$O^*(2^{0.75rw^2+O(rw)})$	$O^*(2^{0.75rw^2+O(rw)})$
rw	$O^*(2^{0.75rw^2+O(rw)})$	$O^*(2^{3rw})$
$\log rw$	$O^*(2^{0.75rw^2+O(rw)})$	$O^*(1)$

Fig. 3. Runtimes achievable for Minimum Dominating Set. Given a decomposition tree (T, δ) of rank-width rw, we know that the boolean-width of (T, δ) lies between $\log rw$ and $0.25rw^2 + O(rw)$.

For an appropriate class of Hsu-grids we are able to show that *any* optimal rank-decomposition will have boolean-width exponential in the optimal boolean-width. This suggests that although we can solve NP-hard problems in polynomial time on Hsu-grids if we use boolean-width as the parameter (as we see in Figure 3) we would get exponential time if we used any of the other graph width parameters.

Finally, we remark that the use of Boolean-sums in the definition of boolean-width (see Section 2) means a new application for the theory of Boolean matrices, *i.e.* matrices with Boolean entries, to the field of algorithms. Boolean matrices already have applications, *e.g.* in switching circuits, voting methods, applied logic, communication complexity, network measurements and social networks [7,15,17,21].

2   Boolean-Width

When applying divide-and-conquer to a graph we first need to divide the graph. A common way to store this information is to use a *decomposition tree* and to evaluate decomposition trees using a *cut function*. The following formalism is standard in graph and matroid decompositions (see, *e.g.*, [10,20,22]).

Definition 1. A decomposition tree of a graph G is a pair (T, δ) where T is a tree having internal nodes of degree three and $n = |V(G)|$ leaves, and δ is a bijection between the vertices of G and the leaves of T. For $A \subseteq V(G)$ let $\overline{A}$ denote the set $V(G) \setminus A$. Every edge of T defines a cut $\{A, \overline{A}\}$ of the graph, i.e. a partition of $V(G)$ in two parts, namely the two parts given, via δ, by the leaves of the two subtrees of T we get by removing the edge. Let $f : 2^V \to \mathbb{R}$ be a symmetric function, i.e. $f(A) = f(\overline{A})$ for all $A \subseteq V(G)$, also called a *cut function*. The f-width of (T, δ) is the maximum value of $f(A)$, taken over all cuts $\{A, \overline{A}\}$ of G given by an edge uv of T. The f-width of G is the minimum f-width over all decomposition trees of G.

The cuts $\{A, \overline{A}\}$ given by edges of the decomposition tree are used in the divide step of a divide-and-conquer approach. For the conquer step we solve the problem recursively, following the edges of the tree T (after choosing a root) in a bottom-up fashion, on the graphs induced by vertices of one side and of the other side of the cuts. In the combine step we must join solutions from the two sides, and this is usually the most costly and complicated operation. The question of what 'solutions' we should store to get an efficient combine step is related to what type of problem we are solving. Let us consider vertex subset or vertex partitioning problems on graphs, and in particular Maximum Independent Set for simplicity[1]. For a cut $\{A, \overline{A}\}$ we note that if two independent sets $X \subseteq A$ and $X' \subseteq A$ have the same set of neighbors in $\overline{A}$ then for any $Y \subseteq \overline{A}$ we have $X \cup Y$ an independent set if and only if $X' \cup Y$ an independent set. This suggests that the following equivalence relation on subsets of A will be useful.

Definition 2. Let G be a graph and $A \subseteq V(G)$. Two vertex subsets $X \subseteq A$ and $X' \subseteq A$ are *neighbourhood equivalent* w.r.t. A, denoted by $X \equiv_A X'$, if $\overline{A} \cap N(X) = \overline{A} \cap N(X')$.

If for each class $[X]_{\equiv_A}$ we store the maximum independent set in $[X]_{\equiv_A}$, and similarly for each class $[Y]_{\equiv_{\overline{A}}}$ we store the maximum independent set in $[Y]_{\equiv_{\overline{A}}}$, then we can perform the combine step in time depending only on the number of such equivalence classes. The same argument can be made for a large class of vertex subset and partitioning problems. Thus, to solve these problems as fast as possible on general graphs by divide-and-conquer we need a decomposition tree minimizing the number of equivalence classes over each cut defined by the tree. This minimum value is given by the boolean-width of the graph.

Definition 3 (Boolean-width). The *cut-bool* $: 2^{V(G)} \to \mathbb{R}$ function of a graph G is

$$cut\text{-}bool(A) = \log_2 |\{S \subseteq \overline{A} : \exists X \subseteq A \wedge S = \overline{A} \cap \bigcup_{x \in X} N(x)\}|$$

---

[1] Minimum Dominating Set is the main example of this paper and solving it by divide-and-conquer is indeed more complicated than solving Maximum Independent Set. Nevertheless, the runtime of our algorithm for Minimum Dominating Set, after employing several tricks, will in fact have a runtime matching what we could get for Maximum Independent Set.

It is known from Boolean matrix theory that *cut-bool* is symmetric [15, Theorem 1.2.3]. Using Definition 1 with $f = cut\text{-}bool$ we define the boolean-width of a decomposition tree, denoted $boolw(T, \delta)$, and the boolean-width of a graph, denoted $boolw(G)$.

Note that we take the logarithm base 2 of the number of equivalence classes simply to ensure that $0 \leq boolw(G) \leq |V(G)|$, which will ease the comparison of boolean-width to other parameters. For a vertex subset A, the value of $cut\text{-}bool(A)$ can also be seen as the logarithm in base 2 of the number of pairwise different vectors that are spanned, via Boolean sum, by the rows of the $A \times \overline{A}$ sub-matrix of the adjacency matrix of G.

3   Values of Boolean-Width Compared to Other Graph Width Parameters

Missing proofs can be found in the appendix. In this section we compare boolean-width to tree-width tw, branch-width bw, clique-width cw and rank-width rw. For any graph, it holds that the rankwidth of the graph is essentially the smallest parameter among the four [19,20,22]: $rw \leq cw$ and $rw \leq bw \leq tw + 1$ (unless $bw = 0$ and $rw = 1$). Accordingly, we focus on comparing boolean-width to rankwidth, and prove that $\log rw \leq boolw \leq \frac{1}{4}rw^2 + \frac{5}{4}rw + \log rw$ with the lower bound being tight to a constant multiplicative factor. We also know that the boolean-width of a graph is never larger than its tree-width plus one [1]. Furthermore, we also prove that $\log cw - 1 \leq boolw \leq cw$ with both bounds being tight to a constant multiplicative factor.

Rank-width was introduced in [18,20] based on the $cut\text{-}rank : 2^{V(G)} \to \mathbb{N}$ function of a graph G, which is the rank over $GF(2)$ of the submatrix of the adjacency matrix of G having rows A and columns $\overline{A}$. To see the connection with boolean-width note that

$$cut\text{-}rank(A) = \log_2 |\{Y \subseteq \overline{A} : \exists X \subseteq A \wedge Y = \overline{A} \cap \triangle_{x \in X} N(x)\}|$$

Here $\triangle$ is the symmetric difference operator. Note that $cut\text{-}rank$ is a symmetric function having integer values. Using Definition 1 with $f = cut\text{-}rank$ will define the rankwidth of a decomposition tree, denoted $rw(T, \delta)$, and the rankwidth of a graph, denoted $rw(G)$. We first investigate the relationship between the $cut\text{-}bool$ and the $cut\text{-}rank$ functions.

Lemma 1. *[5] Let G be a graph and $A \subseteq V(G)$. Let $nss(A)$ be the number of spaces that are $GF(2)$-spanned by the rows (resp. columns) of the $A \times V(G) \setminus A$ submatrix of the adjacency matrix of G. Then, $\log cut\text{-}rank(A) \leq cut\text{-}bool(A) \leq \log nss(A)$. Moreover, it is well-known from linear algebra that $nss(A) \leq 2^{\frac{1}{4}cut\text{-}rank(A)^2 + \frac{5}{4}cut\text{-}rank(A)} cut\text{-}rank(A)$.*

This lemma can be derived from a reformulation of [5, Proposition 3.6]. We now prove that both bounds given in this lemma are tight. For the lower bound we recall the graphs used in the definition of Hsu's generalized join [13]. For all $k \geq 1$, the graph H_k is defined as the bipartite graph having color classes $A(H_k) = \{a_1, a_2, \ldots, a_{k+1}\}$ and $B(H_k) = \{b_1, b_2, \ldots, b_{k+1}\}$ such that $N(a_1) = \emptyset$ and $N(a_i) = \{b_1, b_2, \ldots, b_{i-1}\}$ for all $i \geq 2$. Here, a union of neighborhoods of vertices of $A(H_k)$ is always of the form $\{b_1, b_2, \ldots, b_l\}$, hence,

Lemma 2. *For the above defined graph H_k, it holds that cut-bool$(A(H_k)) = \log k$ and cut-rank$(A(H_k)) = k$.*

For the tightness of the upper bound of Lemma 1 we now recall the graphs used in the characterization of rank-width given in [5]. The graph R_k is defined as a bipartite graph having color classes $A(R_k) = \{a_S, \ S \subseteq \{1, 2, \ldots, k\}\}$ and $B(R_k) = \{b_S, \ S \subseteq \{1, 2, \ldots, k\}\}$ such that a_S and b_T are adjacent if and only if $|S \cap T|$ is odd.

Lemma 3. *For the above defined graph R_k, it holds that cut-bool$(A(R_k)) = \log nss(A(R_k))$ and cut-rank$(A(R_k)) = k$.*

Since Lemma 1 holds for all edges of all decomposition trees, it is clear for every graph G that $\log rw(G) \leq boolw(G) \leq \frac{1}{4}rw(G)^2 + \frac{5}{4}rw(G) + \log rw(G)$. We now address the tightness of this lower bound. A cut $\{A, \overline{A}\}$ is *balanced* if $\frac{1}{3}|V(G)| \leq |A| \leq \frac{2}{3}|V(G)|$. In any decomposition tree of G, there always exists an edge of the tree which induces a balanced cut in the graph. We lift the tightness result on graph cuts given by Lemma 2 to the level of graph parameters in a standard way, by using the structure of a grid. The main idea is that any balanced cut of a grid will contain either a large part of some column of the grid, or it contains a large enough matching. We then add edges to the columns of the grid and fill each of them into a Hsu graph (see below). Note that graphs with a similar definition have also been studied in relation with clique-width in a different context [3].

Definition 4 (Hsu-grid $HG_{p,q}$). Let $p \geq 2$ and $q \geq 2$. The Hsu-grid $HG_{p,q}$ is defined by $V(HG_{p,q}) = \{v_{i,j} \mid 1 \leq i \leq p \ \wedge \ 1 \leq j \leq q\}$ with $E(HG_{p,q})$ being exactly the union of the edges $\{(v_{i,j}, v_{i+1,j}) \mid 1 \leq i < p \ \wedge \ 1 \leq j \leq q\}$ and of the edges $\{(v_{i,j}, v_{i',j+1}) \mid 1 \leq i \leq i' \leq p \ \wedge \ 1 \leq j < q\}$. We say that vertex $v_{i,j}$ is at the ith row and the jth column.

Lemma 4. *For large enough integers p and q, we have that boolw$(HG_{p,q}) \leq \min(2\log p, q)$ and $rw(HG_{p,q}) \geq \min(\lfloor \frac{p}{4} \rfloor, \lfloor \frac{q}{6} \rfloor)$. Moreover, if $q < \lfloor \frac{p}{8} \rfloor$ then any optimal rank decomposition of $HG_{p,q}$ has boolean-width at least $\lfloor \frac{q}{6} \rfloor$.*

Notice that not only the lemma addresses the tightness of the lower bound on boolean-width as a function of rank-width, but also the additional stronger property that for a special class of Hsu-grids any optimal rank decomposition has boolean-width exponential in the optimal boolean-width.

Theorem 1. *For any decomposition tree (T, δ) of any graph G it holds that $\log rw(T, \delta) \leq boolw(T, \delta) \leq \frac{1}{4}rw(T, \delta)^2 + \frac{5}{4}rw(T, \delta) + \log rw(T, \delta)$ and $\log rw(G) \leq boolw(G) \leq \frac{1}{4}rw(G)^2 + \frac{5}{4}rw(G) + \log rw(G)$. For large enough integer k, there are graphs L_k and U_k of rank-width at least k such that boolw$(L_k) \leq 2\log rw(L_k) + 4$ and boolw$(U_k) \geq \lfloor \frac{1}{6}rw(U_k) \rfloor - 1$.*

Remark 1. The inequalities about L_k is a direct application of Lemma 4 for well-chosen values of p and q. The graph U_k are standard $k \times k$ grids.

Remark 2. Let (T, δ) and (T', δ') be such that $rw(G) = rw(T, \delta)$ and $OPT = boolw(G) = boolw(T', \delta')$. We then have from Theorem 1 that $boolw(T, \delta) \leq rw(T, \delta)^2 \leq rw(T', \delta')^2 \leq (2^{OPT})^2$. Hence, any optimal rank-width decomposition of G is also a $2^{2 \cdot OPT}$-approximation of an optimal boolean-width decomposition of G. There is an FPT algorithm to compute an optimal rank-width decomposition of G in $O(f(rw(G)) \times |V(G)|^3)$ time [11].

One of the most important applications of rank-width is to approximate the clique-width $cw(G)$ of a graph by $\log(cw(G) + 1) - 1 \leq rw(G) \leq cw(G)$ [20]. Although we have seen that the difference between rank-width and boolean-width can be quite large, we remark that, w.r.t. clique-width, boolean-width behaves similarly as rank-width, namely

Theorem 2. *For any graph G it holds that $\log cw(G) - 1 \leq boolw(G) \leq cw(G)$. For large enough integer k, there are graphs L_k and U_k of clique-width at least k such that $boolw(L_k) \leq 2 \log cw(L_k) + 4$ and $boolw(U_k) \geq \lfloor \frac{1}{6} cw(U_k) \rfloor - 1$.*

4   Algorithms

Given a decomposition tree (T, δ) of a graph G we will in this section show how to solve a problem on G by a divide-and-conquer (or dynamic programming) approach. We subdivide an arbitrary edge of T to get a new root node r, denoting by T_r the resulting rooted tree, and let the algorithm follow a bottom-up traversal of T_r. With each node w of T_r we associate a table data structure Tab_w, that will store optimal solutions to subproblems related to the cut $\{A, \overline{A}\}$ given by the edge between w and its parent. In Subsection 4.2 we will define the tables used and in particular give the details of the combine step. For the moment it suffices to say that the table indices will be related to the classes of the equivalence relation $\equiv_A$ of Definition 2. Firstly, in Subsection 4.1 we show how to enhance the decomposition tree with information needed to handle these equivalence classes.

4.1   Computing Representatives

We assume a total ordering on the vertex set of G which stays the same throughout the whole paper. If vertex u comes before vertex v in the ordering then we say u is *smaller* than v. Using this ordering we also denote that a vertex set X is lexicographically smaller than vertex set Y by $X \leq_{lex} Y$. Let $\{A, \overline{A}\}$ be a cut given by an edge of the decomposition tree. For each equivalence class of $\equiv_A$ we want to choose one vertex subset as a representative for that class. The representative set for a class will be the lexicographically smallest among the sets in the class with minimum cardinality. More formally we define for $A \subseteq V(G)$ the list LR_A of all representatives of $\equiv_A$.

Definition 5 (List of Representatives). Given a graph G and $A \subseteq V(G)$ we define the list LR_A of representatives of $\equiv_A$ as the unique set of subsets of A satisfying:

1) $\forall X \subseteq A, \exists R \in LR_A : R \equiv_A X$
2) $\forall R \in LR_A :$ if $R \equiv_A X$ then $|R| \leq |X|$
3) $\forall R \in LR_A :$ if $R \equiv_A X$ and $|R| = |X|$ then $R \leq_{lex} X$.

Note that such a list will contain exactly one element for each equivalence class of $\equiv_A$.

Lemma 5. *Let G be a graph and $A \subseteq V(G)$. Let R be an element of the list LR_A of representatives of $\equiv_A$, then for any $X, Y \subseteq R$ s.t. $X \neq Y$, we have $X \not\equiv_A Y$.*

Corollary 1. *Given a graph G and $A \subseteq V(G)$, every element R of the list LR_A of representatives of $\equiv_A$ satisfies $|R| \leq$ cut-bool(A).*

We now describe an algorithm to compute LR_A. It will at the same time compute a list LNR_A containing $N(R) \cap \overline{A}$, for every element R of LR_A. These two lists will be linked together, in such a way that given an element N of LNR_A we can access in constant time the element R of LR_A such that $N = N(R) \cap \overline{A}$, and vice versa. To do this in time depending only on cut-bool(A) we will need the notion of twin classes.

Definition 6. Let G be a graph and let $A \subseteq V(G)$ be a vertex subset. A subset $X \subseteq A$ is a *twin set of A* if, for every $z \in \overline{A}$ and pair of vertices $x, y \in X$, we have x adjacent to z if and only if y adjacent to z. A twin set X is a *twin class of A* if X is a maximal twin set. The set of all twin classes of A forms a partition of A, that we call the *twin class partition of A*. We denote by TC_A the set containing for each twin class of A the smallest vertex of the class.

Note that u and v belong to the same twin class of A if and only if $\{u\} \equiv_A \{v\}$. One consequence is that $|TC_A| \leq 2^{cut\text{-}bool(A)}$. Our algorithm will handle the edges crossing a cut $\{A, \overline{A}\}$ by using the two vertex sets TC_A and $TC_{\overline{A}}$. As a preprocessing step, we will compute TC_A and $TC_{\overline{A}}$ associated to every $A \subseteq V(G)$ that will be needed in our principal dynamic programming algorithm as specified in the lemma below.

Lemma 6. *Let G be a graph and (T, δ) a decomposition tree of G. Then, in $O(n(n + 2^{2boolw(T,\delta)}))$ global runtime we can compute, for every edge uv of T the two vertex sets TC_A and $TC_{\overline{A}}$ for $\{A, \overline{A}\}$ being the 2-partition of $V(G)$ induced by the leaves of the trees we get by removing uv from T. In the same runtime, for every $v \in A$, resp. $\overline{A}$, we compute a pointer to the vertex u in TC_A, resp. $TC_{\overline{A}}$, such that u and v are in the same twin class of A, resp. $\overline{A}$.*

We now focus on a particular cut $\{A, \overline{A}\}$, induced by some edge of the decomposition tree of G. Our algorithms will use the bipartite graph H_A with color-classes TC_A and $TC_{\overline{A}}$ and containing all edges of G crossing the cut $\{A, \overline{A}\}$. The graph H_A can be built in $O(|TC_A| \times |TC_{\overline{A}}|) = O(2^{2cut\text{-}bool(A)})$ time.

Lemma 7. *Given H_A as defined above for any $A \subseteq V(G)$, we can in time $O(2^{3cut\text{-}bool(A)} cut\text{-}bool(A))$ compute the list of representatives LR_A and the sorted list LNR_A of neighborhoods of elements of LR_A.*

Proof. We describe the algorithm. The lists LR_A and LNR_A are initially empty. We will use auxiliary lists *NextLevel*, initially empty, and *LastLevel* which initially will contain the empty set as its single element. We then run the following nested loops.

> **while** LastLevel $!= \emptyset$ **do**
> > **for** R in LastLevel **do**
> > > **for** every vertex v of TC_A **do**
> > > > $R' = R \cup \{v\}$
> > > > compute $N' = N_{H_A}(R')$
> > > > **if** $R' \not\equiv_A R$ and N' is not contained in LNR_A **then**
> > > > > add R' to LR_A and NextLevel, and add N' to LNR_A at proper position
> > > >
> > > > **end if**
> > >
> > > **end for**
> >
> > **end for**
> > set LastLevel = NextLevel, and NextLevel = $\emptyset$
>
> **end while**

Let us first argue for correctness. The first iteration of the while-loop will set $\{v\}$ as representative, for every $v \in TC_A$, and there exist no other representatives of size 1 in LR_A. The algorithm computes all representatives of size i before it moves on to those of size $i + 1$. LastLevel will contain all representatives of size i while NextLevel will contain all representatives of size $i + 1$ found so far. Every representative will be expanded by every possible node and checked against all previously found representatives. The only thing left to prove is that any representative R can be written as $R' \cup \{v\}$ for some representative R'. Assume for contradiction that no R' exists such that $R = R' \cup \{v\}$. Then let v be the lexicographically largest element of R, then $R \setminus \{v\}$ can not be a representative so let R' be the representative of $[R \setminus \{v\}]_{\equiv_A}$. We know that $R' \cup \{v\} \equiv_A R$, we know that $|R' \cup \{v\}| \leq |R|$ and that $R' \cup \{v\}$ comes before R in a lexicographical ordering contradicting that R is a representative.

We now argue for the runtime. Let $k = cut\text{-}bool(A)$. The three loops loop once for each pair of element R (of TC_A) and vertex v (of TC_A). The number of representatives are exactly 2^k, while $|TC_A| \leq 2^k$, hence at most $O(2^{2k})$ iterations in total. Inside the innermost for-loop we need to calculate the neighbourhood of R', from Corollary 1 we get $|R'| \leq k + 1$. Since no node in H_A have degree more than 2^k we can find $N_{H_A}(R')$ in $O(k2^k)$ time. Then to see if $R' \equiv_A R$ we compare the two neighbourhoods in $O(2^k)$ time. Then we want to check if the neighbourhood is contained in the list LNR_A, hence we want LNR_A to be a sorted list, then searching only takes $O(k)$ steps, however for each step comparing two neighbourhoods can take $O(2^k)$ time. Inserting into the sorted list LNR_A takes $O(2^k)$, and in the other lists $O(1)$ time. This means all operations in the inner for-loop can be done in $O(k2^k)$ time, giving a total running-time of $O(k2^{3k})$. $\qquad\square$

Algorithm 1. Initialize datastructure used for finding representative R of $[X]_{\equiv_A}$

> INPUT: Lists LR_A and LNR_A and bipartite graph H_A
> Initialize M to a two dimensional table with $|LR_A| \times |TC_A|$ elements.
> **for** every vertex v of TC_A **do**
> > **for** R in LR_A **do**
> > > $R' = R \cup \{v\}$
> > > find R_U in LR_A that is linked to the neighbourhood $N_{H_A}(R')$ in LNR_A
> > > add a pointer from $M[R][v]$ to R_U
> > **end for**
> **end for**
> OUTPUT: M

Given $X \subseteq A$ we will now address the question of computing the representative R of $[X]_{\equiv_A}$, in other words accessing the entry R of LR_A such that $X \equiv_A R$. The naive way to do this is to binary-search in the list LNR_A for the set $N(X) \cap \overline{A}$ in time $O(2^{cut\text{-}bool(A)} cut\text{-}bool(A))$, but we want to do this in $O(|X|)$ time. To accomplish this we construct an auxiliary data-structure that maps a pair (R, v), consisting of one representative R from LR_A and one vertex from TC_A, to the representative R' of the class $[R \cup \{v\}]_{\equiv_A}$. It will be stored as a two dimensional table, leading to a constant time lookup.

Lemma 8. *Given H_A as defined above for any $A \subseteq V(G)$, we can in time $O(2^{3cut\text{-}bool(A)} cut\text{-}bool(A))$ compute a datastructure allowing, for any $X \subseteq A$, to access in $O(|X|)$ time the entry R of LR_A such that $X \equiv_A R$.*

Proof. Let $k = cut\text{-}bool(A)$. First we need to initialize the datastructure used for finding representatives using Algorithm 1. It goes through 2 for-loops, in total iterating $O(2^{2k})$ times. To find the neighbourhood of R' takes $O(2^k k)$ time. To search LNR_A for the neighbourhood takes $O(2^k k)$ time. All other operations are done in constant time, thus the runtime is $O(2^{3k} k)$.

Given $X \subseteq A$ we find the representative R of $[X]_{\equiv_A}$ as follows. Initially R will be empty. Then we iterate over all elements $u \in X$, first looking up $v \in TC_A$ such that u and v belong to the same twin class of A, and then replacing R by the representative of the class $[R \cup \{v\}]_{\equiv_A}$ (as given by the auxiliary data structure). $\qquad\square$

4.2   Dynamic Programming for Dominating Set

This section is based on the dynamic programming scheme used in [5] to give an algorithm for Minimum Dominating Set parameterized by rankwidth. For example, Lemma 11 is an adaptation from that paper to the current formalism parameterizing by boolean-width. Recall that our algorithm will follow a bottom-up traversal of the tree T_r, computing at each node w of the tree a table Tab_w, that will store optimal solutions to subproblems related to the cut $\{A, \overline{A}\}$ given by the edge between w and its parent. If we were solving Maximum Independent Set then Tab_w would simply be indexed by the equivalence classes of $\equiv_A$. However, unlike the case of independent sets we note that a set of vertices D dominating A will include also vertices of $\overline{A}$ that dominate vertices of A 'from the outside'. This motivates the following definition.

Definition 7. Let G be a graph and $A \subseteq V(G)$. For $X \subseteq A$, $Y \subseteq \overline{A}$, if $A \setminus X \subseteq N(X \cup Y)$ we say that the pair (X, Y) *dominates* A.

The main idea for dealing with this complication is to index the table at w by two sets, one that represents the equivalence class of $D \cap A$ under $\equiv_A$ that dominates 'from the inside', and one that represents the equivalence class of $D \cap \overline{A}$ under $\equiv_{\overline{A}}$ that helps dominate the rest of A 'from the outside'. The subsequent lemma should indicate why this will work.

Lemma 9. *Let G be a graph and $A \subseteq V(G)$. For $X \subseteq A$, $Y, Y' \subseteq \overline{A}$, If (X, Y) dominates A and $Y \equiv_{\overline{A}} Y'$ then (X, Y') dominates A.*

Proof. Since (X, Y) dominates A we have $A \setminus X \subseteq N(X \cup Y)$. Since $Y \equiv_{\overline{A}} Y'$ we have $N(Y) \cap A = N(Y') \cap A$. Then it follows that $A \setminus X \subseteq N(X \cup Y')$, meaning (X, Y') dominates A. □

For a node w of T_r we denote by $\{A_w, \overline{A_w}\}$, the cut given by the edge between w and its parent. In the previous subsection we saw how to compute for every node w of T_r the lists LR_{A_w} of representatives of $\equiv_{A_w}$ and $LR_{\overline{A_w}}$ of representatives of $\equiv_{\overline{A_w}}$.

Definition 8. The two-dimensional table Tab_w will have index set $LR_{A_w} \times LR_{\overline{A_w}}$. For $R_w \in LR_{A_w}$ and $R_{\overline{w}} \in LR_{\overline{A_w}}$ the contents of $Tab_w[R_w][R_{\overline{w}}]$ after updating should be:

$$Tab_w[R_w][R_{\overline{w}}] \overset{def}{=} min_{S \subseteq A_w}\{|S| : S \equiv_{A_w} R_w \text{ and } (S, R_{\overline{w}}) \text{ dominates } A_w\}$$

Note that the table Tab_w will have $2^{2cut\text{-}bool(A_w)}$ entries. For every node w we assume that initially every entry Tab_w is set to ∞. For a leaf l of T_r, since $A_l = \{\delta(l)\}$, note that $\equiv_{A_l}$ has only two equivalence classes: one containing $\emptyset$ and the other containing A_l. For $\overline{A_l}$, we have the same situation with only two equivalence classes: one containing $\emptyset$ and the other containing $\overline{A_l}$. Therefore, we set $Tab_l[\emptyset][\emptyset] := \infty$, and $Tab_l[\{\delta(l)\}][\emptyset] := 1$ and $Tab_l[\{\delta(l)\}][R] := 1$ and $Tab_l[\emptyset][R] := 0$ (where R is the representative of $[\overline{A_l}]_{\equiv_{\overline{A_l}}}$) since the only of the four combinations that does not dominate A_l as in Definition 7 is $(\emptyset, \emptyset)$. Note that there would be a special case if $\delta(l)$ was an isolated vertex, but isolated vertices can easily be removed.

For the updating of internal nodes we have a node w with two children a and b and can assume that the tables Tab_a and Tab_b have been correctly computed. We need to correctly compute the value of $Tab_w[R_w][R_{\overline{w}}]$ for each $R_w \in LR_{A_w}$ and $R_{\overline{w}} \in LR_{\overline{A_w}}$. Each table can have $2^{2boolw(T,\delta)}$ entries. Therefore, the number of pairs of entries, one from each of Tab_a and Tab_b, could be as much as $2^{4boolw(T,\delta)}$. Looping over all such pairs of entries we would in fact spend time $2^{5boolw(T,\delta)}$ since we would have to compute the right entry in Tab_w. Instead we achieve

$2^{3boolw(T,\delta)}$ time by looping only over one half of the entries in each of the three tables, as follows:

for all $R_a \in LR_{A_a}, R_b \in LR_{A_b}, R_{\overline{w}} \in LR_{\overline{A_w}}$ **do**
    find the representative $R_{\overline{a}}$ of the class $[R_b \cup R_{\overline{w}}]_{\equiv_{\overline{A_a}}}$
    find the representative $R_{\overline{b}}$ of the class $[R_a \cup R_{\overline{w}}]_{\equiv_{\overline{A_b}}}$
    find the representative R_w of the class $[R_a \cup R_b]_{\equiv_{A_w}}$
    $Tab_w[R_w][R_{\overline{w}}] = min(\,Tab_w[R_w][R_{\overline{w}}],\, Tab_a[R_a][R_{\overline{a}}] + Tab_b[R_b][R_{\overline{b}}])$
end for

Lemma 10. *For a graph G, let A, B, W be a 3-partitioning of $V(G)$, and let $S_a \subseteq A, S_b \subseteq B$ and $S_w \subseteq W$. $(S_a, S_b \cup S_w)$ dominates A and $(S_b, S_a \cup S_w)$ dominates B iff $(S_a \cup S_b, S_w)$ dominates $A \cup B$.*

Proof. Let $S = S_a \cup S_b \cup S_w$. Clearly, $(S_a, S_b \cup S_w)$ dominates A iff $A \setminus S_a \subseteq N(S)$. Likewise, $(S_b, S_a \cup S_w)$ dominates B iff $B \setminus S_b \subseteq N(S)$. Therefore, $A \setminus S_a \subseteq N(S)$ and $B \setminus S_b \subseteq N(S)$ iff $A \cup B \setminus S_a \cup S_b \subseteq N(S)$ iff $(S_a \cup S_B, S_w)$ dominates $A \cup B$. $\square$

Lemma 11. *The table at node w is updated correctly, namely for any representative $R_w \in LR_{A_w}$ and $R_{\overline{w}} \in LR_{\overline{A_w}}$, if $Tab_w[R_w][R_{\overline{w}}]$ is not ∞ then*

$$Tab_w[R_w][R_{\overline{w}}] = min_{S \subseteq A_w}\{|S| : S \equiv_{A_w} R_w \wedge (S, R_{\overline{w}}) \text{ dominates } A_w\}.$$

Theorem 3. *Given an n-vertex graph G and a decomposition tree (T, δ) of G, the Minimum Dominating Set problem on G can be solved in time $O(n(n + 2^{3boolw(T,\delta)}boolw(T, \delta)))$.*

Proof. As a preprocessing step we compute the twin classes for all cuts induced by the edges of (T, δ) as described in Lemma 6. We then loop over all edges uv of T. Let $\{A, \overline{A}\}$ be the cut of G induced by the leaves of T after removing uv from T. We compute the graph H_A, as well as the lists $LR_A, LR_{\overline{A}}, LNR_A$, and $LNR_{\overline{A}}$ as described in Lemma 7, and also the datastructure for finding a representative of $[X]_{\equiv_A}$ and $[Y]_{\equiv_{\overline{A}}}$ as described in Lemma 8. After this loop we subdivide an arbitrary edge of T by a new root node r to get T_r. We then initialize the table Tab_l for every leaf l of T_r as described after Definition 8. Finally, we scan T_r in a bottom-up traversal and update the table Tab_w for every internal node w as described right before Lemma 10. After this, the optimum solution can be read at the (unique) entry $Tab_r[V(G)][\emptyset]$ of the table at the root of T_r.

The correctness follows from Lemma 11, when applied to $w = r$. The complexity analysis of the computation before setting the root r is a straightforward combination of those given in Lemmas 6, 7 and 8. After this, the initialization at every leaf of T_r takes $O(1)$ time. The update at every internal node w of T_r loops through $2^{3boolw(T,\delta)}$ triplets, and for each of them spend $O(boolw(T, \delta))$ time finding the three representatives and $O(1)$ time updating the value of $Tab_w[R_w][R_{\overline{w}}]$. $\square$

Solving Maximum Independent Set (MIS) is simpler than solving Minimum Dominating Set. The table Tab_w at a node w will then be one-dimensional, indexed by the equivalence classes of $\equiv_{A_w}$, and will store the size of the maximum independent set in that class. In the combine step we loop over all pairs of representatives R_a from Tab_a and R_b from Tab_b and check if there are any edges between R_a and R_b. If not, then we look up the representative R_w of $[R_a \cup R_b]_{\equiv_{A_w}}$ and update $Tab_w[R_w]$ by the maximum of its old value and $Tab_a[R_a] + Tab_b[R_b]$. Combining these ideas we can solve both the Minimum and Maximum Independent Dominating Set problems. The runtimes will be dominated by the computation of representatives.

Corollary 2. *Given an n-vertex graph G and a decomposition tree (T, δ) of G, we can solve the Maximum Independent Set, Minimum Independent Dominating Set and Maximum Independent Dominating Set problems on G in time $O(n(n + 2^{3boolw(T,\delta)}boolw(T, \delta)))$.*

5   Conclusion and Perspectives

There are many questions about boolean-width left unanswered. The foremost concerns possibly its practical applicability. The divide-and-conquer algorithms given here are practical and easy to implement, but we need fast and good heuristics computing decomposition trees of low boolean-width. Research in this direction is underway.

On the theoretical side it would be nice to improve on the $2^{2 \cdot OPT}$-approximation algorithm to an optimal boolean-width decomposition (c.f. Remark 2) we get by applying the algorithm computing an optimal rank-width decomposition [11]. Note that the runtime of that approximation algorithm is FPT when parameterized by boolean-width. The best we can hope for is an FPT algorithm computing a decomposition of optimal boolean-width, but any polynomial approximation would be nice.

The graphs of boolean-width at most one are exactly the graphs of rank-width one, *i.e.* the distance-hereditary graphs. What about the graphs of boolean-width at most two, do they also have a nice characterization? Is there a polynomial-time algorithm to decide if a graph has boolean-width at most two? More generally, is there an alternative characterization of the graphs of boolean-width at most k?

We do not know if the bound $boolw(G) \leq \frac{1}{4}rw(G)^2 + \frac{5}{4}rw(G) + \log rw(G)$ is tight to a multiplicative factor. For most well-known classes we should have $boolw = O(rw)$, but this needs to be investigated. Are there well-known graph classes where $boolw = O(\log rw)$? It has been shown that a $k \times k$ grid has rank-width k [14], and we have seen that its boolean-width lies between $\frac{1}{6}k$ (proof Theorem 1) and $k + 1$ (derived from its clique-width). What is the right value? All these questions should benefit from the connections between boolean-width and the field of Boolean matrix theory.

References

1. Adler, I., Vatshelle, M.: Personal communication
2. Bodlaender, H., Koster, A.: Treewidth Computations I Upper Bounds. Technical Report UU-CS-2008-032, Department of Information and Computing Sciences, Utrecht University (2008)

3. Brandstaedt, A., Lozin, V.V.: On the linear structure and clique-width of bipartite permutation graphs. Ars Combinatoria 67, 719–734 (2003)
4. Bui-Xuan, B.-M., Telle, J.A., Vatshelle, M.: Fast FPT algorithms for vertex subset and vertex partitioning problems using neighborhood unions, http://arxiv.org/abs/0903.4796+
5. Bui-Xuan, B.-M., Telle, J.A., Vatshelle, M.: H-join decomposable graphs and algorithms with runtime single exponential in rankwidth. To appear in DAM: special issue of GROW, http://www.ii.uib.no/~telle/bib/BTV.pdf
6. Corneil, D., Rotics, U.: On the relationship between clique-width and treewidth. SIAM Journal on Computing 34(4), 825–847 (2005)
7. Damm, C., Kim, K.H., Roush, F.W.: On covering and rank problems for boolean matrices and their applications. In: Asano, T., Imai, H., Lee, D.T., Nakano, S.-i., Tokuyama, T. (eds.) COCOON 1999. LNCS, vol. 1627, pp. 123–133. Springer, Heidelberg (1999)
8. Dorn, F.: Dynamic programming and fast matrix multiplication. In: Azar, Y., Erlebach, T. (eds.) ESA 2006. LNCS, vol. 4168, pp. 280–291. Springer, Heidelberg (2006)
9. Ganian, R., Hliněný, P.: On Parse Trees and Myhill-Nerode-type Tools for handling Graphs of Bounded Rank-width (submitted manuscript), http://www.fi.muni.cz/~hlineny/Research/papers/MNtools+dam3.pdf
10. Geelen, J., Gerards, A., Whittle, G.: Branch-width and well-quasi-ordering in matroids and graphs. Journal of Combinatorial Theory, Series B 84(2), 270–290 (2002)
11. Hliněný, P., Oum, S.: Finding branch-decompositions and rank-decompositions. SIAM Journal on Computing 38(3), 1012–1032 (2008); Abstract at ESA 2007.
12. Hliněný, P., Oum, S., Seese, D., Gottlob, G.: Width parameters beyond tree-width and their applications. The Computer Journal 51(3), 326–362 (2008)
13. Hsu, W.-L.: Decomposition of perfect graphs. Journal of Combinatorial Theory, Series B 43(1), 70–94 (1987)
14. Jelínek, V.: The rank-width of the square grid. In: Broersma, H., Erlebach, T., Friedetzky, T., Paulusma, D. (eds.) WG 2008. LNCS, vol. 5344, pp. 230–239. Springer, Heidelberg (2008)
15. Kim, K.H.: Boolean matrix theory and its applications. Marcel Dekker, New York (1982)
16. Kobler, D., Rotics, U.: Edge dominating set and colorings on graphs with fixed clique-width. Discrete Applied Mathematics 126(2-3), 197–221 (2003); Abstract at SODA 2001
17. Nguyen, H.X., Thiran, P.: Active measurement for multiple link failures diagnosis in IP networks. In: Barakat, C., Pratt, I. (eds.) PAM 2004. LNCS, vol. 3015, pp. 185–194. Springer, Heidelberg (2004)
18. Oum, S.: Graphs of Bounded Rank-width. PhD thesis, Princeton University (2005)
19. Oum, S.: Rank-width is less than or equal to branch-width. Journal of Graph Theory 57(3), 239–244 (2008)
20. Oum, S., Seymour, P.: Approximating clique-width and branch-width. Journal of Combinatorial Theory, Series B 96(4), 514–528 (2006)
21. Pattison, P., Breiger, R.: Lattices and dimensional representations: matrix decompositions and ordering structures. Social Networks 24(4), 423–444 (2002)
22. Robertson, N., Seymour, P.: Graph minors. X. Obstructions to tree-decomposition. Journal of Combinatorial Theory, Series B 52(2), 153–190 (1991)
23. Rooij, J., Bodlaender, H., Rossmanith, P.: Dynamic programming on tree decompositions using generalised fast subset convolution. In: Fiat, A., Sanders, P. (eds.) ESA 2009. LNCS, vol. 5757, pp. 566–577. Springer, Heidelberg (2009)

The Complexity of Satisfiability of Small Depth Circuits

Chris Calabro, Russell Impagliazzo*, and Ramamohan Paturi**

Department of Computer Science and Engineering
University of California, San Diego
La Jolla, CA 92093-0404, USA

Abstract. Say that an algorithm solving a Boolean satisfiability problem x on n variables is *improved* if it takes time $\text{poly}(|x|)2^{cn}$ for some constant $c < 1$, *i.e.*, if it is exponentially better than a brute force search. We show an improved randomized algorithm for the satisfiability problem for circuits of constant depth d and a linear number of gates cn: for each d and c, the running time is $2^{(1-\delta)n}$ where the improvement $\delta \geq 1/O(c^{2^{d-2}-1}\lg^{3\cdot 2^{d-2}-2}c)$, and the constant in the big-Oh depends only on d. The algorithm can be adjusted for use with Grover's algorithm to achieve a run time of $2^{\frac{1-\delta}{2}n}$ on a quantum computer.

1 Introduction

All **NP**-complete problems are equivalent as far as the existence of polynomial time algorithms is concerned. However, the exact complexities of these problems vary widely. There are frequently algorithms for **NP**-complete problems that achieve substantial improvement over exhaustive search. This raises the questions: Which problems have such improved algorithms? How much can we improve? Can we provide evidence that no improvement over some known algorithm is possible? Work addressing such questions, both from the algorithmic and complexity theoretic sides, has become known as *exact complexity,* and it is related to the field of parameterized complexity. While significant work has been done, both areas are still fairly new and leave open many problems. In particular, the answers and techniques seem to rely on the exact **NP**-complete problem in question, and there are few unifying techniques. (This is in some ways similar to the situation for the exact approximation ratios achievable for different **NP**-complete problems, which also is problem dependent. However, the use of probabilistically checkable proofs, and the unique games conjecture and related conjectures, provide very general tools for

* This research is supported by supported by the Simonyi Fund, the Bell Company Fellowship and the Fund for Math, and NSF grants DMS-083573, CNS-0716790 and CCF-0832797.
** This research is supported by NSF grant CCF-0947262 from the Division of Computing and Communication Foundations. Any opinions, findings and conclusions or recommendations expressed in this material are those of the authors and do not necessarily reflect the views of the National Science Foundation.

J. Chen and F.V. Fomin (Eds.): IWPEC 2009, LNCS 5917, pp. 75–85, 2009.
© Springer-Verlag Berlin Heidelberg 2009

understanding approximability for a wide variety of problems. We are still looking for similar tools for exact complexity.)

From the viewpoint of exact complexity, the most studied and best understood problems are probably the restricted versions of the satisfiability problem (SAT), in particular, k-SAT, a restriction of SAT to k-CNFs, and CNF-SAT, a restriction to general CNFs. There has been a sequence of highly nontrivial and interesting algorithmic approaches [Sch99, PPZ99, PPSZ05, Sch05, DW05, CIP06] to these problems, where the best known constant factor improvements in the exponent are of the form $1 - 1/O(k)$ for k-SAT and $1 - 1/O(\lg c)$ for CNF-SAT with at most cn clauses. Also, a sequence of papers ([IPZ01, IP01, CIKP08, CIP06]) has shown many nontrivial relationships between the exact complexities of these problems, and helped characterize their hardest instances (under the assumption that they are indeed exponentially hard.) For what other circuit/formula models can we expect to show *improved* exponential-time (*i.e.*, $O(2^{cn})$-time for $c < 1$) algorithms for the satisfiability problem?

1.1   Linear-Size Bounded Depth Circuits

In this paper, we give what we believe is the first improved algorithm for the satisfiability problem for circuits of constant depth and linear size (AC^0 type), which seems significantly harder than k-SAT. (Note that it is trivially possible to give an improved algorithm in terms of the circuit size parameter m if the circuit has fan-in 2, e.g. by the standard reduction to 3-SAT and then applying a 3-SAT solver, but this is no better than exhaustive search once m gets larger than around $4n$ or so. Sergey Nurik [Nur09] has recently communicated a somewhat improved bound along these lines.) For each $c, d > 0$, we give a constant $\delta > 0$ and a randomized algorithm that works in $2^{(1-\delta)n}$ time and solves the satisfiability problem for depth d, size at most cn circuits. (Here, it is significant that circuit size is measured by gates rather than wires.)

For $d = 2$, our algorithm becomes deterministic and matches the current best bound [CIP06], since our algorithm and analysis are generalizations of the ones there. However, randomizing the algorithm also yields the best quantum algorithm for this case, with running time $2^{\frac{1-1/O(\lg c)}{2}n}$. For $d = 3$, this gives $\delta \geq 1/O(c \lg^4 c)$, which, as far as the authors know, is the first improvement achieved for this problem.

There are a few motivations to consider linear-size circuits. One is the question of ideal block cipher design. Block ciphers are carefully constructed to maximize efficiency for a given level of security. Particularly, since we want ciphers to be usable by low-power devices, and to be implemented at the network level, it is often very important to have efficient hardware implementations that make maximum use of parallelism. A typical cipher computes for a small number of "rounds", where in each round, very simple operations are performed (e.g., substitutions from a look-up table called an S box, permutations of the bit positions, or bit-wise $\oplus$ operations). These operations are almost always AC^0 type or even simpler. It is also considered vital to have key sizes that are as small as possible, and an algorithm that breaks the cryptosystem in significantly less time than exhaustive search over keys is considered worrisome. So this raises the question: Can we have an ideal block cipher family (one per key size), *i.e.*, so that the number of rounds remains constant, each round being implementable in constant depth with a linear number of gates, and security is almost that of exhaustive search over keys? Our results rule out such ideal block ciphers, and so give a partial explanation for why the number of rounds needs to increase in new generations of block ciphers. (Block ciphers require average-case security, not worst-case, but worst-case algorithms obviously also rule out average-case security. Our values of δ are vanishingly small for the sizes and depths of real cryptosystems, so our results cannot be used for cryptanalysis of existing block ciphers.)

Another motivation is that linear-size circuits are perhaps the most general class of circuits for which we can expect to show improved upper bounds on their exact complexity. To explain this statement, we need the following notation. Let $s_k = \inf\{c|\exists$ a randomized algorithm for k-SAT with time complexity $\text{poly}(m)2^{cn}$ for k-CNF formulas of size m over n variables$\}$. Let **ETH** denote the Exponential-Time Hypothesis: $s_3 > 0$. We know that the sequence $\{s_k\}$ has a limit and let s_∞ denote this limit. [IP01] proposed the open question whether $s_\infty = 1$, which we will call the *Strong Exponential-Time Hypothesis* (**SETH**). The best known upper bounds for s_k are all of the form $1 - 1/O(k)$, which makes the conjecture **SETH** plausible.

Here is the connection between **SETH** and the complexity of satisfiability of linear-size circuits: Since one can embed k-CNFs for any k into any non-linear size circuit model (in particular, nonconstant density CNF) [CIP06], improved upper bounds for the satisfiability problem for nonlinear-size circuits would imply $s_\infty < 1$. Thus, we are primarily left with the question of the complexity of the satisfiability problem for linear-size circuits if **SETH** holds. The following partial converse shows a further connection between **SETH** and improved bounds for the satisfiability of linear-size circuits. If $s_\infty < 1$, one can easily show using the depth-reduction technique of Valiant [Val77] (see also [Cal08]) that the satisfiability problem for cn-size series-parallel circuits has an improved upper bound of $2^{\delta(c)n}$ where $\delta(c) < 1$.

Yet another motivation is that improved algorithms for SAT for a circuit model $\mathcal{C}$ may reveal structural properties of the solution space of circuits in $\mathcal{C}$. These structural properties may in turn be helpful in proving stronger lower bounds on the size of circuits which are disjunctions of circuits in $\mathcal{C}$. In fact, [PSZ00, PPZ99, PPSZ05, IPZ01] exploit this connection to provide the best known lower bounds of the form $2^{\Delta(k)n/k}$ where $\Delta(k) > 1$ for depth-3 unbounded fan-in circuits with bounded bottom fan-in k. This connection between the hardness of the satisfiability problem for a circuit model and lower bounds of a related circuit model is not surprising since, as a more general circuit can compute more complicated functions, it may be more difficult to invert, *i.e.*, check the satisfiability of these functions.

1.2   Extension to Quantum Computing Model

Since Grover's quantum search algorithm [Gro96] provides a quadratic speed-up, the baseline in the quantum model for improved algorithms is $2^{n/2}$. In other

words, a quantum algorithm is an improvement for the satisfiability problem if
the constant factor in the exponent in the running time is strictly less than $1/2$.
However, it is not clear that every improved algorithm in the classical model can
benefit from a quadratic speed-up in the quantum model. It is known that the
class of algorithms that are exponential iterations of probabilistic polynomial
time algorithms can obtain quadratic speed-up using Grover's technique.

More precisely, [Gro96, BBHT96] show that a probabilistic algorithm running
in time t and with success probability p can be transformed into a quantum algorithm with running time $O(t/\sqrt{p})$ and with constant success probability. The
quadratic speed-up provided by quantum algorithms prompts the following question: Given an algorithm A with exponential running time t, can we transform
it into an exponential iteration of a polynomial time algorithm B with success
probability approximately $1/t$? Such a transformation would prime A for use in
Grover's algorithm and we could reap the full benefit of its quadratic speedup
in the quantum model. We will show that our algorithm for the satisfiability of
bounded-depth linear-size circuits can be sped up quadratically in the quantum
model, *i.e.*, that our algorithm, which runs in time $2^{(1-\delta)n}$ with constant success
probability, can be sped up by transforming it into a probabilistic polynomial
time algorithm that succeeds with probability at least $2^{-(1-\delta)n}$.

For simplicity, we will only describe this probabilistic version of the algorithm.
To convert this back into a backtracking algorithm, simply try both branches
deterministically. Note that either way, the subroutine find_restriction would still
use randomness, though the authors strongly suspect that it can be derandomized. For completeness, we include the backtracking version (without analysis,
which would be essentially the same, just more verbose) at the end in figure 2.

2   The Algorithm

2.1   Definitions

The *inputs* (*outputs*) of a dag are those nodes with indegree 0 (outdegree 0).
A *circuit* F is a dag where each input is labeled with a literal, each non-input
(called a *gate*) is labeled AND or OR, and there is exactly 1 output. A *subgate*
of size k of a gate g is a gate h (not necessarily in F) with the same label as g
and with k of g's inputs. The *depth* d of F is the number of edges in a longest
path in F. The *ith level* of F is the set of gates a distance of i from the output,
e.g. the output is at level 0 and the bottom gates are at level $d - 1$.

2.2   High Level Description

The overall algorithm consists of four subroutines, $A_{d,c}, A_{d,c,k}$, find_restriction
and PPZ_main. We first provide a high level description of the key routines $A_{d,c}$
and $A_{d,c,k}$ which mutually call each other. We then provide an intuitive explanation as to how the algorithm reduces the average number of variables whose

values need to be guessed. In the next subsection, we provide a detailed description of the subroutines accounting for all the parameters. Figure 1 provides a complete description of the routines except for PPZ_main.

The argument F of $A_{d,c}$ and $A_{d,c,k}$ denotes the current circuit after it is simplified as a result of variable assignments. We use the argument V of $A_{d,c}$ and $A_{d,c,k}$ to keep track of the unassigned variables, which is a superset of the remaining variables $\text{var}(F)$. In particular, each time we simplify the circuit by assigning to a group of variables, we will remove them from V. Subsequently simplifying the circuit may result in the removal of other variables from F so that $\text{var}(F) \subseteq V$ may become a proper containment. Because of these simplifying steps, $|V|$ is a measure of the algorithm's progress that will be easier to keep track of than $|\text{var}(F)|$. In order to describe the progress made at various points in $A_{d,c}$ and $A_{d,c,k}$, we use the identifier V to denote the set of unassigned variables at that point whereas we use n to denote the number of unassigned variables at the beginning of an invocation of the routines.

$A_{d,c}$ starts with a circuit F of depth d and a set of unassigned variables V satisfying the condition, $\text{var}(F) \subseteq V$ and $|F| \leq cn$ where $n = |V|$. It reduces the the fan-in of each bottom gate to k by repeatedly selecting a subgate h of size k of any bottom gate g with fan-in greater than k and setting h to either true or false. One of these settings will eliminate k variables from V and the other will eliminate a gate. The one that eliminates k variables we choose with probability q and the other with probability $1 - q$. We continue setting the subgates of bottom gates until we reach one of two cases. In the first case, we've at least halved the number of unassigned variables compared to the number n at the invocation of $A_{d,c}$. In this case, we simply guess an assignment to the unassigned variables.

In the second case we have that each bottom gate has fan-in less than or equal to k and that the number of gates is at most $2c$ times the number $|V|$ of unassigned variables. In this case the control passes to the routine $A_{d,2c,k}$.

$A_{d,c,k}$ takes as input a set V of unassigned variables and a circuit F of depth d, bottom fan-in restricted to k satisfying the condition $\text{var}(F) \subseteq V$ and $|F| \leq cn$ where $n = |V|$. It chooses a random restriction (by invoking the routine find_restriction) to all but a p fraction of the variables of V for some p. This may leave some bottom level gates with more than one unassigned variable. We clean up these gates by randomly setting all the variables in them. By choosing k, p appropriately, with probability at least $\frac{1}{2}$ there will still be $\geq \frac{1}{2}pn$ unassigned variables but the bottom level gates will each have at most one unassigned variable. So we can collapse the circuit to depth $d - 1$ and recurse. If the circuit is already at depth 2, $A_{d,c,k}$ applies PPZ_main, which applies one iteration of the PPZ solver [PPZ99], which takes polynomial time and finds a satisfying assignment with probability at least $2^{-(1-\frac{1}{k})n}$ if one exists.

To see why the algorithm succeeds better than random guessing, observe that the algorithm either preserves a constant fraction of the unassigned variables by the time the circuit reaches depth 2 or sets a large number of variables correctly according to a satisfying assignment without having to guess each one of them independently. If the algorithm produces a circuit of depth 2 with a constant fraction of unassigned variables, PPZ_main guarantees that at most $(1 - \frac{1}{k})$ fraction of the unassigned variables need to be looked at on average to find a satisfying assignment. If the algorithm terminates earlier in $A_{d,c}$ when the number of unassigned variables gets halved, it must be the case that at least half of the variables are assigned in $A_{d,c}$ by setting subgates of size k. The variables of the subgates are set with probability q and each such setting results in an assignment to k variables where k is sufficiently large compared to $\lg \frac{1}{q}$ thus saving a number of bits.

The overall algorithm takes polynomial time and has exponentially small probability s of finding a satisfying assignment given that there is one. By iterating s^{-1} times, we increase the probability of success to a constant.

2.3   Detailed Description

We describe our algorithm in several subroutines:

- $A_{d,c}(F, V)$ seeks a solution of F when F has depth d with V as the set of unassigned variables such that $\text{var}(F) \subseteq V$, $|V| = n$, and $|F| \leq cn$. Although initially $\frac{|F|}{|V|} \leq c$, the algorithm may set variables, increasing the ratio. If it ever exceeds $2c$, $A_{d,c}$ will simply guess an assignment to the remaining variables.
- $A_{d,c,k}(F, V)$ seeks a solution of F when F has depth d, $\text{var}(F) \subseteq V$, $|V| = n$, $|F| \leq cn$, and F has bottom fan-in at most k.
- find_restriction(F, V, p) finds, with probability at least $\frac{1}{2}$, a set of variables $W \subseteq V$ whose complement has size in the interval $[\frac{1}{2}pn, pn]$ and such that if the variables of W are assigned, then each bottom level gate has at most one unassigned variable.
- PPZ_main is 1 iteration of the k-SAT solver (which is the same as a depth 2 circuit solver) from [PPZ99] which takes polynomial time and has success probability $\geq 2^{-(1-\frac{1}{k})n}$. More specifically, PPZ_main assigns the variables, one at a time, in a random order. If a variable about to be assigned appears in a unit clause C (a clause of size 1), then it is assigned so as to satisfy C, otherwise it is assigned uniformly randomly. Note that this algorithm solves depth 1 circuits in polynomial time and with success probability 1.

Our algorithm description is not the most succinct. For example, one could construct an equivalent algorithm containing only one subroutine by eliminating tail recursion, but this would make the analysis obtuse. Below, the choices of k, p, q, c' are unspecified, and are left for the analysis section.

If h is an AND of literals, then $F|(h = 1)$ sets those literals to true and simplifies the circuit by removing true children of AND gates, false children of OR gates, replacing empty AND gates by true, replacing empty OR gates by false; unless h contains contradictory literals, in which case $F|(h = 1)$ is simply false. If h is an OR of literals, $F|(h = 1)$ removes any gate of which h is a subgate and then performs a similar simplification. Also if h is an AND (OR) of literals, then $F|(h = 0)$ can be treated as $F|(h' = 1)$ where h' is the OR (AND) of the negations of those literals.

$A_{d,c}(F, V)$ // F has depth d, $\text{var}(F) \subseteq V$, $|V| = n$, $|F| \le cn$
  choose k, q // as some function of d, c
  while $\exists$ bottom gate g in F of fan-in $> k$
    let h be a subgate of g of size k
$$b \leftarrow \begin{cases} 1 & \text{with probability } 1 - q \\ 0 & \text{with probability } q \end{cases}$$
$$b' \leftarrow \begin{cases} 1 & \text{if } h \text{ is an AND gate} \\ 0 & \text{if } h \text{ is an OR gate} \end{cases}$$
    $F \leftarrow F|(h = b \text{ XOR } b')$
    if $b = 0$, $V \leftarrow V - \text{var}(h)$
    if $|F| > 2c|V|$ // guess assignment
      choose $a \in_u 2^{\text{var}(F)}$
      if $F(a) = 1$, return a
      return "probably not satisfiable"
  return $A_{d,2c,k}(F, V)$

$A_{d,c,k}(F, V)$ // F has depth d, F also has bottom fan-in $\le k$,
$\text{var}(F) \subseteq V$, $|V| = n$
  if $d \le 2$, return PPZ_main(F)
  choose p, c' // as functions of d, c, k
  $W \leftarrow \text{find\_restriction}(F, V, p)$
  choose $a \in_u 2^W$
  // the bottom level gates of $F|a$ are trivial
  $F' \leftarrow F|a$ but collapsing the bottom level
  return $A_{d-1,c'}(F', V - W)$

find_restriction(F, V, p) // $|V| = n$
  $B \leftarrow \{\text{bottom gates of } F\}$
  $U \leftarrow$ random subset of V of size $(1 - p)n$
  $G \leftarrow \{g \in B \mid |\text{var}(g) - U| > 1\}$
  $U' \leftarrow \text{var}(G)$
  if $|U'| \le \frac{1}{2}pn$, return $U \cup U'$
  else die // algorithm fails

Fig. 1. Linear size, constant depth circuit solver

The purpose of all these definitions is so that below, in $A_{d,c}$, the line $F \leftarrow F|(h = b \text{ XOR } b')$ sets h in the way that eliminates k variables with probability q and the other way with probability $1 - q$. See figure 1.

3   Run Time Analysis

Suppose $A_{d,c}$, $A_{d,c,k}$ succeed with probability $\ge 2^{-(1-a_{d,c})n}$, $2^{-(1-a_{d,c,k})n}$, respectively, given that find_restriction succeeds on each call to it – we will eliminate this assumption later. Here n is the number of variables in V at the time $A_{d,c}$ or $A_{d,c,k}$ are invoked. We assume $c \ge 2$ and $\forall d \ge 2, k \ge 4$ $a_{d,c}, a_{d,c,k} \le \frac{1}{4}$.

Lemma 1. $\forall d, c \geq 2$ if $k \geq 4 \lg \frac{4c}{a_{d,2c,k}}$, then $a_{d,c} \geq \frac{1}{2} a_{d,2c,k}$.

Proof. Each iteration of the while loop of $A_{d,c}(F,V)$ eliminates (1) a gate or (2) k variables. (1) occurs $\leq cn$ times and (2) occurs $r \leq \frac{n}{k}$ times. Let a be a solution to F. Exactly one sequence of random choices, say with r choices of type (2), can lead $A_{d,c}(F,V)$ to find a. So the probability that $A_{d,c}(F,V)$ finds a solution given that each call to find_restriction succeeds, is

$$\geq q^r (1-q)^{cn} \min\{2^{-(1-a_{d,2c,k})(n-kr)}, 2^{-\frac{n}{2}}\}.$$

(Note carefully that we are *not* asserting that with at least this probability a is found. This is because PPZ_main may return another solution.)

To lower bound $q^r(1-q)^{cn} 2^{-(1-a_{d,2c,k})(n-kr)}$, take the logarithm, divide by n, and set $r' = \frac{r}{n} \in [0, \frac{1}{k}]$ to get

$$r' \lg q + c \lg(1-q) - (1 - a_{d,2c,k}) + (1 - a_{d,2c,k})kr'$$
$$= -(1 - a_{d,2c,k}) + c\lg(1-q) + r'(\lg q + (1 - a_{d,2c,k})k)$$
$$\geq -(1 - a_{d,2c,k}) - 2cq + r'(\lg q + (1 - a_{d,2c,k})k)$$

$$\left(\text{taking } n = 4, x = nq \text{ in the fact } \forall n \geq 1, x \in [0,1] \left(1 - \frac{x}{n}\right)^{n-1} \geq e^{-x}\right)$$

$$\geq -(1 - a_{d,2c,k}) - 2cq + r'\left(\lg q + \frac{1}{2}k\right) \quad \text{since } a_{d,2c,k} \leq \frac{1}{2},$$

which is $\geq -(1 - \frac{1}{2} a_{d,2c,k})$ if we set $q = \frac{a_{d,2c,k}}{4c}, k \geq -2 \lg q$.

To lower bound $q^r(1-q)^{cn} 2^{-\frac{n}{2}}$, note that if we choose $k \geq -4 \lg q$, then

$$r' \lg q + c \lg(1-q) - \frac{1}{2}$$

$$\geq \frac{1}{k} \lg q - 2cq - \frac{1}{2}$$

$$\geq -\frac{1}{4} - \frac{1}{2} a_{d,2c,k} - \frac{1}{2}$$

$$\geq -\left(1 - \frac{1}{2} a_{d,2c,k}\right) \quad \text{since } a_{d,2c,k} \leq \frac{1}{4}.$$

Lemma 2. *If* $|V| = n$, $|F| \leq cn$, F *has bottom fan-in* $\leq k$, *and we choose* $p = \frac{1}{2ck^3}$, *then the probability that* find_restriction(F, V, p) *dies is* $\leq \frac{1}{2}$.

Proof. Let $g \in B$ and $X = |\text{var}(g) - U|$. g has a fan-in $k' \leq k$ and so X is hypergeometric with parameters n, k', pn. We claim that $Pr(g \in G) = Pr(X \geq 2) \leq \binom{k'}{2}p^2$. To see this, note that the sample points where $X \geq 2$ can be partitioned according to the positions among the k' variables of g of the first 2 free variables (*i.e.*, not in U), and the probability of any of these $\binom{k'}{2}$ events is $\leq p^2$. So

$$E(|U'|) \leq E(|G|)k \leq \binom{k}{2}p^2|B|k \leq \frac{1}{2}k^3 p^2 cn \leq \frac{1}{4}pn.$$

By Markov's inequality,

$$Pr\left(|U'| > \frac{1}{2}pn\right) \le \frac{E(|U'|)}{\frac{1}{2}pn} \le \frac{1}{2}.$$

So the probability that find_restriction(F, V, p) dies is $\le \frac{1}{2}$.

Lemma 3. *If we set* $p = \frac{1}{2ck^3}, c' = 4c^2k^3$, *then* $a_{d,c,k} \ge \frac{1}{4ck^3}a_{d-1,4c^2k^3}$.

Proof. $A_{d,c,k}(F, V)$ leaves $f \in [\frac{1}{2}pn, pn]$ variables of V free and sets the rest. So $\frac{|F'|}{|V-W|} \le \frac{cn}{\frac{1}{2}pn} = 4c^2k^3 = c'$. (The proof of this lemma would have been confounded if we had used var(F) to keep track of variables instead of V.) The probability that $A_{d,c,k}(F, V)$ finds a solution, assuming each call to find_restriction succeeds, is then

$$2^{-(n-f+(1-a_{d-1,c'})f)} = 2^{-(n-a_{d-1,c'}f)} > 2^{-(1-\frac{1}{2}pa_{d-1,c'})n},$$

and the lemma follows.

Lemma 4. $\forall d, c \ge 2,$

$$a_{d,c} \ge 1/O(c^{2^{d-2}-1} \lg^{3 \cdot 2^{d-2}-2} c),$$

where the constant in the big-Oh depends only on d.

Proof. We use induction to show that for each d, k can be chosen to be $O(\lg c)$ so as to satisfy the hypothesis of lemma 1 and $a_{d,c} \ge 1/O(c^{f(d)} \lg^{g(d)} c)$ for some functions f, g where the constants in the big-Ohs depend only on d.

$a_{2,c,k} = \frac{1}{k}$. From lemma 1, we need to choose k such that $k \ge O(\lg \frac{c}{a_{2,2c,k}}) = O(\lg c + \lg k)$. So $k = O(\lg c)$ suffices, and we conclude that $a_{2,c} \ge 1/O(\lg c)$. So $f(2) = 0, g(2) = 1$. This completes the base case.

From the inductive hypothesis and lemma 3,

$$a_{d,c,k} \ge 1/O(ck^3(c^2k^3)^{f(d-1)} \lg^{g(d-1)}(c^2k^3))$$
$$= 1/O(c^{2f(d-1)+1}k^{3f(d-1)+3} \lg^{g(d-1)}(ck)).$$

To use lemma 1, we need to choose k such that

$$k \ge O(\lg(c^{2f(d-1)+2}k^{3f(d-1)+3} \lg^{g(d-1)}(ck)))$$
$$= O(\lg c + \lg k).$$

So $k = O(\lg c)$ suffices, and we conclude that

$$a_{d,c} \ge 1/O(c^{2f(d-1)+1} \lg^{3f(d-1)+3+g(d-1)} c).$$

So we have the recurrence

$$f(d) = 2f(d-1) + 1 \qquad\qquad f(2) = 0$$
$$g(d) = g(d-1) + 3f(d-1) + 3 \qquad\qquad g(2) = 1$$

which has solution $f(d) = 2^{d-2} - 1, g(d) = 3 \cdot 2^{d-2} - 2$.

$A_{d,c}(F, V)$
   if $|F| \geq 2c|V|$ // solve by brute force
      for each $a \in 2^{\mathrm{var}(F)}$
         if $F(a)$, return 1
      return 0
   choose k // as some function of d, c
   if $\exists$ bottom gate g in F of fan-in $> k$ // branch
      let h be a subgate of g of size k
      $(V_0, V_1) \leftarrow \begin{cases} (V, V - \mathrm{var}(h)) & \text{if } h \text{ is an AND gate} \\ (V - \mathrm{var}(h), V) & \text{if } h \text{ is an OR gate} \end{cases}$
      if $A_{d,c}(F|(h = 1), V_1)$, return 1
      if $A_{d,c}(F|(h = 0), V_0)$, return 1
      return 0
   return $A_{d,2c,k}(F, V)$
$A_{d,c,k}(F, V)$
   if $d \leq 2$, return PPZ(F)
   // PPZ solves k-SAT in time poly($|F|$)$2^{(1-\frac{1}{k})n}$
   // with exponentially small error probability
   choose p, c' // as functions of d, c, k
   $W \leftarrow$ find_restriction(F, V, p)   // same subroutine as in figure 1
   for each $a \in 2^W$
      // the bottom level gates of $F|a$ are trivial
      $F' \leftarrow F|a$ but collapsing the bottom level
      if $A_{d-1,c'}(F', V - W)$, return 1
   return 0

Fig. 2. Linear size, constant depth circuit solver, backtracking version

Theorem 1. $\forall d, c \geq 2$, the probability that $A_{d,c}(F, V)$ finds some solution is $\geq 2^{-(1-\alpha)n}$ where

$$\alpha \geq 1/O(c^{2^{d-2}-1} \lg^{3 \cdot 2^{d-2}-2} c),$$

where the constant in the big-Oh depends only on d.

Proof. This is a corollary from the previous lemma together with the following: find_restriction is called $\leq d$ times, each with success probability $\geq \frac{1}{2}$, so the probability that in every call it succeeds is $\geq 2^{-d}$, a penalty that can be absorbed into the big-Oh in the theorem statement.

Again, we are not asserting that every solution is found with this probability, just that some is, and this asymmetric property is inherited from the PPZ algorithm.

4   Open Problems

Can find_restriction be derandomized without too much performance penalty? Can we find a nontrivial algorithm for the case where d grows very slowly with n?

Acknowledgments. We would like to thank Mike Saks for useful discussions.

References

[BBHT96]   Boyer, M., Brassard, G., Høyer, P., Tapp, A.: Tight bounds on quantum searching (May 1996)

[Cal08]   Calabro, C.: A lower bound on the size of series-parallel graphs dense in long paths. In: Electronic Colloquium on Computational Complexity (ECCC), vol. 15(110) (2008)

[CIKP08]   Calabro, C., Impagliazzo, R., Kabanets, V., Paturi, R.: The complexity of unique k-sat: An isolation lemma for k-cnfs. J. Comput. Syst. Sci. 74(3), 386–393 (2008)

[CIP06]   Calabro, C., Impagliazzo, R., Paturi, R.: A duality between clause width and clause density for sat. In: CCC 2006: Proceedings of the 21st Annual IEEE Conference on Computational Complexity, Washington, DC, USA, 2006, pp. 252–260. IEEE Computer Society Press, Los Alamitos (2006)

[DW05]   Dantsin, E., Wolpert, A.: An improved upper bound for sat. In: Bacchus, F., Walsh, T. (eds.) SAT 2005. LNCS, vol. 3569, pp. 400–407. Springer, Heidelberg (2005)

[Gro96]   Grover, L.K.: A fast quantum mechanical algorithm for database search. In: STOC, pp. 212–219 (1996)

[IP01]   Impagliazzo, R., Paturi, R.: On the complexity of k-sat. J. Comput. Syst. Sci. 62(2), 367–375 (2001)

[IPZ01]   Impagliazzo, R., Paturi, R., Zane, F.: Which problems have strongly exponential complexity? J. Comput. Syst. Sci. 63(4), 512–530 (2001)

[Nur09]   Nurik, S.: Personal communication. To appear in ECCC (2009)

[PPSZ05]   Paturi, R., Pudlák, P., Saks, M.E., Zane, F.: An improved exponential-time algorithm for k-sat. J. ACM 52(3), 337–364 (2005)

[PPZ99]   Paturi, R., Pudlák, P., Zane, F.: Satisfiability coding lemma. Chicago Journal of Theoretical Computer Science 115 (December 1999)

[PSZ00]   Paturi, R., Saks, M.E., Zane, F.: Exponential lower bounds for depth 3 boolean circuits. Computational Complexity 9(1), 1–15 (2000); Preliminary version in 29th annual ACM Symposium on Theory of Computing, pp. 96–91 (1997)

[Sch99]   Schöning, U.: A probabilistic algorithm for k-sat and constraint satisfaction problems. In: FOCS, pp. 410–414 (1999)

[Sch05]   Schuler, R.: An algorithm for the satisfiability problem of formulas in conjunctive normal form. J. Algorithms 54(1), 40–44 (2005)

[Val77]   Valiant, L.: Graph-theoretic arguments in low level complexity. In: Gruska, J. (ed.) MFCS 1977. LNCS, vol. 53, Springer, Heidelberg (1977)

On Finding Directed Trees with Many Leaves

Jean Daligault and Stéphan Thomassé

Abstract. The ROOTED MAXIMUM LEAF OUTBRANCHING problem consists in finding a spanning directed tree rooted at some prescribed vertex of a digraph with the maximum number of leaves. Its parameterized version asks if there exists such a tree with at least k leaves. We use the notion of $s - t$ numbering studied in [19, 6, 20] to exhibit combinatorial bounds on the existence of spanning directed trees with many leaves. These combinatorial bounds allow us to produce a constant factor approximation algorithm for finding directed trees with many leaves, whereas the best known approximation algorithm has a $\sqrt{OPT}$-factor [11]. We also show that ROOTED MAXIMUM LEAF OUTBRANCHING admits an edge-quadratic kernel, improving over the vertex-cubic kernel given by Fernau et al [13].

1 Introduction

An *outbranching* of a digraph D is a spanning directed tree in D. We consider the following problem:

ROOTED MAXIMUM LEAF OUTBRANCHING:

Input: A digraph D, an integer k, a vertex r of D.
Output: TRUE if there is an outbranching of D rooted at r with at least k leaves, otherwise FALSE.

This problem is equivalent to finding a Connected Dominating Set of size at most $|V(D)| - k$, connected meaning in this setting that every vertex is reachable by a directed path from r. Indeed, the set of internal nodes in an outbranching correspond to a connected dominating set.

Finding undirected trees with many leaves has many applications in the area of communication networks, see [8] or [24] for instance. An extensive literature is devoted to the paradigm of using a small connected dominating set as a backbone for a communication network.

ROOTED MAXIMUM LEAF OUTBRANCHING is NP-complete, even restricted to acyclic digraphs [2], and MaxSNP-hard, even on undirected graphs [16].

Two natural ways to tackle such a problem are, on the one hand, polynomial-time approximation algorithms, and on the other hand, parameterized complexity. Let us give a brief introduction on the parameterized approach.

An efficient way of dealing with NP-hard problems is to identify a parameter which contains its computational hardness. For instance, instead of asking for a minimum vertex cover in a graph - a classical NP-hard optimization question - one can ask for an algorithm which would decide, in $O(f(k).n^d)$ time for some

J. Chen and F.V. Fomin (Eds.): IWPEC 2009, LNCS 5917, pp. 86–97, 2009.

fixed d, if a graph of size n has a vertex cover of size at most k. If such an algorithm exists, the problem is called *fixed-parameter tractable*, or FPT for short. An extensive literature is devoted to FPT, the reader is invited to read [10], [14] and [21].

Kernelization is a natural way of proving that a problem is FPT. Formally, a *kernelization algorithm* receives as input an instance (I, k) of the parameterized problem, and outputs, in polynomial time in the size of the instance, another instance (I', k') such that: $k' \le g(k)$ for some function g, the size of I' only depends of k, and the instances (I, k) and (I', k') are both true or both false.

The reduced instance (I', k') is called a *kernel*. The existence of a kernelization algorithm clearly implies the FPT character of the problem since one can kernelize the instance, and then solve the reduced instance (G', k') using brute force, hence giving an $O(f(k) + n^d)$ algorithm. A classical result asserts that being FPT is indeed equivalent to having kernelization. The drawback of this result is that the size of the reduced instance G' is not necessarily small with respect to k. A much more constrained condition is to be able to reduce to an instance of polynomial size in terms of k. Consequently, in the zoology of parameterized problems, the first distinction is done between three classes: W[1]-hard, FPT, polykernel.

A kernelization algorithm can be used as a preprocessing step to reduce the size of the instance before applying some other parameterized algorithm. Being able to ensure that this kernel has actually polynomial size in k enhances the overall speed of the process. See [17] for a recent review on kernelization.

An extensive litterature is devoted to finding trees with many leaves in undirected and directed graphs. The undirected version of this problem, MAXIMUM LEAF SPANNING TREE, has been extensively studied. There is a factor 2 approximation algorithm for the MAXIMUM LEAF SPANNING TREE problem [22], and a $3.75k$ kernel [12]. An $O^*(1,94^n)$ exact algorithm was designed in [15]. Other graph theoretical results on the existence of trees with many leaves can be found in [9] and [23].

The best approximation algorithm known for MAXIMUM LEAF OUTBRANCHING is a factor $\sqrt{OPT}$ algorithm [11]. From the Parameterized Complexity viewpoint, Alon et al showed that MAXIMUM LEAF OUTBRANCHING restricted to a wide class of digraphs containing all strongly connected digraphs is FPT [1], and Bonsma and Dorn extended this result to all digraphs and gave a faster parameterized algorithm [4]. Very recently, Kneis, Langer and Rossmanith [18] obtained an $O^*(4^k)$ algorithm for MAXIMUM LEAF OUTBRANCHING, which is also an improvement for the undirected case over the numerous FPT algorithms designed for MAXIMUM LEAF SPANNING TREE (Chen and Liu have a similar algorithm in [5]). Fernau et al [13] proved that ROOTED MAXIMUM LEAF OUTBRANCHING has a polynomial kernel, exhibiting a cubic kernel. They also showed that the unrooted version of this problem admits no polynomial kernel, unless polynomial hierarchy collapses to third level, using a breakthrough lower bound result by Bodlaender et al [3]. A linear kernel for the acyclic subcase of ROOTED MAXIMUM

LEAF OUTBRANCHING and an $O^*(3, 72^k)$ algorithm for ROOTED MAXIMUM LEAF OUTBRANCHING were exhibited in [7].

This paper is organized as follows. In Section 2 we exhibit combinatorial bounds on the problem of finding an outbranching with many leaves. We use the notion of $s - t$ *numbering* introduced in [19]. We next present our reduction rules, which are independent of the parameter, and in the following section we prove that these rules give an edge-quadratic kernel. We finally present a constant factor approximation algorithm in Section 5 for finding directed trees with many leaves.

2   Combinatorial Bounds

Let D be a directed graph. For an arc (u, v) in D, we say that u is an *in-neighbour* of v, that v is an *outneighbour* of u, that (u, v) is an *in-arc* of v and an *out-arc of* u. The *outdegree* of a vertex is the number of its outneighbours, and its *indegree* is the number of its in-neighbours. An outbranching with a maximum number of leaves is said to be *optimal*. Let us denote by maxleaf(D) the number of leaves in an optimal outbranching of D.

Without loss of generality, we restrict ourselves to the following. We exclusively consider loopless digraphs with a distinguished vertex of indegree 0, denoted by r. *We assume that there is no arc (u, r) in D with $u \in V(D)$, and no arc (x, y) with $x \neq r$ and y an outneighbour of r, and that r has outdegree at least 2.* Throughout this paper, we call such a digraph a *rooted digraph*. Definitions will be made exclusively with respect to rooted digraphs, hence the notions we present, like connectivity and resulting concepts, do slightly differ from standard ones. Let D be a rooted digraph with a specified vertex r.

The rooted digraph D is *connected* if every vertex of D is reachable by a directed path starting at r in D. A *cut* of D is a set $S \subseteq V(D) - r$ such that there exists a vertex $z \notin S$ endpoint of no directed path from r in $D - S$. We say that D is *2-connected* if D has no cut of size at most 1. A cut of size 1 is called a *cutvertex*. Equivalently, a rooted digraph is 2-connected if there are two internally vertex-disjoint paths from r to any vertex besides r and its outneighbours.

We will show that the notion of $s - t$ numbering behaves well with respect to outbranchings with many leaves. It has been introduced in [19] for 2-connected undirected graphs, and generalized in [6] by Cheriyan and Reif for digraphs which are 2-connected in the usual sense. We adapt it in the context of rooted digraphs.

Let D be a 2-connected rooted digraph. An $r - r$ numbering of D is a linear ordering σ of $V(D) - r$ such that, for every vertex $x \neq r$, either x is an outneighbour of r or there exist two in-neighbours u and v of x such that $\sigma(u) < \sigma(x) < \sigma(v)$. An equivalent presentation of an $r - r$ numbering of D is an injective embedding f of the graph D where r has been duplicated into two vertices r_1 and r_2, into the $[0, 1]$-segment of the real line, such that $f(r_1) = 0$, $f(r_2) = 1$, and such that the image by f of every vertex besides r_1 and r_2 lies inside the convex hull

of the images of its in-neighbours. Such *convex embeddings* have been defined
and studied in general dimension by Lovász, Linial and Wigderson in [20] for
undirected graphs, and in [6] for directed graphs.

Given a linear order σ on a finite set V, we denote by $\bar{\sigma}$ the linear order on
V which is the reverse of σ. An arc (u, v) of D is a *forward* arc if $u = r$ or if u
appears before v in σ; (u, v) is a *backward* arc if $u = r$ or if u appears after v in
σ. A spanning out-tree T is *forward* if all its arcs are forward. Similar definition
for *backward* out-tree.

The following result and proof is just an adapted version of [6], given here for
the sake of completeness.

Lemma 1. *Let D be a 2-connected rooted digraph. There exists an $r - r$ numbering of D.*

Proof: By induction over D. We first reduce to the case where the indegree of
every vertex besides r is exactly 2. Let x be a vertex of indegree at least 3 in
D. Let us show that there exists an in-neighbour y of x such that the rooted digraph $D - (y, x)$ is 2-connected. Indeed, there exist two internally vertex disjoint
paths from r to x. Consider such two paths intersecting the in-neighbourhood
$N^-(x)$ of x only once each, and denote by D' the rooted digraph obtained from
D by removing one arc (y, x) not involved in these two paths. There are two
internally disjoint paths from r to x in D'. Consider $z \in V(D) - r - x$. Assume
by contradiction that there exists a vertex t which cuts z from r in D'. As t does
not cut z from r in D and the arc (y, x) alone is missing in D', t must cut x and
not y from r in D'. Which is a contradiction, as there are two internally disjoint
paths from r to x in D'. By induction, D' has an $r - r$ numbering, which is also
an $r - r$ numbering for D.

Hence, let D be a rooted digraph, where every vertex besides r has indegree
2. As r has indegree 0, there exists a vertex v with outdegree at most 1 in D
by a counting argument. If v has outdegree 0, then let σ be an $r - r$ numbering
of $D - v$, let u_1 and u_2 be the two in-neighbours of v. Insert v between u_1 and
u_2 in σ to obtain an $r - r$ numbering of D. Assume now that v has a single
outneighbour u. Let w be the second in-neighbour of u. Let D' be the graph
obtained from D by contracting the arc (v, u) into a single vertex uv. As D' is
2-connected, consider by induction an $r - r$ numbering σ of D'. Replace uv by
u. It is now possible to insert v between its two in-neighbours in order to make
it so that u lies between v and w. Indeed, assume without loss of generality that
w is after uv in σ. Consider the in-neighbour t of v smallest in σ. As σ is an
$r - r$ numbering of D', t lies before uv in σ. We insert v just after t to obtain
an $r - r$ numbering of D.   □

Note that an $r - r$ numbering σ of D naturally gives two acyclic covering sub-
digraphs of D, the rooted digraph $D_{|\sigma}$ consisting of the forward arcs of D, and
the rooted digraph $D_{|\bar{\sigma}}$ consisting of the backward arcs of D. The intersection
of these two acyclic digraphs is the set of out-arcs of r.

Corollary 1. *Let D be a 2-connected rooted digraph. There exists an acyclic connected spanning subdigraph A of D which contains at least half of the arcs of $D - r$.*

Let G be an undirected graph. A *vertex cover* of G is a set of vertices covering all edges of G. A *dominating set* of G is a set $S \subseteq V$ such that for every vertex $x \notin S$, x has a neighbour in S. A *strongly dominating set* of G is a set $S \subseteq V$ such that every vertex has a neighbour in S.

Let D be a rooted digraph. A *strongly dominating set* of D is a set $S \subseteq V$ such that every vertex besides r has an in-neighbour in S. We need the following folklore result:

Lemma 2. *Any undirected graph G on n vertices and m arcs has a vertex cover of size $\frac{n+m}{3}$.*

Proof: By induction on $n + m$. If there exists a vertex of degree at least 2 in G, choose it in the vertex cover, otherwise choose any non-isolated vertex.   □

Lemma 3. *Let G be a bipartite graph over $A \cup B$, with $d(a) = 2$ for every $a \in A$. There exists a subset of B dominating A with size at most $\frac{|A|+|B|}{3}$.*

Proof: Let G' be the graph which vertex set is B, and where (b, b') is an arc if b and b' share a common neighbour in A. The result follows from Lemma 2 since G' has $|A|$ arcs and $|B|$ vertices.   □

Corollary 2. *Let D be an acyclic rooted digraph with l vertices of indegree at least 2 and with a root of outdegree $d(r) \geq 2$. Then D has an outbranching with at least $\frac{l+d(r)-1}{3} + 1$ leaves.*

Proof: Denote by n the number of vertices of D. For every vertex v of indegree at least 3, delete incoming arcs until v has indegree exactly 2. Since D is acyclic, it has a vertex s with outdegree 0.

Let Z be the set of vertices of indegree 1 in D, of size $n - 1 - l$. Let Y be the set of in-neighbours of vertices of Z, of size at most $n - 1 - l$. Let A' be the set of vertices of indegree 2 dominated by Y. Let $B = V(D) - Y - s$. Let A be the set of vertices of indegree 2 not dominated by Y. Note that Y cannot have the same size as $Z \cup A'$. Indeed, Z contains the outneighbours of r, and hence Y contains r, which has outdegree at least 2. More precisely, $|Y| + d(r) - 1 \leq |Z \cup A'|$. As $B = V(D) - Y - s$ and $A = V(D) - A' - Z - r$, we have that $|B| \geq |A| + d(r) - 1$. Moreover, as Y has size at most $n - 1 - l$, we have that $|B| \geq l$. Consider a copy A_1 of A and a copy B_1 of B. Let G be the bipartite graph with vertex bipartition (A_1, B_1), and where (b, a), with $a \in A_1$ and $b \in B_1$, is an edge if (b, a) is an arc in D. By Lemma 3 applied to G, there exists a set $X \subseteq B$ of size at most $\frac{|A|+|B|}{3} \leq \frac{2|B|-(d(r)-1)}{3}$ which dominates A in D. The set $C = X \cup Y$ strongly dominates $V(D) - r$ in D, and has at most $|X| + |Y| \leq \frac{2|B|-(d(r)-1)}{3} + |Y| = |B| + |Y| - \frac{|B|+d(r)-1}{3}$. As $|Y| + |B| = n - 1$ and $|B| \geq l$, this yields $|X \cup Y| \leq n - 1 - \frac{l+d(r)-1}{3}$. As D is acyclic, any set

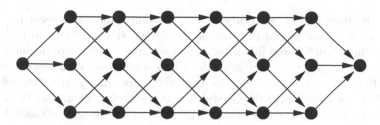

Fig. 1. The "boloney" graph D_6

strongly dominating $V - r$ contains r and is a connected dominating set. Hence there exists an outbranching T of D having a subset of C as internal vertices. T has at least $\frac{l+d(r)-1}{3} + 1$ leaves.   □

This bound is tight up to one leaf. The rooted digraph D_k depicted in Figure 1 is 2-connected, has $3k - 2$ vertices of indegree at least 2, $d(r) = 3$ and maxleaf$(D_k) = k + 2$.

Finally, the following combinatorial bound is obtained:

Theorem 1. *Let D be a 2-connected rooted digraph with l vertices of indegree at least 3. Then $maxleaf(D) \geq \frac{l}{6}$.*

Proof: Apply Corollary 2 to the rooted digraph with the larger number of vertices of indegree 2 among D_σ and $D_{\bar{\sigma}}$.   □

An arc (u, v) of D is said to be a *2-circuit* if (v, u) is also an arc in D. An arc is *simple* if does not belong to a 2-circuit. A vertex v is *nice* if it is incident to a simple in-arc.

The second combinatorial bound is the following:

Theorem 2. *Let D be 2-connected rooted digraph. Assume that D has l nice vertices. Then D has an outbranching with at least $\frac{l}{24}$ leaves.*

Proof: By Lemma 1, we consider an $r - r$ numbering σ of D. For every nice vertex v (incident to some in-arc a) with indegree at least three, delete incoming arcs of v different from a until v has only one incoming forward arc and one incoming backward arc. For every other vertex of indegree at least 3 in D, delete incoming arcs of v until v has only one incoming forward arc and one incoming backward arc. At the end of this process, σ is still an $r - r$ numbering of the digraph D, and the number of nice vertices has not decreased.

Denote by T_f the set of forward arcs of D, and by T_b the set of backward arcs of D. As σ is an $r - r$ numbering of D, T_f and T_b are spanning trees of D which partition the arcs of $D - r$.

The crucial definition is the following: say that an arc uv of T_f (resp. of T_b), with $u \neq r$, is *transverse* if u and v are *incomparable* in T_b (resp. in T_f), that is if v is not an ancestor of u in T_b (resp. in T_f). Observe that u cannot be an ancestor of v in T_b (resp. in T_f) since T_b is backward (resp. T_f is forward) while uv is forward (resp. backward) and $u \neq r$.

Assume without loss of generality that T_f contains more transverse arcs than T_b. Consider now any planar drawing of the rooted tree T_b. We will make use of this drawing to define the following: if two vertices u and v are incomparable in T_b, then one of these vertices is to the left of the other, with respect to our drawing. Hence, we can partition the transverse arcs of T_f into two subsets: the set S_l of transverse arcs uv for which v is to the left of u, and the set S_r of transverse arcs uv for which v is to the right of u. Assume without loss of generality that $|S_l| \geq |S_r|$.

The digraph $T_b \cup S_l$ is an acyclic digraph by definition of S_l. Moreover, it has $|S_l|$ vertices of indegree two since the heads of the arcs of $|S_l|$ are pairwise distinct. Hence, by Corollary 2, $T_b \cup S_l$ has an outbranching with at least $\frac{|S_l|+d(r)-1}{3} + 1$ leaves, hence so does D.

We now give a lower bound on the number of transverse arcs in D to bound $|S_l|$. Consider a nice vertex v in D, which is not an outneighbour of r, and with a simple in-arc uv belonging to, say, T_f. If uv is not a transverse arc, then v is an ancestor of u in T_b. Let w be the outneighbor of v on the path from v to u in T_b. Since uv is simple, the vertex w is distinct from u. No path in T_f goes from w to v, hence vw is a transverse arc. Therefore, we proved that v (and hence every nice vertex) is incident to a transverse arc (either an in-arc, or an out-arc). Thus there are at least $\frac{l-d(r)}{2}$ transverse arcs in D.

Finally, there are at least $\frac{l-d(r)}{4}$ transverse arcs in T_f, and thus $|S_l| \geq \frac{l-d(r)}{8}$. In all, D has an outbranching with at least $\frac{l}{24}$ leaves. $\square$

As a corollary, the following result holds for oriented graphs (digraphs with no 2-circuit):

Corollary 3. *Every 2-connected rooted oriented graph on n vertices has an outbranching with at least $\frac{n-1}{24}$ leaves.*

3   Reduction Rules

We say that $P = \{x_1, \ldots, x_l\}$, with $l \geq 3$, is a *bipath of length* $l-1$ if the set of arcs adjacent to $\{x_2, \ldots, x_{l-1}\}$ in D is exactly $\{(x_i, x_{i+1}), (x_{i+1}, x_i) | i \in \{1, \ldots, l-1\}\}$.

To exhibit a quadratic kernel for ROOTED MAXIMUM LEAF OUTBRANCH-ING, we use the following four reduction rules:

(0) If there exists a vertex not reachable from r in D, then reduce to a trivially FALSE instance.

(1) Let x be a cutvertex of D. Delete vertex x and add an arc (v, z) for every $v \in N^-(x)$ and $z \in N^+(x) - v$.

(2) Let P be a bipath of length 4. Contract two consecutive internal vertices of P.

(3) Let x be a vertex of D. If there exists $y \in N^-(x)$ such that $N^-(x) - y$ cuts y from r, then delete the arc (y, x).

Note that these reduction rules are not parameter dependent. Rule (0) only needs to be applied once.

Observation 1. *Let S be a cutset of a rooted digraph D. Let T be an outbranching of D. There exists a vertex in S which is not a leaf in T.*

Lemma 4. *The above reduction rules are safe and can be checked and applied in polynomial time.*

Proof:

(0) Reachability can be tested in linear time.

(1) Let x be a cutvertex of D. Let D' be the graph obtained from D by deleting vertex x and adding an arc (v, z) for every $v \in N^-(x)$ and $z \in N^+(x) - v$. Let us show that maxleaf(D) = maxleaf(D'). Assume T is an outbranching of D rooted at r with k leaves. By Observation 1, x is not a leaf of T. Let $f(x)$ be the father of x in T. Let T' be the tree obtained from T by contracting x and $f(x)$. T' is an outbranching of D' rooted at r with k leaves.
Let T' be an outbranching of D' rooted at r with k leaves. $N^-(x)$ is a cut in D', hence by Observation 1 there is a non-empty collection of vertices $y_1, \ldots, y_l \in N^-(x)$ which are not leaves in T'. Choose y_i such that y_j is not an ancestor of y_i for every $j \in \{1, \ldots, l\} - \{i\}$. Let T be the graph obtained from T' by adding x as an isolated vertex, adding the arc (y_i, x), and for every $j \in \{1, \ldots, l\}$, for every arc $(y_j, z) \in T$ with $z \in N^+(x)$, delete the arc (y_j, z) and add the arc (x, z). As y_i is not reachable in T' from any vertex $y \in N^-(x) - y_i$, there is no cycle in T. Hence T is an outbranching of D rooted at r with at least k leaves. Moreover, deciding the existence of a cut vertex and finding one if such exists can be done in polynomial time.

(2) Let P be a bipath of length 4. Let u, v, w, x and z be the vertices of P in this consecutive order. Let T be an outbranching of D. Let D' be the rooted digraph obtained from D by contracting v and w. The rooted digraph obtained from T by contracting w with its father in T is an outbranching of D' with as many leaves as T.
Let T' be an outbranching of D'. If the father of vw in T' is x, then $T' - (x, vw) \cup (x, w) \cup (w, v)$ is an outbranching of D with at least as many leaves as T'. If the father of vw in T' is u, then $T' - (u, vw) \cup (u, v) \cup (v, w)$ is an outbranching of D with at least as many leaves as T'.

(3) Let x be a vertex of D. Let $y \in N^-(x)$ be a vertex such that $N^-(x) - y$ cuts y from r. Let D' be the rooted digraph obtained from T by deleting the arc (y, x). Every outbranching of D' is an outbranching of D. Let T be an outbranching of D containing (y, x). There exists a vertex $z \in N^-(x) - y$ which is an ancestor of x. Thus $T - (y, x) \cup (z, x)$ is an outbranching of D' with at least as many leaves as T.   □

We apply these rules iteratively until reaching a *reduced instance*, on which none can be applied.

Lemma 5. *Let D be a reduced rooted digraph with a vertex of indegree at least k. Then D is a TRUE instance.*

Proof: Assume x is a vertex of D with in-neighbourhood $N^-(x) = \{u_1, \ldots, u_l\}$, with $l \geq k$. For every $i \in \{1, \ldots, l\}$, $N^-(x) - u_i$ does not cut u_i from r. Thus there exists a path P_i from r to u_i outside $N^-(x) - u_i$. The rooted digraph $D' = \cup_{i \in \{1,\dots,l\}} P_i$ is connected, and for every $i \in \{1,\dots,l\}$, u_i has outdegree 0 in D'. Thus D' has an outbranching with at least k leaves, and such an outbranching can be extended into an outbranching of D with at least as many leaves.   □

4   Quadratic Kernel

In this section and the following, a vertex of a 2-connected rooted digraph D is said to be *special* if it has indegree at least 3 or if one of its incoming arcs is simple. A non special vertex is a vertex u which has exactly two in-neighbours, which are also outneighbours of u. A *weak bipath* is a maximal connected set of non special vertices. If $P = \{x_1, \dots, x_l\}$ is a weak bipath, then the in-neighbours of x_i, for $i = 2, \dots, l - 1$ in D are exactly x_{i-1} and x_{i+1}. Moreover, x_1 and x_l are each outneighbour of a special vertex. Denote by $s(P)$ the in-neighbour of x_1 which is a special vertex.

This section is dedicated to the proof of the following statement:

Theorem 3. *A digraph D of size at least $(3k - 2)(30k - 2)$ reduced under the reduction rules of previous section has an outbranching with at least k leaves.*

Proof: By Theorem 1 and Theorem 2, if there are at least $6k + 24k - 1$ special vertices, then D has an outbranching with at least k leaves. Assume that there are at most $30k - 2$ special vertices in D.

As D is reduced under Rule (2), there is no bipath of length 4. We can associate to every weak bipath B of D of length t a set A_B of $\lceil t/3 \rceil$ out-arcs toward special vertices. Indeed, let $P = (x_1, \dots, x_l)$ be a weak bipath of D. For every three consecutive vertices x_i, x_{i+1}, x_{i+2} of P, $2 \le i \le l - 3$, $(x_{i-1}, x_i, x_{i+1}, x_{i+2}, x_{i+3})$ is not a bipath by Rule (2), hence there exists an arc (x_j, z) with $j = i, i + 1$ or $i + 2$ and $z \notin P$. Moreover z must be a special vertex as arcs between non-special vertices lie within their own weak bipath. The set of these arcs (x_j, z) has the prescribed size.

By Lemma 5, any vertex in D has indegree at most $k - 1$ as D is reduced under Rule (3), hence there are at most $3(k - 1)(30k - 2)$ non special vertices in D.   □

To sum up, the kernelization algorithm is as follows: starting from a rooted digraph D, apply the reduction rules. Let D' be the obtained reduced rooted digraph. If D has size more than $(3k - 2)(30k - 2)$, then reduce to a trivially TRUE instance. Otherwise, D' is an instance equivalent to D with $O(k^2)$ vertices and $O(k^2)$ edges.

This quadratic bound is tight up to a constant factor with respect to our reduction rules.

5   Approximation

Let us describe our constant factor approximation algorithm for ROOTED MAXIMUM LEAF OUTBRANCHING, being understood that this also gives an

approximation algorithm of the same factor for MAXIMUM LEAF OUTBRANC-
HING as well as for finding an out-tree (not necessarily spanning) with many
leaves in a digraph.

Our reduction rules directly give an approximation algorithm asymptotically
as good as the best known approximation algorithm [11] (see Annex). Let us now
describe our constant factor approximation algorithm. Given a rooted digraph
D'', apply exhaustively Rule (1) of Section 3. The resulting rooted digraph D is
2-connected. By Lemma 4, maxleaf(D'') = maxleaf(D).

Let us denote by D_{ns} the digraph D restricted to non special vertices. Recall
that D_{ns} is a disjoint union of bipaths, which we call *non special components*. A
vertex of outdegree 1 in D_{ns} is called an *end*. Each end has exactly one special
vertex as an in-neighbour in D.

Theorem 4. *Let D be a 2-connected rooted digraph with l special vertices and
h non special components. Then $max(\frac{l}{30}, h - l) \leq maxleaf(D) \leq l + 2h$.*

Proof: The upper bound is clear, as at most two vertices in a given non special
component can be leaves of a given outbranching. The first term of the lower
bound comes from Theorem 1 and Theorem 2. To establish the second term,
consider the digraph D' whose vertices are the special vertices of D and r. For
every non special component of D, add an edge in D' between the special in-
neighbours of its two ends. Consider an outbranching of D' rooted at r. This
outbranching uses $l - 1$ edges in D', and directly corresponds to an out-tree
T in D. Extend T into an outbranching $\tilde{T}$ of D. Every non special component
which is not used in T contributes to at least a leaf in $\tilde{T}$, which concludes the
proof.   □

Consider the best of the three outbranchings of D obtained in polynomial time
by Theorem 1, Theorem 2 and Theorem 4. This outbranching has at least
$max(\frac{l}{30}, h - l)$ leaves. The worst case is when $\frac{l}{30} = h - l$. In this case, the
upper bound becomes: $\frac{92l}{30}$, hence we have a factor 92 approximation algorithm
for ROOTED MAXIMUM LEAF OUTBRANCHING.

6   Conclusion

We have given an edge-quadratic kernel and a constant factor approximation
algorithm for ROOTED MAXIMUM LEAF OUTBRANCHING: reducing the gap
between the problem of finding trees with many leaves in undirected and directed graphs. The gap now essentially lies in the fact that MAXIMUM LEAF
SPANNING TREE has a linear kernel while ROOTED MAXIMUM LEAF
OUTBRANCHING has a quadratic kernel. Deciding whether ROOTED MAXI-
MUM LEAF OUTBRANCHING has a vertex-linear kernel is a challenging question. Whether long paths made of 2-circuits can be dealt with or not might be
key to this respect.

References

[1] Alon, N., Fomin, F., Gutin, G., Krivelevich, M., Saurabh, S.: Parameterized algorithms for directed maximum leaf problems. In: Arge, L., Cachin, C., Jurdziński, T., Tarlecki, A. (eds.) ICALP 2007. LNCS, vol. 4596, pp. 352–362. Springer, Heidelberg (2007)

[2] Alon, N., Fomin, F., Gutin, G., Krivelevich, M., Saurabh, S.: Spanning directed trees with many leaves. SIAM J. Discrete Maths. 23(1), 466–476 (2009)

[3] Bodlaender, H.L., Downey, R.G., Fellows, M.R., Hermelin, D.: On problems without polynomial kernels (Extended abstract). In: Aceto, L., Damgård, I., Goldberg, L.A., Halldórsson, M.M., Ingólfsdóttir, A., Walukiewicz, I. (eds.) ICALP 2008, Part I. LNCS, vol. 5125, pp. 563–574. Springer, Heidelberg (2008)

[4] Paul, S.: Bonsma and Frederic Dorn. An fpt algorithm for directed spanning k-leaf. abs/0711.4052 (2007)

[5] Chen, J., Liu, Y.: On the parameterized max-leaf problems: digraphs and undirected graphs. Technical report, Department of Computer Science, Texas A& M University (2008)

[6] Cheriyan, J., Reif, J.: Directed s-t numberings, rubber bands, and testing digraph k-vertex connectivity. Combinatorica 14(4), 435–451 (1994)

[7] Daligault, J., Gutin, G., Kim, E.J., Yeo, A.: FPT algorithms and kernels for the Directed k-Leaf problem. To appear in Journal of Computer and System Sciences

[8] Dijkstra, E.: Self-stabilizing systems in spite of distributed control. Commun. ACM 17(11), 643–644 (1974)

[9] Ding, G., Johnson, T., Seymour, P.: Spanning trees with many leaves. J. Graph Theory 37(4), 189–197 (2001)

[10] Downey, R.G., Fellows, M.R.: Parameterized complexity. Springer, Heidelberg (1999)

[11] Drescher, M., Vetta, A.: An approximation algorithm for the maximum leaf spanning arborescence problem. To appear in ACM Transactions on Algorithms

[12] Estivill-Castro, V., Fellows, M., Langston, M., Rosamond, F.: Fixed-parameter tractability is polynomial-time extremal structure theory i: The case of max leaf. In: Proc. of ACiD 2005 (2005)

[13] Fernau, H., Fomin, F.V., Lokshtanov, D., Raible, D., Saurabh, S., Villanger, Y.: Kernel(s) for problems with no kernel: On out-trees with many leaves. In: Albers, S., Marion, J.-Y. (eds.) 26th International Symposium on Theoretical Aspects of Computer Science (STACS 2009), Dagstuhl, Germany. Leibniz International Proceedings in Informatics, vol. 3, pp. 421–432. Schloss Dagstuhl - Leibniz-Zentrum fuer Informatik, Germany (2009),
http://drops.dagstuhl.de/opus/volltexte/2009/1843

[14] Flum, J., Grohe, M.: Parameterized Complexity Theory. Springer, Heidelberg (2006)

[15] Fomin, F., Grandoni, F., Kratsch, D.: Solving connected dominating set faster than 2^n. Algorithmica 52(2), 153–166 (2008)

[16] Galbiati, G., Maffioli, F., Morzenti, A.: A short note on the approximability of the maximum leaves spanning tree problem. Inf. Process. Lett. 52(1), 45–49 (1994)

[17] Guo, J., Niedermeier, R.: Invitation to data reduction and problem kernelization. SIGACT News 38(1), 31–45 (2007)

[18] Kneis, J., Langer, A., Rossmanith, P.: A new algorithm for finding trees with many leaves. In: Hong, S.-H., Nagamochi, H., Fukunaga, T. (eds.) ISAAC 2008. LNCS, vol. 5369, pp. 270–281. Springer, Heidelberg (2008)

[19] Lempel, A., Even, S., Cederbaum, I.: An algorithm for planarity testing of graphs. In: Rosenstiehl, P. (ed.) Theory of Graphs: Internat. Sympos.: Rome, pp. 215–232 (1966)

[20] Linial, N., Lovasz, L., Wigderson, A.: Rubber bands, convex embeddings and graph connectivity. Combinatorica 8, 91–102 (1988)

[21] Niedermeier, R.: Invitation to fixed parameter algorithms. Oxford Lectures Series in Mathematics and its Applications, vol. 31. Oxford University Press, Oxford (2006)

[22] Solis-Oba, R.: 2-approximation algorithm for finding a spanning tree with maximum number of leaves. In: Bilardi, G., Pietracaprina, A., Italiano, G.F., Pucci, G. (eds.) ESA 1998. LNCS, vol. 1461, pp. 441–452. Springer, Heidelberg (1998)

[23] Storer, J.A.: Constructing full spanning trees for cubic graphs. Inform Process Lett. 13, 8–11 (1981)

[24] Wu, J., Li, H.: On calculating connected dominating set for efficient routing in ad hoc wireless networks. In: DIALM 1999: Proceedings of the 3rd international workshop on Discrete algorithms and methods for mobile computing and communications, pp. 7–14. ACM Press, New York (1999)

Bounded-Degree Techniques Accelerate Some Parameterized Graph Algorithms

Peter Damaschke

Department of Computer Science and Engineering
Chalmers University, 41296 Göteborg, Sweden
ptr@chalmers.se

Abstract. Many algorithms for FPT graph problems are search tree algorithms with sophisticated local branching rules. But it has also been noticed that using the global structure of input graphs complements the the search tree paradigm. Here we prove some new results based on the global structure of bounded-degree graphs after branching away the high-degree vertices. Some techniques and structural results are generic and should find more applications. First, we decompose a graph by "separating" branchings into cheaper or smaller components wich are then processed separately. Using this idea we accelerate the $O^*(1.3803^k)$ time algorithm for counting the vertex covers of size k (Mölle, Richter, and Rossmanith, 2006) to $O^*(1.3740^k)$. Next we characterize the graphs where no edge is in three conflict triples, i.e., triples of vertices with exactly two edges. This theorem may find interest in graph theory, and it yields an $O^*(1.47^k)$ time algorithm for CLUSTER DELETION, improving upon the previous $O^*(1.53^k)$ (Gramm, Guo, Hüffner, Niedermeier, 2004). CLUSTER DELETION is the problem of deleting k edges to destroy all conflict triples and get a disjoint union of cliques. For graphs where every edge is in $O(1)$ conflict triples we show a nice dichotomy: The graph or its complement has degree $O(1)$. This opens the possibility for future improvements via the above decomposition technique.

1 The Problems and Contributions

A problem is fixed-parameter tractable (FPT) if it can be solved in $O(p(n)f(k))$ time where p is a polynomial and f any function. We assume familiarity with the basic notions of FPT algorithms and their analysis [5,13]. Since we focus on the $f(k)$ factor, we sometimes adopt the $O^*(f(k))$ notation that suppresses polynomial factors. In graphs we usually denote by n the number of vertices. For brevity we say "component" instead of "connected component" of a graph.

Counting Vertex Covers: A vertex cover is a set of vertices with at least one vertex from every edge. The problem of counting all vertex covers of size k in graphs (or more generally, hitting sets of size k in hypergraphs of fixed rank) is very natural as such, but has also interesting applications in combinatorial inference, e.g., in computational biology as proposed in [3]. Briefly, the real problem is to infer a set of substances that produced a given set of indicators, where

J. Chen and F.V. Fomin (Eds.): IWPEC 2009, LNCS 5917, pp. 98–109, 2009.

an indicator may come from several candidate substances. Substances and indicators are modeled as vertices and edges of a hypergraph. Since many different solutions of a certain expected size exist, we count the solutions containing every fixed vertex, in order to evaluate how likely the presence of every substance is. Many vertex cover variants have been studied as well, see, e.g., [11] for pointers.

The VERTEX COVER COUNTING algorithm in [12] first branches on vertices of degree 4 or larger. The branching vector $(1,4)$ has branching number 1.3803, and vertex covers in the residual graphs of degree 3 can be counted in $O^*(1.26^k)$ time by a nontrivial technique. This gives an overall time bound of $O^*(1.3803^k)$, where the $(1,4)$-branching is the bottleneck. In order to take this hurdle we must avoid the situation that only $(1,4)$-branching takes place and cheaper rules are never invoked. Our idea is to reuse the above algorithm but choose the initial $(1,4)$-branchings in a special way that guarantees a "large and easy" residual graph. We either retain one large degree-3 component that can be solved in $O^*(1.26^k)$ time (with the residual k), or we can split the residual graph into several components, process them independently, and finally combine the partial results in polynomial time. Loosely speaking, due to this quasi-parallelization of the remaining branchings, only one component with the largest residual k counts for the exponential term in the complexity. In all cases this drives the overall branching number below 1.3803. The fact that components of a graph can be treated independently is usually not even worth mentioning as it cannot be exploited in a worst-case analysis, but in our approach it becomes essential. The time analysis is not straightworward either. Upon the $(1,4)$-branchings we deduct not only 1 and 4 from parameter k, but also the number of vertices that will be added later to the vertex covers in a cheaper way. On the other hand, we must "charge" the $(1,4)$-branchings for these deferred decisions and increase the size bounds of the search trees after the branching accordingly, as every leaf in the search tree now represents the deferred decisions. But since they are cheaper, this pays off in the end.

This global technique of branching away the hardest parts and splitting the rest into independent pieces seems to be new in this form. The idea is not deep, but this may be a strength. The conceptual simplicity should make it versatile and also applicable to other FPT problems that decompose neatly.

The bound for VERTEX COVER COUNTING we can prove so far is $O^*(1.3740^k)$. Seemingly this is marginal progress, however, the weakness is apparently in the current analysis rather than in the algorithm itself. Our analysis gives something away, hence the true *worst-case* bound is likely to be way better. But it is hard to take advantage of those observations, since different bad cases can appear mixed throughout the search tree. (Cf. also [6,7].)

Cluster Modification Problems: A cluster graph is a disjoint union of cliques, also called clusters. A P_3, or conflict triple, is a path of three vertices (and two edges). Cluster graphs are exactly the graphs without induced P_3. In CLUSTER DELETION we want to delete at most k edges in a graph G so as to obtain a cluster graph. That is, deletions must destroy all induced P_3. CLUSTER EDITING is similarly defined with edge deletions *and* insertions. Both problems are special cases of WEIGHTED CLUSTER EDITING where pairs of vertices have individual edits costs. These NP-hard problems [14] have also applications foremost in computational biology [4,9,14]. CLUSTER DELETION becomes important if only adjacent vertices (representing similar objects) are tolerated in any cluster.

The parameterized complexity of these problems is well studied. In [2] we give problem kernels and algorithms for enumerating *all* solutions to several clustering problems. In [1], WEIGHTED CLUSTER EDITING has been solved in $O(1.82^k + n^3)$ time. This is also the best known time bound for CLUSTER EDITING. The other special case is still somewhat "easier": an $O(1.53^k + n^3)$ time algorithm for CLUSTER DELETION is known from [8]. It was an example of an automatically generated search tree algorithm and improved a "handmade" $O(1.77^k + n^3)$ algorithm [9]. It works with branching rules on subgraphs with at most six vertices. No further progress on CLUSTER DELETION has been made since then.

Here we come back to handicraft and give an $O(1.47^k + n^3)$ time algorithm for CLUSTER DELETION. It starts from the most obvious rule with branching number 1.47: If some edge is in three or more P_3, then *branch on the edge*, i.e., delete the edge, or delete all edges building a P_3 with it. Key to our improvement is a new theorem showing that graphs where this rule is not applicable have simple structures, and no other local branching rules are needed. (Only special cases were already treated in [1, Lemma 4] and in a conference version of [2].)

The result give new indications that local branching rules should be complemented by global structure analysis and techniques. Other global methods like dynamic programming on subsets, bounded treewidth and pathwidth [6,7], and iterative compression [10], have recently shown their great potential. In fact, the $O(1.3803^k)$ algorithm for WEIGHTED VERTEX COVER in [6] is close to the VERTEX COVER COUNTING results, but here we add our separation idea.

Organization of the paper: In Section 2 we state our algorithm for VERTEX COVER COUNTING. It builds upon [12] and is only slightly more complicated. The only new features are that we branch on vertices of degree 4 in some breadth-first order and combine partial results from different components in moderate polynomial time. (The polynomial factor has not been explicitly specified in [12], but we can obviously state that it does not blow up by the routines we add.) In Section 3 we prove a time bound that beats $O^*(1.3803^k)$, and we discuss why the true bound should be even better. Then we turn to some graph theory useful for clustering. In Section 4 we completely characterize the graphs where no edge is in three P_3. In the proof we tried to avoid too many tiresome case distinctions. More generally, for graphs where every edge is in an $O(1)$ number of P_3 we prove a dichotomy: the graph or its complement graph has degree $O(1)$. Section 5 deals with the algorithmic consequences of P_3 structure. We solve CLUSTER DELETION in $O^*(1.47^k)$ time. Improving some details in the polynomial-time parts seems quite possible, however, the more intriguing question is about improved bases of the exponential term. We have to leave this for further research, but, using the above dichotomy we can at least show that the bottleneck case is graphs with bounded degree, and therefore it may be possible to apply the separation technique again. We discuss these possibilities also for CLUSTER EDITING. Due to the page limit, some simpler proofs are omitted, graph-theoretic notions are only briefly reviewed, and there is no space for figures.

2   Vertex Cover Counting Algorithm

Let $c(G, k)$ be the number of vertex covers with exactly k vertices in graph G. The *degree* of a vertex v is the number of vertices adjacent to v. A *degree-d vertex* has degree exactly d. The degree of a graph is the maximum vertex degree. A *subcubic* graph has degree 3. We skip the definition of tree decomposition, because we use the following result only as a "black box": Any subcubic graph has a tree decomposition of width at most $(1/6 + \epsilon)n$, and such a tree decomposition is computable in polynomial time for any fixed $\epsilon > 0$ [7]. By dynamic programming on this tree decomposition one can count the vertex covers of size k in $O^*(2^{(1/3+\epsilon)k}) = O^*(1.26^k)$ time [12], thus we have:

Lemma 1. *In subcubic graphs G one can compute $c(G, k)$ in $O^*(1.26^k)$ time.*

If G is the disjoint union of graphs G_1 and G_2 then, obviously, $c(G, k) = \sum_{j=0}^{k} c(G_1, j) \cdot c(G_2, k - j)$. From this one easily concludes:

Lemma 2. *Once the $c(G', j)$ are known for all components G' of G, and for all $j \leq k$, we can compute $c(G, k)$ in polynomial time.*

Branching on a vertex v means: Either put v or all neighbors of v in the vertex cover, remove the chosen vertices, all incident edges, and isolated vertices.

We refer to a series of such branching decisions on different vertices as a *branch*. Thus, a branch corresponds to a node of the resulting search tree. The *residual graph* in a branch is the graph that remains after the branchings. Note that branching on a vertex divides the family of vertex covers of the current residual graph exactly in two subfamilies of the vertex covers of residual graphs in the two new branches. Hence, when the search tree is completed, we can sum up the numbers of vertex covers (of the proper size) found in the residual graphs at the leaves. Now we describe our algorithm.

Phase 1: preparation. First branch, as long as possible, on degree-d vertices with $d \geq 5$. In every branch consider the residual graph of degree 4 and continue as described below. We will declare certain vertices *roots*.

Phase 2: separation. If a component without a root exists, then fix any vertex in this component as the *active root* and proceed as follows. Layer j contains the vertices at distance j from the active root. As long as degree-4 vertices exist in some layer $j > 2$ of this component, choose one such degree-4 vertex closest to the active root and branch on it. (Note that these branchings may change the layers, and even disconnect some vertices from the active root.) This process stops as soon as no degree-4 vertices remain in the layers $j > 2$. At this moment, set the active root passive. – Iterate.

Phase 3: completion. Branch on the remaining degree-4 vertices near the roots. This results in a subcubic graph. Let $m \leq k$ be the number of vertices that remain to be added to the vertex cover. (Note that m depends on the branch but is exactly known in every branch.) Compute the $c(G', i)$ in all components G' of the residual graph, and for all $i \leq m$, using the algorithm in Lemma 1. Finally combine these results in every residual graph, to compute the number of vertex covers of size k, using the algorithm in Lemma 2. In the last step, the counts from all leaves of the search tree are added.

Some refinement: For phase 2 we refine the rule for selecting the next degree-4 vertex to branch on. This is only a technicality that we insert just because we can prove a better worst-case bound when using it.

Whenever a new layer j for branching is entered (that is, $j > 2$ is the smallest index of a layer where still degree-4 vertices exist), pair up some degree-4 vertices in layer j to siblings: Every degree-4 vertex in layer j is assigned an adjacent *parent vertex* in layer $j - 1$. (If several possible parent vertices exist, then select any one.) *Siblings* are degree-4 vertices with the same parent. It is not hard to see that, for $j > 3$, every degree-4 vertex gets at most one sibling. Now, whenever we have branched on a degree-4 vertex v and added v to the vertex cover (this happens in one branch), and v has a sibling which is still a degree-4 vertex, then branch on this sibling immediately.

3   Analysis

A simple analysis would show that the residual graph after phase 2 (separation) always needs a vertex cover of at least yk vertices for some fraction y, giving a time bound $O^*(1.38028^{k-y}1.26^y)$. However y is very small in branches where often 4 vertices are taken. We got a better result by relating these "cheap" vertices in phase 3 directly to the number of branchings in phase 2.

Theorem 1. *The vertex covers of size k can be counted in $O^*(1.3740^k)$ time.*

Proof. The branching number for branching on degree-d vertices with $d \geq 5$ in phase 1 is the positive root of $x^5 = x^4 + 1$, that is, $x < 1.3248$.

For phase 2 the following observation is crucial: Since we always pick a degree-4 vertex in a layer $j > 2$ closest to the root, and removals can make the distances in the remaining graph only larger, the current distance j from root, of the degree-4 vertices we branch on, can only increase during the process. Hence, if j is the index of the current layer, any vertex being in a layer $i \leq j - 2$ at this moment is never removed later, and it also stays in layer i forever. Accordingly we call these vertices *persistent*. Also note that any vertex being in layer $j - 1$ at this moment can either disappear due to branchings in layer j, or stay in layer $j - 1$ forever, but it cannot slide into a layer with higher index later.

Let j be the current layer where we branch on degree-4 vertices. As already stated in the algorithm, to every degree-4 vertex v in layer j we assign a *parent* $p(v)$ which is a neighbor of v in layer $j - 1$, and a *grandparent* $g(v)$ which is a neighbor of $p(v)$ in layer $j - 2$. Note that $g(v)$ is persistent. Since every vertex in a layer $i > 0$ has at least one edge to a neighbor in layer $i - 1$, and vertex degrees are at most 3 in the layers i, $2 < i < j$, every persistent vertex in a layer

$i > 2$ can have at most 2 children and at most 4 grandchildren. Recall that any two degree-4 vertices v, v' with the same parent are *siblings*. If v is a degree-4 vertex without a sibling v' (a degree-4 vertex with the same parent), we still use the notation v' and simply say that "v' does not exist".

We assign to certain vertices *certificates*: Consider any vertex v and its sibling v' in layer j. *Case (1): Either v' does not exist, or v, v' are adjacent.* Then, when we branch on v we send a certificate to $g(v)$. The branching vector is $(1, 4)$, and after the branching no child of $p(v)$ is a degree-4 vertex anymore, since either v' did not exist, or v' has been removed now, or v has been removed which reduces the degree of v'. *Case (2): v' exists and v, v' are not adjacent.* Then, when we branch on the first sibling v, we send a certificate to $g(v)$. In one branch we have put the 4 neighbors of v in the vertex cover. Otherwise we have put only v in the vertex cover, and then, by the refined rule, we branch on v' which is still a degree-4 vertex. In total we achieve the branching vector $(2, 4, 5)$. After that, no child of $p(v)$ is a degree-4 vertex anymore, since $p(v)$ has been removed which reduces the degree of its children, or v and v' have been removed.

This way, every persistent vertex in layers $i > 2$ receives at most 2 certificates through its (at most 2) children. For each residual graph let r denote the total number of certificates issued in all components. Hence at least $r/2$ persistent vertices exist in layers $i > 2$. Since every such vertex is incident to an edge to layer $i - 1$, at least $r/2$ edges remain after phase 2.

Let us count the branchings on the remaining degree-4 vertices near the roots already in phase 2. Since in each residual graph only one search tree for the components is decisive for the complexity (only the largest one, due to the polynomial-time combination procedure from Lemma 2) and every component has only $O(1)$ degree-4 vertices at that moment, this adds only a constant factor to the complexity. On the other hand, the graphs in phase 3 are now subcubic.

We call the vertices inserted in the vertex cover in phase 2 and 3 expensive and cheap vertices, respectively. Let $T(k, l)$ be the number of leaves of a search tree, when it remains to add $k + l$ vertices at most k of which are expensive and at least l are cheap. Initially, k is the given k, and $l = 0$. In the following, fractional numbers of objects make sense because their sum is finally rounded to the next integer. (They could be avoided by doing an equivalent analysis for blocks of 6 consecutive branchings.) As seen above, for every branching in phase 2, either on a single vertex or on siblings v, v' of degree 4, at least $1/2$ edges need to be covered later in phase 3. Since the maximum degree is 3, at least $1/6$ cheap vertices will be added later. Thus it is safe to transmit $1/6$ from k to l. This yields $T(k, l) \leq T(k - 1 - 1/6, l + 1/6) + T(k - 4 - 1/6, l + 1/6)$ and $T(k, l) \leq T(k - 2 - 1/6, l + 1/6) + T(k - 4 - 1/6, l + 1/6) + T(k - 5 - 1/6, l + 1/6)$ in the two cases. Together with $T(0, l) \leq 1.26^l$ from Lemma 1 we obtain recurrences for $T(k) := T(k, 0)$, the number of leaves in the overall search tree: Since every branching in phase 2 finally incurs another $1.26^{1/6}$ factor in all summands, we have $T(k) \leq (T(k - 1 - 1/6) + T(k - 4 - 1/6))1.26^{1/6}$ and similarly $T(k) \leq (T(k - 2 - 1/6) + T(k - 4 - 1/6) + T(k - 5 - 1/6))1.26^{1/6}$, with numerical solutions 1.3699^k and 1.3740^k, respectively.   □

Our analysis guarantees the existence of at least one persistent vertex, with an edge attached, for any two $(1, 4)$-branchings made. We "certify" only grandparents and ignore further persistent edges, especially in the large component relevant for the complexity. If the analysis could use all persistent vertices, this would almost double the deduction from the parameter. Refinements of the algorithm itself may give further improvements: The worst case in our analysis appears if every vertex in the largest component has two children. But then the component is merely a binary tree, and vertex covers could be counted trivially there. Finally, some clever measure-and-conquer might yield a simpler analysis.

4   Conflict Triple Structure in Graphs

P_n, C_n, K_n denote a chordless path, cycle, and a clique, respectively, of n vertices, and $K_{m,n}$ a complete bipartite graph with m and n vertices in the partite sets. The disjoint union $G + H$ of graphs G and H consists of vertex-disjoint copies of G and H, and pG is the disjoint union of p copies of G. The join $G * H$ is obtained from $G + H$ by inserting all possible edges between the vertices of G and H. The complement G^c of G is obtained by switching all edges into non-edges and vice versa. In an obvious sense, a P_4 has two *inner vertices* forming the *central edge*, and two *outer vertices*, and a P_5 has a *central vertex*.

Let the *score* of an edge be the number of different P_3 the edge belongs to. A graph is score-s if every edge has score at most s. Note that a score-s graph is also score-$(s + 1)$, and an induced subgraph of a score-s graph is score-s. A main goal of this section is to characterize the connected score-2 graphs G. We say that G is *spanning score-2* if G has a spanning tree of score-2 edges. To *add a vertex* to a connected graph G means to introduce a new vertex adjacent to at least one vertex of G. A graph is *maximal score-2* if we cannot add another vertex while keeping the graph score-2 and connected.

Lemma 3. *Suppose that we add a vertex x to a connected score-2 graph G.*
(i) If the extended graph remains score-2, and x is adjacent to vertex u of G, then x is also adjacent to all vertices reachable from u via score-2 edges in G.
(ii) If G is spanning score-2 and the extended graph remains score-2, then x is adjacent to all vertices of G.
(iii) If G is spanning score-2 and has a vertex non-adjacent to at least three other vertices of G, then G is maximal score-2.

Doubling a vertex x of a graph means to insert a new vertex adjacent exactly to x and its neighbors. The result of doubling several vertices does not depend on the order. Now we define several special 6-vertex graphs, with the understanding that only the explicitly mentioned edges exist.

- 3-asterisk: a K_3 where each vertex is adjacent to one further vertex.
- 3-sun: a K_3 where each two vertices are adjacent to one further vertex.
- fat P_4: obtained from a P_4 by doubling both inner vertices.
- fat P_5: obtained from a P_5 by doubling its central vertex.

Theorem 2. *The following graphs (with arbitrarily large positive n, q, p) and their connected induced subgraphs comprise the complete list of connected score-2 graphs: 3-asterisk, 3-sun, fat P_4, fat P_5; C_n ($n \geq 4$); $K_q * C_5$, $K_q * K_3^c$, $K_q * (K_2 + K_2)$; $(qK_1 + pK_2)^c$ ($p \geq 2$).*

Proof. It is easy to check that all listed graphs are score-2. All of them except $(qK_1 + pK_2)^c$ are also spanning score-2. The 3-asterisk, 3-sun, fat P_4, fat P_5, and C_n ($n \geq 6$) also fulfill the conditions of Lemma 3 (iii), hence they are maximal score-2. Adding a vertex to any of $K_q * C_5$, $K_q * K_3^c$, $K_q * (K_2 + K_2)$ while keeping score 2 yields a graph of the same type, with q increased by 1.

The reasoning for $(qK_1 + pK_2)^c$ is slightly more complicated. Graph $(pK_2)^c$ ($p \geq 2$) is spanning score-2, hence, by Lemma 3 (ii), every added vertex is adjacent to all its $2p$ vertices. Moreover, each of the added vertices must be adjacent to all other added vertices except at most one. Thus we obtain only graphs $(qK_1 + pK_2)^c$. It is also impossible to add further vertices adjacent only to vertices in the qK_1 part, as this would create new edges of score above 2.

Thus we have shown that our list contains only score-2 graphs and is closed under vertex addition. Next we prove that no further cases exist, that is, any connected score-2 graph G is mentioned in the Theorem.

Let $N(u)$ denote the set of neighbors of a vertex u in G. In G^c this means, $N(u)$ is the set of all non-neighbors of u. Let $H(u)$ be the subgraph of G^c (!) induced by $N(u)$. If some vertex v has degree larger than 2 in $H(u)$, then the non-edge uv is in three P_3^c in G^c, a contradiction. If $H(u)$ has an induced $2K_2$ then G has an induced $(K_1 + 2K_2)^c$. The other cases are that $H(u)$ has no edges, or the edges in $H(u)$ form one of the induced subgraphs $K_2, P_3, C_3, P_4, C_4, C_5$. We examine these cases one-by-one.

- If $H(u)$ has an induced C_5 then G has an induced $K_1 * C_5$.
- If $H(u)$ has an induced C_4 then G has an induced $K_1 * (K_2 + K_2)$.
- If $H(u)$ has an induced P_4 then G has an induced $K_1 * P_4$. Since the two edges from u to the outer vertices of the P_4 and the central edge of the P_4 have score 2, Lemma 3 (i) gives that any added vertex is adjacent to both inner vertices, or to u and both outer vertices, or to all five vertices. These cases yield an induced 3-sun, $K_1 * C_5$, and $K_2 * P_4$ (induced subgraph of $K_2 * C_5$), respectively. By essentially the same argument, adding further vertices to $K_q * P_4$ can only lead to $K_{q+1} * P_4$ or $K_q * C_5$.
- If $H(u)$ has an induced C_3 then G has an induced $K_{1,3} = K_1 * K_3^c$.

The P_3 case requires some more work. If $H(u)$ has an induced P_3 (but none of P_4, C_4, C_5) then G has an induced subgraph with vertices x, y, u, z and edges xy, xu, yu, uz. Since uz has score 2, any new vertices adjacent to u or z are adjacent to both u and z (Lemma 3 (i)), hence also to x and y and to each other (as no further edges in $H(u)^c = N(u)$ exist by assumption). This yields only graphs of the form $K_q * (K_2 + K_1)$. Now let q be maximum, that is, further vertices that we add are adjacent to x or y only. If some added vertex v is adjacent to x only, then $q = 1$. Since vx, xu, uz have score 2, by Lemma 3 (i) and the assumptions of our case, we can add at most one further vertex, and this one is adjacent to y only, which yields a 3-asterisk. It remains the case that v is adjacent to x and y. First observe $q \leq 2$. If $q = 2$, we have a fat P_4. For $q = 1$, edges xu, yu, uz have score 2, hence any further vertex added is adjacent to v only, and we get a fat P_5.

Now we have settled all cases where $H(u)$ contains more than one edge, for some u. It remains to study connected score-2 graphs G where, for every u, the neighborhood $N(u)$ has at most one non-edge. Clearly, such G cannot have induced $K_{1,3}$. If G also lacks K_3 then the maximum degree in G is 2, hence G is P_n or C_n. Otherwise consider some maximal clique K_q, $q \geq 3$. Any vertex x added to this K_q has some neighbor u in the K_q, hence $x \in H(u)$. In G^c, vertex x is therefore adjacent to exactly one vertex in the K_q^c. It also follows that we can add at most one vertex to the considered K_q. This yields an induced $K_{q-1}*(2K_1)$. Since the same reasoning applies to the other K_q (where x replaces u), u is the only vertex that could be added. Hence $K_{q-1} * (2K_1)$ is already the entire G in this case. □

Theorem 2 is best possible in the sense that already score-3 graphs are not limited to special structures but can be arbitrarily complicated: Subdividing the edges of any graph of degree 3 by further vertices generates a score-3 graph. However, we can still prove an interesting dichotomy for graphs of any fixed score s. Let $N^i(u)$ denote the set of vertices at distance exactly i from u.

Theorem 3. *Let G be any connected graph of score s graph that has more than $(s + 1)^2 (2s + 1)^2 + 1 \sim 4s^4$ vertices. Then G or G^c has degree at most $3s + 1$.*

Proof. Let u be a vertex of maximum degree d in G, hence $N(u)$ has d vertices. Since every edge uv, $v \in N(u)$ has score at most s, every vertex $v \in N(u)$ must be adjacent to at least $d - s - 1$ other vertices in $N(u)$. Let $x \in N^2(u)$, and let $v \in N(u)$ be some neighbor of x. Since vx forms a P_3 with uv, it is in at most $s - 1$ other P_3. It follows that x is adjacent to at least $d - 2s - 2$ of v's $d - s - 1$ neighbors in $N(u)$, provided that $d > 2s - 2$. Next, let $y \in N^3(u)$, and let $x \in N^2(u)$ be some neighbor of y. As shown above, x has at least $d - 2s - 1$ neighbors in $N(u)$, each of which is involved in a P_3 with xy, and xy has score at most s, we get $d - 2s - 1 \leq s$. This shows $d \leq 3s + 1$ or $N^3(u) = \emptyset$.

In case $d \leq 3s + 1$ we are done, so let $N^3(u) = \emptyset$. If $d \leq (s + 1)(2s + 1)$ then, since d is the maximum degree, we have $|N^0(u)| + |N^1(u)| + |N^2(u)| \leq (s + 1)^2 (2s + 1)^2 + 1$. Hence assume $d > (s + 1)(2s + 1)$ in the following. Recall again that every vertex in $N^2(u)$ has at least $d - 2s - 1$ neighbors in $N(u)$, that is, at most $2s + 1$ non-neighbors in $N(u)$. Thus, if $|N^2(u)| \geq s + 1$ then some $v \in N(u)$ is still adjacent to $s + 1$ vertices of $N^2(u)$. But uv has score at most s. This contradiction shows $|N^2(u)| \leq s$. Now we see that the vertices in $N^0(u)$, $N^1(u)$, and $N^2(u)$ have, in G^c, degree at most s, $2s + 1$, and $3s + 1$. □

5   FPT Algorithms Using the Conflict Triple Structure

As usual, $O(1)$ means "bounded by a certain constant".

Theorem 4. CLUSTER DELETION *is solvable in* $O(1.47^k + n^3)$ *time.*

Proof. As long as possible, take an edge e of score larger than 2, and delete e or all edges forming a P_3 with e. The branching number is 1.47. Once the graph is score-2, every component is one of the graphs from Theorem 2. We show how to solve CLUSTER DELETION in polynomial time in these cases. For graphs of size $O(1)$ and for C_n (and P_n) this is evident.

In $(qK_1 + pK_2)^c$ we take one vertex from each K_2^c to form a clique of size p, and the rest is a clique of size $p + q$. This solution needs $p(p + q - 1)$ edge deletions, which is optimal: Any two of the p pairs K_2^c build an induced C_4, hence two edges must be deleted. Since all these C_4 are edge-disjoint, no deletion is counted twice. Thus we must delete at least $p(p - 1)$ edges between these pairs. Moreover, each combination of the p pairs and the q vertices in the $(qK_1)^c$ builds a P_3, and all these P_3 are edge-disjoint. Thus we must delete pq of these edges. The sum is $p(p + q - 1)$.

The other graphs in Theorem 2 consist of one clique K of size q, joined with at most five other vertices. If we first disconnect these extra vertices from K, we always get a solution with at most $5q + 3$ deletions. (Summand 3 is easy to verify.) Assume that some solution disconnects some $r \leq q/2$ vertices from the other $q - r$ vertices of K. This costs already $r(q - r)$ deletions which cannot be optimal unless $r(q - r) \leq 5q + 3$. Since $q - r \geq q/2$, this yields $rq/2 \leq 5q + 3$, hence $r \leq 10 + 6/q$ and finally $r \leq 10$ regardless of q. Since the vertices of K are undistinguishable, we may select any 10 of them as candidate vertices for split-off. Thus there exists an optimal solution that deletes edges only between at most 15 predefined vertices, and we are back to the $O(1)$ size case.

In all cases, the polynomial term in the time bound is dominated by the time needed to enumerate the P_3. Note that, during the process of edge deletions, every triple of vertices becomes a new P_3 at most once.   □

With minor modifications, the algorithm can output a concise enumeration of all solutions (cf. [2]) within the same time bound. Theorem 3 has some interesting algorithmic consequences, too:

Corollary 1. *Let $b > 1.3803$ be any fixed base. If we can solve* CLUSTER DELETION *in* $O^*(b^k)$ *time for graphs of degree $O(1)$, we can do so for general graphs.*

Proof. As long as possible, branch on edges of score 4 or larger. The branching number is 1.3803. It remains a score-3 graph G. Assume G is connected, otherwise we consider the components separately. Theorem 3 gives that G or G^c has degree at most 10 (or G has $O(1)$ size).

In G^c we also observe that the sum of degrees of any two vertices u, v with distance larger than 2 (in G^c) is at most 3, since non-edge uv belongs to at most three P_3^c. Consequently, G^c has at most one non-trivial component H with more than one edge. If the second case of Theorem 3 applies, the degree of G^c is $O(1)$, thus H has size $O(1)$ (or we get again a forbidden pair u, v as above). That means, G has the form $(qK_1 + pK_2 + H)^c$ with a graph H of size $O(1)$, and we can solve this case in polynomial time, similarly as in Theorem 4. If the first case of Theorem 3 holds, the degree of G is $O(1)$.

Hence, either a rule with branching number at most 1.3803 is available, or the instance is solvable in polynomial time, or G has degree $O(1)$. $\square$

Remarkably, Corollary 1 implies that efforts to improve the base 1.47 only need to consider score-3 graphs of $O(1)$ degree. This also suggests that the separation technique from Section 2-3 may be applicable. We have to leave the question how much we can gain in this way for further research. For CLUSTER EDITING we get a similar "reduction to fixed degree" below.

Lemma 4. *For any $b > 1.62$ there exists s such that, if a graph has an edge of score larger than s then a branching rule for* CLUSTER EDITING *with branching number at most b is available.*

Proof. Consider an edge uv with score $s + 1$, and let S be the set of those $s + 1$ vertices that form P_3 with uv. We branch as follows: Either delete uv or keep it. If we keep uv, then for each $w \in S$ we must edit (insert or delete) one of the edges uw, vw. Accordingly, we refer to w as an insertion or deletion vertex. Now decide to make all $w \in S$ deletion vertices, or decide on some insertion vertex $w \in S$. In each of the last $s + 1$ branches continue as follows. Make every $y \in S \setminus \{w\}$ independently an insertion or deletion vertex. In both branches we must edit one of the edges uy, vy, and in one branch we must also edit wy, because any insertion (deletion) vertex must be adjacent (non-adjacent) to w. It is easy to verify that the branching number of this whole branching rule on S, u, v satisfies $x^{2s+1} \le x^{2s} + x^s + (s+1)(x+1)^s$, equivalently $x \le 1 + 1/x^s + (s+1)((x+1)/x^2)^s$. For any $x > 1.62$ we have $(x + 1)/x^2 < 1$, hence the branching number tends to 1.62 as s grows. $\square$

If the score of G is at most our fixed s, then Theorem 3 applies. The case that G^c has degree $O(1)$ is easily settled, similarly as in Theorem 4:

Lemma 5. *In any class of graphs G where G^c has degree d,* CLUSTER EDITING *is solvable in polynomial time.*

Corollary 2. *Let $b > 1.62$ be any fixed base. If we can solve* CLUSTER EDITING *in $O^*(b^k)$ time for graphs of degree $O(1)$ (depending on b), we can also do so for general graphs.*

Proof. Combine Theorem 3 with Lemma 4 and 5. $\square$

We conclude with another observation supporting the conjecture that the separation technique is applicable to CLUSTER EDITING:

Proposition 1. *In graphs of degree d, all clusters in an optimal solution to* CLUSTER EDITING *have at most $2d + 1$ vertices.*

It is also interesting to notice that an optimal solution to CLUSTER EDITING in connected graphs of degree $O(1)$ needs $k = \Theta(n)$ edits, since the cluster size is limited (Proposition 1) and links between the clusters must be removed.

Acknowledgment

This work has been supported by the Swedish Research Council (Vetenskapsrådet), grant 2007-6437, "Combinatorial inference algorithms – parameterization and clustering".

References

1. Böcker, S., Briesemeister, S., Bui, Q.B.A., Truß, A.: Going Weighted: Parameterized Algorithms for Cluster Editing. In: Yang, B., Du, D.-Z., Wang, C.A. (eds.) COCOA 2008. LNCS, vol. 5165, pp. 1–12. Springer, Heidelberg (2008)
2. Damaschke, P.: Fixed-Parameter Enumerability of Cluster Editing and Related Problems. Theory Comp. Systems (to appear)
3. Damaschke, P., Mololov, L.: The Union of Minimal Hitting Sets: Parameterized Combinatorial Bounds and Counting. J. Discr. Alg. (to appear)
4. Dehne, F., Langston, M.A., Luo, X., Pitre, S., Shaw, P., Zhang, Y.: The Cluster Editing Problem: Implementations and Experiments. In: Bodlaender, H.L., Langston, M.A. (eds.) IWPEC 2006. LNCS, vol. 4169, pp. 13–24. Springer, Heidelberg (2006)
5. Downey, R.G., Fellows, M.R.: Parameterized Complexity. Springer, Heidelberg (1999)
6. Fomin, F.V., Gaspers, S., Saurabh, S., Stepanov, A.A.: On Two Techniques of Combining Branching and Treewidth. Algorithmica (to appear)
7. Fomin, F.V., Hoie, K.: Pathwidth of Cubic Graphs and Exact Algorithms. Info. Proc. Letters 97, 191–196 (2006)
8. Gramm, J., Guo, J., Hüffner, F., Niedermeier, R.: Automated Generation of Search Tree Algorithms for Hard Graph Modification Problems. Algorithmica 39, 321–347 (2004)
9. Gramm, J., Guo, J., Hüffner, F., Niedermeier, R.: Graph-Modeled Data Clustering: Fixed-Parameter Algorithms for Clique Generation. Theory Comp. Systems 38, 373–392 (2005)
10. Hüffner, F., Komusiewicz, C., Moser, H., Niedermeier, R.: Fixed-Parameter Algorithms for Cluster Vertex Deletion. In: Laber, E.S., Bornstein, C., Nogueira, L.T., Faria, L. (eds.) LATIN 2008. LNCS, vol. 4957, pp. 711–722. Springer, Heidelberg (2008)
11. Kneis, J., Langer, A., Rossmanith, P.: Improved Upper Bounds for Partial Vertex Cover. In: Broersma, H., Erlebach, T., Friedetzky, T., Paulusma, D. (eds.) WG 2008. LNCS, vol. 5344, pp. 240–251. Springer, Heidelberg (2008)
12. Mölle, D., Richter, S., Rossmanith, P.: Enumerate and Expand: New Runtime Bounds for Vertex Cover Variants. In: Chen, D.Z., Lee, D.T. (eds.) COCOON 2006. LNCS, vol. 4112, pp. 265–273. Springer, Heidelberg (2006)
13. Niedermeier, R.: Invitation to Fixed-Parameter Algorithms. Oxford Lecture Series in Math. and its Appl. Oxford Univ. Press, Oxford (2006)
14. Shamir, R., Sharan, R., Tsur, D.: Cluster Graph Modification Problems. Discr. Appl. Math. 144, 173–182 (2004)

Pareto Complexity of Two-Parameter FPT Problems: A Case Study for Partial Vertex Cover

Peter Damaschke

Department of Computer Science and Engineering
Chalmers University, 41296 Göteborg, Sweden
ptr@chalmers.se

Abstract. We propose a framework for the complexity of algorithms for FPT problems with two separate parameters k, m and with exponential time bounds $O^*(x^k y^m)$ where $x, y > 1$ are constant bases. An optimal combination of bases x, y can be chosen depending on the ratio m/k. As a first illustration we apply the framework to the problem of finding, in a graph, a vertex cover of size k that leaves at most m edges uncovered. We report the best branching rules we could find so far, for all ranges of ratio m/k.

1 Introduction

Parameterized computational problems may have multiple parameters of different nature. To be concrete, this paper will focus on an examplary problem that we call VERTEX COVER WITH MISSED EDGES (VCME): Given a graph G and integers k, m, find a set S of at most k vertices and at most m edges such that every edge either belongs to S or has a vertex in S.

We use it only as a first illustration for the framework, however, VCME (and more generally, partial hitting sets in hypergraphs) is interesting in itself as a formulation of certain error-resilient inference problems: Every vertex is a positive boolean variable, and every edge is a clause, here with two variables. (In the hitting set problem with missed hyperedges, clauses can be larger.) We want to satisfy all clauses by a small number k of true variables. Clauses represent observations and variables are possible alternative "explanations" of them. Due to noise in the data, at most m clauses may be erroneously in the formula and need not be satisfied, but it is unknown which clauses are spurious. Thus we first ask whether there exists a solution with k true variables violating at most m clauses. In the enumeration problem we would like to get all solutions with parameter values at most k and m.

VCME differs from weighted covering problems with one parameter: We may assign weights to vertices and edges and seek a minimum-weight covering. (The problem introduced in [2] is a special case.) Once we have an algorithm for the weighted version, we can get solutions with several ratios m/k by adjusting the weights. However, since the two parameters can have a very different meaning

J. Chen and F.V. Fomin (Eds.): IWPEC 2009, LNCS 5917, pp. 110–121, 2009.

(see above), we want to have them under control separately and explicitly. The "weighting approach" can easily miss solutions with the prescribed parameter values. By treating the parameters separately we can also develop efficient algorithms tuned to specific ranges of k, m. VCME also differs from the PARTIAL VERTEX COVER problem with parameters k, t and the goal to cover at least t edges by k vertices [7,9]. While VCME and PARTIAL VERTEX COVER are equivalent as optimization problems (since $m + t$ is the edge number of the input graph), we are interested in cases where few edges are uncovered. Note that in [9] the time bounds are exponential in t, hence are not directly comparable to ours. In general, complexity results for one parameterization do not say much about others.

By now there is apparently no formal theory of problems that are NP-hard but fixed-parameter tractable (FPT) and have two or more parameters. Usually they are handled in an *ad hoc* way. Some framework is helpful, e.g., when it comes to comparison of two-parameter algorithms.

Algorithms for one-parameter FPT problems often have time bounds $O^*(x^k)$ where $x > 1$ is a constant base. (To keep the discussion simple we leave out polynomial factors and use the O^*-notation which suppresses them. In our concrete examples they are always moderate.) Any two $O^*(x^k)$ algorithms can be directly compared, and the one with smaller base x is superior, at least from sufficiently large k on. In the case of two (or more) parameters k, m, etc. the situation becomes more complicated, even if polynomial factors are neglected.

In this paper we restrict attention to time bounds of the form $O^*(x^k y^m)$ with constants $x, y > 1$. Clearly, time bounds are not necessarily of this form, and we do not insist on it, but for many problems like VCME it naturally comes up and appears as a reasonable and not too narrow generalization of $O^*(x^k)$ (see details later on).

While the complexity of an $O^*(x^k)$ algorithm is mainly characterized by one number x, in the two-parameter case we may combine several available algorithms with different pairs of constant bases (x, y), and either one can be the best for different ranges of parameters k, m. The O^* complexity of such a compound algorithm is then appropriately described by a set of optimal points (x, y) in the quadrant $x, y > 1$ of the (x, y)-plane. (Here, *optimal* is always meant with respect to the given algorithm; there might exist faster but yet unknown algorithms for the problem in question.)

In Section 2 we develop our framework for the complexity of two-parameter problems with time bounds $O^*(x^k y^m)$. In Section 3 we apply this concept and provide some basic two-parameter algorithms for VCME, focussing on the standard technique of bounded search trees. The results cover the whole range of ratios m/k. This is only a starting point. The aim of this paper is to bring forward the approach, whereas the specific results for VCME deserve refinement and improvement. Section 4 points out some questions for further research. The main suggestion is to study other natural two-parameter problems in this framework.

2 Pareto Complexity of Two-Parameter Problems

For a general introduction to FPT algorithms and parameterized complexity we refer to [6,10]. Now we list a number of definitions and facts for two-parameter problems.

Geometric definitions

A point (x_0, y_0) in the (x, y)-plane is said to be *dominated* by point (x_1, y_1) if $x_0 \le x_1$ and $y_0 \le y_1$. Consider any algorithm for a problem with input parameters k, m, and let B be the set of points (x, y) such that $O^*(x^k y^m)$ is an upper bound for its time complexity. Trivially, if $(x, y) \in B$ then every point that dominates (x, y) is also in B.

For any specific problem instance with parameters k and m we want to minimize the algorithm's time bound, that is, choose some $(x, y) \in B$ that minimizes $x^k y^m$. (As said earlier, we can disregard polynomial factors, at least for large enough instances.) Since $x^k y^m = (x y^{m/k})^k$, all parameter pairs (k, m) with the same ratio m/k have the same optimal points $(x, y) \in B$. (We say "points" because the optimal one is not necessarily unique; see the remark below.) We can characterize the optimal points geometrically. Define $X = \log x$ and $Y = \log y$, with any fixed logarithm base. Points in the $x, y > 1$ quadrant of the (x, y)-plane and in the $X, Y > 0$ quadrant of the (X, Y)-plane correspond to each other in the obvious sense. For convenience we do not distinguish them notationally if the reference is clear from context. In particular, B is also considered a set of points in the positive quadrant of the (X, Y)-plane. Let the X- and Y-axis be directed to the right and upwards, respectively. Since $\log(x^k y^m) = kX + mY$, an optimal point for parameter ratio m/k is any point in the (X, Y)-plane where a sweep line with slope $-k/m$ moving upwards meets B first. This gives rise to the following notions in the (X, Y)-plane which we also call the *logarithmic plane*.

Any non-vertical line splits the (X, Y)-plane obviously in a lower and an upper halfplane. A *lower tangent* to B is a line T with negative slope such that T goes through some point of B, and no point of B is in the lower halfplane of T. The *lower convex hull* of B is the intersection of all upper halfplanes of lower tangents to B. The *Pareto curve* of B is the boundary of the lower convex hull of B. It is easy to see that Pareto curves are graphs of monotone decreasing convex functions (with X and Y as the independent and dependent variable, or vice versa). The *maximum of two or more Pareto curves* is defined as the boundary of the intersection of the corresponding lower convex hulls. Equivalently, if we interpret Pareto curves as functions, we take their argument-wise maximum. The *minimum of two or more Pareto curves* is defined similarly, by first taking the union of the lower convex hulls, but after that we build the lower convex hull again, as the union of convex regions is in general not convex. Finally, if B describes a set of upper bounds of a two-parameter FPT algorithm, we simply speak of the *Pareto curve of the algorithm*, with respect to these known time bounds. Any collection of algorithms for the same two-parameter problem, tuned to several ranges of the ratio m/k, can be merged into a new algorithm whose Pareto curve is the minimum of Pareto curves of the single algorithms. Thus we can formulate our goal as follows: Given a problem, devise an algorithm whose Pareto curve is minimal.

Remark: In can be proved that, in the logarithmic plane, points on the straight line segment between two valid bounds are valid bounds, too. However we do not even need this fact. Points in the interior of a straight line segment L on the Pareto curve are futile in the sense that they are optimal only for the same spefific ratio m/k, but then any of the two endpoints of L can be used instead to express the same time bound, as $kX + mY$ is constant on L.

Recurrences

For a problem with parameters k and m, we call the subproblems with $k = 0$ and $m = 0$, respectively, the *marginal problems*. For example, VCME with $m = 0$ is the VERTEX COVER problem, and $k = 0$ is the trivial problem of checking whether at most m edges are left over. The two-parameter problem cannot be easier than its marginal problems, that is, bases x, y in a bound $O^*(x^k y^m)$ are not smaller than the corresponding bases from the algorithms used for the marginal problems.

We can analyze search tree algorithms for two-parameter problems in a similar way as in the one-parameter case, though with some extra twist. The *branching vector* of a branching rule is now a vector of ordered pairs of numbers indicating the reduction of parameters k, m in the different branches. Let $T(k, m)$ be the maximum size, i.e., number of leaves, of a search tree for a problem instance with parameter values k, m. A branching vector with entries $(k_1, m_1), \ldots, (k_s, m_s)$ yields the recurrence $T(k, m) \leq \sum_{i=1}^{s} T(k - k_i, m - m_i)$, where we set $T(0, 0) := 1$. If k or m is already smaller than some k_i or m_i, respectively, the branching rule is no longer applicable. But then we can, in the remaining instance, immediately reduce this parameter to 0 in all possible ways: Even naive exhaustive search costs only polynomial time, since the k_i and m_i are constant. (For example, in VCME this simply means to search for a vertex cover of some constant size.) Afterwards we are left with the marginal problem for $k = 0$ or $m = 0$.

We get solutions of the form $T(k, m) \leq x^k y^m$ with constant bases $x, y > 1$ from the characteristic equation $x^0 y^0 = \sum_{i=1}^{s} x^{-k_i} y^{-m_i}$, which may be multiplied by $x^{k_i} y^{m_j}$ with the largest k_i and m_j, so that we get rid of negative exponents. Any pair of real numbers $x, y > 1$ satisfying the equation yields an upper bound $x^k y^m$ on $T(k, m)$. Formally this is proved by bottom-up induction in the search tree, however the argument is fairly simple: $x^k y^m$ satisfies the recurrence as long as the branching rule applies, and when k or m is already down to a constant then the search tree for the marginal problem is no larger than our y^m or x^k, respectively, where m or k denotes the residual value at this stage. In conclusion, these solution pairs determine the Pareto curve of the branching rule. Accordingly, the Pareto curve of a search tree algorithm working with several branching rules is the maximum of the Pareto curves of the branching rules involved.

An interesting side question is whether the solutions of recurrences of branching rules always form a convex curve in the logarithmic plane. However, if not, then the Pareto curve is just the lower convex hull of the solution curve, that is, nonconvex parts are replaced with lower tangents.

Subgraphs and branching rules

Given any graph $G = (V, E)$, by a *subgraph* we mean a graph formed by a subset of V and E, not necessarily an induced subgraph. The subgraph *spanned* by a set $F \subseteq E$ of edges is the graph consisting of F and the set $V(F)$ of all vertices incident with F, but without the additional edges that might exist inside $V(F)$. For $X \subseteq V$ we denote by $E(X)$ the set of edges inside X.

An FPT algorithm for VCME (or a similar problem) in the bounded search tree paradigm consists of branching rules, each working on a set F of edges. A branching rules decides in every branch which vertices and edges of the subgraph spanned by F shall be added to the solution. We also say that these vertices and edges are *selected*. One has to prove that every graph contains some of the specified subgraphs, i.e., some of the branching rules is applicable, or that the graph is simple in the sense that the problem can be solved in polynomial time.

We call a branching rule on F *exhaustive* if, for every valid solution S to VCME restricted to the subgraph spanned by F, the rule has a branch that selects only vertices and edges from S. Search tree algorithms with exhaustive branching rules is somehow the simplest type of FPT algorithms. Clearly, exhaustiveness guarantees that a rule cannot miss possibles solutions to VCME on the entire graph G. On the other hand, exhaustiveness is not necessary for correctness. The simplest example is the following degree-1 rule on $F = \{uv\}$: If u is a degree-1 vertex and v its only neighbor, either select vertex u or select edge uv. This rule misses solutions containing only u, but u can always be replaced with v without increasing k or m. The additional information that u has no further neighbors outside $V(F)$ implies that u is never needed in an optimal solution. However, exhaustive rules are essential for *enumeration* problems where we want all solutions rather than some arbitrary one.

3   Vertex Cover with Missed Edges

In this section we give a first Pareto curve for VCME. First of all, we cannot expect bounds where $y = 1$, that is, FPT algorithms with parameter k only. This is because VCME and PARTIAL VERTEX COVER are equivalent as optimization problems, and PARTIAL VERTEX COVER with parameter k is $W[1]$-complete [7]. In view of this fact it is pleasing that we can achieve $O^*(x^k y^m)$ time, with some constant x, for any $y > 1$. In fact, a trivial such algorithm is to branch on single edges: Either select the edge or one of its vertices. The characteristic equation $xy = x + 2y$ yields $x = 2y/(y-1) \approx 2/\delta$ for $y = 1 + \delta$ with an arbitrarily small $\delta > 0$. The more interesting question is: What is the best x for any given (small) δ? The considered case corresponds to large ratios m/k.

First we study what can be done with exhaustive branching rules if m/k is large. We can avoid searching for "exotic" branching rules, as the following observation restricts the possibilities:

Proposition 1. *Consider any search tree algorithm with exhaustive branching rules for VCME, such that for every fixed $y = 1 + \delta$ ($\delta > 0$) the algorithm runs in $O^*(x^k y^m)$ time, with some constant x. Then every rule is, without loss of*

*generality, of the following form: For a set F of f edges, either select all edges
of F, or select any one of the g vertices in $V(F)$. (Any other rules make x
worse.) Moreover, for $y = 1 + \delta$, $\delta \to 0$, base x behaves as $x \approx g/f\delta$. Clearly, if
several, but finitely many branching rules are involved, then we have $x \approx g/f\delta$
for the largest g/f.*

Proof. Consider any branching rule working on a set F of edges. Its characteristic
equation can be written as $x^g y^f = x^g p(y) + q(x, y)$, where f, g are defined as
above, p is a polynomial of degree strictly smaller than f, and q is a polynomial
where x appears only with exponents strictly smaller than g. Since the rule must
not enforce that some of the vertices in $V(F)$ be selected, it must possess at least
one branch where only edges are selected. On the other hand, the rule must not
enforce that only edges be selected, hence it must have at least one branch where
also vertices are selected. It follows that neither p nor q is identical to zero.

   Due to the assumptions, the characteristic equation has solutions with $y =
1 + \delta$ for all (arbitrarily small) $\delta > 0$. For ease of presentation we use in the
following some asymptotic arguments and neglect lower-order terms: For small
$\delta > 0$ we may replace terms y^a with $1 + a\delta$, neglecting higher-order powers of δ.
Then our characteristic equation becomes $x^g(1 + f\delta) \approx x^g p(1 + \delta) + q(x, 1 + \delta)$.
Note that $p(1)$ is the sum of coefficients of p. If $p(1) > 1$ then obviously the right-
hand side is too large, and no solution (x, y) can exist for small $\delta > 0$. Since the
coefficients of p are positive integers, the only remaining possibility is $p(1) = 1$.
Hence $p(y) = y^{f-h} = (1 + \delta)^{f-h}$ for some integer $h > 0$, corresponding to the
unique branch where a set $H \subseteq F$ of h edges is selected. (All other branches
have to select also vertices.) Thus, our characteristic equation asymptotically
simplifies to $x^g h\delta \approx q(x, 1 + \delta)$. Since $q(x, 1) > 0$ for any $x > 0$, and polynomial
q is a continuous function, we can even write $x^g h\delta \approx q(x, 1)$. Since $\lim_{\delta \to 0} x = \infty$,
only the term with highest degree in $q(x, 1)$ determines the asymptotic behaviour
of x as a function of δ. Denote this dominating term by ax^{g-c}, with some integers
$a > 0$ and $c > 0$. Hence $x^g h\delta \approx ax^{g-c}$. This finally allows us to express x
explicitly as $x \approx \sqrt[c]{a/h\delta}$.

   For exhaustive rules we claim that necessarily $c = 1$. To see this, recall that
some branch selects H, and all other branches have to select also vertices. Consider any $v \in V(H)$. Since some solution exists that includes v as the only vertex
from $V(F)$, some branch must select v and no other vertices (but perhaps some
edges along with v), showing that $c = 1$. In particular we get $x \approx a/h\delta$.

   On the other hand, the branching rule where one branch selects H and, for
every $v \in V(H)$, some branch selects only v, is already an exhaustive branching
rule on H. Further branches would only increase a and thus x, and selecting edges
along with a vertex would also increase the right-hand side of the characteristic
equation and thus x. Hence we can restrict the possible branching rules to this
form, and redefine $F := H$, $f := h$, $g := a$. Finally note that $x \approx g/f\delta$.   □

For an algorithm of the form as in Proposition 1 we would like to utilize edge sets F
with g/f as small as possible. One obvious idea is to choose a set F of three edges
incident to one vertex. Since VCME is straightforwardly solvable in polynomial

time if all vertices have degree at most 2, this yields an $O^*((4/3\delta)^k(1+\delta)^m)$ time algorithm. However, we can get considerably better x:

Theorem 1. *For every fixed positive integer c, VCME is solvable in time $O^*(((1+1/c)/\delta)^k(1+\delta)^m)$, for all sufficiently small $\delta > 0$.*

Proof. Fix any positive integer c. Find a cycle C in the input graph, which is possible in polynomial time. If C has at most $c+1$ vertices, let F denote its edge set, and apply the branching rule on F as specified in Proposition 1. Using the earlier denotations we have $g/f = 1$ in this case. If C has at least $c+2$ vertices, let F be the edge set of a subpath of C with $c+1$ vertices and c edges. Again, apply the branching rule on F as specified in Proposition 1, now with $g/f = 1 + 1/c$. As soon as all cycles are destroyed, the remaining graph is a forest, and VCME can be solved straightforwardly in polynomial time, by dynamic programming bottom-up in the trees. (We omit the tedious details.)   □

We remark that search tree algorithms based on exhaustive branching rules and with $x \le 1/\delta$ are unlikely to exist: Since the largest g/f determines the constant factor in $x = \Theta(1/\delta)$, they would need a collection of "dense" subgraphs with $g/f < 1$ such that VCME is polynomial-time solvable on graphs that are free of such subgraphs. However, since $g > f$ if F forms a forest, every subgraph F with $g/f \le 1$ must include a cycle. On the other hand, by subviding the edges of arbitrary graphs we can avoid any fixed-length cycles, but still the graphs remain arbitrarily complicated, so that VCME probably remains NP-complete there.

We needed to fix c in Theorem 1, as we can take the maximal $x \approx g/f\delta$ only from a finite collection of branching rules. However, we may choose the optimal c depending on δ as follows. Inspecting the algorithm from Theorem 1, observe that we used branching rules with $x = by^b/(y^b - 1)$, $b \le c$, and $x = (c + 1)y^c/(y^c - 1)$. We can approximate x by $(1 + b\delta)/\delta \le (1 + c\delta)/\delta$ and $(1 + 1/c)(1 + c\delta)/\delta$, respectively. The bound on x is minimized if $1/c + c\delta$ is minimized, that is, $c \approx 1/\sqrt{\delta}$, rounded to an integer.

Corollary 1. *VCME is solvable in $O^*(((1+2\sqrt{\delta}+\delta)/\delta)^k(1+\delta)^m)$ time, for all sufficiently small $\delta > 0$.*   □

Finally we can choose the optimal δ for the given ratio m/k. Straightforward calculation gives $\delta \sim k/m$ and:

Corollary 2. *VCME is solvable in $O^*((em/k)^k)$ time, where e denotes Euler's number.*   □

Recall that this might be already the best possible bound achievable with exhaustive branching rules of fixed size. Due to Stirling's formula, $\binom{m}{k}$ for large m/k behaves roughly as $(em/k)^k$, hence exhaustive search for the vertices in S on a kernel with fewer than m vertices would yield a faster algorithm in the case of large m/k.

Here we finish our discussion of the case when y is close to 1. For increasing y and decreasing x we can afford some branching on the choice of edges while we must make the choice of vertices more efficient. The best algorithm that we found until now for the range $y \leq 1.6$ is stated in:

Theorem 2. *VCME is solvable in $O^*(x^k y^m)$ time, for all (x, y) that fulfill the equation $x^3 y^4 = 2x^2 y^4 + (x + y)^3$.*

Proof. Let F be a subgraph with vertices r, s, t, u, v and edges rs, st, tu, tv. First we show that every graph contains such a subgraph, or VCME is solvable in polynomial time. In all connected components that are not just paths or cycles, there exists a vertex t with at least three neighbors s, u, v. Assume that, for every such t, none of s, u, v has further neighbors outside $\{t, s, u, v\}$. If t has degree exactly 3, then $\{t, s, u, v\}$ forms already a connected component. Thus consider t with degree 4 or larger. If some edge connects two of t's neighbors, then obviously a (non-induced) subgraph isomorphic to F exists. Otherwise, the connected component of t is merely a star, that is, all edges therein are incident with t. Thus we have shown that "F-free" graphs have only trivial connected components where VCME is easy to solve.

On F we branch as follows. Select s or t or st, and in the last case select, for each of the other three edges independently, either the edge or the other vertex distinct from s, t. The characteristic equation is easy to establish.   □

The table shows some points on the Pareto curve:

y	1.02	1.05	1.10	1.15	1.20	1.30	1.40	1.50	1.60
x	63.98	26.49	14.00	9.84	7.77	5.70	4.67	4.06	3.66

For medium y and x we then detected a better algorithm which does not follow the search tree paradigm:

Theorem 3. *VCME is solvable in $O^*(2^k 2^m)$ time.*

Proof. If a graph $G = (V, E)$ has a solution S, then G also has a vertex cover of size at most $k + m$ (take the k vertices of S and any vertex from each of the m edges in S). We cannot recover a VCME solution S directly from any vertex cover, however we can use it in the following way. Compute once some vertex cover C of size $k + m$, here we may even apply a naive $O^*(2^{k+m})$ time algorithm. Then decide on the set $C \cap S$, which can be done in 2^{k+m} ways. (Actually we need to consider only the subsets of C of size at most k, but this would not improve the bases in the final bound.) For any $C \cap S$, the residual problem is solvable in polynomial time: An edge $e \in E(C)$ must be added to S if and only if e is not indident to $C \cap S$. Let k' and m' be the number of vertices of C and edges of $E(C)$, respectively, selected for S. Observe that, since C is a vertex cover, $V \setminus C$ is an independent set. Hence it remains to select $k - k'$ vertices from $V \setminus C$ that leave at most $m - m'$ edges between C and $V \setminus C$ uncovered. Clearly, it is optimal to choose those $k - k'$ vertices from $V \setminus C$ that have the largest numbers of neighbors in $C \setminus S$, ties are broken arbitrarily.   □

For larger y it becomes advantageous to use fixed-size branching rules again but slide to "more vertex-efficient" branching, compared to Theorem 2. We found that the following rule is superior to Theorem 3 from ca. $y = 5.0$ on.

Theorem 4. *VCME is solvable in $O^*(x^k y^m)$ time, for all (x, y) that fulfill the equation $x^3 y^3 = x^2 y^3 + (x + y)^3$.*

Proof. Let F be a set of three edges incident to some vertex t. Graphs without such subgraphs are trivial. If F exists, then either we select t, or we select, for each of the other three edges independently, either the edge or the other vertex distinct from t.   □

For example, we obain (x, y) pairs $(5.00, 1.79)$ and $(6.00, 1.73)$. Note that the rule in Theorem 4 converges for growing y to the simple $x \approx 1.47$ rule for VERTEX COVER with characteristic equation $x^3 = x^2 + 1$: take a vertex of degree at least 3 or all its neighbors. In particular, x is limited from below by 1.47, for any y.

It arises the question how small we can make x for instances with small m/k. Since VCME is a proper generalization of VERTEX COVER, a smooth transition would be desirable: Given a known $O^*(b^k)$ time algorithm for VERTEX COVER, we would like to have an $O^*(x^k y^m)$ algorithm for VCME where x tends to b if m/k goes to 0. We do not have a method that takes an arbitrary VERTEX COVER algorithm as a black box and generates such an algorithm for VCME. We state the existence of such a method as an open problem. However, for search tree algorithms that use only exhaustive (and also certain non-exhaustive) branching rules we can achieve the desired behaviour in a generic way, as we outline below.

First, we can transform every exhaustive branching rule for VERTEX COVER working on an edge set F into a branching rule for VCME: Every branch of the original rule is preserved, i.e., the specified vertices are selected, and for every edge in F we add a branch where only this edge is selected. Correctness is evident. For some non-exhaustive rules we can proceed in the same way, but correctness must be proved separately. For example, consider again the degree-1 rule: If u is a degree-1 vertex and v its only neighbor, we can erase u and put v in the vertex cover. The only role of u in a vertex cover would be to cover the edge uv, hence u can always be replaced with v. The corresponding rule for VCME selects either v or uv. This is still correct, as we discussed earlier.

For the analysis, consider any branching rule in the given VERTEX COVER algorithm, working on a set F of f edges. Let $x^d = p(x)$ be its characteristic equation, where p is a polynomial of degree smaller than d, with nonnegative coefficients and, in particular, with a positive coefficient of x^0. Let b be the branching number of the rule, that is, $b^d = p(b)$. In the following we use some known basic properties of such equations (we refer to, e.g., [4]): $x^d = p(x)$ has a unique positive root b, and it holds $x^d < p(x)$ for $x < b$, and $x^d > p(x)$ for $x > b$. It also follows that $db^{d-1} - p'(b) > 0$, since if $x^d - p(x)$ had derivative zero at $x = b$ then b would be a root with multiplicity 2 or higher, and a slight perturbation of the coefficients would yield another polynomial of the considered type with more than one positive root, which contradicts the mentioned properties. The branching rule for VCME obtained by the scheme above has the characteristic equation $x^d y = p(x)y + fx^d$. Since $fx^d > 0$, any solution (x, y) must satisfy $x > b$, thus $x^d > p(x)$. It follows $y = fx^d/(x^d - p(x))$. For $x = b + \delta$ with small δ we obviously get $y \approx fb^d/(db^{d-1} - p'(b))\delta$. That is, we can bring x as close to b as desired, still keeping a finite base y for each δ, although with $y \to \infty$ for $x \to b$. This reasoning holds in particular for the branching rule with the largest b that determined the $O^*(b^k)$ complexity of the VERTEX COVER algorithm we started from. Since y goes to infinity, for all x sufficiently close to b the Pareto curves of all other rules are below the Pareto curve of this rule with maximum b. Hence this rule also dominates the complexity of the VCME algorithm, for δ sufficiently small.

Now we apply this machinery to the VERTEX COVER algorithm of [1]. It has been subsequently improved several times, but its base 1.325 is already fairly close to the currently best branching number 1.2738 for VERTEX COVER [5]. Moreover, we can easily apply our transformation and analysis.

Theorem 5. *VCME is solvable in $O^*((1.325 + \delta)^k (3.54/\delta)^m)$ time, for all sufficiently small $\delta > 0$.*

Proof. We only sketch the proof and refer to [1] for details of the rules. The VERTEX COVER algorithm consists of twelve branching rules. One can easily check them one by one and verify that they are of the form described above and thus extendible to branching rules for VCME. As stated in [1], Rule 3 and 7 have the maximum branching number $b < 1.325$ in this algorithm. Rule 3 works on a subgraph with $f \le 6$ edges (while rule 7 involves only 5 edges), which yields the characteristic equation $x^3 y = xy + y + 6x^3$ with the claimed solution.   □

For more sophisticated algorithms using other techniques one has to check whether our assumptions on branching rules are still valid, or weaken the assumptions and enrich the rule transformation method. However, we doubt that the use of more complicated VERTEX COVER algorithms with slighlty improved b is of much practical value also for VCME: If x approaches b, then y goes up quickly, so that we need to relax the base x anyway. That is, the smallest possible b and x does not seem to be the main concern in the VCME context. For very small ratios m/k we would in practice rather select m edges exhaustively on a kernel, and then simply apply any VERTEX COVER algorithm to the remaining instances. This type of algorithm does not enjoy constant bases y but is expected to be simpler and faster for very small m/k.

Related to this discussion, we finally give a simple quadratic kernel for VCME by straightforward generalization of the $O(k^2)$ kernelization for VERTEX COVER. Due to kernel results for VERTEX COVER we conjecture that VCME has actually an $O(k + m)$ size kernel.

Theorem 6. *A kernel for VCME with $k^2 + km + m$ edges can be computed in polynomial time.*

Proof. Any vertex of degree larger than $k + m$ must be selected, since otherwise we have to select more than k other vertices or more than m edges. After removal of the enforced vertices and all incident edges, there remains a graph of maximum degree $k + m$. Now k vertices can cover at most $k(k + m)$ edges, hence at most $k(k + m) + m$ edges remained, or there is no solution. $\square$

4   Conclusions

We proposed the framework of Pareto complexity for FPT problems with two parameters k, m, where we want time complexities $O^*(x^k y^m)$ with constant bases x, y. As an illustration and starting point we gave some basic algorithms for VCME, the problem of finding vertex covers with k vertices and m missed edges. The algorithms are tailored to several ranges of m/k. We have not studied minor technical issues like convexity of their Pareto curves in the logarithmic plane (see Section 2). There are some natural questions for further research, besides the ones already brought up in the technical sections:

- Get improvements: Find VCME algorithms with lower Pareto curves. Better branching rules for some ranges of m/k are likely to exist.
- Despite our discussion of small m/k: Give algorithms for VCME with $x < 1.325$, possibly with x matching the best known base for VERTEX COVER algorithms.
- Study the Pareto complexity of the corresponding enumeration and counting problems: How many solutions with k vertices and m edges exist, etc.?
- Extend the approach to hitting sets in hypergraphs of fixed rank $3, 4, 5 \ldots$
- Study other problems. For instance, some variants of the *Cluster Editing* and *Cluster Vertex Deletion* problem naturally have two parameters: (a) number of edge deletions and edge insertions, (b) number of vertex deletions and edge edits, (c) number of vertex deletions and resulting cliques. (The latter problem was studied in [8], however not in the "Pareto framework".)

Finally, the principal discussion whether $O^*(x^k y^m)$ is the "right" type of time bound may be continued.

Acknowledgments

The work is supported by the Swedish Research Council (Vetenskapsrådet), grant 2007-6437, "Combinatorial inference algorithms – parameterization and clustering". The author thanks the anonymous referees for the points raised in their reports on the first version, which also inspired some substantial improvements.

References

1. Balasubramanian, R., Fellows, M.R., Raman, V.: An Improved Fixed-Parameter Algorithm for Vertex Cover. Info. Proc. Letters 65, 163–168 (1998)
2. Bar-Yehuda, R., Hermelin, D., Rawitz, D.: An Extension of the Nemhauser-Trotter Theorem to Generalized Vertex Cover with Applications. In: WAOA 2009. LNCS, vol. 5893. Springer, Heidelberg (to appear, 2010)

3. Böcker, S., Briesemeister, S., Bui, Q.B.A., Truß, A.: Going Weighted: Parameterized Algorithms for Cluster Editing. In: Yang, B., Du, D.-Z., Wang, C.A. (eds.) COCOA 2008. LNCS, vol. 5165, pp. 1–12. Springer, Heidelberg (2008)
4. Chen, J., Kanj, I.A., Xia, G.: A Note on Search Trees. Tecnical Report TR05-006.pdf (2005), facweb.cs.depaul.edu/research/TechReports/
5. Chen, J., Kanj, I.A., Xia, G.: Simplicity is Beauty: Improved Upper Bounds for Vertex Cover. Technical Report TR05-008.pdf (2005),
   cdm.depaul.edu/research/Documents/TechnicalReports/2005/
6. Downey, R.G., Fellows, M.R.: Parameterized Complexity. Springer, Heidelberg (1999)
7. Guo, J., Niedermeier, R., Wernicke, S.: Parameterized Complexity of Generalized Vertex Cover Problems. In: Dehne, F., López-Ortiz, A., Sack, J.-R. (eds.) WADS 2005. LNCS, vol. 3608, pp. 36–48. Springer, Heidelberg (2005)
8. Hüffner, F., Komusiewicz, C., Moser, H., Niedermeier, R.: Fixed-Parameter Algorithms for Cluster Vertex Deletion. In: Laber, E.S., Bornstein, C., Nogueira, L.T., Faria, L. (eds.) LATIN 2008. LNCS, vol. 4957, pp. 711–722. Springer, Heidelberg (2008)
9. Kneis, J., Langer, A., Rossmanith, P.: Improved Upper Bounds for Partial Vertex Cover. In: Broersma, H., Erlebach, T., Friedetzky, T., Paulusma, D. (eds.) WG 2008. LNCS, vol. 5344, pp. 240–251. Springer, Heidelberg (2008)
10. Niedermeier, R.: Invitation to Fixed-Parameter Algorithms. Oxford Lecture Series in Math. and its Appl. Oxford Univ. Press, Oxford (2006)

What Makes Equitable Connected Partition Easy

Rosa Enciso[1], Michael R. Fellows[2], Jiong Guo[3], Iyad Kanj[4],
Frances Rosamond[2], and Ondřej Suchý[5,*]

[1] School of Electrical Engineering and Computer Science
University of Central Florida, Orlando, FL
renciso@cs.ucf.edu
[2] University of Newcastle, Newcastle, Australia
{michael.fellows,frances.rosamond}newcastle.edu.au
[3] Institut für Informatik, Friedrich-Schiller-Universität Jena,
Ernst-Abbe-Platz 2, D-07743 Jena, Germany
jiong.guo@uni-jena.de
[4] School of Computing, DePaul University
243 S. Wabash Ave, Chicago, IL 60604
ikanj@cs.depaul.edu
[5] Department of Applied Mathematics and Institute for Theoretical Computer
Science, Charles University,
Malostranské nám. 25, 118 00 Praha, Czech Republic
suchy@kam.mff.cuni.cz

Abstract. We study the EQUITABLE CONNECTED PARTITION problem:
partitioning the vertices of a graph into a specified number of classes,
such that each class of the partition induces a connected subgraph, so
that the classes have cardinalities that differ by at most one. We examine the problem from the parameterized complexity perspective with respect to various (aggregate) parameterizations involving such secondary
measurements as: (1) the number of partition classes, (2) the treewidth,
(3) the pathwidth, (4) the minimum size of a feedback vertex set, (5)
the minimum size of a vertex cover, (6) and the maximum number of
leaves in a spanning tree of the graph. In particular, we show that the
problem is W[1]-hard with respect to the first four combined, while it
is fixed-parameter tractable with respect to each of the last two alone.
The hardness result holds even for planar graphs. The problem is in XP
when parameterized by treewidth, by standard dynamic programming
techniques. Furthermore, we show that the closely related problem of
EQUITABLE COLORING (equitably partitioning the vertices into a specified number of independent sets) is FPT parameterized by the maximum
number of leaves in a spanning tree of the graph.

* Work partially supported by the ERASMUS program and by the DFG, project
NI 369/4 (PIAF) while visiting Friedrich-Schiller-Universität Jena (October 2008–
March 2009), by grant 201/05/H014 of the Czech Science Foundation and by grant
1M0021620808 of the Czech Ministry of Education.

J. Chen and F.V. Fomin (Eds.): IWPEC 2009, LNCS 5917, pp. 122–133, 2009.

1 Introduction and Preliminaries

Let $G = (V, E)$ be an undirected graph. We say that $V_1, V_2, \ldots, V_r$ is *a partition of V* if and only if $\bigcup_{i=1}^{r} V_i = V$, and $\forall i, j, 1 \leq i < j \leq r : V_i \cap V_j = \emptyset$. A partition is *equitable* if $\forall i, j, 1 \leq i < j \leq r : ||V_i| - |V_j|| \leq 1$. In this paper we consider the following two NP-hard problems:

EQUITABLE CONNECTED PARTITION (ECP)
Instance: A simple graph $G = (V, E)$, and a positive integer $r \in \mathbb{N}$.
Question: Is there an equitable partition of V into r classes $V_1, V_2, \ldots, V_r$, such that each class of the partition induces a connected subgraph?

We consider parameterizations based on various combinations of:

- the treewidth of the input graph $tw(G)$,
- the pathwidth of the input graph $pw(G)$,
- the minimum size of a feedback vertex set in the input graph $fvs(G)$,
- the minimum size of a vertex cover in the input graph $vc(G)$,
- the maximum number of leaves in a spanning tree of the input graph $ml(G)$, and
- the number of partitions r

We show that ECP is W[1]-hard when parameterized by $pw(G)$, $fvs(G)$, and the number of partition classes r combined. We show that this result holds true even for planar graphs. On the positive side, we show that ECP becomes *fixed-parameter tractable* (FPT) when parameterized by $vc(G)$, or by $ml(G)$. We also show that ECP, parameterized by treewidth, is in XP.

EQUITABLE COLORING
Instance: A simple graph $G = (V, E)$, the number of partitions $r \in \mathbb{N}$.
Question: Is there an equitable partition $V_1, V_2, \ldots, V_r$ of the vertex set V such that each partition induces an independent set?

For the EQUITABLE COLORING problem, we show that the problem is FPT when parameterized by $ml(G)$.

The first problem we study (ECP) arises in computational social choice in the subject of *redistricting* [1]. It is known to be NP-complete, even for planar graphs or for fixed $r \geq 2$ [4,10]. Similar problems were studied for example by Ito et al. in [11,12].

It was shown in [6] that EC is W[1]-hard when parameterized by $tw(G)$ and r combined. More recently, it was shown in [8] that EC is FPT when parameterized by $vc(G)$.

For the background and terminologies on graphs, we refer the reader to West [13], and for that on parameterized complexity, we refer the reader to Downey and Fellows' book [3].

In the rest of the paper, we will denote by $l := (n \mod r)$ the number of partition classes whose size is larger by one than the size of the other classes, i.e., we have $r - l$ classes of size $s := \lfloor n/r \rfloor$ and l classes of size $s + 1$.

2 Hardness Results

This section is mainly devoted to proving the following theorem:

Theorem 1. EQUITABLE CONNECTED PARTITION *is W[1]-hard with respect to the pathwidth* $pw(G)$, *the minimum size of a feedback vertex set* $fvs(G)$ *and the number of partition classes* r *combined.*

Proof. We will provide a parameterized reduction from the $W[1]$-complete MULTI-COLORED CLIQUE (MCC) problem [7]. In MCC we are given an undirected graph that is properly colored by k colors and the question is whether there is a size-k clique in this graph consisting of exactly one vertex from each color class. The parameter is k.

A basic building block of our construction is an *anchor*. It is a vertex (the *root* of the anchor) with many neighbors of degree one (see Fig. 1). As the prescribed class size will be much greater than one, the pendant degree one vertices must belong to the same partition class as the root of the anchor. The number of degree one vertices of an anchor will be chosen so that two anchors cannot belong to the same class. We create exactly as many anchors as the specified number of classes r of the equitable partition; thus there must be exactly one anchor in each class. The situation can be viewed intuitively as if each class is "started" with one single anchor and then some more vertices are later to be added. The number of vertices that need to be added to the class started by a particular anchor will differ for different anchors and is forced by the number of pendant vertices that the anchor is "missing" relative to the prescribed class size. Anchors will be denoted by uppercase letters and by connecting something to an anchor we mean connecting it to the root of the anchor.

We interconnect the anchors using a building block gadget called a *choice*. If $A = \{a_1, \ldots, a_t\}, 0 \le a_1 < a_2 < a_3 < \ldots < a_t$ is a set of integers and $b \ge a_t$, then an (A, b)-*choice* is a path with $t+1$ vertices $v_1, \ldots, v_{t+1}$, where each vertex of the path can have some degree-one vertices pendant on it (see Fig. 1). In particular, vertex v_1 has a_1 degree one vertices pendant on it, vertex v_{t+1} has $b - a_t$ pendant vertices, and for all $i, 2 \le i \le t$, vertex v_i has $a_i - a_{i-1} - 1$ pendant

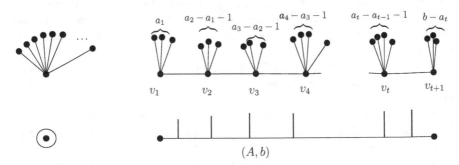

Fig. 1. Basic building blocks of our construction: Anchor (left), (A, b)-choice (right) and the way they are depicted in further figures (bellow)

vertices. Observe again that the pendant vertices must always fall into the same class as their unique neighbor.

Now if an anchor X is connected to an anchor Y by an (A, b)-choice (which is done by simply identifying vertex v_1 and v_{t+1} with the root of anchor X and Y, respectively), then there is an i, with $1 \leq i \leq t$, such that the vertices $v_1, \ldots, v_i$ (and their pendant vertices) fall into the partition class of X, while the vertices $v_{i+1}, \ldots, v_{t+1}$ fall into the class of Y. The vertices v_1 and v_{t+1} are identified with the respective anchors, and hence we do not count them. Thus the number of vertices from this choice that fall into the class of X is $a_1 + \sum_{j=2}^{i}((a_j - a_{j-1} - 1) + 1) = a_i \in A$, while the number of vertices falling into the class of Y is $b - a_t + \sum_{j=i+1}^{t}((a_j - a_{j-1} - 1) + 1) = b - a_i$. Note that the vertices pendant on v_1 and v_{t+1} are in fact pendant on the roots of the appropriate anchors, but we still consider them as a part of this choice.

The construction is based on sending "signals" between anchors. Let A and B be two anchors connected by a choice. The vertices of the choice have to fall into the two partition classes corresponding to A and B. The more vertices that fall into the class of A the less vertices that will fall into the one of B. However, since the sizes of the two classes differ by at most one, the class of B must include vertices from somewhere else in the construction. This generates the signal. The choices allow us to control the signal sent.

Now suppose that $G = (V, E), k, c : V \rightarrow \{1, \ldots, k\}$ is an instance of MCC. Also suppose that, for each i, there are n_i vertices of color i denoted by $v_p^i, 1 \leq p \leq n_i$. Furthermore, we give each edge an integer ID, i.e., there is a bijective labeling $l : E \rightarrow \{1, \ldots, |E|\}$. Our construction has a selection gadget for each vertex color, which ensures both the selection of the vertex of this color and the selection of edges from the selected vertex to the vertices of the other colors. This is sometimes called an *edge representation strategy* for the reduction gadgeteering. The selection gadgets are interconnected in a way that an equitable partition is only possible if the IDs of the "selected" edges match.

The selection gadget for color i is formed by $2(k - 1)$ anchors $N_j^i, P_j^i, 1 \leq j \leq k, j \neq i$, connected into a cycle, each having a connection outside the gadget. The vertex selection is represented by a "big" signal that the outgoing connections are unable to handle, and hence is forced to run along the cycle without a change. The selection of an edge going from the selected vertex to the vertices of color j is then done between N_j^i and P_j^i via a "small" signal that equals the label of the selected edge. This signal is then sent to the anchors N_i^j, P_i^j of the selection gadget for color j, from anchor P_j^i to N_i^j and from N_j^i to P_i^j (in opposite directions).

Now we present the selection gadget more formally. First let $Z_0 := 2|E| + 10$. The "big" signal is formed by the order of the vertex selected times the number Z_0, i.e., the possible signal states are $A_0^i := \{p \cdot Z_0 \mid 1 \leq p \leq n_i\}$. As we mentioned, the small signal is formed by the edge IDs. Between the anchors N_j^i and P_j^i both the big and the small signal is sent, and a particular small signal can only be used with an appropriate big signal. Thus, between N_j^i and P_j^i, the possible signal states are $A_j^i := \{p \cdot Z_0 + l(uv_p^i) \mid c(u) = j \text{ and } uv_p^i \in E\}$. To

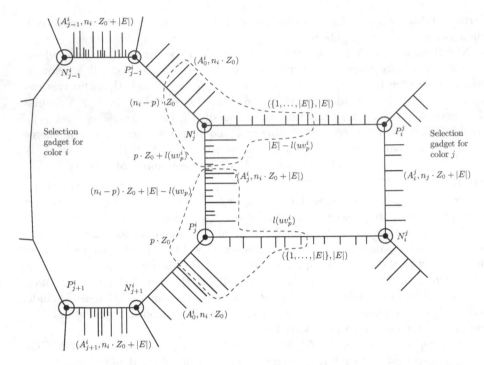

Fig. 2. A part of the selection gadget with possible partition shown by dashed line

catch the order of the anchors along the cycle, we introduce the notion of a *successor*. For each $j, 1 \leq j \leq k$ set $succ(j) := j + 1$ for $j \neq k$ and $j \neq (i-1)$, set $succ(k) := 1$ and $succ(i-1) := succ(i)$. Now for each $j, 1 \leq j \leq k, j \neq i$, the anchor P_j^i is connected to the anchor $N_{succ(j)}^i$ by an $(A_0^i, n_i \cdot Z_0)$-choice, the anchor N_j^i is connected to the anchor P_j^i by an $(A_j^i, n_i \cdot Z_0 + |E|)$-choice, the anchor P_i^i to the anchor N_i^j, and the anchor P_i^j to the anchor N_j^i by two $(\{1, \ldots, |E|\}, |E|)$-choices (see Fig. 2).

Now the class sizes are set such that each of the anchors in this selection gadget needs to get $n_i \cdot Z_0 + |E|$ vertices from the choices that it is incident with. Hence, if the anchor P_j^i takes $p \cdot Z_0$ vertices from the choice connecting it to the anchor $N_{succ(j)}^i$, then the anchor $N_{succ(j)}^i$ gets $(n_i - p) \cdot Z_0$ vertices from this choice, and it must get $p \cdot Z_0 + |E|$ vertices from the two remaining choices that it is incident with. Since it can take at most $|E|$ vertices from the connection to the selection gadget for color $succ(j)$, it must take at least $p \cdot Z_0$ (but at most $p \cdot Z_0 + |E| < (p+1) \cdot Z_0$) vertices from the connection to $P_{succ(j)}^i$. Hence, it has to take $p \cdot Z_0 + l(uv_p^i)$ vertices for some $c(u) = succ(j)$ and $uv_p^i \in E$ from this connection and $|E| - l(uv_p^i)$ vertices from the connection to the other selection gadget. Hence $P_{succ(j)}^i$ gets $(n_i - p) \cdot Z_0 + |E| - l(uv_p)$ vertices from the connection from $N_{succ(j)}^i$, and by a similar reasoning it is forced to take $p \cdot Z_0$ vertices from the connection to $N_{succ(succ(j))}^i$ and $l(uv_p)$ vertices from the connection to the other selection gadget. Thus the anchor $N^i_{succ(succ(j))}$ is again forced to select some edge incident with vertex v^i_p, etc.

The anchor N^i_j is connected to anchor P^j_i by a $(\{1, \ldots, |E|\}, |E|)$-choice, and the number of vertices it takes into its class out of this choice is $|E| - l(uv^i_p)$, where v^i_p is the vertex selected in the selection gadget for color i, $c(u) = j$ and $uv^i_p \in E$. The number of vertices that the anchor P^j_i takes out of this choice is $l(wv^j_q)$ where v^j_q is the vertex selected in the selection gadget for color j, $c(w) = i$ and $wv^j_q \in E$. Since the $|E|$ vertices of the choice must be partitioned into the classes of its endpoints, it follows that $l(wv^j_q) = l(uv^i_p)$ and $wv^j_q = uv^i_p = v^j_q v^i_p$ is an edge of G. Hence a solution for the constructed graph is possible if and only if the selected vertices form a multi-colored clique in the graph G.

Now to determine the right size of the anchors, it is enough to ensure that each class is more than half full, once the starting anchor is added. The maximum demand (the number of vertices that should be added to its class except for itself and vertices pendant on it) of any anchor is less than $n \cdot Z_0 + |E|$, hence it is enough to set the desired class size to $s := (2n + 1) \cdot Z_0 = (2n + 1) \cdot (2|E| + 10)$.

The number of partition classes r is equal to the number of anchors. Since there are k selection gadgets, each containing $2(k-1)$ anchors, we have $2k(k-1)$ anchors in total. Now observe that if we delete all roots of the anchors, the resulting graph consists of paths with pendant vertices. Hence, the roots form a feedback vertex set in the graph, and the pathwidth of the graph is also bounded by the number of roots plus one. The construction can be clearly carried out in polynomial time; in particular, the graph has $r \cdot s = O(k^2 \cdot n^3)$ vertices.   □

Corollary 1. EQUITABLE CONNECTED PARTITION *is W[1]-hard for planar graphs with respect to the pathwidth* $pw(G)$, *the minimum size of a feedback vertex set* $fvs(G)$ *and the number of partition classes r combined.*

Proof. The graph H' constructed in Theorem 1 is in general not planar. But there is a drawing of this graph such that only the edges of the choices connecting two different selection gadgets cross. Moreover, we can assume that each pair of them crosses at most once, and only in the edges of their paths, not in the edges connecting the pendant vertices. We replace each such crossing one by one by a planar *crossing gadget*, such that the resulting planar graph H has a solution if and only if the graph H' does.

Suppose that in our drawing of H' the $(\{1, \ldots, |E|\}, |E|)$-choices between anchors A and B and between C and D cross. The crossing gadget is formed by four anchors R, S, X, Y such that the anchor R is connected to A, X to C, S to B, Y to D and X to Y by a $(\{1, \ldots, |E|\}, |E|)$-choice, respectively. X is connected to both R and S by $(\{z \cdot (Z_0 + 1) \mid 1 \le z \le |E|\}, |E| \cdot (Z_0 + 1))$-choices and Y is connected to both R and S by $(\{z \cdot Z_0 \mid 1 \le z \le |E|\}, |E| \cdot Z_0)$-choices (see Fig. 3). The anchors R, S and Y need $|E| \cdot (Z_0 + 1)$ vertices to be added to their respective classes, while X needs $|E| \cdot (Z_0 + 2)$ vertices.

If X gets $p \cdot (Z_0 + 1)$ vertices from the choice connecting it to R, then it can get between 2 and $2 \cdot |E|$ vertices from the connections to Y and C, and hence it must take between $|E| \cdot (Z_0 + 2) - p \cdot (Z_0 + 1) - 2 < (|E| - p + 1) \cdot (Z_0 + 1)$

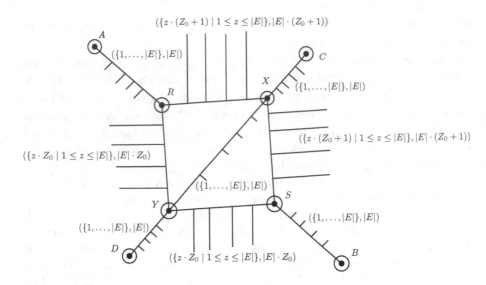

Fig. 3. The crossing gadget

and $|E| \cdot (Z_0 + 2) - p \cdot (Z_0 + 1) - 2 \cdot |E| > (|E| - p - 1) \cdot (Z_0 + 1)$ and thus $(|E| - p) \cdot (Z_0 + 1)$ vertices out of the choice connecting it to S. The anchor Y works in a similar way. Thus, if C takes q vertices from the connection to X, then X gets $|E| - q$ vertices, and takes q vertices from the connection to Y; Y does similarly, and thus D gets $|E| - q$ vertices, as if it were connected to C directly.

Counting the number of vertices inside the crossing gadget and the demands of the anchors, it follows that the number of vertices A get and the number of vertices B get must also sum up to $|E|$ as if they were connected directly.

It is easy to check, that the resulting graph H is planar, with $pw(G)$ a $fvs(H)$ bounded in terms of k. Moreover it can be constructed in polynomial time and has a solution if and only if the graph H' constructed in Theorem 1 has. We defer further details to the full version of the paper. □

Since the treewidth of a graph is never greater than its pathwidth, we immediately get also the following corollary:

Corollary 2. EQUITABLE CONNECTED PARTITION *is W[1]-hard for planar graphs with respect to the treewidth* $tw(G)$.

3   Algorithmic Results

Before we give the FPT results, we note that the EQUITABLE CONNECTED PARTITION problem is in XP, parameterized by the treewidth. This can be proved using standard techniques for problems on graphs of bounded treewidth. We defer the proof to the full version of the paper.

Theorem 2. EQUITABLE CONNECTED PARTITION *is in XP with respect to the treewidth* $tw(G)$.

Now we present two FPT results for ECP:

Theorem 3. EQUITABLE CONNECTED PARTITION *is in FPT with respect to the minimum size of a vertex cover* $vc(G)$.

Proof. Assume that we are given a vertex cover $C \subseteq V$ of size $c := vc(G)$. If not, we can compute it in time $O(1.2738^c + cn)$ by [2]. Each class that contains at least 2 vertices must contain some vertex of C; otherwise, it would not be connected. Hence, if the minimum size of each class s is at least 2, then either $r \leq c$ or (G, r) is a no-instance. If $s = 1$, then we have l classes of size 2 and $r - l$ of size 1. Since a size-2 class contains two vertices connected by an edge, such a partition is in fact a matching of size l in G. The existence of such a matching can be decided in polynomial time. The case of $s = 0$ is trivial and yields a yes-instance. Hence, in what follows we can assume that $s \geq 2$ and $r \leq c$.

We search for an equitable partition such that the first l classes are the larger ones and the last $r - l$ are the smaller ones. We start by trying all the possibilities of partitioning the vertices of C into r (not necessarily connected) non-empty classes $V_1^C, \ldots, V_r^C$. For each such partition, and each disconnected class V_i^C, we try the possibilities of adding at most $|V_i^C| - 1$ vertices of $V \setminus \bigcup_{i=1}^r V_i^C$ into V_i^C to make it connected. But we do not try all the vertices. Instead, each vertex tried must have a different neighborhood. We try all such possibilities with different neighborhoods. It remains to distribute the remaining vertices among the classes so that the partition becomes equitable. We construct a network such that there is a flow of certain size in it if and only if the vertices can be distributed among the classes.

Let us denote by $D := V \setminus \bigcup_{i=1}^r V_i^C$ the set of vertices that are not used yet. The network consists of three intermediate layers, in addition to the source z and the target t. There are r vertices in the first layer, denoted $a_1, \ldots, a_r$. Vertex a_i is connected to the source z by an arc of capacity equal to the number of vertices that should still be added to class i, i.e., $s - |V_i^C|$ if $i > l$ and $s + 1 - |V_i^C|$ if $i \leq l$. The second layer is formed by the vertices of C, and for each i, there are arcs of capacity ∞ from a_i to each vertex in $V_i^C \cap C$. The third layer is formed by vertices b_J, $J \subseteq C$, and there is an arc between $v \in C$ and b_J if and only if $v \in J$. Such arcs have also infinite capacity. Finally each b_J is connected to t by an arc with capacity equal to the number of vertices in D with neighborhood J.

The flow on arcs between the vertices of C and the vertices of type b_J directly shows how many vertices of the particular type should be put into the same class as a vertex of C. Hence it is easy to see that there is a flow of size $|D|$ in the constructed network if and only if the vertices can be distributed among the classes. Concerning the running time of the algorithm, there are at most r^c different colorings of C; for each of them we try adding at most $c - 1$ vertices to the classes, each of at most 2^c types. Hence there are at most $O(2^{c^2})$ possibilities to do so. The network can be constructed in time linear in the number of edges of the original graph and the number of vertex types. The flow can be found in the time cubic in the number of vertices of the network, i.e., $O((2^c + 2c + 2)^3)$. Hence the overal running time of the algorithm is bounded by $O(2^{c^2 + c \cdot \log c + 4c} \cdot n^2)$.   □

Theorem 4. EQUITABLE CONNECTED PARTITION *(ECP) is in FPT with respect to the maximum number of leaves in a spanning tree $ml(G)$.*

Proof. We construct an instance of Integer Linear Programing (ILP) whose number of variables is a function of $ml(g)$ for ECP. It is known that G is a subdivision of some graph H on (at most) $4k$ vertices, for $ml(G) = k$ [5]. Such a graph H can be easily found in linear time. We say that a class is *simple* if it contains no vertex of H. For an edge uv of H we use P_{uv} to denote the unique path in G having as endpoints u and v, and whose internal vertices are in $G \setminus H$; let $|P_{uv}|$ denote the number of its internal vertices.

We first branch on all possibilities of partitioning the vertices of H into at most $4k$ classes. Note that quite possibly $4k < r$. We construct an ILP instance as follows. On a given branch, assume that the classes that are assigned vertices of H, according the the partition of the branch, are $V_1, \ldots, V_p$. Recall that $s := \lfloor n/r \rfloor$. If a path P_{uv} has at least s internal vertices, then it cannot be fully contained in one class. Hence, it must be split into several parts. The first part is put into the same class as the vertex u, the last part into the same class as v and the rest is divided into several simple classes. Since the order of the simple classes on the path does not matter, we only have to know the number of classes having size s, and the number of classes having size $s + 1$. Hence, for such an edge uv of H, we introduce four variables: $t_{u,uv}$ and $t_{v,uv}$ representing the number of internal vertices of the path to be placed in the same class as u and v, respectively, and a_{uv} and b_{uv} representing the number of simple classes of size s and $s + 1$ on P_{uv}, respectively.

If a path P_{uv} contains less than s internal vertices, then there is no simple class on this path, and each vertex of the path is in the same class as one of the endpoints. In particular, if u and v are in the same class, then the whole path P_{uv} is in that class. If u and v are in different classes, then we introduce two variables $t_{u,uv}$ and $t_{v,uv}$ for that path, with the same meaning as in the previous case. To simplify the equations, we denote by E_1 the set $\{uv \in E(H) \mid |P_{uv}| < s$ and u and v lie in different classes$\}$, E_2 the set $\{uv \in E(H) \mid |P_{uv}| \geq s\}$, $E_3^i := \{uv \in E(H) \mid |P_{uv}| < s$ and $u, v \in V_i\}$ and $E_3 := \bigcup_{i=1}^{p} E_3^i$. The class V_i is connected if and only if the graph (V_i, E_3^i) is connected. We check this before we call the procedure to solve the ILP. Finally, we introduce a $\{0, 1\}$-variable c_i, for each $1 \leq i \leq p$, so that the size of V_i in the final partition will be $s + c_i$. The variables introduced are subject to following constraints:

$$\forall uv \in E_1 \cup E_2 : 0 \leq t_{u,uv}, t_{v,uv},$$
$$\forall uv \in E_2 : 0 \leq a_{uv}, b_{uv},$$
$$\forall i, 1 \leq i \leq p : 0 \leq c_i \leq 1,$$

The above formalize obvious matters pertaining to our approach. The following constraints do the main work:

$$\forall uv \in E_1 : t_{u,uv} + t_{v,uv} = |P_{uv}|, \tag{1}$$

$$\forall uv \in E_2 : t_{u,uv} + t_{v,uv} + s \cdot a_{uv} + (s+1) \cdot b_{uv} = |P_{uv}|, \tag{2}$$

$$\forall i, 1 \le i \le p : \sum_{v \in V(H) \cap V_i} (1 + \sum_{uv \in E_1 \cup E_2} t_{v,uv}) + \sum_{uv \in E_3^i} |P_{uv}| = s + c_i, \tag{3}$$

$$\sum_{uv \in E_2} a_{uv} + \sum_{i=1}^{p} (1 - c_i) = r - l, \tag{4}$$

$$\sum_{uv \in E_2} b_{uv} + \sum_{i=1}^{p} c_i = l. \tag{5}$$

Equation 1 ensures that the paths corresponding to the edges in E_1 are correctly divided. Equation 2 ensures the same for the edges in E_2. Equation 3 ensures that the classes containing some vertices of H have the right size, and the last two equations (4 and 5) ensure that there are the right numbers of large and small classes. It is easy to see that there is a solution to this ILP instance if and only if there is an equitable connected partition of the vertices of G with r classes, that extends the initial assignments made according to the branch (partition of $V(H)$) being explored. Since there are at most $4 \cdot \binom{4k}{2} + 4k \le 32k^2$ variables, each of them used at most three times, the overall size of the instance is bounded by $O(k^2)$ and it can be solved in $O((32k^2)^{2.5 \cdot 32k^2 + o(k^2)} \cdot k^2)$ time [9], which yields a running time of $O(m \cdot 2^{160k^2 \log k + o(k^2 \log k)})$ for the whole algorithm. $\qquad\square$

4   Equitable Coloring

In this section we show that EQUITABLE COLORING is in FPT with respect to the maximum number $ml(G)$ of leaves in a spanning tree of G. Since a graph with bounded $ml(G)$ contains a lot of induced paths, we first examine the situation on the paths. There is a nice characterisation lemma for that case (the inductive proof of this Lemma is deferred to the full version of the paper):

Lemma 1. *Let $k \ge 2$ be an integer. Let P be a path with endpoints possibly colored by one of the colors $1, \ldots, k$. Let n be the number of uncolored vertices on the path, and $\forall i$ let $t(i) \in \{0, 1, 2\}$ be the number of endpoints colored by color i. Then P can be properly colored by the colors $1, \ldots, k$ such that there are $n_i + t(i)$ vertices of color i if and only if $\forall i : 0 \le n_i \le \lceil \frac{1}{2}(n - t(i)) \rceil$ and $\sum_i n_i = n$.*

Theorem 5. *EQUITABLE COLORING is in FPT with respect to the maximum number of leaves in a spanning tree $ml(G)$.*

Proof. First we show that if there are many classes and the graph is big, then we have a yes-instance; otherwise, we construct an instance of Integer Linear Programing (ILP) for EQUITABLE COLORING. It is known that G is a subdivision of some graph H on (at most) $4k$ vertices, for $ml(G) = k$ [5]. Such a graph H can be easily found in linear time. For an edge uv of H we use P_{uv} to denote the unique path in G having as endpoints u and v, and whose internal vertices are in $G \setminus H$. Let $|P_{uv}|$ denote the number of its internal vertices.

Claim. If $r \geq 16k$ and $n := |V(G)| \geq 32k^2$ then (G, r) is a yes-instance.

Proof (of Claim). We color the vertices of H arbitrarily by at most $4k$ colors. Then we fill these color classes with some (at most $4k(\lceil n/r \rceil - 1) \leq 4kn/r \leq n/4$) vertices. To avoid conflicts, we do not use the (at most $4k(4k-1)$) vertices that are neighbors in G of the H-vertices to fill the at most $4k$ classes. Since the rest of the graph is just a collection of paths, we can color at least each second vertex by one of the at most $4k$ colors used for the H-vertices. Hence there are at least $(n - 16k^2)/2 \geq n/4$ available vertices which are at least as many as we need. Now we are left with the task of equitably coloring the collection of paths with at least $12k$ colors, which is possible due to Lemma 1.

If the graph has at most $32k^2$ vertices, then we can solve the instance by brute force. If this is not the case, we can assume, due to the claim, that the number of colors r is less than $16k$. Now we try all the possibilities $c : V(H) \to \{1, \ldots r\}$ to color the vertices of H. For each such possibility, we construct an instance of ILP, which will have a variable q_{uv}^i for each combination of color i and an edge uv of H. This variable expresses the number of the vertices of color i on the path P_{uv}. They are subject to the constraints given by Lemma 1 and the constraints that enforce the classes to have the right number of vertices. In the following formal description of the constraints, for a logical formulae ϕ the expression $[\phi]$ is 1 if ϕ is true and 0 otherwise. Note that these expressions as well as the ceilings only appear on the constant sides of the equations.

$$\forall uv \in E(H), 1 \leq i \leq r : 0 \leq q_{uv}^i \leq \left\lceil \frac{1}{2}(|P_{uv}| - [c(u) = i] - [c(v) = i]) \right\rceil,$$

$$\forall uv \in E(H) : \sum_{i=1}^{r} q_{uv}^i = |P_{uv}|,$$

$$\forall i, 1 \leq i \leq l : \sum_{uv \in E(H)} q_{uv}^i = s + 1 - \sum_{v \in V(H)} [c(v) = i],$$

$$\forall i, l+1 \leq i \leq r : \sum_{uv \in E(H)} q_{uv}^i = s - \sum_{v \in V(H)} [c(v) = i].$$

Clearly, there is a solution for EQUITABLE COLORING if there is a solution to the ILP for one of the colorings c. Since an instance X of ILP with t variables can be solved in time $O(t^{2.5t + o(t)} \cdot |X|)$ [9], the overall running time is at most $O((32k^2)^{32k^2} + (256k^3)^{2.5 \cdot 256k^3 + o(k^3)} poly(n))$, where the polynomial is independent of k.   □

Some Open Questions. We wonder whether ECP might be in FPT for 3-connected planar graphs, and for 3-connected graphs of bounded treewidth.

References

1. Altman, M.: Is automation the answer? the computational complexity of automated redistricting. Rutgers Computer and Technology Law Journal 23, 81–142 (2007)
2. Chen, J., Kanj, I.A., Xia, G.: Improved parameterized upper bounds for vertex cover. In: Královič, R., Urzyczyn, P. (eds.) MFCS 2006. LNCS, vol. 4162, pp. 238–249. Springer, Heidelberg (2006)
3. Downey, R.G., Fellows, M.R.: Parameterized Complexity. Springer, Heidelberg (1999)
4. Dyer, M.E., Frieze, A.M.: A partitioning algorithm for minimum weighted euclidean matching. Inf. Process. Lett. 18(2), 59–62 (1984)
5. Estivill-Castro, V., Fellows, M.R., Langston, M.A., Rosamond, F.A.: FPT is P-Time Extremal Structure I. In: Hajo Broersma, M.J., Szeider, S. (eds.) ACiD. Texts in Algorithmics, vol. 4, pp. 1–41. King's College, London (2005)
6. Fellows, M.R., Fomin, F.V., Lokshtanov, D., Rosamond, F.A., Saurabh, S., Szeider, S., Thomassen, C.: On the complexity of some colorful problems parameterized by treewidth. In: Dress, A.W.M., Xu, Y., Zhu, B. (eds.) COCOA. LNCS, vol. 4616, pp. 366–377. Springer, Heidelberg (2007)
7. Fellows, M.R., Hermelin, D., Rosamond, F.A., Vialette, S.: On the parameterized complexity of multiple-interval graph problems. Theor. Comput. Sci. 410(1), 53–61 (2009)
8. Fiala, J., Golovach, P.A., Kratochvíl, J.: Parameterized complexity of coloring problems: Treewidth versus vertex cover. In: Chen, J., Cooper, S.B. (eds.) TAMC 2009. LNCS, vol. 5532, pp. 221–230. Springer, Heidelberg (2009)
9. Frank, A., Tardos, É.: An application of simultaneous diophantine approximation in combinatorial optimization. Combinatorica 7(1), 49–65 (1987)
10. Garey, M.R., Johnson, D.S.: Computers and Intractability: A Guide to the Theory of NP-Completeness. W.H. Freeman, New York (1979)
11. Ito, T., Goto, K., Zhou, X., Nishizeki, T.: Partitioning a multi-weighted graph to connected subgraphs of almost uniform size. IEICE Transactions 90-D(2), 449–456 (2007)
12. Ito, T., Zhou, X., Nishizeki, T.: Partitioning a weighted graph to connected subgraphs of almost uniform size. In: Hromkovič, J., Nagl, M., Westfechtel, B. (eds.) WG 2004. LNCS, vol. 3353, pp. 365–376. Springer, Heidelberg (2004)
13. West, D.: Introduction to graph theory. Prentice Hall Inc., Upper Saddle River, NJ (1996)

Improved Induced Matchings in Sparse Graphs

Rok Erman[1], Łukasz Kowalik[2], Matjaž Krnc[1], and Tomasz Waleń[2]

[1] Department of Mathematics, University of Ljubljana, Slovenia
[2] Institute of Informatics, University of Warsaw, Poland

Abstract. An induced matching in graph G is a matching which is an induced subgraph of G. Clearly, among two vertices with the same neighborhood (called *twins*) at most one is matched in any induced matching, and if one of them is matched then there is another matching of the same size that matches the other vertex. Motivated by this, Kanj, Pelsmajer, Schaefer and Xia [10] studied induced matchings in twinless graphs. They showed that any twinless planar graph contains an induced matching of size at least $\frac{n}{40}$ and that there are twinless planar graphs that do not contain an induced matching of size greater than $\frac{n}{27} + O(1)$. We improve both these bounds to $\frac{n}{28} + O(1)$, which is tight up to an additive constant. This implies that the problem of deciding an whether a planar graph has an induced matching of size k has a kernel of size at most $28k$. We also show for the first time that this problem is FPT for graphs of bounded arboricity.

Kanj et al. presented also an algorithm which decides in $O(2^{159\sqrt{k}} + n)$-time whether an n-vertex planar graph contains an induced matching of size k. Our results improve the time complexity analysis of their algorithm. However, we show also a more efficient, $O(2^{25.5\sqrt{k}} + n)$-time algorithm. Its main ingredient is a new, $O^*(4^l)$-time algorithm for finding a maximum induced matching in a graph of branch-width at most l.

1 Introduction

An induced matching in graph G is a matching which is an induced subgraph of G. It was introduced by Stockmeyer and Vazirani [17] and motivated as the "risk-free" marriage problem (decide whether there exist at least k pairs such that each married person is compatible with no married person except the one he or she is married to). In this paper we study induced matchings in planar, or more generally bounded arboricity graphs, both from combinatorial and computational perspective.

1.1 Combinatorial Perspective

It is a natural and heavily researched area in extremal graph theory to establish lower bounds on the size of various structures in selected graph classes. For example, Nishizeki and Baybars [15] and later Biedl et al. [3] showed tight lower bounds on the size of matching in subclasses of planar graphs, Alon, Mubayi and Thomas [1] show a lower bound on the size of induced forest in sparse graphs.

J. Chen and F.V. Fomin (Eds.): IWPEC 2009, LNCS 5917, pp. 134–148, 2009.

Kanj, Pelsmajer, Schaefer and Xia [10] were first to consider the size of induced matchings in planar graphs. Graphs like $K_{1,n}$, $K_{2,n}$ show that general planar graphs have no nontrivial lower bound on the induced matching size. Kanj et al. observed that among two vertices with the same neighborhood (called *twins*) at most one is matched in any induced matching, and if one of them is matched then there is another matching of the same size that matches the other vertex. In particular, after removing one of two twins from a graph, the size of maximum induced matching does not change. Motivated by this, Kanj et al. studied induced matchings in twinless graphs. They showed that any twinless planar graph contains an induced matching of size at least $\frac{n}{40}$ and that there are twinless planar graphs that do not contain an induced matching of size greater than $\frac{n}{27} + O(1)$.

In this paper we improve both these bounds to $\frac{n}{28} + O(1)$, which is tight up to an additive constant. The lower bound is also generalized to bounded genus graphs, i.e. we show that any twinless graph of genus g contains an induced matching of size at least $\frac{2(n-10g+9)}{7(7+\sqrt{1+48g})}$. This improves an earlier bound $\frac{2(n-10g+10)}{13(7+\sqrt{1+48g})}$ of Kanj et al. [10].

Kanj et al. showed also that any planar graph of minimum degree 3 contains an induced matching of size $(n+8)/20$. We note that results of Nishizeki and Baybars [15] imply a better bound of $(n+2)/12$ for these graphs, as well as some better bounds for planar graphs of minimum degree 4 and 5.

Finally, we consider graphs of bounded arboricity, i.e. graphs whose edges set can be partitioned into $O(1)$ forests. For example, planar graphs have arboricity 3. Intuitively, graphs of bounded arboricity are uniformly sparse, since this class is equal to the class of graphs of bounded maximum density, where maximum density of a graph G is defined as $d^* = \max_{J \subseteq V, J \neq \emptyset} \frac{|E(G[J])|}{|J|}$ (see e.g. [11] for some relations between classes of sparse graphs). We show that any n-vertex twinless graph of arboricity c contains an induced matching of size $\Omega(\frac{1}{c}n^{1/c})$.

1.2   Computational Perspective

It was shown by Yannakakis [18] that deciding whether a planar graph contains an induced matching is NP-complete. Although the optimization problem is APX-complete in general [8], for planar graphs, and more generally for graphs that do not contain K_5 or $K_{3,3}$ as a minor, there is a PTAS working in $2^{O(1/\epsilon)}n$ time due to Baker [2] and Chen [4]. The PTAS (though with a worse running-time bound) can be generalized to H-minor-free graphs due to Demaine et al. [7].

In the area of parameterized complexity, one asks whether there is an algorithm for the induced matching problem which verifies whether an n-vertex graph contains an induced matching of size k in time $n^{O(1)}f(k)$. If so, then the problem is *fixed parameter tractable* (FPT in short). It is known that the problem is $W[1]$-hard in general [13], which means that most likely the induced matching problem is not FPT. However, there is a $2^{O(\sqrt{k})}n^{O(1)}$-time parameterized algorithm for H-minor-free graphs [6] due to Demaine, Fomin, Hajiaghayi and Thilikos. For the (smaller) class of planar graphs, Moser and Sikdar [12] showed that the problem has a linear *kernel*, which means that one can reduce the problem in polynomial time to the same problem but on instance of size $O(k)$. The result of Kanj et al. mentioned in Section 1.1 implies that the size of the kernel is bounded by $40k$. Our results improve the bound further to $28k$.

We show, using the concept of eliminating twins, that the induced matching problem has a polynomial kernel for graphs of bounded arboricity. This implies that for such graphs there is an FPT algorithm with time complexity of the form $O(n + f(k))$. Since H-minor-free graphs have bounded arboricity, this generalizes the result of Demaine et al. [6] for the special case of the induced matching problem (the results in [6] are stated for all so-called bidimensional problems). This is also particularly interesting because there are classes of bounded arboricity graphs (like 4-regular graphs) for which the problem is APX-hard.

By using the linear kernel and planar separator technique, Kanj et al. showed an $O(2^{159\sqrt{k}} + n)$-time parameterized algorithm. Our lower bound of the size of induced matching in twinless planar graphs improves the time complexity analysis of their algorithm to $O(2^{133\sqrt{k}} + n)$. However, we show also a more efficient, $O(2^{25.5\sqrt{k}} + n)$-time algorithm based on the branch-width decomposition.

We also note that the proof of the lower bound gives a very practical and easy-to-implement algorithm for finding large induced forests in planar graphs (see Section 2.4).

2   Lower Bounds

In this section we present some lower bounds on the size of induced matching in subclasses of planar graphs and some classes of twinless sparse graphs. Our general approach is the same as that of Kanj et al., who used the following lemma. (We give a simple proof for completeness).

Lemma 1 (Kanj et al. [10]). *Let $\mathcal{G}$ be a minor-closed family of graphs and let c be a constant such that any graph in G is c-colorable. Moreover, let G be a graph from $\mathcal{G}$ and let M be a matching in G. Then G contains an induced matching of size at least $|M|/c$.*

Proof. Let M be a matching in G. We obtain graph G' by removing all unmatched vertices and contracting all edges of matching M. Then $G' \in \mathcal{G}$. Color the vertices of G' in c colors. The largest color class in $V(G')$ is an independent set of size at least $|M|/c$. It corresponds to an induced matching in G of size at least $|M|/c$.

It follows that a lower bound on the size of a matching in a subclass $\mathcal{H}$ of a minor-closed graph family implies a bound on the size of an induced matching in $\mathcal{H}$.

2.1   Planar Graphs of Large Minimum Degree

Kanj et al. showed that a planar graph of minimum degree 3 contains a matching of size at least $(n + 8)/5$. Using this with Lemma 1 and the Four Color Theorem they obtained that any planar graph of minimum degree 3 contains an induced matching of size at least $(n+8)/20$. However, this bound can be easily improved by using the following tight bounds for the size of matchings due to Nishizeki and Baybars.

Theorem 1 (Nishizeki and Baybars [15]). *Let G be an n-vertex planar graph of minimum degree δ and let M be a maximum cardinality matching in G. Then,*

(i) *if $\delta = 3$ and $n \geq 10$, then $|M| \geq \frac{n+2}{3}$,*
(ii) *if $\delta = 4$ and $n \geq 16$, then $|M| \geq \frac{2n+3}{5}$,*
(iii) *if $\delta = 5$ and $n \geq 34$, then $|M| \geq \frac{5n+6}{11}$.*

Corollary 1. *Let G be an n-vertex planar graph of minimum degree δ and let M be a maximum cardinality induced matching in G. Then,*

(i) *if $\delta = 3$ and $n \geq 10$, then $|M| \geq \frac{n+2}{12}$,*
(ii) *if $\delta = 4$ and $n \geq 16$, then $|M| \geq \frac{2n+3}{20}$,*
(iii) *if $\delta = 5$ and $n \geq 34$, then $|M| \geq \frac{5n+6}{44}$.* $\qquad\square$

Let us note that the above bound $|M| \geq \frac{n+2}{12}$ is tight (up to an additive constant), as we show in Section 3. Let us also note that the paper of Nishizeki and Baybars contains also tight lower bounds on the matching size in graphs of minimum degree 3, 4 and 5 and *vertex connectivity* 1, 2, 3 and 4 and the corresponding bounds for induced matchings can be obtained.

2.2   Twinless Graphs of Bounded Genus

In this section we present an improved lower bound for the size of induced matchings in twinless graph of bounded genus. To this end, we are going to establish a lower bound on the size of a maximum cardinality matching in such graphs, and apply Lemma 1.

We begin with two simple observations.

Lemma 2. *Let uv be an edge in a maximum cardinality matching M in graph G and let I the set of unmatched vertices. If $N(u) \cap I \neq \emptyset$ and $N(v) \cap I \neq \emptyset$, then there is a vertex $x \in I$ such that $N(u) \cap I = N(v) \cap I = \{x\}$. In particular u, v and x form a triangle.*

Proof. Follows from the maximality of M.

Lemma 3. *For any twinless graph G there exists a maximum cardinality matching such that all 1-vertices of G are matched.*

Proof. Let M be a maximum cardinality matching in G and let I be the set of unmatched vertices. Suppose I contains a 1-vertex v. Let y be the sole neighbor of v. Then y is matched for otherwise M is not maximal. Let x be the vertex matched with y by M. Since G is twinless, x has degree at least two. We can now replace the edge xy with the edge vy in matching M and hereby decrease the number of vertices of degree 1 in I, without changing the size of M. After applying the above procedure to all 1-vertices we get the desired matching.

Now we are ready to show a lower bound on the size of a matching in twinless graphs of bounded genus.

Theorem 2. *Every n-vertex twinless graph G of genus g contains a matching of size $\frac{n+10(1-g)-1}{7}$.*

Proof. We will show that if G has no isolated vertices then G contains a matching of size $\frac{n+10(1-g)}{7}$. Since a twinless graph contains at most one isolated vertex then the claimed bound will follow. In what follows, M denotes the matching described in Lemma 3 and $I = V \setminus V(M)$. Note that I is an independent set, by the maximality of M. In what follows, we show a lower bound on $|M|$.

Let $M^\triangle \subset M$ be the set of edges in matching M that form triangles with vertices in I. Similarly, let $I^\triangle \subset I$ be the set of vertices in I that form triangles with edges of M. Let $I_2 \subset I$ denote the vertices of degree two in I and I_{3+} the vertices with degree three or more.

First note that Lemma 2 implies that

$$|M^\triangle| \geq |I^\triangle|. \tag{1}$$

Hence it suffices to bound $|M \setminus M^\triangle|$ from below. Let R be the set of vertices in $M \setminus M^\triangle$ that are adjacent to I. Note that by Lemma 2 each edge of $M \setminus M^\triangle$ has at most one endpoint in R, so

$$|M \setminus M^\triangle| \geq |R|. \tag{2}$$

Now we bound R from below in terms of $|I_2 \setminus I^\triangle|$. Let G_2 be a graph on the vertices R such that G_2 contains an edge uv when there is a vertex $x \in I_2 \setminus I^\triangle$ adjacent to both u and v. Observe that G_2 has genus at most g, because after subdividing its edges we get a subgraph of G. Hence, by Euler's Formula, $|E(G_2)| \leq 3|V(G_2)| - 6 + 6g$. Since $|E(G_2)| = |I_2 \setminus I^\triangle|$ and $V(G_2) = R$, we get $|R| \geq \frac{|I_2 \setminus I^\triangle| + 6 - 6g}{3}$. By (2),

$$|M \setminus M^\triangle| \geq \frac{|I_2 \setminus I^\triangle| + 6 - 6g}{3}. \tag{3}$$

Now we bound $|R|$ from below in terms of $|I_{3+} \setminus I^\triangle|$.

Let G_3 be the bipartite subgraph of G, on the vertices $R \cup (I_{3+} \setminus I^\triangle)$ and with edges incident with $I_{3+} \setminus I^\triangle$. Since G_3 is bipartite its embedding on an orientable surface of genus g has no triangles and we get the following bound on the number of its edges by Euler's Formula :

$$|E(G_3)| \leq 2|V(G_3)| - 4 + 4g.$$

By combining it with the fact that vertices in I_{3+} have degree at least 3, we can bound $|E(G_3)|$ as follows.

$$3 \cdot |I_{3+} \setminus I^\triangle| \leq |E(G_3)| \leq 2 \cdot (|R| + |I_{3+} \setminus I^\triangle|) - 4 + 4g.$$

It gives us $|R| \geq \frac{|I_{3+}\backslash I^{\triangle}|+4-4g}{2}$ so with (2) we get

$$|M \setminus M^{\triangle}| \geq \frac{|I_{3+} \setminus I^{\triangle}| + 4 - 4g}{2}. \tag{4}$$

Now we are going to merge the bounds (3) and (4) into the following bound:

$$|M \setminus M^{\triangle}| \geq \frac{|I \setminus I^{\triangle}|}{5} + 2(1 - g). \tag{5}$$

When $|I_2\backslash I^{\triangle}| \geq \frac{3}{5}|I\backslash I^{\triangle}|$, we get (5) from (3). Similarly, when $|I_2\backslash I^{\triangle}| \leq \frac{3}{5}|I\backslash I^{\triangle}|$ we get (5) from (4) by replacing $|I_{3+} \setminus I^{\triangle}|$ by $|I \setminus I^{\triangle}| - |I_2 \setminus I^{\triangle}|$.

By combining (1) and (5) we get

$$|M| = |M \setminus M^{\triangle}| + |M^{\triangle}| \geq \frac{|I \setminus I^{\triangle}|}{5} + 2(1-g) + |I^{\triangle}| =$$

$$= \frac{|I|}{5} + 2(1-g) + \frac{4}{5}|I^{\triangle}| \geq \frac{|I|}{5} + 2(1-g).$$

Since $I = n - 2|M|$, we get $|M| \geq \frac{n+10(1-g)}{7}$, as desired.

By using Lemma 1, Four Color Theorem, and Heawood's Theorem (which states that any graph of genus $g > 0$ is $\lfloor(7 + \sqrt{1 + 48g})/2\rfloor$-colorable), we get the following corollaries.

Corollary 2. *Every n-vertex twinless graph of genus g contains an induced matching of size $(2n + 20(1 - g) - 2)/(49 + 7\sqrt{1 + 48g})$.*

Corollary 3. *Every n-vertex twinless planar graph contains an induced matching of size $\frac{n+9}{28}$.*

2.3   Twinless Sparse Graphs

In this section we focus on graphs of bounded arboricity. Let $arb(G)$ and $d^*(G)$ denote arboricity and the maximum density of graph G, respectively.

Theorem 3. *Any n-vertex twinless graph of maximum density d^* contains a matching of size $\Omega(n^{1/\lceil d^* \rceil})$.*

Proof. Let G be an n-vertex twinless graph of maximum density d^* and let M be a maximum cardinality matching in G. Denote $d = \lceil d^* \rceil$. Let I denote the independent set $V(G) \setminus V(M)$. Let us partition I into vertices of degree at least $d + 1$ and vertices of degree at most d, denoted by $I_{d+1\uparrow}$ and $I_{d\downarrow}$ respectively.

Let $E(V(M), I_{d+1\uparrow})$ denote set the edges between $V(M)$ and $I_{d+1\uparrow}$. Then

$$(d + 1)|I_{d+1\uparrow}| \leq |E(V(M), I_{d+1\uparrow})| \leq d(2|M| + |I_{d+1\uparrow}|),$$

where the second inequality follows from the fact that $E(V(M), I_{d+1\uparrow})$ induces a graph of maximum density at most d^*. By rearranging we get

$$|M| = \Omega(|I_{d+1\uparrow}|/d). \tag{6}$$

On the other hand, since G is twinless,

$$|I_{d\downarrow}| \leq \sum_{i=0}^{d} \binom{2|M|}{i} = O((2|M|)^d).$$

Hence, $|M| = \Omega(|I_{d\downarrow}|^{1/d})$. Together with (6) we get the claimed bound.

Theorem 4. *Any n-vertex twinless graph G of maximum density d^* contains an induced matching of size $\Omega(\frac{1}{d^*} \cdot n^{1/\lceil d^* \rceil})$.*

Proof. Let M be a maximum cardinality matching in G. Similarly as in Lemma 1 we consider graph G' which is obtained from G by removing all unmatched vertices and contracting all edges of matching M. Consider any set of vertices $S' \subseteq V(G')$. Then S' corresponds to a set $S \subseteq V(G)$, i.e. S' is obtained from S by identifying endpoints of edges of M. Then $|E(G'[S'])| \leq |E(G[S])|$ and $|S'| = |S|/2$, hence $\frac{|E(G'[S'])|}{|S'|} \leq 2\frac{|E(G[S])|}{|S|} \leq 2d^*(G)$. It follows that $G'[S']$ contains a vertex of degree at most $4d^*$. Since S' was chosen arbitrarily we infer that G' is $\lfloor 4d^* \rfloor$-degenerate and hence $(\lfloor 4d^* \rfloor + 1)$-vertex-colorable (by a simple algorithm which chooses a vertex v with the smallest degree, removes it from the graph, colors the resulting graph recursively and assigns to v the smallest color which is unused by v's neighbors). By choosing the subset of M corresponding to the biggest color class in G' we obtain an induced matching of size $|M|/(\lfloor 4d^* \rfloor + 1) = \Omega(|M|/d^*)$. Since $|M| = \Omega(n^{1/\lceil d^* \rceil})$ by Theorem 3, the claim follows[1].

Although it is more convenient to prove the above result refering to maximum density, we feel that arboricity is more often used as a measure of graph sparsity. However, we can easily reformulate Theorem 4 using the following lemma, which follows from the Nash-Williams Theorem [14].

Lemma 4. *For any graph G with at least one edge, $\lceil d^*(G) \rceil < \text{arb}(G)$.*   □

Corollary 4. *Any n-vertex twinless graph G of arboricity c contains an induced matching of size $\Omega(\frac{1}{c} n^{1/c})$.*

Now, if we want to decide whether an n-vertex graph of arboricity bounded by a constant c contains an induced matching of size k, we begin by eliminating twins in linear time (see [10]). Let H be the resulting graph. From Theorem 4 we know that H contains an induced matching of size $\alpha \cdot |V(H)|^{1/c}$, for some constant α. Hence if $k \leq \alpha \cdot |V(H)|^{1/c}$ we answer "yes", and otherwise we know that $|V(H)| = O(k^c)$ and hence $|E(H)| = O(ck^c) = O(k^c)$. Since we can find a maximum induced matching in H by the exhaustive search, the overall algorithm runs in time $O(n + \exp(k^c))$. We note that one can also *find* induced matchings of size k within this time bound (see Section 4). We summarize it with the following corollary.

---

[1] Independently, Kanj et al. [10] in the journal version observed that any matching M in a graph of maximum density d^* contains an induced matching of size at least $|M|/(4d^* - 1)$.

Corollary 5. *The induced matching problem for graphs of arboricity bounded by $c = O(1)$ has kernel of size $O(k^c)$. In particular, this problem is fixed parameter tractable for these graphs.* □

2.4 A Practical Algorithm for Planar Graphs

The discussion in Section 1.2 shows that if we want to find large induced matching in a planar graph *in practice*, then most likely we should use PTAS of Baker or Chen, since they are linear-time (for any fixed approximation ratio) and their time complexities do not hide large constants. However, these algorithms are still very complicated and hard to implement.

Here we want to note that the proof technique of Corollary 3 (introduced by Kanj et al. [10]) can be turned into the following algorithm. Given an input graph G, remove twins, find a maximum matching M, remove the unmatched vertices, contract the edges from M, color the resulting graph and choose the subset of M which corresponds to the biggest color class.

Eliminating twins can be easily done in linear time (see [10]). Finding a maximum matching using Hopcroft-Karp algorithm works in $O(n^{3/2})$-time for planar graphs and is implemented in many libraries. Since so far there is no fast and simple algorithm for 4-coloring planar graphs, we use 5 colors instead and then the coloring can be found by a simple linear-time algorithm (see e.g. [5]). Because of using 5 colors instead of 4 the constant 28 increases to 35. Then we get a $O(n^{3/2})$-time algorithm which always finds an induced matching of size at least $n'/35$, where n' is the number of pairwise different vertex neighborhoods in G. If one insists on linear-time, a maximal matching can be used instead of maximum matching M. (Then the constant 35 doubles because any maximal matching has size at least $|M|/2$.)

3 An Upper Bound

In this section we show that the bound in Corollary 3 is tight, up to an additive constant. Namely, we show the following.

Theorem 5. *For any $n_0 \in \mathbb{N}$ there is an n-vertex twinless planar graph G such that $n > n_0$ and any induced matching in G is of size at most $\frac{n}{28} + O(1)$.*

Proof. In what follows we describe an n-vertex planar graph with maximum induced matching of size at most $\frac{n}{28} + O(1)$. It will be clear from our construction that the number of vertices can be made arbitrarily large.

We begin with a graph T_k, which consists of k copies of K_4 and some additional edges. We obtain T_k from the graph drawn in Fig 1 by identifying vertex v_1 with w_1, v_2 with w_2 and so on. It is easy to see that the resulting graph is still planar, since the cylinder is homeomorphic to a subset of the plane. Also, T_k is twinless.

Note that T_k has $4k$ vertices, $8k + O(1)$ triangular faces and $12k + O(1)$ edges. Now, we build a new graph G_k by extending T_k, as follows:

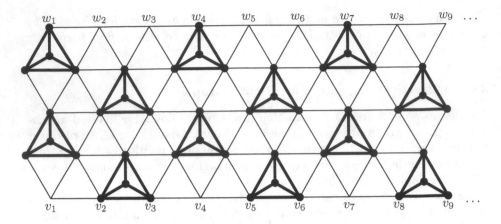

Fig. 1. Building T_k: arranging k copies of K_4 in 4 layers of triangles

(*i*) For each 3-face xyz of T_k add a 3-vertex v adjacent to x, y and z,
(*ii*) For each edge xy of T_k add a 2-vertex v adjacent to x and y.
(*iii*) For each vertex x of T_k add a 1-vertex v adjacent to x.

Note that by adding vertices like this we do not introduce twins and the graph stays planar. It is clear that G_k has $4k + 8k + 12k + 4k + O(1) = 28k + O(1)$ vertices. Moreover, every edge of G_k is incident with a vertex of one of the k copies of K_4. On the other hand, if M is an induced matching in G_k, vertices of each copy of K_4 are incident with at most one edge of M. It follows that $|M| \leq k$, so $|M| \leq |V(G_k)|/28 + O(1)$.

In a very similar way, we get that Corollary 1 (i) is also tight.

Corollary 6. *For any $n_0 \in \mathbb{N}$ there is an n-vertex planar graph G of minimum degree 3 such that $n > n_0$ and any induced matching in G is of size at most $\frac{n}{12} + O(1)$.*

Proof. Just remove the 1- and 2-vertices from the graphs constructed in the proof of Theorem 5.

4   An Algorithm Based on Branch-Width

In this section we discuss an algorithm that, given a planar graph G on n vertices and an integer k, either computes a induced matching of size $\geq k$, or concludes that there is no such induced matching. The algorithm requires $O(n + 2^{25.5\sqrt{k}})$ time.

4.1   Preliminaries

A *branch decomposition* of a graph G is a pair (T, r), where T is a tree with vertices of degree 1 or 3 and τ is a bijection from $E(G)$ to the set of leaves of

T. The *order function* $\omega : E(T) \to 2^{V(G)}$ of a branch decomposition maps every edge e of T to a subset of vertices $\omega(e) \subseteq V(G)$ as follows. The set $\omega(e)$ consists of all vertices of $V(G)$ such that for every vertex $v \in \omega(e)$ there exist two edges $f_1, f_2 \in E(G)$ that are incident with v and the leaves $\tau(f_1), \tau(f_2)$ are in different components of $T - \{e\}$. The *width* of (T, τ) is equal to $\max_{e \in E(T)} |\omega(e)|$ and the branch-width of G, $bw(G)$, is the minimum width over all branch decomposition of G. A set $D \subseteq V(G)$ is a *dominating set* in a graph G if every vertex in $V(G) - D$ is adjacent to a vertex in D.

Now we will introduce a few lemmas, that will connect induced matching problem with branch-width decomposition.

Lemma 5. *In any graph without isolated vertices if D is a minimum dominating set and M is a maximum cardinality matching, then $|D| \le |M|$.*

Proof. Let $V(M)$ be the set of the vertices of edges from M. We will describe a dominating set of size $|M|$. Let us observe that every vertex is adjacent to some vertex from the matching M, for otherwise the matching M is not maximal. Hence $V(M)$ is dominating. However, it is sufficient to choose just one endpoint for each edge of M. Then clearly all vertices of $V(M)$ are dominated, but we need to be careful about which endpoint we choose to dominate the unmatched vertices. Namely, for each edge of $uv \in M$ we choose its endpoint which has unmatched neighbors (or any endpoint if both have only matched neighbors). It may happen that both endpoints have unmatched neighbors but then by Lemma 2, $N(u) - V(M) = N(v) - V(M) = \{x\}$ for some x, so it does not matter whether we choose u or v.

Lemma 6 ([9]). *For any planar graph G with dominating set D,*

$$bw(G) \le 3\sqrt{4.5 \cdot |D|}.$$

Lemma 7. *For any planar graph G with maximum induced matching I,*

$$bw(G) \le 3\sqrt{18 \cdot |I|} \cong 12.7 \cdot \sqrt{|I|}.$$

Proof. From Lemma 1 we know that the maximum cardinality matching M of G has size $|M| \le 4|I|$. Combining lemmas 5 and 6 we get

$$bw(G) \le 3\sqrt{18 \cdot |I|} \cong 12.7 \cdot \sqrt{|I|}.$$

4.2 Algorithm Outline

Let G be the input planar graph on n vertices and let k be the size of induced matching we look for. As long as there is a pair of twins in G we remove one of them. This can be implemented in $O(n)$ time (see [10]). Let n' be the number of vertices of the resulting graph H. Now we describe an algorithm which *decides* whether H has an induced matching of size k.

Step 1. If $n' > 28k$ we can answer *True*, since the induced matching of size at least k exists as a consequence of Corollary 3. Otherwise we proceed with the next step and we can assume our graph has $O(k)$ vertices.

Step 2. Compute the optimal branch-decomposition of graph H. Using algorithm of Seymour and Thomas [16] this step requires $O(k^4)$ time. If $bw(G) \geq 12.7\sqrt{k}$ then as a consequence of Lemma 7 we can return answer *True*. Otherwise we proceed with the next step.

Step 3. Use the dynamic programming approach for finding a maximum cardinality induced matching in graph G. In Section 4.3 we present an algorithm that solves this problem on graphs with branch-decomposition of width $\leq l$ in $O(m \cdot 4^l)$ time where m is the number of edges in a graph. This step requires $O(k \cdot 4^{12.7\sqrt{k}}) = O(2^{25.5\sqrt{k}})$ time, since $l \leq 12.7\sqrt{k}$.

If we want to *find* the matching, in Step 1, we check whether $n' > 70k$ and if so we find an induced matching of size k by the linear-time algorithm from Section 2.4. Otherwise, we know that our graph has $O(k)$ vertices. Then we find the matching using the self-reducibility approach. Let $T(n')$ denote the time complexity of the decision algorithm described above. First, using the decision problem we determine the size s of the maximum induced matching in H. Then we can test in time $O(k+T(n'))$ whether a chosen edge e belongs to some induced matching of size s: just remove e and the adjacent vertices and test whether there is an induced matching of size $s-1$. If that is the case, find the induced matching of size $s-1$ recursively, and otherwise we put back the removed vertices (and their incident edges) and we test another edge, which has not been excluded so far. Clearly this procedure takes overall $O(|E(H)|(k+T(n')))$ time, which is $O(k^2 \cdot 4^{12.7\sqrt{k}}) = O(2^{25.5\sqrt{k}})$.

Theorem 6. *For any planar graph G on n vertices and an integer k, there is an $O(n + 2^{25.5\sqrt{k}})$-time algorithm which finds in G an induced matching of size k if and only if such a matching exists.*

4.3   Dynamic Programming on Graphs of Bounded Branch-Width

Our approach here is based on the algorithm for dominating set proposed by Fomin and Thilikos in [9]. We closely follow the notation and presentation from their paper.

Let (T', τ) be a branch decomposition of a graph G with m edges, let $\omega' : E(T') \to 2^{V(G)}$ be the order function of (T', τ). The tree T' is unrooted, so we build its rooted version T, by choosing an edge xy in T', putting new vertex v of degree 2 on this edge and making v adjacent to new vertex r, which is the new root of tree T. For every edge $f \in E(T) \cap E(T')$ we define $\omega(f) = \omega'(f)$, and for edges $\omega(xv) = \omega(vy) = \omega'(xy)$ and $\omega(rv) = \emptyset$.

For an edge f of T we define E_f (V_f) as the set of edges (vertices) of tree T that are "below" f, i.e. the set of all edges (vertices) g such that every path containing g and vr in T contains f. Every edge f of T that is not incident to a leaf has two children that are edges of E_f incident to f.

For every edge f of T we color the vertices of $\omega(f)$ in three colors $\{0, 1, 2\}$. We say that an induced matching M is *valid* for a coloring $c : \omega(f) \rightarrow \{0, 1, 2\}$ when for every $x \in \omega(f)$:

- if $c(x) = 2$, then $x \in V(M)$,
- if $c(x) = 1$, then $x \notin V(M)$, but it can be adjacent to some vertex of the matching,
- if $c(x) = 0$, then $x \notin V(M)$, and x is not adjacent to a vertex of the matching (for all $y \in N(x)$, $y \notin V(M)$).

For every edge f of T we use a mapping:

$$A_f : \{0, 1, 2\}^{\omega(f)} \rightarrow \mathbb{N} \cup \{-\infty\}$$

For a coloring $c \in \{0, 1, 2\}^{\omega(f)}$, the value $A_f(c)$ denotes the largest cardinality of an induced matching in the subgraph G_f of G that is defined by inducing the edge set:

$$\{\tau^{-1}(x) : x \in V_f \text{ and } x \text{ is a leaf of } T'\}$$

subject to the condition that the matching is valid with coloring c.

We define $A_f(c) = -\infty$ if there is no valid induced matching in G_f with coloring c.

Let f be a non-leaf edge of T and let f_1, f_2 be the children of f. Define $X_1 = \omega(f) - \omega(f_2)$, $X_2 = \omega(f) - \omega(f_1)$, $X_3 = \omega(f) \cap (\omega(f_1) \cap \omega(f_2))$, $X_4 = (\omega(f_1) \cap \omega(f_2)) - \omega(f)$.

Let us note that for $1 \le i \ne j \le 4$ we have $X_i \cap X_j = \emptyset$. Moreover, $\omega(f) = X_1 \cup X_2 \cup X_3$, $\omega(f_1) = X_1 \cup X_3 \cup X_4$, and $\omega(f_2) = X_2 \cup X_3 \cup X_4$.

We say that a coloring c of $\omega(f)$ is *formed* from coloring c_1 of $\omega(f_1)$ and coloring c_2 of $\omega(f_2)$ if

(F1) For every $x \in X_1$, $c(x) = \max(\{c_1(x)\} \cup \{c_2(y) - 1 : y \in \omega(f_2) \cap N(x)\})$,
(F2) For every $x \in X_2$, $c(x) = \max(\{c_2(x)\} \cup \{c_1(y) - 1 : y \in \omega(f_1) \cap N(x)\})$,
(F3) For every $x \in X_3$, $c(x) = \max\{c_1(x), c_2(x)\}$,
(F4) For every $x \in X_3 \cup X_4$, $c_1(x) + c_2(x) \le 2$.

If coloring c of $\omega(f)$ is formed from colorings c_1 of $\omega(f_1)$ and coloring c_2 of $\omega(f_2)$, then $A_f(c) \ge A_{f_1}(c_1) + A_{f_2}(c_2)$.

We compute functions A_f from leaves of T by bottom-up fashion.

For a leaf edge $f \in E(T)$, and its leaf node $v \in V(T)$ corresponding to an edge $xy \in E(G)$ we define function A_f as follows:

- if $c(x) \le 1$ and $c(y) \le 1$, then $A_f(c) = 0$,
- if $c(x) = c(y) = 2$ then $A_f(c) = 1$,
- otherwise $A_f(c) = -\infty$.

For non-leaf edges f of T we can compute function A_f as follows (f_1, f_2 denote the children of f):

$$A_f(c) = \max\{A_{f_1}(c_1) + A_{f_2}(c_2) \mid c_1, c_2 \text{ forms } c\}$$

If coloring c can not be formed from colorings c_1, c_2 of f_1, f_2, then we define $A_f(c) = -\infty$.

Let x_i denote $|X_i|$. The number of pairs (c_1, c_2) of colorings that can form a coloring c, can be bounded by

$$3^{x_1+x_2} \cdot 6^{x_3+x_4}$$

since there are three possible colorings of vertices $u \in X_1 \cup X_2$, and six pairs of colorings of vertices $u \in X_3 \cup X_4$, that is:

$$(c_1(u), c_2(u)) \in \{(0,0), (0,1), (1,0), (1,1), (2,0), (0,2)\}.$$

We can observe that if for some coloring c, $A_f(c) \neq -\infty$, we change coloring c into c' by replacing the color of a vertex x with $c(x) = 0$, to a new color $c'(x) = 1$, then $A_f(c') \neq -\infty$, and $A_f(c') \geq A_f(c)$. This leads us to an observation, that during computation of function A_f, instead of investigating pairs of colorings (from sets X_3, X_4) $\{(0,1), (1,0), (1,1)\}$, it is sufficient to check only one pair, namely $(1,1)$.

We can compute function A_f using a slightly modified formula:

$$A_f(c) = \max\{A_{f_1}(c_1) + A_{f_2}(c_2) \mid c_1, c_2 \text{ satisfies (F1), (F2), (F3) and (F4')}\}$$

where condition (F4') is defined as follows:

(F4') For every $x \in X_3 \cup X_4$, $(c_1(x), c_2(x)) \in \{(0,0), (1,1), (2,0), (0,2)\}$.

The complexity of computing A_f, with this optimization, can be bounded by:

$$3^{x_1+x_2} \cdot 4^{x_3+x_4}$$

Let $l = bw(G)$, and $x_i = |X_i|$, the values x_i are bounded by following inequalities:

$$x_1 + x_2 + x_3 = |\omega(f)| \leq l$$
$$x_1 + x_3 + x_4 = |\omega(f_1)| \leq l$$
$$x_2 + x_3 + x_4 = |\omega(f_2)| \leq l$$

The maximum value of linear functions $\log_4 3(x_1 + x_2) + x_3 + x_4$ subject to constraints on x_i is l (which is achieved for $x_1 = x_2 = 0$, $x_3 = x_4 = 0.5l$). Hence the cost of computing function A_f for a single edge f can be bounded by $O(4^l)$. Since we have to compute function A_f for each edge of tree T, the total time complexity is $O(m \cdot 4^l)$. The size of the maximum induced matching is stored in $A_{vr}(\epsilon)$, (where ϵ is the coloring of the empty set). The matching itself can be easily retrieved using standard methods. This gives us the following theorem.

Theorem 7. *For a graph G on m edges and with given a branch-decomposition of width l, the maximum induced matching of G can be computed in $O(m \cdot 4^l)$ time.*

We finish this section by noting that there is also an $O^*(4^t)$-time algorithm by Moser and Sikdar [12], where t denotes the tree-width of the input graph. It follows that our algorithm improves on this result, since for any graph G of at least 3 edges, $bw(G) \leq tw(G) + 1 \leq \frac{3}{2}bw(G)$ and the existing algorithms for finding optimal branch-decomposition are regarded as more practical than those for finding optimal tree-decomposition.

Acknowledgments

We are grateful to Fedor Fomin for helfpul hints. We are indebted to anonymous IWPEC referees for careful reading and numerous helpful comments.

The work was partially supported by Polish-Slovenian project "Graph colorings and their applications". Ł. Kowalik and T. Waleń were also supported by a grant from the Polish Ministry of Science and Higher Education, projects N N206 355636 and N206 004 32/0806.

References

1. Alon, N., Mubayi, D., Thomas, R.: Large induced forests in sparse graphs. J. Graph Theory 38(3), 113–123 (2001)
2. Baker, B.S.: Approximation algorithms for NP-complete problems on planar graphs. J. ACM 41(1), 153–180 (1994)
3. Biedl, T.C., Demaine, E.D., Duncan, C.A., Fleischer, R., Kobourov, S.G.: Tight bounds on maximal and maximum matchings. Discrete Mathematics 285(1-3), 7–15 (2004)
4. Chen, Z.-Z.: Efficient approximation schemes for maximization problems on $K_{3,3}$-free graphs. J. Algorithms 26(1), 166–187 (1998)
5. Chiba, N., Nishizeki, T., Saito, N.: A linear algorithm for five-coloring a planar graph. J. Algorithms 2, 317–327 (1981)
6. Demaine, E.D., Fomin, F.V., Hajiaghayi, M.T., Thilikos, D.M.: Subexponential parameterized algorithms on bounded-genus graphs and H-minor-free graphs. J. ACM 52(6), 866–893 (2005)
7. Demaine, E.D., Hajiaghayi, M.T., Kawarabayashi, K.: Algorithmic graph minor theory: Decomposition, approximation, and coloring. In: Proc. FOCS 2005, pp. 637–646. IEEE Computer Society Press, Los Alamitos (2005)
8. Duckworth, W., Manlove, D., Zito, M.: On the approximability of the maximum induced matching problem. J. Discrete Algorithms 3(1), 79–91 (2005)
9. Fomin, F.V., Thilikos, D.M.: Dominating sets in planar graphs: Branch-width and exponential speed-up. SIAM J. Comput. 36(2), 281–309 (2006)
10. Kanj, I.A., Pelsmajer, M.J., Xia, G., Schaefer, M.: On the induced matching problem. In: Proc. STACS 2008, pp. 397–408 (2008); Journal version to appear in J. Comput. Sys. Sci.
11. Kowalik, Ł.: Approximation scheme for lowest outdegree orientation and graph density measures. In: Asano, T. (ed.) ISAAC 2006. LNCS, vol. 4288, pp. 557–566. Springer, Heidelberg (2006)
12. Moser, H., Sikdar, S.: The parameterized complexity of the induced matching problem in planar graphs. Discrete Applied Mathematics 157, 715–727 (2009)

13. Moser, H., Thilikos, D.M.: Parameterized complexity of finding regular induced subgraphs. In: Broersma, H., Dantchev, S.S., Johnson, M., Szeider, S. (eds.) Proc. ACiD 2006. Texts in Algorithmics, vol. 7, pp. 107–118. King's College, London (2006)
14. Nash-Williams, C.S.J.A.: Decomposition of finite graphs into forests. Journal of the London Mathematical Society 39, 12 (1964)
15. Nishizeki, T., Baybars, I.: Lower bounds on the cardinality of the maximum matchings of planar graphs. Discrete Mathematics 28(3), 255–267 (1979)
16. Seymour, P.D., Thomas, R.: Call routing and the ratcatcher. Combinatorica 14(2), 217–241 (1994)
17. Stockmeyer, L.J., Vazirani, V.V.: NP-completeness of some generalizations of the maximum matching problem. Inf. Process. Lett. 15(1), 14–19 (1982)
18. Yannakakis, M.: Node- and edge-deletion NP-complete problems. In: Proc. STOC 1978, pp. 253–264. ACM Press, New York (1978)

Well-Quasi-Orders in Subclasses of Bounded Treewidth Graphs

Michael R. Fellows[1], Danny Hermelin[2,*], and Frances A. Rosamond[1]

[1] School of Electrical Engineering and Computer Science,
The University of Newcastle, Calaghan NSW 2308 - Australia
{mike.fellows,frances.rosamond}@cs.newcastle.edu.au
[2] Department of Computer Science, University of Haifa,
Mount Carmel, Haifa 31905 - Israel
danny@cri.haifa.ac.il

Abstract. We show that three subclasses of bounded treewidth graphs are well-quasi-ordered by refinements of the minor order. Specifically, we prove that graphs with bounded feedback-vertex-set are well-quasi-ordered by the topological-minor order, graphs with bounded vertex-covers are well-quasi-ordered by the subgraph order, and graphs with bounded circumference are well-quasi-ordered by the induced-minor order. Our results give an algorithm for recognizing any graph family in these classes which is closed under the corresponding minor order refinement.

1 Introduction

The treewidth parameter is one of the most commonly used structural parameterizations in parameterized complexity [7,10,14]. The reason for this being that many natural graph problems turn out to be fixed-parameter tractable when parameterized by the treewidth of the input graph. Indeed, various algorithmic methodologies such as tree-decomposition dynamic programming [1,2,4] and Courcelle's Theorem [5] provide a single framework to a vast multitude of different combinatorial problems in bounded treewidth graphs.

With that being said, there are still quite a few problems which are impregnable by any of the algorithmic methodologies for bounded treewidth graphs. For instance, vertex ordering problems such as BANDWIDTH or COALITION WIDTH have no known fixed-parameter algorithm when the treewidth is taken as a parameter. There is thus room for more algorithmic methodologies, perhaps by imposing more structure on the input than bounded treewidth. In this paper we suggest the method of well-quasi-ordering as a means towards this aim. Using this method, we are able to prove the following algorithmic result concerning subclasses of bounded treewidth graphs:

Theorem 1. *Let k be some fixed positive integer. There is a linear-time algorithm for recognizing:*

---

* Supported by the Adams Fellowship of the Israel Academy of Sciences and Humanities.

J. Chen and F.V. Fomin (Eds.): IWPEC 2009, LNCS 5917, pp. 149–160, 2009.

- *Any family of graphs with vertex-cover at most k that is closed under subgraphs.*
- *Any family of graphs with feedback-vertex-set at most k that is closed under topological minors.*
- *Any family of graphs with circumference at most k that is closed under induced-minors.*

We recall that a vertex cover in a graph is a set of vertices which covers all edges in the graph, a feedback-vertex-set is a set of vertices which covers all cycles in the graph, and the circumference of a graph is the length of its maximum cycle. Bounding each of these parameters results in a bound in the treewidth as well. By closed under subgraphs (resp. topological minors, induced minors), we mean that whenever a graph belongs to the family, then all of its subgraphs (resp. topological minors, induced-minors) also belong to the family. We mention that the first item of Theorem 1 was already shown indirectly by Ding [6] (see Section 3).

To see how our theorem applies to fixed-parameter algorithms, lets us consider some examples. Given a graph G, the *bandwidth* of G is the minimum bandwidth of all vertex-orderings $\pi : V(G) \rightarrow \{1, \ldots, |V(G)|\}$, where the bandwidth of a given vertex-ordering π is defined as $\max_{\{u,v\} \in E(G)} |\pi(v) - \pi(u)|$. The BANDWIDTH problem is the problem of computing the bandwidth of a given graph. It is known to be W[t]-hard for all $t > 1$ when parameterized by the bandwidth of the input graph [3], and not known to be in FPT when parameterized by the treewidth of the graph. However, observe that for each $\ell \in \mathbb{N}$, the family of graphs with bandwidth at most ℓ is closed under subgraphs. Thus, by the first item of Theorem 1, we get:

Corollary 1. *For any $k, \ell \in \mathbb{N}$, there is an $f(k + \ell) \cdot n$ time algorithm which determines whether a given graph G with n vertices and vertex-cover at most k, has bandwidth at most ℓ.*

A *coalition* in a graph is a subset of vertices pairwise connected by vertex-disjoint paths, or in other words, a topological clique minor. Given a vertex-ordering π of G, let us denote by $G_\pi^+(v)$ the graph induced by $\{v, \pi^{-1}(\pi(v)+1), \pi^{-1}(\pi(v)+2), \ldots, \pi^{-1}(|V(G)|)\}$, and by $N_\pi^+(v)$ the set of neighbors v has in $G_\pi^+(v)$. The *coalition-width* of a given graph G is defined as the minimum coalition-width over all vertex-orderings of G, where the coalition-width of a given vertex-ordering π is defined as $\max_{v \in V(G)} |\{|K| : K \subseteq N_\pi^+(v), K \text{ forms a coalition in } G_\pi^+(v)\}|$. We do not know whether the corresponding COALITION WIDTH problem is in FPT when parameterized by the treewidth of the input graph. However, observe that for each $\ell \in \mathbb{N}$, the family of all graphs with coalition-width at most ℓ is closed under topological minors, and so according to second item of Theorem 1 we get:

Corollary 2. *For any $k, \ell \in \mathbb{N}$, there is an $f(k + \ell) \cdot n$ time algorithm which determines whether a given graph G with n vertices and feedback-vertex-set at most k, has coalition-width at most ℓ.*

The reader should observe that all algorithms implied by the corollaries above are non-uniform by nature: For each k and ℓ we get a different algorithm. However, using the techniques by Fellows and Langston [8,9], the above results along with many other natural examples can be transformed into uniform algorithms. We refer the reader for more details also to [7].

The remainder of this paper is devoted to proving Theorem 1. We begin by briefly reviewing the fundamentals behind the method of well-quasi-ordering, and how it applies to bounded treewidth graphs. In Section 3, we provide the general framework for proving Theorem 1 by devising what we call well-quasi-order identification tools. The correctness of these tools is proved in Sections 4 and 5.

2   The WQO Method in Bounded Treewidth Graphs

Let us begin with some fundamental terminology from well-quasi-order theory. A *quasi-order* on a set X is a reflexive transitive subset of $X \times X$. That is, if $\preceq$ is a quasi-order on X, then $x \preceq y$ and $y \preceq z$ implies $x \preceq z$ for all $x, y, z \in X$, and $x \preceq x$ for all $x \in X$. We write $x \succ y$ if $x \succeq y$ and $y \npreceq x$. An infinite sequence $x_1, x_2, x_3 \cdots$ is called *strictly descending* if $x_1 \succ x_2 \succ x_3 \cdots$, and *good* if it is a *good pair* – a pair (x_i, x_j) with $x_i \preceq x_j$ and $i < j$. A *bad* sequence is an infinite sequence which is not good. A *well-founded quasi-order* is a quasi order with no infinite strictly descending sequences. A *well-quasi-order* (*wqo*) is a well-founded quasi-order with no infinite bad sequences. Equivalently, a well-quasi-order is a well-founded quasi-order with no infinite antichain.

Given a quasi-ordered set $\langle X, \preceq \rangle$, a subset $X' \subseteq X$ is said to be *closed under* $\preceq$, if for all $x, y \in X$ we have $x \in X'$ whenever $x \preceq y$ and $y \in X'$. Closed subsets of wqo sets have a property which is very interesting in our context: Consider the set

$$\mathrm{Forb}(X') := \{y \in X \setminus X' : z \nprec y \text{ for all } z \in X \setminus X'\}.$$

This set has the property that $x \in X'$ iff $y \npreceq x$ for all $y \in \mathrm{Forb}(X')$, and thus it is called a *forbidden characterization* of X'. Furthermore, since $\preceq$ is wqo, $\mathrm{Forb}(X')$ is necessarily finite as it constitutes an anti-chain w.r.t $\preceq$. Thus, every closed subset of a wqo set has a finite forbidden characterization.

Theorem 2 (WQO Recognition Theorem). *Let $\langle X, \preceq \rangle$ be a quasi-ordered set. If:*

(i) $\preceq$ is a wqo on X, and
(ii) for any $x, y \in X$, one can determine whether $y \preceq x$ in $f(|y|) \cdot |x|^c$ time, for some $c \in \mathbb{N}$.

Then one can recognize in $O(n^c)$ time any subset $X' \subseteq X$ that is closed under $\preceq$.

Proof. We describe an algorithm for recognizing an arbitrary subset $X' \subseteq X$ that is closed under $\preceq$. Since X' is closed under $\preceq$, the set $\mathrm{Forb}(X')$ defined above is a forbidden characterization of X'. According to the first condition in the theorem, this set is finite, and so our algorithm can have all elements of Forb(X') "hardwired" into it. On input $x \in X$, our algorithm checks whether $y \preceq x$ for each $y \in$ Forb(X'), using the order testing procedure promised by the second condition in the theorem. It determines that $x \in X'$ iff $y \not\preceq x$ for all $y \in$ Forb(X'). Correctness of this algorithm follows from the fact Forb(X') is a forbidden characterization of X'. Furthermore, its running-time can be bounded by $O(n^c)$, $n := |x|$, since the number and sizes of the elements in Forb(X') depends only on X', and is constant with respect to $|x|$. $\qquad \Box$

The WQO Recognition Theorem encapsulates the two main ingredients behind the method of well-quasi-ordering. Probably the best known application of this method is the astonishing result implied by Robertson and Seymour's graph minor project: For any graph family $\mathcal{G}$ closed under minors, there is an $O(n^3)$ time algorithm for recognizing $\mathcal{G}$. This result is proved in an ongoing series of over twenty papers, where the two items of the theorem above are shown to apply to graph minors: The set of all graphs is wqo by the minor order, and one can test whether a k-vertex graph is a minor of an n-vertex graph in $f(k) \cdot n^3$ time. Combining these two extremely complex results together gives one of the deepest result in graph theory yielding polynomial-time algorithms for many problems previously not known to be even decidable.

The minor order on graphs is typically defined via graph operations: A graph H is *minor* of a graph G if H can be obtained in G (via isomorphism) by vertex and edge deletions, and by edge contractions. A *contraction* of an edge $\{u, v\}$ in a graph G is the operation that replaces u and v by a new vertex which is adjacent to all neighbors of u and v (and removing all resulting multiple edges and self loops). We can therefore consider orders that are defined by a subset of these operations, or by applying restrictions on them. For example, a graph H is *topological-minor* of a graph G if H can be obtained in G by vertex and edge deletions, and by topological contractions, where a *topological contraction* (or *subdivision removal*) is a contraction of an edge incident to at least one vertex of degree 2. The *induced minor* order is similar to the minor order but without edge-deletions, and the *subgraph order* is the well-known order defined by vertex and edge deletions alone.

In contrast to the graph minor order, none of the above orders is a wqo, not even in the very restrictive universe of bounded treewidth graphs [15]. However, in bounded treewidth graphs we have Courcelle's Theorem [5] which states that for any monadic-second-order formula ϕ there is an $f(|\phi| + k) \cdot n$ time algorithm for determining whether a given graph on n vertices and treewidth at most k satisfies ϕ. Since order testing for any of the above refinements of the graph minor order can be expressed in monadic-second-order logic, we get the second ingredient in the WQO method for free in bounded treewidth graphs, due to Courcelle's Theorem.

Lemma 1 (Bounded Treewidth Order-Testing Lemma). *For any graph H on ℓ vertices, there is an $f(\ell + k) \cdot n$ time algorithm for determining whether H is a subgraph (resp. topological-minor, induced-minor) of a given graph G on n vertices and treewidth at most k.*

The setting should be clear by now. In order to obtain the recognition algorithm promised in Theorem 1, we need to show that each subclass of bounded treewidth graphs in the theorem is wqo by its corresponding order. This along with the Bounded Treewidth Order-Testing Lemma will give us both conditions in the WQO Recognition Theorem, which in turn will give us Theorem 1 as a direct corollary. In the following sections of this chapter we will prove that:

Lemma 2. *Let k be any fixed positive integer. Then:*

- *The set of all graphs with vertex-cover at most k is wqo by subgraphs.*
- *The set of all graphs with feedback-vertex-set at most k is wqo by topological-minors.*
- *The set of all graphs with circumference at most k is wqo by induced-minors.*

Proof (of Theorem 1 assuming Lemma 2). Consider a family of graphs $\mathcal{G}$ with vertex-cover at most k. Then according to Lemma 2, $\mathcal{G}$ satisfies the first requirement of the WQO Recognition Theorem, and since each graph in $\mathcal{G}$ has treewidth at most $f(k)$ for some function $f()$, the second condition is satisfied according to the Bounded Treewidth Order Testing Lemma. Thus, any subset of $\mathcal{G}$ that is closed under subgraphs can be recognized in $O(n)$ time. The second and third items of the theorem can be proven similarly. $\square$

Thus, what remains to prove is Lemma 2. For this we develop in the next section two tools for determining when a given graph order is a wqo on a specific graph class.

3   Two Tools for Identifying WQOs

In this section we develop two tools that will help us in proving Lemma 2. These tools allow us to reduce the question of whether a given graph family is wqo by a particular order, to the question of whether a simpler family is wqo by some "colored variant" on that order. To specify these colored variants precisely, it will be convenient to speak of the graph orders we study in terms of embeddings. Let H and G be two given graphs:

- A *subgraph embedding* of H in G is an injection $f : V(H) \rightarrow V(G)$ with $\{u, v\} \in E(H) \Rightarrow \{f(u), f(v)\} \in E(G)$.
- A *topological-minor embedding* of H in G is an injection $f : V(H) \rightarrow V(G)$ where there exist vertex disjoint paths in G between $f(u)$ and $f(v)$ for every $\{u, v\} \in E(H)$.
- An *induced-minor embedding* of H in G is a injective mapping $f : V(H) \rightarrow 2^{V(G)}$ with $f(v)$ connected in G for all $v \in V(H)$, $f(u) \cap f(v) = \emptyset$ for all $u \neq v \in V(H)$, and $\{u, v\} \in E(H) \iff \exists x \in f(u)$ and $\exists y \in f(v)$ with $\{x, y\} \in E(G)$.

We write $H \subseteq G$ (resp. $H \trianglelefteq G$, $H \sqsubseteq G$) if there exists a subgraph (resp. topological-minor, induced-minor) embedding of H in G. It is easy to see that

$H \subseteq G$ (resp. $H \trianglelefteq G$, $H \sqsubseteq G$) iff H is a subgraph (resp. topological-minor, induced-minor) of G as defined in the previous section. We will also use $H \subseteq^* G$ to denote that there is an induced-subgraph embedding of H in G, where an *induced subgraph embedding* is an injection $f : V(H) \to V(G)$ with the condition that $\{u, v\} \in E(H) \Leftrightarrow \{f(u), f(v)\} \in E(G)$. Finally, we write $H \cong G$ to denote that H and G are *isomorphic*, i.e. that $H \subseteq G$ and $G \subseteq H$.

We will speak of graph universes, where by a universe $\mathcal{U}$ we mean an infinite set of graphs which is closed under vertex deletions, i.e. $G \in \mathcal{U} \Longrightarrow G - V \in \mathcal{U}$ for all $V \subseteq V(G)$. Let $\mathcal{U}$ be some graph universe. A *labeling* of $\mathcal{U}$ is a set $\{\sigma_G : G \in \mathcal{U}\}$, where each σ_G is a *labeling* of the vertices of G by a set of labels Σ_G, i.e. $\sigma_G : V(G) \to \Sigma_G$. The set $\Sigma = \bigcup_{G \in \mathcal{U}} \Sigma_G$ is the set of labels assigned by σ to $\mathcal{U}$. If Σ is wqo by some quasi order $\preceq$, we say that σ is a *wqo labeling w.r.t* $\preceq$. Well-quasi-ordered labelings of $\mathcal{U}$ allow us to refine the subgraph, topological minor, and induced minor orders on $\mathcal{U}$ in a natural manner. Given a wqo labeling $\sigma = \{\sigma_G : G \in \mathcal{U}\}$ w.r.t $\preceq$, and a pair of graphs $H, G \in \mathcal{U}$, we will write $H \subseteq_\sigma G$ (resp. $H \subseteq_\sigma^* G$, $H \trianglelefteq_\sigma G$) if there is a subgraph (induced subgraph, topological-minor) embedding of H in G with $\sigma_H(v) \preceq \sigma_G(f(v))$ for all $v \in V(H)$. We write $H \cong_\sigma G$ whenever $H \subseteq_\sigma G$ and $G \subseteq_\sigma H$. Also, we extend this definition to the induced-minor order, and write $H \sqsubseteq_\sigma G$ whenever there exists an induced-minor embedding of H in G where for each $v \in V(H)$ there is some $x \in f(v)$ with $\sigma(v) \preceq \sigma(x)$.

Let us next give two important examples of well-quasi-ordered graph families, that we will use later on. The first is due to Kruskal, and is known as the famous Labeled Forests Theorem, and the second is due to Ding:

Theorem 3 (Kruskal's Labeled Forests Theorem [11]). *The universe of all forests is wqo by $\trianglelefteq_\sigma$ for any wqo labeling σ.*

Theorem 4 (Ding's Bounded Paths Theorem [6]). *For any $k \in \mathbb{N}$, the universe of all graphs with no paths of length greater than k is wqo by $\subseteq_\sigma^*$, for any wqo labeling σ.*

We are now in position to describe our first wqo identification tool. This tool is especially suited for universes consisting of graphs which have a small subset of vertices whose removal leaves a very simple structured graph, e.g. graphs with bounded vertex-cover or bounded feedback-vertex-set. Given a graph universe $\mathcal{U}$, and a natural k, let us denote by $\mathcal{U}_k$ is the universe of all graphs G which have a subset of k vertices V with $G - V \in \mathcal{U}$.

Theorem 5 (WQO Identification Tool 1). *If a universe $\mathcal{U}$ is wqo by $\subseteq_\sigma$ (resp. $\trianglelefteq_\sigma$) for any finite wqo labeling σ, then $\mathcal{U}_k$ is wqo under $\subseteq$ (resp. $\trianglelefteq$).*

Our second identification tool is concerned with the induced-minor order and 2-connected graphs. In general, a connected graph G is called *2-connected* if it has at least two vertices, and no removal of less than two vertices leaves G disconnected (the empty graph is assumed to be disconnected). The second identification tool is especially suited for graph universes which have 2-connected graphs with very simple structure:

Theorem 6 (WQO Identification Tool 2). *If the subset of all 2-connected graphs in some universe $\mathcal{U}$ is wqo by $\sqsubseteq_\sigma$ for any wqo labeling σ, then $\mathcal{U}$ itself is wqo by $\sqsubseteq$.*

The next two sections are devoted each to proving Theorem 5 and Theorem 6. But for now, let us next see how these two identification tools easily imply Lemma 2 of the previous section:

Proof (of Lemma 2 assuming Theorem 5 and Theorem 6). We prove the first two items of the lemma using Theorem 5, and the last item using Theorem 6:

- Let $\mathcal{U}$ denote the set of all graphs with no edges. Then for any $k \in \mathbb{N}$, $\mathcal{U}_k$ is the universe of all graphs with vertex-cover at most k by definition. According to Kruskal's Labeled Forests Theorem, we know that $\mathcal{U}$ is wqo by $\trianglelefteq_\sigma$ for any wqo labeling σ, since $\mathcal{U}$ includes only forests. Moreover, if H is a topological-minor of G for $H, G \in \mathcal{U}$, then H is also a subgraph of G, since graphs in $\mathcal{U}$ have no edges. This implies that $\mathcal{U}$ is also wqo by $\sqsubseteq_\sigma$ for any wqo labeling σ. Plugging this into Theorem 5 gives us that graphs with vertex-cover at most k are wqo by subgraphs.
- For graphs with bounded feedback-vertex-set the argument is similar to the above. Note that if $\mathcal{U}$ is the set of all forests, then for any $k \in \mathbb{N}$, $\mathcal{U}_k$ is the universe of all graphs with feedback-vertex-set at most k by definition. Due to Kruskal's Labeled Forests Theorem we get by Theorem 5 that graphs with feedback-vertex-set at most k are wqo by topological-minors.
- Let $\mathcal{U}$ denote the set of all graphs with circumference at most k, and let $\mathcal{U}'$ denote the subset of 2-connected graphs in $\mathcal{U}$. Since any two vertices in a 2-connected graph are connected by at least two paths (according to Menger's Theorem [12]), and thus belong together to some cycle, we get that 2-connected graphs in $\mathcal{U}'$ have no paths of length greater than k, due to the bound on the circumference of graphs in $\mathcal{U}$. Therefore, according to Ding's Theorem, we get that $\mathcal{U}'$ is wqo by $\subseteq_\sigma^*$ for any wqo labeling σ, and so it is also wqo by $\sqsubseteq_\sigma^*$ for any wqo labeling σ. Plugging this into Theorem 6 gives us that graphs with circumference at most k are wqo by induced-minors. □

4 Correctness of the First Tool

In this section we prove the correctness of Theorem 5. We will specify only the proof for the $\subseteq$ order, as the proof for the $\trianglelefteq$ order follows the same lines. To start with, we will assume we have a positive integer k, and a graph universe $\mathcal{U}$ which is wqo by $\subseteq_\sigma$ for any wqo labeling σ. We will show that any infinite sequence of graphs in $\mathcal{U}_k$ is good, *i.e.* it has graph which is a subgraph of another graph succeeding it in the sequence.

Let $\{G_i\}_{i=1}^\infty$ be any infinite sequence in $\mathcal{U}_k$. By definition, each graph G_i in this sequence has a subset of k vertices U_i with $G_i - U_i \in \mathcal{U}$. Let V_i denote the subset of vertices $V(G_i) \setminus U_i$. We construct a labeling $\sigma = \{\sigma_i : i \in \mathbb{N}\}$ on $\{G_i : i \in \mathbb{N}\}$ in a way that codifies the adjacency of vertices in U_i with vertices

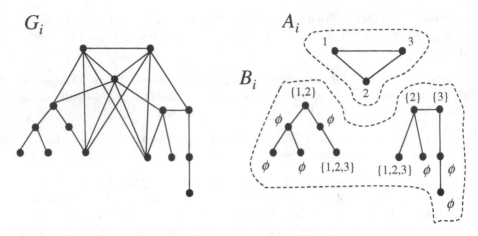

Fig. 1. The labeling used in proving the first identification tool

of V_i, for each $i \in \mathbb{N}$. For this, σ_i first assigns each vertex $u \in U_i$ an arbitrary distinct label $\sigma_i(u) \in \{1, \ldots, k\}$, and then it assigns a label in $2^{\{1, \ldots, k\}}$ to each $v \in V_i$ by

$$\sigma_i(v) := \big\{ x \ : \ \exists u \in U_i \text{ with } \{u, v\} \in E(G_i) \text{ and } \sigma_i(u) = x \big\}$$

(see Fig. 1 for an example). Observe that since the set of labels Σ assigned by σ is finite, it is wqo by equality, and σ is a wqo labeling on $\mathcal{U}$ with respect to $=$.

Now, for each $i \in \mathbb{N}$, let A_i denote the graph $G_i - V_i$, and let B_i denote $G_i - U_i$. Then $B_i \in \mathcal{U}$ for all $i \in \mathbb{N}$. Since there are only finitely many graphs A_i under isomorphism, and only finitely many ways to label the vertices of these graphs with distinct labels in $\{1, \ldots, k\}$, there must be an infinite subsequence $G_{i_1}, G_{i_2}, \ldots$ in $\{G_i\}_{i=1}^{\infty}$ with $A_{i_1} \cong_\sigma A_{i_2} \cong_\sigma \cdots$. By our assumption, the family of graphs $\{B_{i_j} : j \in \mathbb{N}\}$ is wqo by $\subseteq_\sigma$, and as this set is infinite, there must be a pair B_{i_x} and B_{i_y}, $x < y$, with $B_{i_x} \subseteq_\sigma B_{i_y}$. Write $i = i_x$ and $j = i_y$. We argue that:

$$A_i \cong_\sigma A_j \text{ and } B_i \subseteq_\sigma B_j \text{ implies } G_i \subseteq G_j$$

Let f_A denote the isomorphic embedding showing that $A_i \cong_\sigma A_j$, and let f_B denote the isomorphic embedding of B_i in B_j. We argue that the mapping $g = f_A \cup f_B$ is an isomorphic embedding of G_i in G_j. Clearly, for all edges $\{u, v\} \in E(G_i)$ with either $u, v \in A_i$ or $u, v \in B_i$, we have $\{g(u), g(v)\} \in E(G_i)$ by our assumptions on f_A and f_B. For $\{u, v\} \in E(G_i)$ with $u \in U_i$ and $v \in V_i$, we have $\sigma_i(u) = \sigma_j(g(u))$ and $\sigma_i(v) = \sigma_j(g(v))$. Thus, by construction of σ, we get $\{u, v\} \in E(G_i) \implies \sigma_i(u) \in \sigma_i(v) \implies \sigma_j(g(u)) \in \sigma_j(g(v)) \implies \{g(u), g(v)\} \in E(G_j)$.

It follows that $\{G_i\}_{i=1}^{\infty}$ is a good sequence, and as this sequence was chosen arbitrarily, this implies that $\mathcal{U}_k$ does not contain any bad sequences. This completes the proof of Theorem 5.

5   Correctness of the Second Tool

In this section we prove the correctness of Theorem 6. We start with the following lemma which allows us to restrict our attention w.l.o.g. to universes containing only connected graphs.

Lemma 3. *If the subset $\mathcal{U}' \subseteq \mathcal{U}$ of all connected graphs in some graph universe $\mathcal{U}$ is wqo by $\sqsubseteq$, then $\mathcal{U}$ is also wqo by $\sqsubseteq$.*

Proof. Since $\mathcal{U}$ is closed under vertex deletions, any component of a graph in $\mathcal{U}$ is a graph in $\mathcal{U}'$. Define the family of edge-less graphs $\mathcal{U}^* := \{G^* : G \in \mathcal{U}\}$, where G^* is the graph obtained by contracting each connected component of G into a single vertex. Next define a labeling σ for $\mathcal{U}$ that assigns each vertex of a graph $G^* \in \mathcal{U}$, the graph of $\mathcal{U}'$ which was contracted into this vertex. Applying Kruskal's Labeled Forests Theorem on $\mathcal{U}^*$, we get that $\mathcal{U}^*$ is wqo by $\trianglelefteq_\sigma$, and in fact also by $\subseteq_\sigma^*$ since graphs in $\mathcal{U}^*$ have no edges. Since by our construction, $\mathcal{U}$ is wqo by $\sqsubseteq$ iff $\mathcal{U}^*$ is wqo by $\subseteq_\sigma^*$, the lemma is proven. $\square$

We next introduce some additional terminology: A *rooted graph* is a pair (G, v) where G is a graph and v is a single distinguished vertex v of G referred to as its *root*. Thus two rooted graphs with the same vertex and edge set, but with different roots, are considered different. Apart from the following definition, we will omit the parentheses notation and simply state that G is a rooted graph with $root(G) = v$.

Definition 1 (Rooted Closure). *The rooted closure of a universe $\mathcal{U}$, denoted $\mathcal{U}_r$, is defined as the universe of rooted graphs $\mathcal{U}_r = \{(G, v) : G \in \mathcal{U}, v \in V(G)\}$.*

We say that an induced-minor embedding f of a rooted graph H in a rooted graph G *preserves roots* if $root(G) \in f(root(H))$, and we will write $H \sqsubseteq G$ (and say that H is an induced minor of G) only when there exists a root-preserving minor-embedding of H in G. Our main interest in rooted graphs lies in the above refinement of minor embeddings, and in the obvious fact that $\mathcal{U}$ is wqo under $\sqsubseteq$ whenever $\mathcal{U}_r$ is wqo under $\sqsubseteq$.

Another important notion we need to introduce before beginning the proof of Theorem 6 is the notion of minimal bad sequences, a concept first introduced by Nash-Williams [13], and later also used by Kruskal in proving his Labeled Forests Theorem:

Definition 2 (Minimal Bad Sequence). *A bad sequence $G_1, G_2, \ldots$ is minimal if for every bad sequence $H_1, H_2, \ldots$, whenever $|V(H_j)| < |V(G_j)|$ for some j, there is always some $i < j$ such that $|V(G_i)| < |V(H_i)|$.*

Let $\mathcal{U}$ be a universe of connected graphs whose subset of 2-connected graphs are wqo by $\sqsubseteq_\sigma$, for any wqo labeling σ (recall that we can assume that all graphs in $\mathcal{U}$ are connected due to Lemma 3). We can also assume w.l.o.g. that $\mathcal{U}$ contains no forests, by observing that Kruskal's Labeled Forests Theorem actually applies also for the induced-topological-minor order. To prove the theorem, we assume

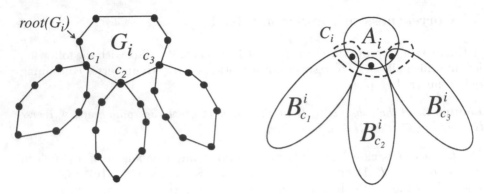

Fig. 2. The notation used in proving the second identification tool

that $\mathcal{U}$ is not wqo by $\sqsubseteq$, which implies that $\mathcal{U}_r$ is also not wqo by $\sqsubseteq$, and arrive at a contradiction by showing that in $\mathcal{U}_r$ contains no bad sequences.

For this, let $G_1, G_2, \ldots$ be a minimal bad sequence in $\mathcal{U}_r$. A *block* in a graph G_i, $i \in \mathbb{N}$, is a maximal 2-connected induced subgraph of G. For each $i \in \mathbb{N}$, select a block A_i in G_i which contains $root(G_i)$, and let C_i denote the set of cutvertices of G_i that are included in A_i. For each cutvertex $c \in C_i$, let B_c^i denote the connected component in $G_i - (V(A_i) \setminus C_i)$ including the vertex c and made into a rooted graph by setting $root(B_c^i) = c$ (see Fig. 2). Observe that for any $c \in C_i$, we have $B_c^i \sqsubseteq G_i$ by the induced-minor root-preserving embedding f that maps every non-root vertex $v \neq c$ of B_c^i to itself, and has $f(c) = A_i \ni root(G_i)$. We argue that:

The family of rooted graphs $\mathcal{B} = \{B_c^i : c \in C_i,\ i \in \mathbb{N}\}$ is wqo by $\sqsubseteq$.

To see this, let $\{H_j\}_{j=1}^\infty$ be any sequence in $\mathcal{B}$, and for every $j \in \mathbb{N}$, choose an $i(j)$ for which $H_j = B_c^i$ for some $c \in C_i$. Pick a j with smallest $i(j)$, and consider the sequence

$$G_1, \ldots, G_{i(j)-1}, H_j, H_{j+1}, \ldots$$

Then this sequence is good by the minimality of $\{G_i\}_{i=1}^\infty$, and by our selection of j, and so it contains a good pair (G, G'). Now, G cannot be among the first $i(j) - 1$ elements of this sequence, since otherwise $G' = H_{j'}$ for some $j' \geq j$, and we will have

$$G \sqsubseteq G' = H_{j'} = B_c^{i(j')} \sqsubseteq G_{i(j')},$$

implying that $(G, G_{i(j')})$ is a good pair in the bad sequence $\{G_i\}_{i=1}^\infty$. Thus, (G, G') must be a good pair in $\{H_j\}_{j=1}^\infty$, and so $\{H_j\}_{j=1}^\infty$ is good.

We next use the above to show that $\{G_i\}_{i=1}^\infty$ has a good pair, bringing us to our desired contradiction. For this, we will label the graph family $\mathcal{A} = \{A_i : i \in \mathbb{N}\}$ so that each cutvertex c of a graph A_i gets labeled by their corresponding connected component B_c^i of G_i, and the roots are preserved under this labeling. More precisely, for each A_i we define a labeling σ_i that assigns a pair of labels $(\sigma_i^{(1)}(v), \sigma_i^{(2)}(v))$ to every vertex $v \in V(G_i)$, where the labelings $\sigma_i^{(1)}$ and $\sigma_i^{(2)}$ are defined by:

- $\sigma_i^{(1)}(v) = 1$ if $v = root(G_i)$, and otherwise $\sigma^{(1)}(v) = 0$.
- $\sigma_i^{(2)}(v) = B_v^i$ if $v \in C_i$, and otherwise $\sigma_i^{(2)}(v) = 0$.

The labeling σ of $\mathcal{A}$ is then $\{\sigma_i : i \in \mathbb{N}\}$. We define a quasi-ordering $\preceq$ on the set of labels Σ assigned by σ. For two labels $(\varsigma_a^{(1)}, \varsigma_a^{(2)}), (\varsigma_b^{(1)}, \varsigma_b^{(2)}) \in \Sigma$, we define

$$(\varsigma_a^{(1)}, \varsigma_a^{(2)}) \preceq (\varsigma_b^{(1)}, \varsigma_b^{(2)}) \iff \varsigma_a^{(1)} = \varsigma_b^{(1)} \text{ and } \varsigma_a^{(2)} \sqsubseteq \varsigma_b^{(2)}.$$

Observe that the $\sqsubseteq$ order above is between rooted graphs. Also, we allow 0 to be $\sqsubseteq$-comparable only to itself. It is not difficult to see that since $\sqsubseteq$ is a wqo on $\mathcal{B}$, the $\preceq$ order is wqo on Σ. Thus, σ is wqo labeling on $\mathcal{A}$ w.r.t. σ. By the assumptions in the theorem, we know that $\mathcal{A}$ is wqo by $\sqsubseteq_\sigma$. It follows that there is a pair of graphs $A_i, A_j \in \mathcal{A}$ with $A_i \sqsubseteq_\sigma A_j$. To complete the proof we will show that:

$$A_i \sqsubseteq_\sigma A_j \Rightarrow G_i \sqsubseteq G_j.$$

To see this, let f be the induced-minor embedding of A_i in A_j. Then for each cutvertex $c \in C_i$, $f(c)$ contains a vertex $d \in C_j$ with $B_c^i \sqsubseteq B_d^j$. Let f_c denote the induced-minor root-preserving embedding of B_c^i in B_d^j. We construct an embedding $g : V(G_i) \to 2^{V(G_j)}$ defined by

$$g(v) = \begin{cases} f(v) & : \quad v \text{ is a vertex of } A_i \text{ and } v \notin C_i, \\ f_c(v) & : \quad v \text{ is a vertex of } B_c^i \text{ and } v \neq c, \\ f(v) \cup f_v(v) & : \quad v \in C_i. \end{cases}$$

We argue that g is an induced minor embedding of G_i in G_j.

To see this, first note that by definitions of f and each f_c, we have $g(u) \cap g(v) = \emptyset$ for any pair of distinct vertices u and v in G_i. Moreover, for any edge $\{u, v\}$ of G_i there is a vertex $x \in g(u)$ and a vertex $y \in g(v)$ with $\{x, y\}$ and edge in G_j. Thus what remains to be shown is that $g(u)$ is connected in G_j for every vertex u of G_i. This is obviously true when $u \notin C_i$, again by the definitions of f and each f_c. If $u \in C_i$, then $f(u)$ contains a vertex $v \in C_j$ for which $B_u^i \sqsubseteq B_v^j$, and v is also contained in $f_v(v)$ since f_v preserves roots. Thus, $g(u)$ is connected also when $u \in C_i$. Noting also that the labeling σ ensures that $root(G_j) \in g(root(G_i))$, we establish that $G_i \sqsubseteq G_j$. This completes the proof of Theorem 6.

References

1. Arnborg, S.: Efficient algorithms for combinatorial problems on graphs with bounded decomposability. A survey. BIT Numerical Mathematics 25(1), 2–23 (1985)
2. Arnborg, S., Proskurowski, A.: Linear time algorithms for NP-hard problems restricted to partial k-trees. Discrete Applied Mathematics 23, 11–24 (1989)
3. Bodlaender, H.L., Fellows, M.R., Hallett, M.T.: Beyond NP-completeness for problems of bounded width: Hardness for the W-hierarchy. In: Proceedings of the 26th annual ACM Symposium on Theory of Computing (STOC), pp. 449–458 (1994)

4. Corneil, D.G., Keil, J.M.: A dynamic programming approach to the dominating set problem on k-trees. SIAM Journal on Algebraic and Discrete Methods 8(4), 535–543 (1987)
5. Courcelle, B.: The monadic second-order logic of graphs I. Recognizable sets of finite graphs. Information and Computation 85(1), 12–75 (1990)
6. Ding, G.: Subgraphs and well-quasi-ordering. Journal of Graph Theory 16(5), 489–502 (1992)
7. Downey, R., Fellows, M.: Parameterized Complexity. Springer, Heidelberg (1999)
8. Fellows, M.R., Langston, M.A.: Fast self-reduction algorithms for combinatorial problems of VLSI design. In: Proc. of the 3rd Aegean Workshop On Computing (AWOC), pp. 278–287 (1988)
9. Fellows, M.R., Langston, M.A.: On well-paritial-order theory and its application to combinatorial problems of VLSI design. SIAM Journal on Discrete Mathematics 5 (1992)
10. Flum, J., Grohe, M.: Parameterized Complexity Theory. Springer, Heidelberg (2006)
11. Kruskal, J.B.: Well-quasi-ordering, the tree theorem, and Vazsonyi's conjecture. Transactions of the American Mathematical Society 95, 210–225 (1960)
12. Menger, K.: Zur allgemeinen kurventheorie. Fundamenta Mathematicae 10, 96–115 (1927)
13. Nash-Williams, C.S.J.A.: On well-quasi-ordering finite trees. Mathematical Proceedings of the Cambridge Philosophical Society 59(4), 833–835 (1963)
14. Niedermeier, R.: Invitation to Fixed-Parameter Algorithms. Oxford University Press, Oxford (2006)
15. Thomas, R.: Well-quasi-ordering infinite graphs with forbidden finite planar minor. Transactions of the American Mathematical Society 312(1), 67–76 (1990)

An Exact Algorithm for the
Maximum Leaf Spanning Tree Problem

Henning Fernau[1], Joachim Kneis[2], Dieter Kratsch[3], Alexander Langer[2],
Mathieu Liedloff[4], Daniel Raible[1], and Peter Rossmanith[2]

[1] Universität Trier, FB 4—Abteilung Informatik, D-54286 Trier, Germany
{fernau,raible}@uni-trier.de
[2] Department of Computer Science, RWTH Aachen University, Germany*
{kneis,langer,rossmani}@cs.rwth-aachen.de
[3] Laboratoire d'Informatique Théorique et Appliquée,
Université Paul Verlaine - Metz, 57045 Metz Cedex 01, France
kratsch@univ-metz.fr
[4] Laboratoire d'Informatique Fondamentale d'Orléans, Université d'Orléans,
45067 Orléans Cedex 2, France
liedloff@univ-orleans.fr

Abstract. Given an undirected graph G with n nodes, the MAXIMUM
LEAF SPANNING TREE problem asks to find a spanning tree of G with as
many leaves as possible. When parameterized in the number of leaves k,
this problem can be solved in time $O(4^k \text{poly}(n))$ using a simple branching
algorithm introduced by a subset of the authors [13]. Daligault, Gutin,
Kim, and Yeo [6] improved this branching algorithm and obtained a
running time of $O(3.72^k \text{poly}(n))$. In this paper, we study the problem
from an exact exponential time point of view, where it is equivalent to
the CONNECTED DOMINATING SET problem. For this problem Fomin,
Grandoni, and Kratsch showed how to break the $\Omega(2^n)$ barrier and
proposed an $O(1.9407^n)$ time algorithm [10]. Based on some properties
of [6] and [13], we establish a branching algorithm whose running time
of $O(1.8966^n)$ has been analyzed using the Measure-and-Conquer technique. Finally we provide a lower bound of $\Omega(1.4422^n)$ for the worst case
running time of our algorithm.

1 Introduction

The MAXIMUM LEAF SPANNING TREE (MLST) problem, which asks to find a
spanning tree of the input graph with as many leaves as possible, is one of the
classical NP-hard problems [12]. Ongoing research on this topic is motivated by
the fact that variants of this problem occur frequently in real life applications.
For example, some broadcasting problems in network design ask to minimize the
number of broadcasting nodes, which must be connected to a single root. This
translates nicely to finding a spanning tree with many leaves and few internal
nodes. There are many other results dealing with this topics, e.g., [5,14,16,18].

* Partially supported by the DFG under grant RO 927/7.

J. Chen and F.V. Fomin (Eds.): IWPEC 2009, LNCS 5917, pp. 161–172, 2009.
© Springer-Verlag Berlin Heidelberg 2009

The MAXIMUM LEAF SPANNING TREE problem is equivalent to the CONNECTED DOMINATING SET problem, which asks to find a connected set of nodes which dominates the whole graph: it is easy to see that the internal nodes of a spanning tree with k leaves form a connected dominating set of size $|V| - k$ and vice versa.

Known results. In the field of exact exponential time algorithms, there is only the paper by Fomin, Grandoni, and Kratsch [10]. They present an exact algorithm of running time of $O(1.9407^n)$. This result was the first that improved over the trivial enumeration algorithm of running time $\Theta^*(2^n)$. However there is a long research history for this problem in the field of parameterized complexity, see [1,7,9,3,8,2,4]. The currently fastest published algorithm has running time $O^*(4^k)$ and is due to Kneis, Langer, and Rossmanith [13]. This has been further improved to $O^*(3.72^k)$ by Daligault, Gutin, Kim, and Yeo in a recent technical report [6]. The ideas behind these improvements are also used in our exact algorithm. Moreover, their algorithm directly leads to an $O^*(1.9973^n)$ exact algorithm, based on observations due to Raman, Saurabh and Sikdar [17].

Our results. In the next sections we present an exact algorithm solving the MLST problem in time $O(1.8966^n)$; improving upon the result of [10]. The algorithm is based on a parameterized one [13], which basically repeatedly branches on leaves of the current subtree of the graph into two subproblems: either this node remains a leaf or it becomes an internal node. When analyzing the running time as a function of n, the number of nodes of the input graph, one observes that branching on nodes of small degree (with two possible successors) is the worst case; forcing a too large running time. This resembles the worst case of the parameterized algorithm, and the changes in [6] aim at improving exactly this case. We use a similar approach in our exact algorithm. We mark nodes as leaves as early as possible even when they are not yet attached to an internal node of the current subtree. In the Measure-and-Conquer analysis, this balances the bad cases against the better cases, i.e., "the better cases lend some running time to the bad cases" for an overall improvement. This approach requires a rather complicated measure and a sophisticated analysis.

2   Preliminaries

Let $G = (V, E)$ be a simple and undirected graph. We denote by n the number of its nodes and by m the number of its edges. Given a node $v \in V$, the set of its neighbors is defined by $N(v) = \{ u \in V \mid \{u, v\} \in E \}$. The closed neighborhood of v is $N[v] = \{v\} \cup N(v)$. Given a subset $S \subset V$, we define $N(S)$ as the set $\bigcup_{v \in S} N(v) \setminus S$ and for a $X \subset V$, we define $N_X(S) = N(S) \cap X$. We write $H \subseteq G$ if H is a subgraph of G. A node whose removal increases the number of connected components is called a *cut vertex*.

A tree $T = (V_T, E_T)$ is a *subtree* of G or a *tree in G* if T is a subgraph of G. A tree T in G is a *spanning tree* of G if $V_T = V$. As usual, a node of degree 1 in a tree T is called a *leaf* and all other nodes are called *internal* nodes. Wlog, we assume that each spanning tree contains at least one internal node. Once we fix some arbitrary node, wlog an internal node, as the root of the tree, we can also speak of *parents* of nodes within this tree. A spanning tree of G is a *maximum leaf spanning tree* (MLST), also called optimal spanning tree in this paper, if there is no spanning tree of G with a larger number of leaves.

In the following, we identify trees $T = (V_T, E_T)$ with the bipartition of V_T into the sets of internal nodes and leaves, denoted as internal(T) and leaves(T), respectively. Although there might be multiple subtrees of G sharing the same set of internal nodes and leaves, either both are subtrees of some optimal solution of MLST or none of them is (recall that G is undirected).

For more information on branching algorithms, branching vectors, and the Measure-and-Conquer approach we refer to [11,15].

3   A New Exact Algorithm

Our algorithm partitions the set of nodes of the input graph G into the sets of *free nodes* (Free), *floating leaves* (FL), *branching nodes* (BN), *leaf nodes* (LN), and *internal nodes* (IN), where the latter three form the nodes of some subtree T of G. Initially, all nodes are in the set Free, i.e., the tree is empty.

The key idea of the branching algorithm is to recursively build a subtree T of G such that $V_T = \text{IN} \cup \text{BN} \cup \text{LN}$, internal($T$) = IN and leaves($T$) = BN $\cup$ LN, which might in some *branch* of the algorithm turn into a spanning tree T' of G.

Definition 1. *Let $G = (V, E)$ be a graph, let* IN, BN, LN, FL $\subseteq V$ *be disjoint sets of nodes, and let $x_1, \ldots, x_l \in V$. By $x_1 \to X_1, \ldots, x_l \to X_l$, where each X_i is one of* IN, BN, LN *or* FL, *we denote the* branch *that corresponds to moving each x_i to the respective set X_i, and additionally, if $X_i = \text{IN}$, all $y \in N_{\text{Free}}(x_i)$ to* BN *and all $y \in N_{\text{FL}}(x_i)$ to* LN. *The notation is extended to sets $Y \to X$ in a straightforward manner. The recursive* branching *over multiple branches is denoted by*

$$\langle x_1 \to X_1, \ldots, x_l \to X_l \parallel \ldots \parallel x_1' \to X_1', \ldots, x_{l'}' \to X_{l'}' \rangle.$$

In particular, whenever Algorithm $\mathcal{M}$ decides that some node $x \in V$ becomes an internal node, all of its neighbors are directly attached to the tree, which is never worse than connecting these neighbors through some other nodes [13]. However, nodes of T belonging to LN will always remain leaves in subsequent calls, whereas the status of a node in BN is still subject to change. Similarly, nodes in FL $\subseteq V \setminus V_T$ will be leaves in the spanning tree T', but their parents in T' have not yet been determined. This is formally defined as follows.

Definition 2. *Let $G = (V, E)$ be a graph, and let* IN, BN, LN, FL $\subseteq V$ *be disjoint sets of nodes and $T \subseteq G$ be a tree. We say T extends* (IN, BN, LN, FL) *iff* IN $\subseteq$ internal(T), LN $\subseteq$ leaves(T), BN $\subseteq$ internal(T) $\cup$ leaves(T), *and* FL $\cap$ internal(T) = $\emptyset$.

If $N(\text{internal}(T)) \subseteq \text{internal}(T) \cup \text{leaves}(T)$, we call T an inner-maximal *tree. A node $v \in \text{Free} \cup \text{FL}$ is* unreachable, *if there is no path $uv_1 \ldots v_t v$, where $t \geq 0$, $u \in \text{BN}$ and $v_i \in \text{Free}$ for all $1 \leq i \leq t$.*

For any $v \in V \setminus (\text{IN} \cup \text{LN})$, we define its *degree* $d(v)$ as $d(v) = |N(v) \cap (\text{Free} \cup \text{FL})|$ if $v \in \text{BN}$, as $d(v) = |N(v) \cap (\text{Free} \cup \text{FL} \cup \text{BN})|$ if $v \in \text{Free}$, and as $d(v) = |N(v) \cap (\text{Free} \cup \text{BN})|$ if $v \in \text{FL}$.

Our algorithm uses the following reduction rules.

Definition 3. *Let $G = (V, E)$ be a graph and let $\text{IN} \cup \text{BN} \cup \text{LN} \cup \text{FL} \cup \text{Free}$ be a partition of V.*

(R1) *If there exist two adjacent nodes $u, v \in V$ such that $u, v \in \text{FL}$ or $u, v \in \text{BN}$, then remove the edge $\{u, v\}$.*

(R2) *If there exists a node $v \in \text{BN}$ with $d(v) = 0$, then move v into LN.*

(R3) *If there exists a node $v \in \text{Free}$ with $d(v) = 1$, then move v into FL.*

(R4) *If there exists a node $v \in \text{Free}$ with no neighbors in $\text{Free} \cup \text{FL}$, then move v into FL.*

(R5) *If there exists a triangle $\{x, y, z\}$ with $d(x) = 2$ and $x \in \text{Free}$, then move x into FL.*

(R6) *If there exists a node $u \in BN$ which is a cut vertex, then apply $u \to \text{IN}$.*

(R7) *If there exists adjacent nodes $u, v \in V$ such that $u \in \text{LN}$ and $v \in V \setminus \text{IN}$, then remove the edge $\{u, v\}$.*

The correctness of the reduction rules is easy to prove and details are not given in this extended abstract due to space limitations.

The halting and branching rules are described in Algorithm $\mathcal{M}$ (see Figure 1). Their correctness is shown in the following Section. The running time analysis is provided in Section 5.

4   Correctness of the Algorithm

The following lemma will ease the forthcoming correctness proof. It enables us to turn some nodes into additional floating leaves (in some special cases). A similar technique has already been used in [6].

Lemma 1. *Let $G = (V, E)$ be a graph, T a tree in G and $v \in leaves(T)$ such that $N(v) \setminus V(T) = \{x_1, x_2\}$. If every optimal spanning tree $T' \supseteq T$ is such that v is an internal node and each x_i is a leaf in T', then there is also some optimal spanning tree where additionally each $w \in N(\{x_1, x_2\}) \setminus (\text{internal}(T) \cup \{v\})$ is a leaf.*

Lemma 2. *Algorithm $\mathcal{M}$ solves the* MAXIMUM LEAF SPANNING TREE *problem if called with $\text{BN} = \{r\}$ and $\text{IN} = \text{LN} = \text{FL} = \emptyset$, where r is the root of some optimal spanning tree.*

Proof. The reduction rules update a partition $\mathcal{P} = (\text{Free}, \text{IN}, \text{BN}, \text{LN}, \text{FL})$ to a partition $\mathcal{P}' = (\text{Free}', \text{IN}', \text{BN}', \text{LN}', \text{FL}')$ so that any maximum leaf spanning tree T' that extends $\mathcal{P}'$ has at least as many leaves as any spanning tree T extending $\mathcal{P}$. Note that given some disjoint subsets IN, BN, LN, FL, the subset Free is uniquely determined by $V \setminus (\text{IN} \cup \text{BN} \cup \text{LN} \cup \text{FL})$. Thus, we omit the explicit notion of the set Free.

Algorithm $\mathcal{M}$
Input: A graph $G = (V, E)$, IN, BN, LN, FL $\subseteq V$

Reduce G according to the reduction rules
if there is some unreachable $v \in$ Free $\cup$ FL **then** return 0
if $V =$ IN $\cup$ LN **then** return $|$LN$|$
Choose a node $v \in$ BN of maximum degree
if $d(v) \geq 3$ **or** $(d(v) = 2$ and $N_{FL}(v) \neq \emptyset)$ **then**
$\qquad \langle v \to$ LN $\| v \to$ IN$\rangle$ (B1)
else if $d(v) = 2$ **then**
$\qquad$ Let $\{x_1, x_2\} = N_{Free}(v)$ such that $d(x_1) \leq d(x_2)$
$\qquad$ **if** $d(x_1) = 2$ **then**
$\qquad\qquad$ Let $\{z\} = N(x_1) \setminus \{v\}$
$\qquad\qquad$ **if** $z \in$ Free **then**
$\qquad\qquad\qquad \langle v \to$ LN $\| v \to$ IN, $x_1 \to$ IN $\| v \to$ IN, $x_1 \to$ LN$\rangle$ (B2)
$\qquad\qquad$ **else if** $z \in$ FL **then** $\langle v \to$ IN$\rangle$
$\qquad$ **else if** $(N(x_1) \cap N(x_2)) \setminus$ FL $= \{v\}$ **and** $\forall z \in (N_{FL}(x_1) \cap N_{FL}(x_2))$,
$\qquad\qquad d(z) \geq 3$ **then**
$\qquad\qquad \langle v \to$ LN $\| v \to$ IN, $x_1 \to$ IN $\| v \to$ IN, $x_1 \to$ LN, $x_2 \to$ IN $\|$ (B3)
$\qquad\qquad\quad v \to$ IN, $x_1 \to$ LN, $x_2 \to$ LN, $N_{Free}(x_1, x_2) \to$ FL, $N_{BN}(x_1, x_2) \to$ LN$\rangle$
$\qquad$ **else** $\langle v \to$ LN $\| v \to$ IN, $x_1 \to$ IN $\| v \to$ IN, $x_1 \to$ LN, $x_2 \to$ IN$\rangle$ (B4)
else if $d(v) = 1$ **then**
$\qquad$ Let $P = (v = v_0, v_1, \ldots, v_k)$ be a maximum path such that
$\qquad\qquad d(v_i) = 2$, $1 \leq i \leq k$, $v_1, \ldots, v_k \in$ Free
$\qquad$ Let $z \in N(v_k) \setminus V(P)$
$\qquad$ **if** $z \in$ FL **and** $d(z) = 1$ **then** $\langle v_0, \ldots, v_k \to$ IN, $z \to$ LN$\rangle$
$\qquad$ **else if** $z \in$ FL **and** $d(z) > 1$ **then** $\langle v_0, \ldots, v_{k-1} \to$ IN, $v_k \to$ LN$\rangle$
$\qquad$ **else if** $z \in$ BN **then** $\langle v \to$ LN$\rangle$
$\qquad$ **else if** $z \in$ Free **then** $\langle v_0, \ldots, v_k \to$ IN, $z \to$ IN $\| v \to$ LN$\rangle$ (B5)

Fig. 1. An algorithm for MAXIMUM LEAF SPANNING TREE. The notation $\langle v \to$
IN $\| v \to$ LN$\rangle$ denotes the corresponding recursive branches, e.g., in this case v
either becomes an internal node or a leaf (see Definition 1).

In the following, (IN $\cup$ BN $\cup$ LN $\cup$ FL)$_{x_1 \to X_1, \ldots, x_l \to X_l}$ denotes the partition
(Free$'$, IN$'$, BN$'$, LN$'$, FL$'$) obtained from (Free, IN, BN, LN, FL) by the algorithm
in the $x_1 \to X_1, \ldots, x_l \to X_l$ branch. In particular, whenever Algorithm $\mathcal{M}$
decides that some nodes $X \subseteq$ BN $\cup$ Free become internal nodes, all nodes in
$N(X) \cap$ Free become new branching nodes (BN) and all nodes in $N(X) \cap$ FL
become leaves (LN). Hence, Algorithm $\mathcal{M}$ always computes an inner-maximal
tree. It thus remains to show that if there is some spanning tree T with k leaves
that extends the current (IN, BN, LN, FL), then Algorithm $\mathcal{M}$ calls itself with an
new (IN$'$, BN$'$, LN$'$, FL$'$) such that there is some spanning tree T' with k leaves
that extends (IN$'$, BN$'$, LN$'$, FL$'$) as well.

We prove this by induction. For the base step, any spanning tree extends
(IN, BN, LN, FL) with BN $= \{r\}$ and IN $=$ LN $=$ FL $= \emptyset$, where the root r
is the only branching node. Now let T be a spanning tree with k leaves that
extends (IN, BN, LN, FL), and let $v \in$ BN be of maximum degree.

- If $d(v) \geq 3$ or $d(v) = 2$ and $N_{\mathrm{FL}}(v) \neq \emptyset$, then Algorithm $\mathcal{M}$ calls itself recursively in (B1). Since v is either an internal node or a leaf in any spanning tree, T extends either $(\mathrm{IN}, \mathrm{BN}, \mathrm{LN}, \mathrm{FL})_{v \to \mathrm{IN}}$ or $(\mathrm{IN}, \mathrm{BN}, \mathrm{LN}, \mathrm{FL})_{v \to \mathrm{LN}}$.
- If $d(v) = 2$, $N_{\mathrm{Free}}(v) = \{x_1, x_2\}$ and $N(x_1) \setminus \{v\} = \{z\}$, such that $z \in \mathrm{FL}$, we do not need to branch, since x_1 must somehow be connected to the tree in any solution extending $(\mathrm{IN}, \mathrm{BN}, \mathrm{LN}, \mathrm{FL})$, and v is the only choice. If otherwise $z \in \mathrm{Free}$, then T either extends $(\mathrm{IN}, \mathrm{BN}, \mathrm{LN}, \mathrm{FL})_{v \to \mathrm{LN}}$, or $(\mathrm{IN}, \mathrm{BN}, \mathrm{LN}, \mathrm{FL})_{v \to \mathrm{IN}, x_1 \to \mathrm{LN}}$, or $(\mathrm{IN}, \mathrm{BN}, \mathrm{LN}, \mathrm{FL})_{v \to \mathrm{IN}, x_1 \to \mathrm{IN}}$, because if v is not a leaf in T, then it is an internal node and x_1 is either leaf or internal node.
- In the case where $d(v) = 2$, $3 \leq d(x_1) \leq d(x_2)$ and $N(x_1) \cap N(x_2) \cap (\mathrm{Free} \cup \mathrm{BN}) = \{v\}$, the algorithm branches on all possibilities whether v, x_1 and x_2 are internal nodes or leaves. If there is some $z \in (N_{\mathrm{FL}}(x_1) \cap N_{\mathrm{FL}}(x_2))$ with $d(z) \leq 2$, not both x_1 and x_2 can be leaves and we skip the last branch (which yields (B4)). Otherwise, Lemma 1 guarantees that in the last branch where v must be an internal node and x_1 and x_2 are leaves, we can assume that all other nodes neighbors of x_1 and x_2 are leaves in some optimal solution as well. Hence there is a tree that extends either $(\mathrm{IN}, \mathrm{BN}, \mathrm{LN}, \mathrm{FL})_{v \to \mathrm{LN}}$, or $(\mathrm{IN}, \mathrm{BN}, \mathrm{LN}, \mathrm{FL})_{v \to \mathrm{IN}, x_1 \to \mathrm{IN}}$, or $(\mathrm{IN}, \mathrm{BN}, \mathrm{LN}, \mathrm{FL})_{v \to \mathrm{IN}, x_1 \to \mathrm{LN}, x_2 \to \mathrm{IN}}$, or $(\mathrm{IN}, \mathrm{BN}, \mathrm{LN}, \mathrm{FL})_{v \to \mathrm{IN}, x_1 \to \mathrm{LN}, x_2 \to \mathrm{LN}, N_{\mathrm{Free}}(x_1, x_2) \to \mathrm{FL}, N_{\mathrm{BN}}(x_1, x_2) \to \mathrm{LN}}$.
- In the case where $d(v) = 2$, $3 \leq d(x_1) \leq d(x_2)$ and $N(x_1) \cap N(x_2) \cap (\mathrm{Free} \cup \mathrm{BN}) \neq \{v\}$ we can assume that if v is an internal node in every optimal solution, either x_1 or x_2 is an internal node as well. Otherwise we could connect x_1 and x_2 to $z \in (N(x_1) \cap N(x_2)) \setminus \mathrm{FL}$ instead of v, which might destroy the leaf z, that is connected somehow else, but yields the new leaf v. Since z is either a branching node or a free node, this is still allowed. Hence, there is also some optimal solution that extends $(\mathrm{IN}, \mathrm{BN}, \mathrm{LN}, \mathrm{FL})_{v \to \mathrm{LN}}$, or $(\mathrm{IN}, \mathrm{BN}, \mathrm{LN}, \mathrm{FL})_{v \to \mathrm{IN}, x_1 \to \mathrm{IN}}$, or $(\mathrm{IN}, \mathrm{BN}, \mathrm{LN}, \mathrm{FL})_{v \to \mathrm{IN}, x_1 \to \mathrm{LN}, x_2 \to \mathrm{IN}}$.
- Finally, if $d(v) = 1$, let $P = (v = v_0, v_1, \dots, v_k)$ be a maximum path such that $d(v_i) = 2$, $1 \leq i \leq k$, $v_1, \dots, v_k \in \mathrm{Free}$ and let $z \in (N(v_k) \setminus V(P))$, as described in Algorithm $\mathcal{M}$. If $z \in \mathrm{FL}$ and $d(z) = 1$, all nodes in P must be internal nodes in any spanning tree that extends $(\mathrm{IN}, \mathrm{BN}, \mathrm{LN}, \mathrm{FL})$, because there is no other way to connect z. If otherwise $d(z) > 1$, there is always an inner-maximal solution where v_k is a leaf by a simple exchange argument. If on the other hand $z \in \mathrm{BN}$, then the nodes in P must either be connected through v or through z, and hence we can just decide to make v a leaf, again by a simple exchange argument.
  Now assume $z \in \mathrm{Free}$. Since T is inner-maximal we know by [13], that there is some inner-maximal T' that extends either $(\mathrm{IN}, \mathrm{BN}, \mathrm{LN}, \mathrm{FL})_{v \to \mathrm{LN}}$, or $(\mathrm{IN}, \mathrm{BN}, \mathrm{LN}, \mathrm{FL})_{v, v_1, \dots, v_k, z \to \mathrm{IN}}$ in this case.

Since this concludes a complete distinction of all possible values of $d(v)$, the claim follows by induction. $\qquad\square$

5   Analysis of the Running Time

To analyze the time via Measure-and-Conquer, we use the following measure:

$$\mu(G) = \sum_{i=1}^{n} \epsilon_i^{\mathrm{BN}} |\mathrm{BN}_i| + \sum_{i=2}^{n} \epsilon_i^{\mathrm{Free}} |\mathrm{Free}_i| + \sum_{i=2}^{n} \epsilon_i^{\mathrm{FL}} |\mathrm{FL}_i|,$$

where BN_i (resp. Free_i and FL_i) denotes the set of node in BN (resp. Free and FL) with degree i, and the values of the ϵ's are chosen in $[0,1]$ so that $\mu(G) \leq n$, more precisely:

- $\epsilon_0^{\mathrm{Free}} = \epsilon_1^{\mathrm{Free}} = 0$, $\epsilon_2^{\mathrm{Free}} = 0.731975$, $\epsilon_3^{\mathrm{Free}} = 0.946609$, and $\epsilon_i^{\mathrm{Free}} = 1$ for all $i \geq 4$;
- $\epsilon_0^{\mathrm{BN}} = 0$, $\epsilon_1^{\mathrm{BN}} = 0.661662$, $\epsilon_i^{\mathrm{BN}} = 0.730838$ for all $i \geq 2$;
- $\epsilon_0^{\mathrm{FL}} = \epsilon_1^{\mathrm{FL}} = 0$, $\epsilon_2^{\mathrm{FL}} = 0.331595$, $\epsilon_3^{\mathrm{FL}} = 0.494066$, and $\epsilon_i^{\mathrm{FL}} = 0.628886$ for all $i \geq 4$.

Lemma 3. Let $G = (V, E)$ be a graph and let $\mathrm{Free} \cup \mathrm{BN} \cup \mathrm{LN} \cup \mathrm{FL} \cup \mathrm{IN}$ be a partition of V. Moreover let $v \in \mathrm{BN}$ such that $d(v) \geq 3$ or $d(v) = 2$ and there is some $u \in N_{\mathrm{FL}}(v)$. Then branching according to (B1) yields a branching number less than 1.8966.

Proof. By the reduction rules (R3) and (R6), we have $d(u) \geq 2$ for all $u \in \mathrm{Free} \cup \mathrm{FL}$.

1. In the first branch, v is added to the internal nodes. Thus, all nodes in $N_{\mathrm{Free}}(v)$ are added to the branching nodes. This reduces the degree of all these nodes by at least one, since the edge to v is not counted anymore. Moreover, all nodes in $N_{\mathrm{FL}}(v)$ are now leaf nodes. Thus, the measure decreases by at least

$$\Delta_1 = \epsilon_{d(v)}^{\mathrm{BN}} + \sum_{x \in N_{\mathrm{Free}}(v)} (\epsilon_{d(x)}^{\mathrm{Free}} - \epsilon_{d(x)-1}^{\mathrm{BN}}) + \sum_{y \in N_{\mathrm{FL}}(v)} \epsilon_{d(y)}^{\mathrm{FL}}.$$

2. In the second branch, v becomes a leaf. Therefore, the degree of all nodes in $N_{\mathrm{Free} \cup \mathrm{FL}}(v)$ decreases by one, as the edge to v is removed. This implies a change in the measure of at least

$$\Delta_2 = \epsilon_{d(v)}^{\mathrm{BN}} + \sum_{x \in N_{\mathrm{Free}}(v)} (\epsilon_{d(x)}^{\mathrm{Free}} - \epsilon_{d(x)-1}^{\mathrm{Free}}) + \sum_{y \in N_{\mathrm{FL}}(v)} (\epsilon_{d(y)}^{\mathrm{FL}} - \epsilon_{d(y)-1}^{\mathrm{FL}}).$$

Since higher degrees only imply a higher change, it is now sufficient to test all combinations where $d(v) = 3$ or $d(v) = 2$ and there is some $u \in N_{\mathrm{FL}}(v)$. For all other nodes $u \in N_{\mathrm{Free} \cup \mathrm{FL}}(v)$, we can similarly assume $2 \leq d(u) \leq 5$. The worst case (branching vector $(1.538324, 0.730838)$, branching number less than 1.8966) occurs at $d(v) = 3$, where v has three free neighbors of degree at least five. $\square$

Lemma 4. *Let $G = (V, E)$ be a graph and let* Free $\cup$ BN $\cup$ LN $\cup$ FL $\cup$ IN *be a partition of V. Moreover let $v \in$ BN such that $d(v) = 2$ and there is some $x_1 \in N_{\text{Free}}(v)$ with $d(x_1) = 2$ and the remaining $z \in N(x_1) \setminus \{v\}$ is contained in* Free*. Then branching according to (B2) yields a branching number less than 1.8966.*

Proof. By the reduction rule (R5), we know that $z \neq x_2$. Moreover, (R3) implies $d(z) \geq 2$.

1. Again, v becomes leaf in the first branch. Similar to Lemma 3, this implies a change in the measure of at least

$$\Delta_1 = \epsilon_2^{\text{BN}} + (\epsilon_2^{\text{Free}} - \epsilon_1^{\text{FL}}) + (\epsilon_{d(x_2)}^{\text{Free}} - \epsilon_{d(x_2)-1}^{\text{Free}})$$
$$= \epsilon_2^{\text{BN}} + \epsilon_2^{\text{Free}} + (\epsilon_{d(x_2)}^{\text{Free}} - \epsilon_{d(x_2)-1}^{\text{Free}}),$$

   because x_1 becomes a floating leaf of degree one and the degree of x_2 decreases by one.
2. In the second branch, both v and x_1 become internal nodes, which implies that z and x_2 become branching nodes. Again, $d(z)$ and $d(x_2)$ decrease by one. The measure decreases by at least

$$\Delta_2 = \epsilon_2^{\text{BN}} + \epsilon_2^{\text{Free}} + (\epsilon_{d(z)}^{\text{Free}} - \epsilon_{d(z)-1}^{\text{BN}}) + (\epsilon_{d(x_2)}^{\text{Free}} - \epsilon_{d(x_2)-1}^{\text{BN}}).$$

3. In the third branch, v becomes an internal node and x_1 becomes a leaf connected to v. Thus, x_2 is now a branching node and $d(x_2)$ decreases. Moreover, $d(z)$ decreases by one as well. This implies that the measure is reduced by at least

$$\Delta_3 = \epsilon_2^{\text{BN}} + \epsilon_2^{\text{Free}} + (\epsilon_{d(z)}^{\text{Free}} - \epsilon_{d(z)-1}^{\text{Free}}) + (\epsilon_{d(x_2)}^{\text{Free}} - \epsilon_{d(x_2)-1}^{\text{BN}}).$$

Since $d(v) = d(x_1) = 2$, we need to try all possible combinations of $d(z)$ and $d(x_2)$, both between 2 and 5. Here, the worst case is $d(z) = d(x_2) = 5$ (1.8965 for branching vector $(1.462813, 1.731975, 2.001137)$). $\qquad\square$

Lemma 5. *Let $G = (V, E)$ be a graph and let* Free $\cup$ BN $\cup$ LN $\cup$ FL $\cup$ IN *be a partition of V. Moreover let $v \in$ BN such that $N_{\text{Free}}(v) = \{x_1, x_2\}$ with $3 \leq d(x_1) \leq d(x_2)$ and let $(N(x_1) \cap N(x_2)) \setminus$ FL $= \{v\}$. Finally, let $x_1 \notin N(x_2)$. Then branching according to (B3) yields a branching number less than 1.8966.*

Proof. 1. In the first branch, v becomes a leaf, which yields

$$\Delta_1 = \epsilon_2^{\text{BN}} + (\epsilon_{d(x_1)}^{\text{Free}} - \epsilon_{d(x_1)-1}^{\text{Free}}) + (\epsilon_{d(x_2)}^{\text{Free}} - \epsilon_{d(x_2)-1}^{\text{Free}}).$$

2. In the second branch, v and x_1 become internal nodes. As a consequence, x_2 becomes a branching leaf and its degree decreases by one. Furthermore, the degree of all nodes in $N_{\text{Free}\cup\text{FL}}(x_1)$ decreases by one. We gain at least

$$\Delta_2 = \epsilon_2^{BN} + \epsilon_{d(x_1)}^{Free} + (\epsilon_{d(x_2)}^{Free} - \epsilon_{d(x_2)-1}^{BN}) + \sum_{x \in N_{Free}(x_1)} (\epsilon_{d(x)}^{Free} - \epsilon_{d(x)-1}^{BN})$$

$$+ \sum_{y \in N_{FL}(x_1)} \epsilon_{d(y)}^{FL} + \sum_{z \in N_{BN}(x_1) \setminus \{v\}} (\epsilon_{d(z)}^{BN} - \epsilon_{d(z)-1}^{BN}).$$

3. In the third branch, v and x_2 become internal nodes, while x_1 becomes a leaf. Thus, the degree decreases by one for all nodes in $N_{Free \cup FL}(x_1)$ as well as for all nodes in $N_{BN}(x_2)$. Moreover, all nodes in $N_{Free}(x_2)$ become branching nodes and all nodes in $N_{FL}(x_2)$ become leaves. Since $(N(x_1) \cap N(x_2)) \setminus FL = \emptyset$, the measure decreases by at least

$$\Delta_3 = \epsilon_2^{BN} + \epsilon_{d(x_1)}^{Free} + \epsilon_{d(x_2)}^{Free} + \sum_{x \in N_{Free}(x_1)} (\epsilon_{d(x)}^{Free} - \epsilon_{d(x)-1}^{Free})$$

$$+ \sum_{y \in N_{FL}(x_1) \setminus N(x_2)} (\epsilon_{d(y)}^{FL} - \epsilon_{d(y)-1}^{FL}) + \sum_{z \in N_{BN}(\{x_1,x_2\}) \setminus \{v\}} (\epsilon_{d(z)}^{BN} - \epsilon_{d(z)-1}^{BN})$$

$$+ \sum_{x' \in N_{Free}(x_2)} (\epsilon_{d(x')}^{Free} - \epsilon_{d(x')-1}^{BN}) + \sum_{y' \in N_{FL}(x_2)} \epsilon_{d(y')}^{FL}.$$

4. In the last branch, v becomes an internal node, x_1 and x_2 become leaves, and all nodes in $N_{Free}(\{x_1, x_2\})$ become floating leaves. Moreover, all nodes in $N_{BN}(\{x_1, x_2\})$ become leaves as well and finally, the degree decreases by at least one for all $u \in N_{FL}(\{x_1, x_2\})$. This implies that the measure decreases by at least

$$\Delta_4 = \epsilon_2^{BN} + \epsilon_{d(x_1)}^{Free} + \epsilon_{d(x_2)}^{Free} + \sum_{x \in N_{Free}(\{x_1,x_2\})} (\epsilon_{d(x)}^{Free} - \epsilon_{d(x)-1}^{FL})$$

$$+ \sum_{y \in N_{FL}(\{x_1,x_2\}) \setminus (N(x_1) \cap N(x_2))} (\epsilon_{d(y)}^{FL} - \epsilon_{d(y)-1}^{FL})$$

$$+ \sum_{y \in FL \cap N(x_1) \cap N(x_2)} (\epsilon_{d(y)}^{FL} - \epsilon_{d(y)-2}^{FL})$$

$$+ \sum_{z \in N_{BN}(\{x_1,x_2\}) \setminus \{v\}} \epsilon_{d(z)}^{BN}.$$

Again, we have to compute all possible neighborhoods. This requires us to test all $3 \leq d(x_1) \leq d(x_2) \leq 5$, all $1 \leq d(u) \leq 2$ for all $u \in N_{BN}(\{x_1, x_2\})$, all $2 \leq d(u) \leq 5$ for each $u \in N_{FL}(\{x_1, x_2\})$ and finally all $2 \leq d(u) \leq 5$ for all $u \in N_{Free}(\{x_1, x_2\})$. Note that it is sufficient to assume that all floating leaves are of degree at least two. Otherwise, some of the branches yield new instances that will be solved in polynomial time, because they are obvious "No" instances. Thus, the exponential parts of the runtime only depend on the other branches, which yields a much better runtime bound, even if some floating leaves are of degree one. Similarly, we can assume that floating leaves of degree two are not contained in $N(x_1) \cap N(x_2)$, because otherwise the last branch (both, x_1 and x_2 are in LN) is found to be a "No" instance in polynomial time.

It turns out that the largest branching number in this case is smaller than 1.8506 with a branching vector $(0.730838, 2.476690, 3.216207, 8.218955)$ for the case $d(x_1) = d(x_2) = 5$, $N_{\text{Free}}(x_1) = \{u\}$ with $d(u) = 5$, $N_{\text{FL}}(x_1) = \emptyset$, $N_{\text{BN}}(x_1) = \{u_1, u_2, u_3\}$ with $d(u_1) = d(u_2) = d(u_3) = 2$, $N_{\text{Free}}(x_2) = \{w\}$ with $d(w) = 2$, $N_{\text{FL}}(x_2) = \emptyset$, and $N_{\text{BN}}(x_2) = \{w_1, w_2, w_3\}$ with $d(w_1) = d(w_2) = d(w_3) = 2$. $\qquad\square$

Lemma 6. *Let $G = (V, E)$ be a graph and let $\text{Free} \cup \text{BN} \cup \text{LN} \cup \text{FL} \cup \text{IN}$ be a partition of V. Moreover let $v \in \text{BN}$ such that $N_{\text{Free}}(v) = \{x_1, x_2\}$ with $3 \leq d(x_1) \leq d(x_2)$ and let $(N(x_1) \cap N(x_2)) \setminus \text{FL} = \{v\}$. Finally, let $x_1 \in N(x_2)$. Then branching according to (B3) yields a branching number less than 1.8966.*

The proof is very similar to the previous lemma, we only need to make sure that the edge between x_1 and x_2 is not counted twice.

Lemma 7. *Let $G = (V, E)$ be a graph and let $\text{Free} \cup \text{BN} \cup \text{LN} \cup \text{FL} \cup \text{IN}$ be a partition of V. Moreover let $v \in \text{BN}$ such that $N_{\text{Free}}(v) = \{x_1, x_2\}$ with $3 \leq d(x_1) \leq d(x_2)$. Then branching according to (B4) yields a branching number less than 1.8966.*

Proof. Similar to Lemma 5 and Lemma 6, x_1 and x_2 can possibly be neighbors.

1. In the first branch, v becomes a leaf. Similar to above, we obtain at least
$$\Delta_1 = \epsilon_2^{\text{BN}} + (\epsilon_{d(x_1)}^{\text{Free}} - \epsilon_{d(x_1)-1}^{\text{Free}}) + (\epsilon_{d(x_2)}^{\text{Free}} - \epsilon_{d(x_2)-1}^{\text{Free}}).$$

2. In the second branch, v and x_1 become internal nodes. As a consequence, the degree decreases for all nodes in $N_{\text{Free}\cup\text{FL}}(\{v, x_1\})$ and these nodes turn into branching nodes or leaves, respectively. The measure decreases by at least
$$\Delta_2 = \epsilon_2^{\text{BN}} + \epsilon_{d(x_1)}^{\text{Free}} + (\epsilon_{d(x_2)}^{\text{Free}} - \epsilon_{d(x_2)-1}^{\text{BN}}) + \sum_{x \in N_{\text{Free}}(x_1)\setminus\{x_2\}} (\epsilon_{d(x)}^{\text{Free}} - \epsilon_{d(x)-1}^{\text{BN}})$$
$$+ \sum_{y \in N_{\text{FL}}(x_1)} \epsilon_{d(y)}^{\text{FL}} + \sum_{z \in N_{\text{BN}}(x_1)\setminus\{v\}} (\epsilon_{d(z)}^{\text{BN}} - \epsilon_{d(z)-1}^{\text{BN}}).$$

Note that when $x_2 \in N(x_1)$, $d(x_2)$ decreases even more. However, this estimation is good enough to obtain the claimed bounds.

3. In the last branch, v and x_2 become internal nodes and x_1 becomes a leaf. As usual, the measure decreases by at least
$$\Delta_3 = \epsilon_2^{\text{BN}} + \epsilon_{d(x_1)}^{\text{Free}} + \epsilon_{d(x_2)}^{\text{Free}} + \sum_{x \in N_{\text{Free}}(x_1)\setminus\{x_2\}} (\epsilon_{d(x)}^{\text{Free}} - \epsilon_{d(x)-1}^{\text{Free}})$$
$$+ \sum_{y \in N_{\text{FL}}(x_1)} (\epsilon_{d(y)}^{\text{FL}} - \epsilon_{d(y)-1}^{\text{FL}}) + \sum_{z \in N_{\text{BN}}(x_1)\setminus\{v\}} (\epsilon_{d(z)}^{\text{BN}} - \epsilon_{d(z)-1}^{\text{BN}}).$$

Similar to previous lemmas, we can safely assume that $d(u) \geq 2$ for all floating leaves $u \in N(x_1)$.

In order to compute all possible branching vectors, we need to test all $3 \leq d(x_1) \leq d(x_2) \leq 5$. Furthermore, we need to try all $1 \leq d(u) \leq 2$ for all $u \in N_{BN}(x_1)$, all $2 \leq d(u) \leq 5$ for each $u \in N_{FL}(x_1)$ and finally all $2 \leq d(u) \leq 5$ for all $u \in N_{Free}(x_1)$.

The worst case of 1.8966 (branching vector $(0.730838, 2.407514, 2.869190)$) occurs when $d(x_1) = d(x_2) = 5$, $N_{Free}(x_1) = \{u_1, u_2\}$ with $d(u_1) = d(u_2) = 5$, $N_{FL}(x_1) = \emptyset$, and $N_{BN}(x_1) = \{u_1' u_2'\}$ with $d(u_1') = d(u_2') = 2$. □

Lemma 8. *Let $G = (V, E)$ be a graph and let* Free $\cup$ BN $\cup$ LN $\cup$ FL $\cup$ IN *be a partition of V. Moreover let $v \in$ BN such that $d(v) = 1$. Then branching according to (B5) yields a branching number less than 1.8966.*

Proof. Let $v_1, \ldots, v_k$ and $z \in V$ as described in Algorithm $\mathcal{M}$ and recall that $d(z) \geq 3$ and $z \in$ Free.

1. In the first branch, v becomes an internal node as well as all $v_1, \ldots, v_k$ and z. This implies that the measure decreases by at least

$$\Delta_1 = \epsilon_1^{BN} + k\epsilon_2^{Free} + \epsilon_{d(z)}^{Free}.$$

2. In the other branch, v becomes a leaf. If $k = 0$, then the degree of z will decrease and otherwise the node v_1 becomes a floating leaf of degree one. Therefore, we gain at least

$$\Delta_2 = \epsilon_1^{BN} + \min(\epsilon_{d(z)}^{Free} - \epsilon_{d(z)-1}^{Free}, \epsilon_2^{Free}).$$

The worst case occurs when $d(z) = 5$ and $k = 0$ with a branching vector of $(1.661662, 0.661662)$ and a branching number less than 1.8966. □

From the above lemmas as well as from Lemma 2, which guarantees the correctness of our algorithm, we can conclude our main result.

Theorem 1. *The given algorithm solves the* MAXIMUM LEAF SPANNING TREE *problem in time $O(1.8966^n)$.*

It is known that the current time analysis, even when based on Measure-and-Conquer, seems to produce upper bounds overestimating the worst case running time of the algorithm. The following Theorem provides a lower bound on the worst case running time of our algorithm. We recall that Fomin et al. [10] present an algorithm solving the problem in worst case running time (upper bounded by) $O(1.9407^n)$ and they also provide a lower bound of $\Omega(1.3195^n)$.

Theorem 2. *We can give a lower bound of $\Omega(3^{n/3}) = \Omega(1.4422^n)$ for the worst case running time of our algorithm.*

To conclude let us mention that our algorithm is based on ideas of a parameterized algorithm that also work for directed graphs. Thus it would be interesting to know whether one may obtain an exact exponential-time algorithm of competitive running time for directed graphs.

References

1. Bodlaender, H.L.: On linear time minor tests with depth-first search. J. Algorithms 14(1), 1–23 (1993)
2. Bonsma, P.: Sparse cuts, matching-cuts and leafy trees in graphs. PhD thesis, University of Twente, the Netherlands (2006)
3. Bonsma, P.S., Brueggemann, T., Woeginger, G.J.: A faster FPT algorithm for finding spanning trees with many leaves. In: Rovan, B., Vojtáš, P. (eds.) MFCS 2003. LNCS, vol. 2747, pp. 259–268. Springer, Heidelberg (2003)
4. Bonsma, P.S., Zickfeld, F.: Spanning trees with many leaves in graphs without diamonds and blossoms. In: Laber, E.S., Bornstein, C., Nogueira, L.T., Faria, L. (eds.) LATIN 2008. LNCS, vol. 4957, pp. 531–543. Springer, Heidelberg (2008)
5. Dai, F., Wu, J.: An extended localized algorithm for connected dominating set formation in ad hoc wireless networks. IEEE Transactions on Parallel and Distributed Systems 15(10), 908–920 (2004)
6. Daligault, J., Gutin, G., Kim, E.J., Yeo, A.: FPT Algorithms and Kernels for the Directed k-Leaf Problem. CoRR abs/0810.4946 (2008); also: J. Comput. System Sci. (2009), doi:10.1016/j.jcss.2009.06.005
7. Downey, R.G., Fellows, M.R.: Parameterized computational feasibility. In: Feasible Mathematics II, pp. 219–244. Birkhäuser, Boston (1995)
8. Estivill-Castro, V., Fellows, M.R., Langston, M.A., Rosamond, F.A.: FPT is P-time extremal structure I. In: Proc. of 1st ACiD, pp. 1–41. College Publications (2005)
9. Fellows, M.R., McCartin, C., Rosamond, F.A., Stege, U.: Coordinatized kernels and catalytic reductions: An improved FPT algorithm for max leaf spanning tree and other problems. In: Kapoor, S., Prasad, S. (eds.) FST TCS 2000. LNCS, vol. 1974, pp. 240–251. Springer, Heidelberg (2000)
10. Fomin, F.V., Grandoni, F., Kratsch, D.: Solving connected dominating set faster than 2^n. Algorithmica 52(2), 153–166 (2008)
11. Fomin, F.V., Grandoni, F., Kratsch, D.: A measure & conquer approach for the analysis of exact algorithms. J. ACM 56(5) (2009)
12. Garey, M., Johnson, D.: Computers and Intractability: A Guide to the Theory of NP-completeness. Freeman, San Francisco (1979)
13. Kneis, J., Langer, A., Rossmanith, P.: A new algorithm for finding trees with many leaves. In: Hong, S.-H., Nagamochi, H., Fukunaga, T. (eds.) ISAAC 2008. LNCS, vol. 5369, pp. 270–281. Springer, Heidelberg (2008)
14. Liang, W.: Constructing minimum-energy broadcast trees in wireless ad hoc networks. In: Proc. of 3rd MOBIHOC, pp. 112–122 (2002)
15. Niedermeier, R.: Invitation to Fixed Parameter Algorithms. Oxford University Press, Oxford (2006)
16. Park, M.A., Willson, J., Wang, C., Thai, M., Wu, W., Farago, A.: A dominating and absorbent set in a wireless ad-hoc network with different transmission ranges. In: Proc. of 8th MOBIHOC, pp. 22–31. ACM, New York (2007)
17. Raman, V., Saurabh, S., Sikdar, S.: Improved exact exponential algorithms for vertex bipartization and other problems. In: Coppo, M., Lodi, E., Pinna, G.M. (eds.) ICTCS 2005. LNCS, vol. 3701, pp. 375–389. Springer, Heidelberg (2005)
18. Thai, M., Wang, F., Liu, D., Zhu, S., Du. Connected, D.Z.: dominating sets in wireless networks with different transmission ranges. IEEE Trans. Mobil. Comp. 6(7), 721–730 (2007)

An Exponential Time 2-Approximation Algorithm for Bandwidth

Martin Fürer[1],*, Serge Gaspers[2],**, and Shiva Prasad Kasiviswanathan[3]

[1] Computer Science and Engineering, Pennsylvania State University
furer@cse.psu.edu
[2] CMM, Universidad de Chile
sgaspers@dim.uchile.cl
[3] Los Alamos National Laboratory
kasivisw@gmail.com

Abstract. The bandwidth of a graph G on n vertices is the minimum b such that the vertices of G can be labeled from 1 to n such that the labels of every pair of adjacent vertices differ by at most b.

In this paper, we present a 2-approximation algorithm for the Bandwidth problem that takes worst-case $\mathcal{O}(1.9797^n) = \mathcal{O}(3^{0.6217n})$ time and uses polynomial space. This improves both the previous best 2- and 3-approximation algorithms of Cygan *et al.* which have an $\mathcal{O}^*(3^n)$ and $\mathcal{O}^*(2^n)$ worst-case time bounds, respectively. Our algorithm is based on constructing bucket decompositions of the input graph. A bucket decomposition partitions the vertex set of a graph into ordered sets (called *buckets*) of (almost) equal sizes such that all edges are either incident on vertices in the same bucket or on vertices in two consecutive buckets. The idea is to find the smallest bucket size for which there exists a bucket decomposition. The algorithm uses a simple divide-and-conquer strategy along with dynamic programming to achieve this improved time bound.

1 Introduction

Let $G = (V, E)$ be a graph on n vertices and b be an integer. The Bandwidth problem asks whether the vertices of G can be labeled from 1 to n such that the labels of every pair of adjacent vertices differ by at most b. The Bandwidth problem is a special case of the Subgraph Isomorphism problem, as it can be formulated as follows: Is G isomorphic to a subgraph of P_n^b? Here, P_n^b denotes the graph obtained from P_n (the path on n vertices) by adding an edge between every pair of vertices at distance at most b in P_n.

A typical scenario in which the Bandwidth problem arises is that of minimizing the bandwidth of a symmetric matrix M to allow for more efficient storing and manipulating procedures [11]. The bandwidth of M is b if all its non-zero entries are at a distance of at most b from the diagonal. Applying permutations on the rows and columns to reduce the bandwidth of M corresponds then to reordering the vertices of a graph whose adjacency matrix corresponds to M by replacing all non-zero entries by 1.

* Visiting EPFL Lausanne and Universität Zürich. Research supported in part by NSF Grant CCF-0728921.
** Partially supported by the Research Council of Norway (NFR) and by the GRAAL project ANR-06-BLAN-0148 of the French National Research Agency (ANR).

J. Chen and F.V. Fomin (Eds.): IWPEC 2009, LNCS 5917, pp. 173–184, 2009.
© Springer-Verlag Berlin Heidelberg 2009

The Bandwidth problem is NP-hard [19], even for trees of maximum degree at most three [14] and caterpillars with hair length at most three [17]. Even worse, approximating the bandwidth within a constant factor is NP-hard, even for caterpillars of degree three [21]. Further, it is known that the problem is hard for every fixed level of the W-hierarchy [3] and unlikely to be solvable in $f(b)n^{o(b)}$ time [4].

Faced with this immense intractability, several approaches have been proposed in the literature for the Bandwidth problem. The first (polynomial time) approximation algorithm with a polylogarithmic approximation factor was provided by Feige [10]. Later, Dunagan and Vempala gave an $\mathcal{O}(\log^3 n\sqrt{\log\log n})$-approximation algorithm. The current best approximation algorithm achieves an $\mathcal{O}(\log^3 n(\log\log n)^{1/4})$-approximation factor [16]. For large b, the best approximation algorithm is the probabilistic algorithm of Blum et al. [2] which has an $\mathcal{O}(\sqrt{n/b}\log n)$-approximation factor.

Super-polynomial time approximation algorithms for the Bandwidth problem have also been widely investigated [5,8,9,12]. Feige and Talwar [12], and Cygan and Pilipczuk [8] provided subexponential time approximation schemes for approximating the bandwidth of graphs with small treewidth. For general graphs, a 2-approximation algorithm with a running time of $\mathcal{O}^*(3^n)$[1] is easily obtained by combining ideas from [11] and [12] (as noted in [5]). Further, Cygan et al. [5] provide a 3-approximation algorithm with a running time of $\mathcal{O}^*(2^n)$, which they generalize to a $(4r-1)$-approximation algorithm (for any positive integer r) with a running time of $\mathcal{O}^*(2^{n/r})$.

Concerning exact exponential time algorithms, the fastest polynomial space algorithm is still the elegant $\mathcal{O}^*(10^n)$ time algorithm of Feige [11]. When allowing exponential space, this bound is improved in a sequence of algorithms by Cygan and Pilipczuk; their $\mathcal{O}^*(5^n)$ time algorithm uses $\mathcal{O}^*(2^n)$ space [6], their $\mathcal{O}(4.83^n)$ time algorithm uses $\mathcal{O}^*(4^n)$ space [7], and their $\mathcal{O}(4.473^n)$ time algorithm uses $\mathcal{O}(4.473^n)$ space [8]. The most practical of these algorithms is probably the $\mathcal{O}^*(5^n)$ time algorithm as the space requirements of the other ones seems forbiddingly large for practical applications. The Bandwidth problem can also be solved exactly in $\mathcal{O}(n^b)$ time using dynamic programming [18,20].

Another recent approach to cope with the intractability of Bandwidth is through the concept of *hybrid algorithms*, introduced by Vassilevska et al. [22]. They gave an algorithm that after a polynomial time test, either computes the minimum bandwidth of a graph in $\mathcal{O}^*(4^{n+o(n)})$ time, or provides a polylogarithmic approximation ratio in polynomial time. This result was recently improved by Amini et al. [1] who give an algorithm which, after a polynomial time test, either computes the minimum bandwidth of a graph in $\mathcal{O}^*(4^n)$ time, or provides an $\mathcal{O}(\log^{3/2} n)$-approximation in polynomial time.

Our Results. Our main result is a 2-approximation algorithm for the Bandwidth problem that takes worst-case $\mathcal{O}(1.9797^n)$ time (Theorem 1). This improves the $\mathcal{O}^*(3^n)$ time bound achieved by Cygan et al. [5] for the same approximation ratio. Also, the previous best 3-approximation algorithm of Cygan and Pilipczuk [8] has an $\mathcal{O}^*(2^n)$ time bound. Therefore, our 2-approximation algorithm is also faster than the previous best 3-approximation algorithm.

---

[1] $\mathcal{O}^*(f(n))$ denotes $n^{O(1)}O(f(n))$.

Our algorithm is based on constructing bucket decompositions of the input graph. A bucket decomposition partitions the vertex set of a graph into ordered sets (called *buckets*) of (almost) equal sizes such that all edges are either incident on vertices in the same bucket or on vertices in two consecutive buckets. The idea is to find the smallest bucket size for which there exists a bucket decomposition. This gives a 2-approximation for the Bandwidth problem (Lemmas 2 and 1). The algorithm uses a simple divide-and-conquer strategy along with dynamic programming to achieve this improved time bound.

2   Preliminaries

Let $G = (V, E)$ be a graph on n vertices. A *linear arrangement* of G is a bijective function $L : V \rightarrow [n] = \{1, \ldots, n\}$, that is a numbering of its vertices from 1 to n. The *stretch* of an edge (u, v) is the absolute difference between the numbers assigned to its endpoints $|L(u) - L(v)|$. The *bandwidth* of a linear arrangement is the maximum stretch over all the edges of G and the *bandwidth* of a graph is the minimum bandwidth over all linear arrangements of G.

A *bucket arrangement* of G is a placement of its vertices into buckets such that for each edge, its endpoints are either in the same bucket or in two consecutive buckets [12]. The buckets are linearly ordered and numbered from left to right. A *capacity vector* C is a vector of positive integers. The *length* of a capacity vector $C = (C[1], \ldots, C[k])$ is k and its *size* is $\sum_{i=1}^{k} C[i]$. Given a capacity vector C of size n, a *C-bucket arrangement* of G is a bucket arrangement in which exactly $C[i]$ vertices are placed in bucket i, for each i. For integers n and ℓ with $\ell < n/2$, an *(n, ℓ)-capacity vector* is a capacity vector

$$(a, \underbrace{\ell, \ell, \ldots, \ell}_{\lceil \frac{n}{\ell} \rceil - 2 \text{ times}}, b)$$

of size n such that $a, b \leq \ell$. We say that an (n, ℓ)-capacity vector is *left-packed* if $a = \ell$ and *balanced* if $|a - b| = 1$.

Let $X \subseteq V$ be a subset of the vertices of G. We denote by $G[X]$ the subgraph of G induced on X, and by $G \setminus X$ the subgraph of G induced on $V \setminus X$. The *open neighborhood* of a vertex v is denoted by $N_G(v)$ and the *open neighborhood* of X is $N_G(X) := (\bigcup_{v \in X} N_G(v)) \setminus X$.

3   Exponential Time Algorithms for Approximating Bandwidth

We first establish two simple lemmas that show that constructing a bucket arrangement can approximate the bandwidth of a graph.

Lemma 1. *Let G be a graph on n vertices, and let C be an (n, ℓ)-capacity vector. If there exists a C-bucket arrangement for G then the bandwidth of G is at most $2\ell - 1$.*

Proof. Given a C-bucket arrangement for G, create a linear arrangement respecting the bucket arrangement (if u appears in a smaller numbered bucket than v, then $L(u) < L(v)$), where vertices in the same bucket are numbered in an arbitrary order. As the capacity of each bucket is at most ℓ and each edge spans at most two consecutive buckets, the maximum edge stretch in the constructed linear arrangement is at most $2\ell - 1$.   □

Lemma 2. *Let G be a graph on n vertices, and let C be an (n, ℓ)-capacity vector. If there exists no C-bucket arrangement for G then the bandwidth of G is at least $\ell + 1$.*

Proof. Suppose there exists a linear arrangement L of G of bandwidth at most ℓ. Construct a bucket arrangement placing the first $C[1]$ vertices of L into the first bucket, the next $C[2]$ vertices of L into the second bucket, and so on. In the resulting bucket arrangement, no edge spans more than two consecutive buckets. Therefore, a C-bucket arrangement exists for G, a contradiction.   □

We will use the previous fastest 2-approximation algorithm of Cygan *et al.* [5] as a subroutine. For completeness, we describe this simple algorithm here.

Proposition 1 ([5]). *There is a polynomial space 2-approximation algorithm for the Bandwidth problem that takes $\mathcal{O}^*(3^n)$ time on connected graphs with n vertices.*

Proof. Let G be a connected graph on n vertices. For ℓ increasing from 1 to $\lceil n/2 \rceil$, the algorithm does the following. Let C be an (n, ℓ)-capacity vector. The algorithm goes over all the $k = \lceil \frac{n}{\ell} \rceil$ choices for assigning the first vertex to some bucket. The algorithm then chooses an unassigned vertex u which has at least one neighbor that has already been assigned to some bucket. Assume that a neighbor of u is assigned to the bucket i. Now there are at most three choices of buckets ($i - 1$, i, and $i + 1$) for assigning vertex u. Some of these choices may be invalid either because of the capacity constraints of the bucket or because of the previous assignments of (other) neighbors of u. If the choice is valid, the algorithm recurses by assigning u to that bucket. Let ℓ' be the smallest integer for which the algorithm succeeds, in some branch, to place all vertices of G into buckets in this way. Then, by Lemma 1, G has bandwidth at most $2\ell' - 1$ and by Lemma 2, G has bandwidth at least ℓ'. Thus, the algorithm outputs $2\ell' - 1$, which is a 2-approximation for the bandwidth of G. As the algorithm branches into at most 3 cases for each of the n vertices (except the first one), and all other computations only contribute polynomially to the running time of the algorithm, this algorithm runs in worst-case $\mathcal{O}^*(3^n)$ time using only polynomial space.   □

We now show another simple algorithm based on a divide-and-conquer strategy that given an (n, ℓ)-capacity vector C, decides whether a C-bucket arrangement exists for a connected graph G.

Proposition 2. *Let G be a connected graph on n vertices and C be an (n, ℓ)-capacity vector with $\ell < n/2$. There exists an algorithm that can decide if G has a C-bucket arrangement in $\mathcal{O}^*\left(\binom{n}{\ell} \cdot \binom{n/2}{\ell} \cdot 2^{4\ell} \cdot 3^{n/4} \right)$ time.*

Proof. Let $k = \lceil \frac{n}{\ell} \rceil$ be the number of buckets in the C-bucket arrangement. Number the buckets from 1 to k from left to right according to the bucket arrangement. Select a bucket index i such that the sum of the capacities of the buckets numbered strictly smaller than i and the one for the buckets numbered strictly larger than i are both at most $n/2$.

The algorithm goes over all possible $\binom{n}{\ell}$ choices of filling bucket i with ℓ vertices. Let X be a set of ℓ vertices assigned to the bucket i. Given a connected component of $G \setminus X$, note that all the vertices of this connected component must be placed either only in buckets 1 to $i - 1$ or buckets $i + 1$ to k. Note that each connected component of $G \setminus X$ contains at least one vertex that is adjacent to a vertex in X (as G is connected). Therefore, for each connected component of $G \setminus X$, at least one vertex is placed into the bucket $i - 1$ or $i + 1$. As the capacity of each bucket is at most ℓ, $G \setminus X$ has at most 2ℓ connected components, otherwise there is no $\mathcal{C}$-bucket arrangement where X is assigned to the bucket i. Thus, there are at most $2^{2\ell}$ choices for assigning connected components of $G \setminus X$ to the buckets 1 to $i-1$ and $i+1$ to k. Some of these assignments might be invalid as they might violate the capacity constraints of the buckets. We discard these invalid assignments.

For each choice of X and each valid assignment of the connected components of $G \setminus X$ to the left or right of bucket i, we have now obtained two independent subproblems: one subproblem for the buckets $\{1, \ldots, i - 1\}$ and one subproblem for the buckets $\{i + 1, \ldots, k\}$. These subproblems have sizes at most $n/2$. Consider the subproblem for the buckets $\{1, \ldots, i - 1\}$ (the other one is symmetric) and let Y be the set of vertices associated to these buckets. Let $Z \subseteq Y$ be the set of vertices in Y that have at least one neighbor in X. Now, add edges to the subgraph $G[Y]$ such that Z becomes a clique. This does not change the problem, as all the vertices in Z must be assigned to the bucket $i - 1$, and $G[Y]$ becomes connected. This subproblem can be solved recursively, ignoring those solutions where vertices in Z are not all assigned to the bucket $i - 1$.

The algorithm performs the above recursion until it reaches subproblems of size at most $n/4$, which corresponds to two levels in the corresponding search tree. On instances of size at most $n/4$, the algorithm invokes the algorithm of Proposition 1, which takes worst-case $\mathcal{O}^*(3^{n/4})$ time.

Let $T(n)$ be the running time needed for the above procedure to check whether a graph with n vertices has a bucket arrangement for an (n, ℓ)-capacity vector. Then,

$$T(n) \leq \binom{n}{\ell} \cdot 2^{2\ell} \cdot \binom{n/2}{\ell} \cdot 2^{2\ell} \cdot 3^{n/4} \cdot n^{\mathcal{O}(1)} = \mathcal{O}^*\left(\binom{n}{\ell} \cdot \binom{n/2}{\ell} \cdot 2^{4\ell} \cdot 3^{n/4} \right).$$

This completes the proof of the proposition.   □

Combining Proposition 2 with Lemmas 1 and 2, we have the following corollary for 2-approximating the bandwidth of a graph.

Corollary 1. *There is an algorithm that, for a connected graph G on n vertices and an integer $\ell \leq n$ can decide whether the bandwidth of G is at least $\ell + 1$ or at most $2\ell - 1$ in $\mathcal{O}^*\left(\binom{n}{\ell} \cdot \binom{n/2}{\ell} \cdot 2^{4\ell} \cdot 3^{n/4} \right)$ time.*

Proof. If $\ell \geq n/2$, the bandwidth of G is at most $2\ell - 1$. Otherwise, use Proposition 2 with G and some (n, ℓ)-capacity vector $\mathcal{C}$ to decide if there exists a $\mathcal{C}$-bucket arrangement for G. If so, then the bandwidth of G is at most $2\ell - 1$ by Lemma 1. If not, then the bandwidth of G is at least $\ell + 1$ by Lemma 2.   □

The running time of the algorithm of Corollary 1 is interesting for small values of ℓ. Namely, if $\ell \leq n/26$, the running time is $\mathcal{O}(1.9737^n)$. In the remainder of this section, we improve Proposition 2. We now concentrate on the cases where $k = \lceil n/\ell \rceil \leq 26$.

Let $\mathcal{C}$ be an (n, ℓ)-capacity vector. A *partial $\mathcal{C}$-bucket arrangement* of an induced subgraph G' of G is a placement of vertices of G' into buckets such that: (a) each vertex in G' is assigned to a bucket or to a union of two consecutive buckets, (b) the endpoints of each edge in G' are either in the same bucket or in two consecutive buckets, and (c) at most $\mathcal{C}[i]$ vertices are placed in each bucket i. Let $\mathcal{B}$ be a partial $\mathcal{C}$-bucket arrangement of an induced subgraph G'. We say that a bucket i is *full* in $\mathcal{B}$ if the number of vertices that have been assigned to it equals its capacity ($= \mathcal{C}[i]$). We say that two consecutive buckets i and $i + 1$ are *jointly full* in $\mathcal{B}$ if a vertex subset Y of cardinality equal to the sum of the capacities of i and $i+1$ have been assigned to these buckets (i.e., each vertex $v \in Y$ is restricted to belong to the union of buckets i or $i + 1$, but which among these two buckets v belongs is not fixed). We say that a bucket is *empty* in $\mathcal{B}$ if no vertices have been assigned to it.

Proposition 3. *Let G be a graph on n vertices and $\mathcal{C}$ be a capacity vector of size n and length k, where k is an integer constant. Let $\mathcal{B} = \mathcal{B}(G')$ be a partial $\mathcal{C}$-bucket arrangement of some induced subgraph G' of G such that in $\mathcal{B}$ some buckets are full, some pairs of consecutive buckets are jointly full, and all other buckets are empty. If in $\mathcal{B}$ no 3 consecutive buckets are empty, then it can be decided if $\mathcal{B}$ can be extended to a $\mathcal{C}$-bucket arrangement in polynomial time.*

Proof Outline. Let $G = (V, E)$ and $G' = (V', E')$. Let r be the number of connected components of $G \setminus V'$ (the graph induced on $V \setminus V'$), and let V_l represent the set of vertices in the lth connected component of $G \setminus V'$.

If the bucket i is full in $\mathcal{B}$, let X_i denote the set of vertices assigned to it. If the buckets i and $i + 1$ are jointly full in $\mathcal{B}$, let $X_{i,i+1}$ denote the set of vertices assigned to the union of buckets i and $i+1$. We use dynamic programming to start from a partial bucket arrangement satisfying the above conditions to construct a $\mathcal{C}$-bucket arrangement. During its execution, the algorithm assigns vertices to the buckets which are empty in $\mathcal{B}$. We only present an outline of the dynamic programming algorithm here. The dynamic programming algorithm constructs a table $T[\ldots]$, which has the following indices.

- An index p, representing the subproblem on the first p connected components of $G \setminus V'$.
- For every empty bucket i in $\mathcal{B}$ such that both the buckets $i - 1$ and $i + 1$ are full, it has an index s_i, representing the number of vertices assigned to the bucket i.
- For every two consecutive empty buckets i and $i + 1$ in $\mathcal{B}$, it has indices $t_{i,i+1}$, x_i, and x_{i+1}. The index $t_{i,i+1}$ represents the total number of vertices assigned to the buckets i and $i + 1$. The index x_i represents the number of vertices assigned to the buckets i and $i + 1$ that have at least one neighbor in the bucket $i - 1$. The index x_{i+1} represents the number of vertices assigned to the buckets i and $i + 1$ that have at least one neighbor in the bucket $i + 2$.
- For every two consecutive buckets $i, i + 1$ which are jointly full in $\mathcal{B}$, it has indices f_i and f_{i+1} representing the number of vertices assigned to these buckets that have at least one neighbor in the bucket $i - 1$ (f_i) or in the bucket $i + 2$ (f_{i+1}).

Table $T[\ldots]$ is initialized to false everywhere, except for the entry corresponding to all-zero indices, which is initialized to true. The rest of the table is built by increasing values of p as described below. Here, we only write those indices that differ in the looked-up table entries and the computed table entry (i.e., indices in the table that play no role in a given recursion are omitted). We also ignore the explicit checking of the invalid indices in the following description. The algorithm looks at the vertices which are neighbors (in G) of the vertices in V_p and have already been assigned.

If the vertices in V_p have at least one neighbor in each of the full buckets $i - 1$ and $i + 1$, have no neighbors in any other buckets, and bucket i is empty in $\mathcal{B}$, then

$$T[p, s_i, \ldots] = T[p - 1, s_i - |V_p|, \ldots].$$

If the vertices in V_p have at least one neighbor in the full buckets $i - 1$ and $i + 2$, have no neighbors in any other buckets, and the buckets i and $i + 1$ are both empty in $\mathcal{B}$, then

$$T[p, t_{i,i+1}, x_i, x_{i+1}, \ldots] =$$
$$\begin{cases} \text{false} & \text{if } N_G(X_{i-1}) \cap N_G(X_{i+2}) \neq \emptyset, \\ T[p - 1, t_{i,i+1} - |V_p|, x_i - |V_p \cap N_G(X_{i-1})|, \\ \quad x_{i+1} - |V_p \cap N_G(X_{i+2})|, \ldots] & \text{otherwise.} \end{cases}$$

If the vertices in V_p have at least one neighbor in the jointly full buckets $i - 2$ and $i - 1$, and at least one neighbor in the jointly full buckets $i+1$ and $i+2$, but have no neighbors in any other buckets, and bucket i is empty in $\mathcal{B}$, then

$$T[p, s_i, f_{i-1}, f_{i+1}, \ldots] = T[p - 1, s_i - |V_p|, f_{i-1} - |N_G(V_p) \cap X_{i-2,i-1}|,$$
$$f_{i+1} - |N_G(V_p) \cap X_{i+1,i+2}|, \ldots].$$

The recursion for the other possibilities where V_p has neighbors in two distinct buckets can now easily be deduced. We now consider the cases where V_p has only neighbors in one bucket. Again, we only describe some key-cases, from which all other cases can easily be deduced.

If the vertices in V_p have only neighbors in the full bucket $i - 1$, and the buckets $i - 2$ and i are both empty in $\mathcal{B}$, but the buckets $i - 3$ and $i + 1$ are either full or non-existing, then

$$T[p, s_{i-2}, s_i, \ldots] = T[p - 1, s_{i-2} - |V_p|, s_i, \ldots] \vee T[p - 1, s_{i-2}, s_i - |V_p|, \ldots].$$

If the vertices in V_p have only neighbors in the full bucket $i - 1$, and the buckets $i - 3$, $i - 2$, i, and $i + 1$ are all empty in $\mathcal{B}$, then

$$T[p, t_{i-3,i-2}, x_{i-2}, t_{i,i+1}, x_i, \ldots] =$$
$$T[p - 1, t_{i-3,i-2} - |V_p|, x_{i-2} - |V_p \cap N_G(X_{i-1})|, t_{i,i+1}, x_i, \ldots]$$
$$\vee T[p - 1, t_{i-3,i-2}, x_{i-2}, t_{i,i+1} - |V_p|, x_i - |V_p \cap N_G(X_{i-1})|, \ldots].$$

If the vertices in V_p have only neighbors in the jointly full buckets i and $i + 1$, and the buckets $i - 1$ and $i + 2$ are both empty in $\mathcal{B}$, but the buckets $i - 2$ and $i + 3$ are either full in $\mathcal{B}$ or non-existing, then

$$T[p, s_{i-1}, s_{i+2}, f_i, f_{i+1}, \ldots] =$$
$$T[p-1, s_{i-1} - |V_p|, s_{i+2}, f_i - |N_G(V_p) \cap X_{i,i+1}|, f_{i+1}, \ldots]$$
$$\vee \ T[p-1, s_{i-1}, s_{i+2} - |V_p|, f_i, f_{i+1} - |N_G(V_p) \cap X_{i,i+1}|, \ldots].$$

The final answer (true or false) produced by the algorithm is a disjunction over all table entries whose indices are as follows: $p = r$, $s_i = C[i]$ for every index s_i, $t_{i,i+1} = C[i] + C[i+1]$ for every index $t_{i,i+1}$, $x_i \leq C[i]$ for every index x_i, and $f_i \leq C[i]$ for every index f_i.   □

Remark 1. The dynamic programming algorithm in Proposition 3 can easily be modified to construct a C-bucket arrangement (from any partial bucket arrangement $\mathcal{B}$ satisfying the stated conditions), if one exists.

If the number of buckets is a constant, the following proposition will be crucial in speeding up the procedure for assigning connected components to the right or the left of a bucket filled with a vertex set X. Denote by $\mathsf{sc}(G)$ the set of all connected components of G with at most $\sqrt{n}$ vertices and by $\mathsf{lc}(G)$ the set of all connected components of G with more than $\sqrt{n}$ vertices. Let $V(\mathsf{sc}(G))$ and $V(\mathsf{lc}(G))$ denote the set of all vertices which are in the connected components belonging to $\mathsf{sc}(G)$ and $\mathsf{lc}(G)$, respectively. We now make use of the fact that if there are many small components in $G \setminus X$, several of the assignments of the vertices in $V(\mathsf{sc}(G \setminus X))$ to the buckets are equivalent.

Let C be a capacity vector of size n (i.e., $\sum_i C[i] = n$) and let $\mathcal{B}$ be a partial C-bucket arrangement of an induced subgraph G' of G. Let C' be the capacity vector obtained from C by decreasing the capacity $C[i]$ of each bucket i by the number of vertices assigned to the bucket i in $\mathcal{B}$. We say that $\mathcal{B}$ *produces* the capacity vector C'.

Proposition 4. *Let $G = (V, E)$ be a graph on n vertices. Let C be a capacity vector of size n and length k, where k is an integer constant. Let j be a bucket and $X \subseteq V$ be a subset of $C[j]$ vertices. Consider all capacity vectors which are produced by the partial C-bucket arrangements of $G[V(\mathsf{sc}(G \setminus X)) \cup X]$ where the vertices in X are always assigned to the bucket j. Then, there exists an algorithm which runs in $\mathcal{O}^*(3^{\sqrt{n}})$ time and takes polynomial space, and enumerates all (distinct) capacity vectors produced by these partial C-bucket arrangements.*

Proof. Let V_l be the vertex set of the lth connected component in $\mathsf{sc}(G \setminus X)$. Let $\mathcal{L}_p$ denote the list of all capacity vectors produced by the partial C-bucket arrangements of $G[\bigcup_{1 \leq l \leq p} V_l \cup X]$ where the vertices in X are always assigned to the bucket j. Note that since k is a constant, the number of distinct vectors in $\mathcal{L}_p$ is polynomial (at most n^k). Then, $\mathcal{L}_1$ can be obtained by executing the algorithm of Proposition 1 on the graph $G[V_1]$ with a capacity vector C' which is the same as C except that $C'[i] = 0$. In general, $\mathcal{L}_p$ can be obtained from $\mathcal{L}_{p-1}$ by executing the algorithm of Proposition 1 on the graph $G[V_p]$ for every capacity vector in $\mathcal{L}_{p-1}$. As the size of each connected component in $\mathsf{sc}(G \setminus X)$ is at most $\sqrt{n}$, the resulting running time is $\mathcal{O}^*(3^{\sqrt{n}})$.   □

3.1   Exponential Time 2-Approximation Algorithm for Bandwidth

Let $G = (V, E)$ be the input graph. Our algorithm tests all bucket sizes ℓ from 1 to $\lceil n/2 \rceil$ until it finds an (n, ℓ)-capacity vector C such that G has a C-bucket arrangement.

For a given ℓ, let $k = \lceil \frac{n}{\ell} \rceil$ denote the number of buckets. Our algorithm uses various strategies depending on the value of k. The case of $k = 1$ is trivial. If $\ell = \lceil n/2 \rceil$, we have at most two buckets and any partition of the vertex set of G into sets of sizes ℓ and $n - \ell$ is a valid $\mathcal{C}$-bucket arrangement. If $k \geq 27$, Corollary 1 gives a running time of $\mathcal{O}(1.9737^n)$. For all other values of k, we will obtain running times in $\mathcal{O}(1.9797^n)$.

Let I_k be the set of all integers lying between $n/(k - 1)$ and n/k. The basic idea (as illustrated in Proposition 2) is quite simple. The algorithm tries all possible ways of assigning vertices to the middle bucket. Once the vertex set X assigned to the middle bucket is fixed and the algorithm has decided for each connected component of $G \setminus X$ if the connected component is to be assigned to the buckets to the left or to the right of the middle bucket, the problem breaks into two independent subproblems on buckets which are to the left and to right of the middle bucket. To get the claimed running time, we build upon this idea to design individualized techniques for different ks (between 3 and 26). For each case, if G has at least one $\mathcal{C}$-bucket arrangement for an (n, ℓ)-capacity vector $\mathcal{C}$, then one such arrangement is constructed. We know that if G has no $\mathcal{C}$-bucket arrangement for an (n, ℓ)-capacity vector $\mathcal{C}$ then the bandwidth of G is at least $\ell + 1$ (Lemma 2), and if it has one then its bandwidth is at most $2\ell - 1$ (Lemma 1). If $k = 8, 10$, or 12, the algorithm uses a left-packed (n, ℓ)-capacity vector $\mathcal{C}$, and otherwise, the algorithm uses a balanced (n, ℓ)-capacity vector $\mathcal{C}$.

$k = 3$. The algorithm goes over all subsets $X \subseteq V$ of cardinality $|X| = \mathcal{C}[3] \leq \lceil (n - \ell)/2 \rceil$ with $\ell \in I_3$. X is assigned to the bucket 3. If the remaining vertices can be assigned to the buckets 1 and 2 in a way such that all vertices which are neighbors of the vertices in X (in G) are assigned to the bucket 2, then G has a $\mathcal{C}$-bucket arrangement where $\mathcal{C}$ has length 3. The worst-case running time for this case is $\max_{\ell \in I_3} \mathcal{O}^*\left(\binom{n}{|X|}\right)$.

$k = 4$ or $k = 5$. The algorithm goes over all subsets $X \subseteq V$ with $|X| = \ell$ and $\ell \in I_k$. X is assigned to the bucket 3. Then, we can conclude using the dynamic programming algorithm outlined in Proposition 3 (see also the remark following it). The worst-case running time for these cases are $\max_{\ell \in I_k} \mathcal{O}^*\left(\binom{n}{\ell}\right)$.

$k = 6$. If $k = 6$, the algorithm goes through all subsets $X \subseteq V$ with $|X| = 2\ell$ and $\ell \in I_6$. X is assigned to the union of buckets 3 and 4 (i.e., some non-specified ℓ vertices from X are assigned to the bucket 3, and the remaining vertices of X are assigned to the bucket 4). Then, we can again conclude by the algorithm outlined in Proposition 3. The worst-case running time for this case is $\max_{\ell \in I_6} \mathcal{O}^*\left(\binom{n}{2\ell}\right)$.

$k = 7$. The algorithm goes through all subsets $X \subseteq V$ with $|X| = \ell$ and $\ell \in I_7$. X is assigned to the bucket 4. For each such X, the algorithm uses Proposition 4 to enumerate all possible capacity vectors produced by the partial $\mathcal{C}$-bucket arrangements of $G[V(\mathsf{sc}(G \setminus X)) \cup X]$ (with X assigned to the bucket 4). This step can be done in $\mathcal{O}^*(3^{\sqrt{n}})$ time. There are only polynomially many such (distinct) capacity vectors. For each of these capacity vector $\mathcal{C}'$, the algorithm goes through all choices of assigning each connected component in $\mathsf{lc}(G \setminus X)$ to the buckets 1 to 3 or to the buckets 5 to 7. Thus, we obtain two independent subproblems on the buckets 1 to 3 and on the buckets 5 to 7. As the number of number of components in $\mathsf{lc}(G \setminus X)$ is at most $\sqrt{n}$ (as each connected component has at least $\sqrt{n}$ vertices), going through all possible ways of assigning each connected component in $\mathrm{lc}(G \setminus X)$ to the buckets numbered smaller or larger than 4 takes $\mathcal{O}^*(2^{\sqrt{n}})$ time. Some of these assignments may turn out to be invalid. For each valid assignment, let V_1 denote the vertex set assigned to the buckets 1 to 3. Then, the vertices of V_1 are assigned to the buckets 1 to 3 as described in the case with 3 buckets with the capacity vector $(\mathcal{C}'[1], \mathcal{C}'[2], \mathcal{C}'[3])$ and with the additional restriction that all vertices in V_1 which are neighbors of the vertices in X need to be assigned to the bucket 3. The number of vertices in V_1 is at most $\lceil (n - \ell)/2 \rceil$ (as $\mathcal{C}$ is balanced). Now the size of bucket 1 is $\mathcal{C}'[1] \leq \lceil (n - 5\ell)/2 \rceil$. Let $n_1 = \lceil (n - \ell)/2 \rceil$ and $\ell_1 = \lceil (n - 5\ell)/2 \rceil$. If V_1 has at least one valid bucket arrangement into 3 buckets (with vertices in V_1 neighboring the vertices in X assigned to the bucket 3), then the above step will construct one in worst-case $\mathcal{O}^*(\binom{n_1}{\ell_1})$ time. The algorithm uses a similar approach for $V_2 = V \setminus (V_1 \cup X)$ with the buckets 5 to 7. Since, the algorithm tries out every subset X for bucket 4, the worst-case running time for this case is

$$\max_{\ell \in I_7} \mathcal{O}^* \left(\binom{n}{\ell} \cdot \left(3^{\sqrt{n}} + 2^{\sqrt{n}} \cdot \binom{n_1}{\ell_1} \right) \right) = \max_{\ell \in I_7} \mathcal{O}^* \left(\binom{n}{\ell} \cdot 2^{\mathcal{O}(\sqrt{n})} \cdot \binom{n_1}{\ell_1} \right).$$

k = 8. The algorithm uses a left-packed (n, ℓ)-capacity vector $\mathcal{C}$ for this case. The algorithm goes through all subsets $X \subseteq V$ with $|X| = \ell$ and $\ell \in I_8$. X is assigned to the bucket 4. The remaining analysis is similar to the case with 7 buckets. Buckets 1 to 3 have a joint capacity of 3ℓ (as $\mathcal{C}$ is left-packed) and the buckets 5 to 8 have a joint capacity of $n - 4\ell$. The worst-case running time for this case is

$$\max_{\ell \in I_8} \mathcal{O}^* \left(\binom{n}{\ell} \cdot 2^{\mathcal{O}(\sqrt{n})} \cdot \max \left\{ \binom{3\ell}{\ell}, \binom{n - 4\ell}{\ell} \right\} \right).$$

The terms in the max expression come from the cases with 3 and 4 buckets.

k = 9 or k = 11. The algorithm goes through all subsets $X \subseteq V$ with $|X| = \ell$ and $\ell \in I_k$. X is assigned to the bucket $\lceil k/2 \rceil$. As in the previous two cases, Proposition 4 is invoked for $G[V(\mathrm{sc}(G \setminus X)) \cup X]$ (with X assigned to the bucket $\lceil k/2 \rceil$). For each capacity vector generated by Proposition 4, the algorithm looks at every possible way of assigning each connected component in $\mathrm{lc}(G \setminus X)$ to the buckets 1 to $\lceil k/2 \rceil - 1$ or to the buckets $\lceil k/2 \rceil + 1$ to k. Each assignment gives rise to two independent subproblems — one on vertices V_1 assigned to the buckets 1 to $(k-1)/2$, and one on vertices V_2 assigned to the buckets $(k + 3)/2$ to k (with vertices in V_1 and V_2 neighboring the vertices in X assigned to the buckets $(k - 1)/2$ and $(k + 3)/2$, respectively). The algorithm solves these subproblems recursively as in the cases with 4 or 5 buckets. Let $n_1 = \lceil (n - \ell)/2 \rceil$. Then, the worst-case running times are $\max_{\ell \in I_k} \mathcal{O}^*(\binom{n}{\ell} \cdot 2^{\mathcal{O}(\sqrt{n})} \cdot \binom{n_1}{\ell})$.

k = 10 or k = 12. The algorithm uses a left-packed (n, ℓ)-capacity vector $\mathcal{C}$ for these cases. The algorithm goes through all subsets $X \subseteq V$ with $|X| = \ell$ and $\ell \in I_k$. X is assigned to the bucket $k/2$. The remaining analysis is similar to the previous cases. For $k = 10$, the worst-case running time is $\max_{\ell \in I_{10}} \mathcal{O}^*(\binom{n}{\ell} \cdot 2^{\mathcal{O}(\sqrt{n})} \cdot \binom{n/2}{\ell})$. For $k = 12$, the worst-case running time is $\max_{\ell \in I_{12}} \mathcal{O}^*(\binom{n}{\ell} \cdot 2^{\mathcal{O}(\sqrt{n})} \cdot \max\{\binom{5\ell}{\ell}, \binom{n - 6\ell}{2\ell}\})$.

13 ≤ k ≤ 26. The algorithm enumerates all subsets $X \subseteq V$ with $|X| = \ell$ and $\ell \in I_k$. X is assigned to the bucket $\lceil k/2 \rceil$. As in the previous cases, Proposition 4 is invoked for $G[V(\mathrm{sc}(G \setminus X)) \cup X]$. For each capacity vector generated by Proposition 4, the algorithm looks at every possible way of assigning each connected component in $\mathrm{lc}(G \setminus X)$ to the buckets 1 to $\lceil k/2 \rceil - 1$ or to the buckets $\lceil k/2 \rceil + 1$ to k. Each assignment gives rise to two independent subproblems. For each of these two subproblems, the algorithm proceeds recursively until reaching subproblems with at most 2 consecutive empty buckets, which can be solved by Proposition 3 in polynomial time. If $k \leq 23$, this recursion has depth 3, giving a running time of

$$\max_{\ell \in I_k} \mathcal{O}^* \left(\binom{n}{\ell} \cdot 2^{\mathcal{O}(\sqrt{n})} \cdot \binom{n/2}{\ell} \cdot 2^{\mathcal{O}(\sqrt{n})} \cdot \binom{n/4}{\ell} \cdot 2^{\mathcal{O}(\sqrt{n})} \right) .$$

If $24 \leq k \leq 26$, the recursion has depth 4, giving a running time of

$$\max_{\ell \in I_k} \mathcal{O}^* \left(\binom{n}{\ell} \cdot 2^{\mathcal{O}(\sqrt{n})} \cdot \binom{n/2}{\ell} \cdot 2^{\mathcal{O}(\sqrt{n})} \cdot \binom{n/4}{\ell} \cdot 2^{\mathcal{O}(\sqrt{n})} \cdot \binom{n/8}{\ell} \cdot 2^{\mathcal{O}(\sqrt{n})} \right) .$$

k ≥ 27. By Proposition 2 the running time of the algorithm is bounded in this case by

$$\max_{\ell \in I_k} \mathcal{O}^* \left(\binom{n}{\ell} \cdot \binom{n/2}{\ell} \cdot 2^{4\ell} \cdot 3^{n/4} \right) .$$

Main Result. Putting together all the above arguments and using numerical values (see [13] for the complete details) we get our main result (Theorem 1). The running time is dominated by the cases where $k = 7$ and $k = 8$. The algorithm outputs $2\ell - 1$, where ℓ is the smallest integer such that G has a bucket arrangement with an (n, ℓ)-capacity vector. The algorithm requires only polynomial space.

If G is disconnected, the algorithm finds for each connected component $G_i = (V_i, E_i)$ the smallest ℓ_i such that G_i has a bucket arrangement corresponding to a $(|V_i|, \ell_i)$-capacity vector and outputs $2\ell_m - 1$, where $\ell_m = \max_i \{\ell_i\}$.

Theorem 1 (Main Theorem). *There is a polynomial space 2-approximation algorithm for the Bandwidth problem that takes $\mathcal{O}(1.9797^n)$ time on graphs with n vertices.*

4   Conclusion

For finding exact solutions, it is known that many problems (by subexponential time preserving reductions) do not admit subexponential time algorithms under the Exponential Time Hypothesis [15] (a stronger hypothesis than P ≠ NP). The Exponential Time Hypothesis supposes that there is a constant c such that 3-SAT cannot be solved in time $\mathcal{O}(2^{cn})$, where n is the number of variables of the input formula. We conjecture that the Bandwidth problem has no subexponential time 2-approximation algorithm, unless the Exponential Time Hypothesis fails.

References

1. Amini, O., Fomin, F.V., Saurabh, S.: Counting Subgraphs via Homomorphisms. In: Proceedings of ICALP 2009, pp. 71–82 (2009)
2. Blum, A., Konjevod, G., Ravi, R., Vempala, S.: Semi-Definite Relaxations for Minimum Bandwidth and other Vertex-Ordering problems. Theor. Comput. Sci. 235(1), 25–42 (2000)
3. Bodlaender, H.L., Fellows, M.R., Hallett, M.T.: Beyond NP-completeness for Problems of Bounded Width: Hardness for the W-hierarchy. In: Proceedings of STOC 1994, pp. 449–458 (1994)
4. Chen, J., Huang, X., Kanj, I.A., Xia, G.: Linear FPT Reductions and Computational Lower Bounds. In: Proceedings of STOC 2004, pp. 212–221 (2004)
5. Cygan, M., Kowalik, L., Pilipczuk, M., Wykurz, M.: Exponential-time Approximation of Hard Problems, Technical Report abs/0810.4934, arXiv, CoRR (2008)
6. Cygan, M., Pilipczuk, M.: Faster exact bandwidth. In: Broersma, H., Erlebach, T., Friedetzky, T., Paulusma, D. (eds.) WG 2008. LNCS, vol. 5344, pp. 101–109. Springer, Heidelberg (2008)
7. Cygan, M., Pilipczuk, M.: Even Faster Exact Bandwidth, Technical Report abs/0902.1661, arXiv, CoRR (2009)
8. Cygan, M., Pilipczuk, M.: Exact and approximate Bandwidth. In: Proceedings of ICALP 2009, pp. 304–315 (2009)
9. Dunagan, J., Vempala, S.S.: On euclidean embeddings and bandwidth minimization. In: Goemans, M.X., Jansen, K., Rolim, J.D.P., Trevisan, L. (eds.) RANDOM 2001 and APPROX 2001. LNCS, vol. 2129, pp. 229–240. Springer, Heidelberg (2001)
10. Feige, U.: Approximating the Bandwidth via Volume Respecting Embeddings. J. Comput. Syst. Sci. 60(3), 510–539 (2000)
11. Feige, U.: Coping with the NP-Hardness of the Graph Bandwidth Problem. In: Halldórsson, M.M. (ed.) SWAT 2000. LNCS, vol. 1851, pp. 10–19. Springer, Heidelberg (2000)
12. Feige, U., Talwar, K.: Approximating the bandwidth of caterpillars. In: Chekuri, C., Jansen, K., Rolim, J.D.P., Trevisan, L. (eds.) APPROX 2005 and RANDOM 2005. LNCS, vol. 3624, pp. 62–73. Springer, Heidelberg (2005)
13. Fürer, M., Gaspers, S., Kasiviswanathan, S.P.: An Exponential Time 2-Approximation Algorithm for Bandwidth, Technical Report abs/0906.1953, arXiv, CoRR (2009)
14. Garey, M.R., Graham, R.L., Johnson, D.S., Knuth, D.E.: Complexity Results for Bandwidth Minimization. SIAM J. Appl. Math. 34(3), 477–495 (1978)
15. Impagliazzo, R., Paturi, R.: On the Complexity of k-SAT. J. Comput. Syst. Sci. 62(2), 367–375 (2001)
16. Lee, J.R.: Volume Distortion for subsets of Euclidean Spaces. Discrete Comput. Geom. 41(4), 590–615 (2009)
17. Monien, B.: The Bandwidth Minimization Problem for Caterpillars with Hair Length 3 is NP-complete. SIAM J. Alg. Disc. Meth. 7(4), 505–512 (1986)
18. Monien, B., Sudborough, I.H.: Bandwidth Problems in Graphs. In: Proceedings of Allerton Conference on Communication, Control, and Computing 1980, pp. 650–659 (1980)
19. Papadimitriou, C.: The NP-completeness of the Bandwidth Minimization Problem. Computing 16, 263–270 (1976)
20. Saxe, J.: Dynamic Programming Algorithms for Recognizing Small-bandwidth Graphs in Polynomial Time. SIAM J. Alg. Disc. Meth. 1, 363–369 (1980)
21. Unger, W.: The Complexity of the Approximation of the Bandwidth Problem. In: Proceedings of FOCS 1998, pp. 82–91 (1998)
22. Vassilevska, V., Williams, R., Woo, S.L.M.: Confronting Hardness using a Hybrid Approach. In: Proceedings of SODA 2006, pp. 1–10 (2006)

On Digraph Width Measures
in Parameterized Algorithmics

Robert Ganian[1], Petr Hliněný[1], Joachim Kneis[2], Alexander Langer[2],
Jan Obdržálek[1], and Peter Rossmanith[2]

[1] Faculty of Informatics, Masaryk University, Brno, Czech Republic
{xganian1,hlineny,obdrzalek}@fi.muni.cz
[2] Theoretical Computer Science, RWTH Aachen University, Germany
{kneis,langer,rossmani}@cs.rwth-aachen.de

Abstract. In contrast to undirected width measures (such as tree-width or clique-width), which have provided many important algorithmic applications, analogous measures for digraphs such as DAG-width or Kelly-width do not seem so successful. Several recent papers, e.g. those of Kreutzer–Ordyniak, Dankelmann–Gutin–Kim, or Lampis–Kaouri–Mitsou, have given some evidence for this. We support this direction by showing that many quite different problems remain hard even on graph classes that are restricted very beyond simply having small DAG-width. To this end, we introduce new measures K-width and DAG-depth. On the positive side, we also note that taking Kanté's directed generalization of rank-width as a parameter makes many problems fixed parameter tractable.

1 Introduction

The very successful concept of graph *tree-width* was introduced in the context of the Graph Minors project by Robertson and Seymour [RS86, RS91], and it turned out to be very useful for efficiently solving graph problems. Tree-width is a property of *undirected graphs*. In this paper we will be interested in *directed graphs* or *digraphs*.

Naturally, a width measure specifically tailored to digraphs with all the nice properties of tree-width would be tremendously useful. The properties of such a measure should include at least the following:

i) The width measure is small on many interesting instances.
ii) Many hard problems become easy if the width measure is bounded.

Obviously, there is a conflict between these goals, and consequently we can expect some trade-off. On the search for such a digraph measure, several suggestions were made, starting with directed tree-width [JRST01], and being complemented recently with several new approaches including DAG-width [Obd06, BDHK06], Kelly-width [HK08], entanglement [BG04], D-width [Saf05], directed path-width [Bar06] (defined by Reed, Seymour, and Thomas), and — although quite different — bi-rank-width [Kan08] (see Section 2).

J. Chen and F.V. Fomin (Eds.): IWPEC 2009, LNCS 5917, pp. 185–197, 2009.

Some positive results were encouraging: The Hamiltonian path problem can be solved in polynomial time (XP) if the directed tree width, the DAG-width, or the Kelly-width are bounded by a constant [JRST01]. More recently, it has been shown that parity games can be solved in polynomial time on digraphs of bounded DAG-width [BDHK06] and Kelly-width [HK08].

Are more results just waiting around the corner and do we just have to wait until we get more familiar with these digraph measures? It is the aim of this paper to answer this question, at least partially.

Unfortunately, as encouraging as the first positive results are, there is also the negative side. Hamiltonian path is W[2]-hard on digraphs of bounded DAG-width [LKM08], and some other natural problems even remain NP-hard on digraphs of low widths [KO08, DGK08, LKM08]. One of the main goals of this paper is to show that not only many problems are hard on DAGs, but rather that they remain hard even if we very severely further restrict the graphs structure.

We introduce two digraph measures for this purpose: K-width and DAG-depth. While K-width (Section 2.3) restricts the number of different simple paths between pairs of vertices, DAG-depth (Definition 2.6) is the directed analog of tree-depth [NdM06]. K-width and DAG-depth are very restrictive digraph measures; at least as high as DAG-width, and often much higher.

The problems we consider in this paper (and formally define in Section 3) are Hamiltonian path (HAM), Disjoint paths (k-PATH), Directed Dominating Set (DIDS), unit cost Directed Steiner Tree (DISTP), Directed Feedback Vertex Set (DFVS), Kernel (KERNEL), Maximum Directed Cut (MAXDICUT), Oriented Colouring (OCN), MSO_1 model checking (ϕ-MSO_1MC), solving Parity Games (PARITY) and LTL-model checking (ϕ-LTLMC). See Table 1 in Section 3.

It turns out that most of the aforementioned problems are not only hard for DAG-width, but even for constant K-width and DAG-depth, or on DAGs. This can be seen as a strong indication that DAG-width or related measures are not yet the right parameters for dealing with standard digraph problems.

On the other hand, one width measure that fares much better in Table 1 is bi-rank-width (Definition 2.4), a width measure generalizing the rank-width of undirected graphs [Kan08]. Nearly all of our problems are fixed parameter tractable or at least in XP with respect to this parameter. Even better, unlike as for DAG-width or Kelly-width, finding an optimal bi-rank-decomposition is known to be in FPT [HO08, Kan08].

2   Digraph Width Measures

The first wave of directed measures to appear shared the following features:

  i) On bidirected orientations of graphs they coincided with the tree-width.
 ii) These measures were strongly based on some variant of the directed *cops-and-robber game* on a digraph: There are k cops and a robber. Each cop can either occupy a vertex, or move around in a helicopter, and the robber occupies a vertex. The robber can, however, see the helicopter landing, and can move at a great speed along a cop-free directed path to another vertex.

The objective of the cops is to capture the robber by landing on the vertex currently occupied by him, the objective of the robber is to avoid capture.

iii) Point (ii) implied that DAGs and other graphs where vertices could be ordered in such a way that edges between them point mainly in one direction, and only a few point backwards, have a very low width.

iv) The last feature (iii) also made the algorithms to be XP, instead of FPT, because of the need to remember the partial results for all vertices with incoming edges from the outside, of which there could be $|V|$.

Directed tree-width. The first explicit directed measure was that of *directed tree-width (dtw)* [JRST01]. In the related cops-and-robber game the robber has to stay in the same cop-free strongly connected component, however the relationship between the number of cops needed and the directed tree-width is not strict. [JRST01] also contains XP algorithms for solving the Hamiltonian cycle, k-path, and related problems on graphs of bounded directed tree-width.

DAG-width. First defined in [Obd06] and, independently, in [BDHK06], *DAG-width (dagw)* was the next attempt to come up with a directed tree-width counterpart. This time the robber does not have to stay in the SCC, but the cop strategy has to be monotone, i.e., a cop cannot be placed on a previously vacated vertex. This game fully characterizes DAG-width. Note that monotone and non-monotone strategies are not equivalent [KO08].

Theorem 2.1 ([Obd06, BDHK06]). *For any graph G, there is a DAG-decomposition of G of width k if, and only if, the cop player has a monotone winning strategy in the k-cops-and-robber game on G.*

Kelly-width. Defined a year later, *Kelly-width (kellyw)* [HK08] aimed to solve an existing problem with DAG-decompositions: the number of nodes can be polynomially larger then the number of vertices in the original graph (the size depends on the width). The idea of Kelly-decompositions is based on the elimination ordering for tree-width, and therefore the size of the decomposition is linear in the size of the graph. The game characterizing Kelly-width is as for DAG-width, but with two important differences: 1) the cops cannot see the robber, and 2) the robber can move only when a cop is about to land on his vertex.

Cycle rank. This is perhaps the oldest definition of a digraph connectivity measure, given in 60's by Eggan and Büchi [Egg63].

Definition 2.2 (Cycle rank). The *cycle rank* $cr(G)$ of a digraph G is defined inductively as follows: For DAGs, $cr(G) = 1$. If G is strongly connected and $E(G) \neq \emptyset$, then $cr(G) = 1 + \min\{ cr(G - v) : v \in V(G) \}$. Otherwise, $cr(G)$ is the maximum over the cycle rank of the strongly connected components of G.

Measure comparison. All the measures presented above are closely related to each other. The following theorem in a summary shows that if a problem is hard for graphs of bounded cycle rank, then it is hard for all the other measures.

Theorem 2.3. *Let G be a digraph. Then (dpw [Bar06] is the directed path-width):*

$$1/3(dtw(G) - 1) \leq_{[BDHK06]} dagw(G) \leq dpw(G) \leq_{[Gru08]} cr(G)$$
$$1/6(dtw(G) + 2) \leq_{[HK08]} kellyw(G) \leq dpw(G) \leq_{[Gru08]} cr(G)$$

Moreover, when DAG-width is bounded, so is Kelly-width [HO06].

2.1 Directed Rank-Width

The rank-width of undirected graphs was introduced by Oum and Seymour in relation to graph clique-width. While the definition of clique-width works "as is" also on digraphs, the following straightforward generalization of rank-width to digraphs (related to clique-width again) has been proposed by Kanté [Kan08].

Definition 2.4 (Bi-rank-width). Consider a digraph G, and vertex subsets $X \subseteq V(G)$ and $Y = V(G) \setminus X$. Let A_X^+ denote the $X \times Y$ $0,1$-matrix with the entries $a_{i,j} = 1$ $(i \in X, j \in Y)$ iff $(i,j) \in E(G)$, and let $A_X^- = (A_Y^+)^T$. The *bi-cutrank* function of G is defined as the sum of the ranks of these two matrices $brk_G(X) = rk(A_X^+) + rk(A_X^-)$ over the binary field $GF(2)$. The *bi-rank-width* $brwd(G)$ of G then equals the branch-width of this bi-cutrank function brk_G.

We remind the readers that the *branch-width* [RS91] of an arbitrary symmetric submodular function $\lambda : 2^E \to \mathbb{N}$ is defined as the minimum width over all branch-decompositions of λ over E, where a *branch-decomposition* is a pair T, τ satisfying the following: T is a tree of degree at most three, and τ is a bijection from E to the leaves of T. If f is an edge of T, then let $X_f \subseteq V(T)$ be the vertex set of one of the two connected components of $T - f$, and let the *width* of f be $\lambda(\tau^{-1}(X_f))$. The width of T, τ is the largest width over all edges of T.

Importantly, as proved by Kanté [Kan08], the rank-decomposition algorithm of [HO08] can also be used to find an optimal bi-rank-decomposition of a digraph.

Theorem 2.5 ([HO08] and [Kan08]). *Let $t \in \mathbb{N}$ be constant. There exists an algorithm that in time $O(n^3)$, for a given n-vertex graph (digraph) G, either outputs a rank-decomposition (bi-rank-decomposition, respectively) of G of width at most t, or certifies that the rank-width (bi-rank-width) is more than t.*

A rank-decomposition is, actually, not so suitable for designing dynamic programming algorithms. Yet, there is an efficient alternative characterization of a rank-decomposition via algebraic terms (or *parse trees*) over the bilinear graph product, which has been proposed by Courcelle and Kanté [CK07] and further extended towards algorithmic applications by [GH08] (see also an independent similar approach of [BXTV08]). As shown in [Kan08], an analogous "dynamic programming friendly" parse-tree view (of bi-rank-width) exists for digraphs, and we will apply this later, e.g. in Theorems 3.7 and 3.12.

2.2 DAG-Depth

This part is inspired by the tree-depth notion of Nešetřil and Ossona de Mendez. [NdM06, Lemma 2.2] gives an inductive definition of the *tree-depth* $td(G)$ of undirected G as follows (compare to Def. 2.2). If G has one vertex, then $td(G) = 1$. If G is connected, then $td(G) = 1 + \min\{\, td(G - v) : v \in V(G) \,\}$. Otherwise, $td(G)$ equals the maximum over the tree-depth of the components of G.

We propose a new "directed" generalization of this definition. For a digraph G and any $v \in V(G)$, let G_v denote the subdigraph of G induced by the vertices reachable from v. The maximal elements of the poset $\{\, G_v : v \in V(G) \,\}$ in the graph-inclusion order are called *reachable fragments* of G. Notice that reachable fragments in the undirected case coincide with connected components.

Definition 2.6 (DAG-depth). The *DAG-depth $ddp(G)$* of a digraph G is inductively defined: If $|V(G)| = 1$, then $ddp(G) = 1$. If G has a single reachable fragment, then $ddp(G) = 1 + \min\{\, ddp(G - v) : v \in V(G) \,\}$. Otherwise, $ddp(G)$ equals the maximum over the DAG-depth of the reachable fragments of G.

Comparing Definitions 2.2 and 2.6, one can see that DAG-depth equals cycle rank on bidirected orientations of graphs. Furthermore, the following useful game characterization of this new measure can be proved along Definition 2.6.

Theorem 2.7. *The DAG-depth of a digraph G is at most k if, and only if, the cop player has a "lift-free" winning strategy in the k-cops and robber game on G, i.e., a strategy that never moves a cop from a vertex once he has landed.*

Corollary 2.8 (cf. Theorem 2.1, Def. 2.2). *For any digraph G, the DAG-depth of G is greater than or equal to the DAG-width and the cycle rank of G.* □

Another claim tightly relates our new measure to directed paths in a digraph.

Proposition 2.9. *Consider a digraph G of DAG-depth t, and denote by ℓ the number of vertices of a longest directed path in G. Then $\lfloor \log_2 \ell \rfloor + 1 \leq t \leq \ell$.*

2.3 K-width

Moreover, applications in various "directed path" problems, see e.g. Section 3.1, inspired the following width measure: The *K-width* (a shortcut of "Kenny width") of a digraph G is the maximum number of distinct (not necessarily disjoint) simple s–t paths in G over all pairs of distinct vertices $s, t \in V(G)$.

Similarly to DAG-depth in Proposition 2.9, K-width can be arbitrarily large on DAGs. By giving a suitable search strategy for the cop player in a digraph G based on a DFS tree of G, we show that K-width is lower-bounded by DAG-width, but K-width is generally incomparable with cycle-rank which is unbounded on bidirected paths.

Theorem 2.10 (cf. Theorem 2.1). *For any digraph G, the K-width of G is greater or equal to the DAG-width of G minus one.*

Furthermore, an easy algorithm enumerating all paths leads to:

Proposition 2.11. *The K-width k of a given digraph G can be computed in time $k \cdot \mathrm{poly}(|V(G)|)$.*

Table 1. Old and new (in boldface) complexity results on digraph measures (*-marked results assume a decomposition is given in advance; p-NPC is a shortcut for the complexity class para-NPC; and c and ϕ are fixed parameters of the respective problems)

Problem	K-width	DAG-depth	DAG-width	Cycle-rank	DAG	Bi-rank-width
HAM	**FPT**	**FPT**	XP[a] * W[2]-hard[c]	XP[a] *	P	XP[b] W[2]-hard[d]
c-PATH	**FPT**	**FPT**	XP[a] *	XP[a] *	P[a]	**FPT**
k-PATH	**p-NPC**	**p-NPC**	NPC	NPC	NPC	**p-NPC**[e]
DiDS	**p-NPC**	**p-NPC**	NPC	NPC	NPC	**FPT**
DiSTP	**p-NPC**	**p-NPC**	NPC	NPC	NPC	**FPT**
MaxDiCut	p-NPC[c]	p-NPC[c]	NPC[c]	NPC[c]	NPC[c]	**XP**
c-OCN	**p-NPC**	**p-NPC**	NPC[f]	NPC[f]	NPC[f]	**FPT**
DFVS	*open*	*open*	p-NPC[g]	p-NPC[g]	P	**FPT**
KERNEL	p-NPC[h]	p-NPC[h]	p-NPC[g,h]	p-NPC[g,h]	P	**FPT**
ϕ-MSO$_1$MC	**p-NPH**	**p-NPH**	NPH	NPH	NPH	**FPT**[i]
ϕ-LTLMC	**p-coNPH**	**p-coNPH**	coNPH	coNPH	coNPC	**p-coNPH**
PARITY	XP[j]	XP[j]	XP[j] *	XP[j] *	P	XP[k]

References [a][JRST01] [b][GH09] [c][LKM08] [d][FGLS09] [e][GW06] [f][CD06] [g][KO08] [h][vL76] [i][CMR00] [j][BDHK06] [k][Obd07] . Refer to the respective following sections for details and the new results.

3   Summary of Complexity Results

3.1   Hamiltonian Path (HAM) and Disjoint Paths (k-PATH)

The classical NP-hard *Hamiltonian Path* (HAM) problem [GJ79] is to find a directed path that visits each vertex of a digraph exactly once. A natural generalization of HAM is the *Longest Path* problem (LONGEST PATH), where one is asked to find the longest simple path in a given digraph.

It is easy to see that HAM can be solved on DAGs in polynomial time. When using the parameter DAG-width, HAM belongs to XP [JRST01], but was also proven to be W[2]-hard [LKM08]. We prove our new FPT results for the parameters K-width and DAG-depth on more general LONGEST PATH. Using a simple enumeration of all distinct paths in the case of bounded K-width, or applying Proposition 2.9 and any FPT-algorithm for LONGEST PATH in the standard parameterization (e.g. [CKL+09]) when DAG-depth is bounded, we get:

Theorem 3.1. *There is a fixed parameter tractable algorithm solving the* LONGEST PATH *problem on a digraph G*

a) in time $O\big(t \cdot |V(G)| \cdot |E(G)|\big)$ if G is of K-width at most t;

b) in time $O\big(4^{2^t + O(t^3)} \cdot |V(G)| \cdot |E(G)|\big)$ if G is of DAG-depth at most t.

Another well-known problem is *Disjoint Paths* (k-PATH); given a digraph and k pairs of nodes (s_i, t_i), $1 \le i \le k$, the task is to find pairwise disjoint directed paths from each s_i to the respective t_i. This problem is NP-complete [FHW80] even when k is bounded by any constant $c \ge 2$ (c-PATH). Moreover, a "mixed" generalization of c-PATH remains NP-complete [BJK09] even on DAGs, and k-PATH is NP-complete [GW06] even on digraphs of bounded bi-rank-width.

If the digraph of an instance of k-PATH has K-width ≤ 2, then it can be expressed as a 2-SAT formula, and if DAG-depth is ≤ 2, then it is equivalent to an SDR instance (system of distinct representatives). If, however, we slightly relax the restrictions as follows, the problem becomes NP-complete again.

Theorem 3.2. *The k-PATH problem (with k as part of input)*
a) can be solved in polynomial time on graphs of K-width or DAG-depth 2;
b) is NP-complete on DAGs of K-width 3 and DAG-depth 4.

Finally, since one can express an instance of c-PATH for any fixed c in MSO_1 logic (Section 3.6), it follows from Theorem 3.12 that this problem is fixed parameter tractable on digraphs of bi-rank-width t with parameters c and t. The c-PATH problem however also becomes easier for the other new measures:

Theorem 3.3. *There is a fixed parameter tractable algorithm (for constant c) solving the c-PATH problem on a digraph G*
a) in time $O(t^c \cdot |E(G)|)$ if G is of K-width at most t;
b) in time $O\big((2c)^{ct4^t} \cdot |E(G)|^2\big)$ if G is of DAG-depth at most t.

3.2 Directed Dominating Set (DiDS) and Steiner Tree (DiSTP)

The well-known NP-hard *Dominating Set* (DS) and *Steiner Tree* (STP) problems both allow for natural directed counterparts. We consider them in their un-weighted variants for simplicity. The *Directed Dominating Set* problem (DiDS) asks for a minimum cardinality vertex set X in a digraph G such that every vertex of G not in X is an outneighbour of X. The *Directed Steiner Tree* problem (DiSTP) [HRW92], given a digraph G and $T \subseteq V(G)$, $r \in V(G)$, asks for a minimum size tree in G spanning $\{r\} \cup T$ with all arcs oriented away from r.

While it is folklore that both of these problems are NP-hard in general, we show (with a simple reduction from VERTEX COVER) that the same holds even on very restricted graph classes.

Theorem 3.4. *DiDS and DiSTP problems are NP-complete on a digraph G even if G is restricted to be a DAG of K-width 2 and DAG-depth 3.*

Applying the MSO_1 optimization framework described in Section 3.6 we get:

Proposition 3.5 (Theorem 3.12). *The (unit cost) DiDS and DiSTP problems are fixed parameter tractable when parameterized by bi-rank-width.*

3.3 Maximum Directed Cut (MaxDiCut)

Maximum directed cut (MaxDiCut) is an extensively studied problem on digraphs. Given a digraph G, the goal is to partition the vertex set $V(G)$ into V_0 and V_1 such that the cardinality of $\{(u,v) \in E(G) : u \in V_0, v \in V_1\}$ is maximized. This problem is often stated with edge weights, but we consider only the unweighted (cardinality MAXDICUT) variant in our paper.

It is well known that the MAXDICUT optimization problem is NP-hard, and it has been shown that MAXDICUT stays NP-hard even on DAGs [LKM08]. A closer, yet quite nontrivial look, at the reduction reveals the resulting graph to have also bounded DAG-depth and K-width.

Theorem 3.6 ([LKM08]). *The* MAXDICUT *problem is NP-hard on a digraph G even if G is restricted to be a DAG of K-width 4608 and DAG-depth 11.*

The only new efficiently solvable case among our measures is the following:

Theorem 3.7. *The unweighted* MAXDICUT *problem on a digraph G of bi-rank-width t is polynomially solvable for every fixed t (i.e. it belongs to the class XP).*

3.4   Oriented Colouring (OCN)

A natural directed generalization of the ordinary graph colouring problem can be obtained as follows: The chromatic number $\chi(G)$ of a graph G equals the minimum c such that G has a homomorphism into the complete graph K_c. The *Oriented Chromatic Number* (OCN) $\chi_o(G)$ of a digraph G is defined as the minimum c such that G has a homomorphism into some(!) orientation of K_c.

In other words, $\chi_o(G)$ equals minimum c such that the vertex set of G can be partitioned into c independent sets such that, between each pair of the sets, all arcs have the same direction. For instance, $\chi_o = 5$ for the directed 5-cycle.

It has been shown [KM04] that checking $\chi_o(G) \leq 3$ is easy, but determining whether $\chi_o(G) \leq 4$ is already NP-complete. Subsequently, [CD06] have shown that the problem $\chi_o(G) \leq 4$ remains NP-complete even on acyclic digraphs. Using a simpler and more powerful reduction than [CD06], we prove:

Theorem 3.8. *The problem (4-OCN) to decide whether a digraph G satisfies $\chi_o(G) \leq 4$ is NP-complete even if G is a DAG of K-width 3 and DAG-depth 5.*

On the other hand, it follows from the general framework of Theorem 3.12:

Proposition 3.9. *The problem (c-OCN) to decide $\chi_o(G) \leq c$ on an input digraph G of bi-rank-width t is fixed parameter tractable with parameters c and t.*

3.5   Directed Feedback Vertex Set (DFVS) and Kernel (KERNEL)

The *directed feedback vertex set* (DFVS) problem is to find a minimum cardinality set S of vertices of a digraph G whose removal leaves $G \setminus S$ acyclic. This problem is trivial for acyclic digraphs, and it is FPT with the parameter $k = |S|$. We hence consider only the optimization variant of DFVS with unbounded k.

Kreutzer and Ordyniak [KO08] gave a reduction showing NP-hardness of the DFVS optimization problem on digraphs of DAG-width 4. A closer look at this reduction reveals that all the produced graphs are moreover of cycle rank 4, but they have unbounded K-width and DAG-depth.

The *kernel* of a digraph G is defined as an independent set $S \subseteq V(G)$ such that for every $x \in V(G) \setminus S$ there is an arc from x into S. Notice that a kernel may not always exist. However, on acyclic digraphs, a kernel can be easily found. Having a closer look at the NP-completeness reduction of van Leeuwen [vL76], one discovers the following claim (cf. also [KO08]).

Theorem 3.10 (van Leeuwen [vL76]). *It is NP-complete to decide whether a digraph G has a kernel, even if G is restricted to have (all at once) DAG-width and K-width 2, cycle rank also 2, and DAG-depth 4.*

Finally, by Example 3.11 and Theorem 3.12, both the KERNEL and DFVS problems are fixed parameter tractable on digraphs of bounded bi-rank-width.

3.6 MSO$_1$ Model Checking (ϕ-MSO$_1$MC)

Monadic second order (MSO) logic is a language often used for description of combinatorial algorithmic problems. When applied to a one-sorted relational graph structure (i.e. to a set V with a symmetric relation $edge(u, v)$), this language is abbreviated as MSO$_1$. We use the same abbreviation MSO$_1$ also for digraphs with a relation $arc(u, v)$.

Example 3.11. The following properties are expressible in MSO$_1$ on digraphs
- a directed dominating set X as $\forall z (z \in X \vee \exists x \in X \, arc(x, z))$,
- the existence of a kernel S as $\exists S \forall x [x \notin S \leftrightarrow (\exists y \in S \, arc(x, y))]$, or
- a feedback vertex set Z as $\forall X [X \cap Z = \emptyset \rightarrow (\exists x \in X \, \forall y \in X \, \neg arc(x, y))]$.

On the other hand, MSO$_1$ cannot express Hamiltonian cycle, for instance.

The MSO$_1$ *model checking problem* (ϕ-MSO$_1$MC), where ϕ is a fixed formula, is FPT on (undirected) graphs of bounded clique-width or rank-width [CMR00, CK07]. Not surprisingly, this extends to digraphs parameterized by bi-rank-width. More generally, the *LinEMSO$_1$ optimization framework* includes all problems which can be expressed as maximization of a linear evaluational term over all tuples of sets $X_1, \ldots, X_j$ satisfying $\psi(X_1, \ldots, X_j)$ where ψ is an MSO$_1$ formula — see [CMR00] for details. Analogously to [CMR00] (or [GH08]) we get:

Theorem 3.12 (cf. [CMR00], and [Kan08, GH08])
Every ψ-LinEMSO$_1$ optimization problem is fixed parameter tractable when restricted to digraphs of bi-rank-width t, with parameters t and ψ.

Theorem 3.12 particularly implies that the problems listed in Example 3.11 (and many others) are FPT on digraphs of bi-rank-width t. No analogous results, however, seem possible for our other directed width measures since one can interpret ϕ-MSO$_1$MC of arbitrary undirected graphs via subdividing each edge and giving the two new edges opposite orientations, leading to:

Proposition 3.13. *The ϕ-MSO$_1$MC problem is NP-hard even when restricted to DAGs that are of K-width 1 and DAG-depth 2.*

3.7 LTL Model Checking (ϕ-LTLMC) and Parity Games (PARITY)

Another useful language that allows to express properties of digraphs is *Linear Temporal Logic* (LTL) — see, e.g., [BK08]. LTL model checking remains hard for a fixed formula ϕ and all of the directed width measures we considered here, including bi-rank-width (as opposed to MSO$_1$ model checking).

Theorem 3.14. *The ϕ-LTLMC problem is coNP-hard even when the input digraph is restricted to have K-width 1, DAG-depth 4, and bi-rank-width 2.*

Theorem 3.15. *The ϕ-LTLMC problem is coNP-complete on DAGs.*

Parity games — see e.g. [GTW02] for a reference, play an important role in the field of model-checking and formal verification. There are many reasons for this. First, solving parity games is equivalent to model-checking the modal μ-calculus, an important modal logic subsuming many other logics (e.g. CTL). Moreover, the modal μ-calculus is a bisimulation invariant fragment of MSO$_1$.

Second, the exact complexity of solving a parity game is a long-standing open problem. It is known to be in NP $\cap$ co-NP, and widely believed to be in P. It is trivially in P for acyclic digraphs. Moreover, it was shown that solving a parity game is in XP for digraphs of bounded tree-width [Obd03], bounded DAG-width [BDHK06] (hence also on bounded K-width, DAG-depth, and cycle rank) and bounded Kelly-width [HK08], and of bounded clique-width [Obd07] (implying the same for bi-rank-width).

4 Conclusion

Table 1, and the related results in this paper, have left several interesting open problems and questions. Just to specifically mention a few:

1) We suggest there exist FPT algorithms solving the DFVS problem for bounded K-width or DAG-depth (two of the open table entries).
2) For some entries in the table, we neither expect an FPT algorithm, nor have an NP-hardness estimate. E.g., MAXDICUT or k-PATH for bi-rank-width, or c-PATH for cycle rank. Can we then, at least, show a W-hardness result?
3) While we have given FPT and XP, respectively, algoritms solving the unit-cost variants of DISTP and MAXDICUT, these problems are usually considered in their weighted variants and then we expect their complexity to be higher. We, however, have no further results in this direction.
4) Some suggest that the DFVS number (see in Section 3.5) perhaps can be a good directed width measure. However, since majority of our sample problems in Table 1 remain hard even on DAGs, there is not much room left for applications of the DFVS parameter. Interestingly though, KERNEL becomes FPT when parametrized by DFVS.

Theorem 4.1. *If a digraph G is given with a directed feedback vertex set of size k, then the KERNEL problem can be solved in time $O(2^k \cdot |V(G)|^2)$.*

Finally, we try to formulate the overall impression coming from Table 1: Robber-and-cops based width measures do not seem to be very useful for parameterized algorithms on digraphs. One reason might be that cops "give" good graph separators in the undirected case, but that does not work any more for digraphs. Considering the DFVS number as a width parameter does not seem to help either. We perhaps need something new to move on. At this moment, bi-rank-width seems like a good alternative.

Acknowledgments. This work has been supported by a Czech–German bilateral grant of GAČR and DFG (201/09/J021 and RO 927/9). Moreover, P. Hliněný has been supported by the Czech research grant GAČR 201/08/0308.

References

[Bar06]   Barát, J.: Directed path-width and monotonicity in digraph searching. Graphs and Combinatorics 22(2), 161–172 (2006)

[BDHK06]   Berwanger, D., Dawar, A., Hunter, P., Kreutzer, S.: DAG-width and parity games. In: Durand, B., Thomas, W. (eds.) STACS 2006. LNCS, vol. 3884, pp. 524–536. Springer, Heidelberg (2006)

[BG04]   Berwanger, D., Grädel, E.: Entanglement – a measure for the complexity of directed graphs with applications to logic and games. In: Baader, F., Voronkov, A. (eds.) LPAR 2004. LNCS (LNAI), vol. 3452, pp. 209–223. Springer, Heidelberg (2005)

[BJK09]   Bang-Jensen, J., Kriesell, M.: Disjoint directed and undirected paths and cycles in digraphs. Technical Report PP-2009-03, University of South Denmark (2009)

[BK08]   Baier, C., Katoen, J.-P.: Principles of Model Checking. The MIT Press, Cambridge (2008)

[BXTV08]   Bui-Xuan, B.-M., Telle, J., Vatshelle, M.: H-join and algorithms on graphs of bounded rank-width (submitted) (November 2008)

[CD06]   Culus, J.-F., Demange, M.: Oriented coloring: Complexity and approximation. In: Wiedermann, J., Tel, G., Pokorný, J., Bieliková, M., Štuller, J. (eds.) SOFSEM 2006. LNCS, vol. 3831, pp. 226–236. Springer, Heidelberg (2006)

[CK07]   Courcelle, B., Kanté, M.: Graph operations characterizing rank-width and balanced graph expressions. In: Brandstädt, A., Kratsch, D., Müller, H. (eds.) WG 2007. LNCS, vol. 4769, pp. 66–75. Springer, Heidelberg (2007)

[CKL⁺09]   Chen, J., Kneis, J., Lu, S., Mölle, D., Richter, S., Rossmanith, P., Sze, S., Zhang, F.: Randomized divide-and-conquer: Improved path, matching, and packing algorithms. SIAM Journal on Computing 38(6), 2526–2547 (2009)

[CMR00]   Courcelle, B., Makowsky, J.A., Rotics, U.: Linear time solvable optimization problems on graphs of bounded clique-width. Theory Comput. Syst. 33(2), 125–150 (2000)

[DGK08]   Dankelmann, P., Gutin, G., Kim, E.: On complexity of minimun leaf out-branching. arXiv:0808.0980v1 (August 2008)

[Egg63]   Eggan, L.: Transition graphs and the star-height of regular events. Michigan Mathematical Journal 10(4), 385–397 (1963)

[FGLS09]   Fomin, F., Golovach, P., Lokshtanov, D., Saurab, S.: Clique-width: On the price of generality. In: SODA 2009, pp. 825–834. SIAM, Philadelphia (2009)

[FHW80]   Fortune, S., Hopcroft, J.E., Wyllie, J.: The directed subgraph homeomorphism problem. Theor. Comput. Sci. 10, 111–121 (1980)

[GH08]   Ganian, R., Hliněný, P.: Automata approach to graphs of bounded rank-width. In: IWOCA 2008, pp. 4–15 (2008)

[GH09]   Ganian, R., Hliněný, P.: Better polynomial algorithms on graphs of bounded rank-width. In: Fiala, J., Kratochvíl, J., Miller, M. (eds.) IWOCA 2009. LNCS, vol. 5874, pp. 266–277. Springer, Heidelberg (2009)

[GJ79]   Garey, M., Johnson, D.: Computers and Intractability: A Guide to the Theory of NP-completeness. W.H. Freeman, New York (1979)

[Gru08]   Gruber, H.: Digraph complexity measures and applications in formal language theory. In: MEMICS 2008, pp. 60–67 (2008)

[GTW02]   Grädel, E., Thomas, W., Wilke, T. (eds.): Automata, Logics, and Infinite Games. LNCS, vol. 2500. Springer, Heidelberg (2002)

[GW06]   Gurski, F., Wanke, E.: Vertex disjoint paths on clique-width bounded graphs. Theor. Comput. Sci. 359(1-3), 188–199 (2006)

[HK08]   Hunter, P., Kreutzer, S.: Digraph measures: Kelly decompositions, games, and orderings. Theor. Comput. Sci. 399(3), 206–219 (2008)

[HO06]   Hliněný, P., Obdržálek, J.: Escape-width: Measuring "width" of digraphs. Presented at Sixth Czech-Slovak International Symposium on Combinatorics, Graph Theory, Algorithms and Applications (2006)

[HO08]   Hliněný, P., Oum, S.: Finding branch-decomposition and rank-decomposition. SIAM J. Comput. 38, 1012–1032 (2008)

[HRW92]   Hwang, F., Richards, D., Winter, P.: The Steiner Tree Problem. Annals of Discrete Mathematics. North-Holland, Amsterdam (1992)

[JRST01]   Johnson, T., Robertson, N., Seymour, P.D., Thomas, R.: Directed tree-width. Journal of Combinatorial Theory, Series B 82(1), 138–154 (2001)

[Kan08]   Kanté, M.: The rank-width of directed graphs. arXiv:0709.1433v3 (March 2008)

[KM04]   Klostermeyer, W., MacGillivray, G.: Homomorphisms and oriented colorings of equivalence classes of oriented graphs. Discrete Mathematics 274, 161–172 (2004)

[KO08]   Kreutzer, S., Ordyniak, S.: Digraph decompositions and monotonicity in digraph searching. In: Broersma, H., Erlebach, T., Friedetzky, T., Paulusma, D. (eds.) WG 2008. LNCS, vol. 5344, pp. 336–347. Springer, Heidelberg (2008)

[LKM08]   Lampis, M., Kaouri, G., Mitsou, V.: On the algorithmic effectiveness of digraph decompositions and complexity measures. In: Hong, S.-H., Nagamochi, H., Fukunaga, T. (eds.) ISAAC 2008. LNCS, vol. 5369, pp. 220–231. Springer, Heidelberg (2008)

[NdM06]   Nešetřil, J., Ossona de Mendez, P.: Tree-depth, subgraph coloring and homomorphism bounds. European J. Combin. 27(6), 1024–1041 (2006)

[Obd03]   Obdržálek, J.: Fast mu-calculus model checking when tree-width is bounded. In: Hunt Jr., W.A., Somenzi, F. (eds.) CAV 2003. LNCS, vol. 2725, pp. 80–92. Springer, Heidelberg (2003)

[Obd06]   Obdržálek, J.: DAG-width – connectivity measure for directed graphs. In: SODA 2006, pp. 814–821. ACM-SIAM, New York (2006)

[Obd07]   Obdržálek, J.: Clique-width and parity games. In: Duparc, J., Henzinger, T.A. (eds.) CSL 2007. LNCS, vol. 4646, pp. 54–68. Springer, Heidelberg (2007)

[RS86]   Robertson, N., Seymour, P.D.: Graph minors. II. Algorithmic aspects of tree-width. Journal of Algorithms 7(3), 309–322 (1986)

[RS91]   Robertson, N., Seymour, P.D.: Graph minors. X. Obstructions to tree-decomposition. J. Comb. Theory B 52(2), 153–190 (1991)

[Saf05]   Safari, M.: D-width: A more natural measure for directed tree-width. In: Jedrzejowicz, J., Szepietowski, A. (eds.) MFCS 2005. LNCS, vol. 3618, pp. 745–756. Springer, Heidelberg (2005)

[vL76]   van Leeuwen, J.: Having a Grundy-numbering is NP-complete. Technical Report 207, The Pennsylvania State University (September 1976)

The Parameterized Complexity of Some Geometric Problems in Unbounded Dimension

Panos Giannopoulos*, Christian Knauer, and Günter Rote

Institut für Informatik, Freie Universität Berlin, Takustraße 9, D-14195 Berlin, Germany
{panos,knauer,rote}@inf.fu-berlin.de

Abstract. We study the parameterized complexity of the following fundamental geometric problems with respect to the dimension d:
  i) Given n points in $\mathbb{R}^d$, compute their minimum enclosing cylinder.
  ii) Given two n-point sets in $\mathbb{R}^d$, decide whether they can be separated by two hyperplanes.
  iii) Given a system of n linear inequalities with d variables, find a maximum size feasible subsystem.
We show that (the decision versions of) all these problems are W[1]-hard when parameterized by the dimension d. Our reductions also give a $n^{\Omega(d)}$-time lower bound (under the Exponential Time Hypothesis).

Keywords: parameterized complexity, geometric dimension, lower bounds, minimum enclosing cylinder, maximum feasible subsystem, 2-linear separability.

1 Introduction

We study the parameterized complexity of the following three fundamental geometric problems with respect to the dimension of the underlying space: minimum enclosing cylinder of a set of points in $\mathbb{R}^d$, 2-linear separation of two point sets in R^d, and maximum-size feasible subsystem of a system of linear inequalities with d variables. All these problems are NP-hard when the dimension d is unbounded and all known exact algorithms run in $n^{O(d)}$ time (basically, using brute force), where n is the total number of objects in the input sets. As with many other geometric problems in d dimensions, it is widely conjectured that the dependence on d cannot be removed from the exponent of n. However, no evidence of this has been given so far.

In terms of parameterized complexity theory the question is whether any of these problems is fixed-parameter tractable with respect to d, i. e., whether there exists an algorithm that runs in $O(f(d)n^c)$ time, for some computable function f and some constant c independent of d. Proving a problem to be W[1]-hard with respect to d, gives a strong evidence that such an algorithm is not possible, under standard complexity theoretic assumptions. We summarize our results bellow.

* This research was supported by the German Science Foundation (DFG) under grant Kn 591/3-1.

J. Chen and F.V. Fomin (Eds.): IWPEC 2009, LNCS 5917, pp. 198–209, 2009.
© Springer-Verlag Berlin Heidelberg 2009

Results. We study the following decision problems:

i) Given n unit balls $\mathbb{R}^d$, decide whether there is a line that stabs all the balls. (Note that since the balls are unit, this is the decision version of the problem of computing the minimum enclosing cylinder of a set of n points.)
ii) Given two n-point sets in $\mathbb{R}^d$, decide whether they can be separated by two hyperplanes.
iii) Given a system of n linear inequalities with d variables and an integer l, decide whether there is a solution satisfying l of the inequalities.

We prove that all three problems are W[1]-hard with respect to d. This is done by fpt-reductions from the k-independent set (or clique) problem in general graphs, which is W[1]-complete [9]. As a side-result, we also show that, when restricted to equalities, problem (iii) is W[1]-hard with respect to both l and d. The reductions for problems (i) and (ii) are based on a technique pioneered in Cabello et al. [7], see next section. With the addition of these two problems this technique shows a generic trait and its potential as a useful tool for proving hardness of geometric problems with respect to the dimension.

In all three reductions the dimension is linear in the size k of the independent set (or clique), hence an $n^{o(d)}$-time algorithm for any of the problems implies an $n^{o(d)}$-time algorithm for the parameterized k-clique problem, which in turn implies that n-variable 3SAT can be solved in $2^{o(n)}$-time. The Exponential Time Hypothesis (ETH) [11] conjectures that no such algorithm exists.

Related work. The dimension of geometric problems is a natural parameter for studying their parameterized complexity. However, there are only few results of this type: Langerman and Morin [12] gave fixed-parameter tractability results for the problem of covering points with hyperplanes, while the 'dual' parameterization of the maximum-size feasible subsystem problem, where parameter l is now the smallest number of inequalities one has to remove to make the system feasible is fixed-parameter tractable with respect to both l and d [4]. As for hardness results, the problems of covering points with balls and computing the volume of the union of axis parallel boxes have been shown to be W[1]-hard by Cabello et al. [7] and Chan [8] respectively. We refer the reader to Giannopoulos et al. [10] for a survey on parameterized complexity results for geometric problems.

The problem of stabbing balls in $\mathbb{R}^d$ with one line was shown to be NP-hard when d is part of the input by Megiddo [14]. This problem is equivalent to the minimum enclosing cylinder problem for points, see Varadarajan et al. [15]. Exact and approximation algorithms for the latter problem can be found, for example, in Bădoiu et al. [5].

Megiddo [13] showed that the problem of separating two point sets in $\mathbb{R}^d$ by two hyperplanes is NP-hard. He also showed that the general problem of separating two point sets by l hyperplanes can be solved in polynomial time when both d and l are fixed.

The complexity of the maximum-size feasible subsystem problem was studied in Amaldi and Kann [1]. Several results on the hardness of approximability can also be found in this paper, as well as in Arora et al. [3]. For exact and approximation algorithms for this and several related problems see Aronov and Har-Peled [2].

2   Preliminaries

2.1   Methodology

As mentioned above, all three hardness results use a reduction from the k-independent set (or clique) problem. Using the technique in [7], we construct of a *scaffolding* structure that restricts the solutions to n^k combinatorially different solutions, which can be interpreted as potential k-cliques in a graph with n vertices. Additional *constraint* objects will then encode the edges of the input graph.

The main ideas are the following. We construct geometric instances which lie in Euclidean space whose dimension depends only on k. Note that the lower the dependence on k, the better the lower bound we get from the hardness result. In our case the dependence is linear. The scaffolding structure is highly symmetric. It is composed of k symmetric subsets of a linear (in n) number of objects that lie in orthogonal subspaces. Orthogonality together with the specific geometric properties of each problem allows us to restrict the solutions to n^k combinatorially different solutions. The way of placing the constraint objects is crucial: each object lies in a 4-dimensional subspace and cancels an exponential number of solutions.

Model of computation. The geometry of the constructions in Sections 3, 4 will be described as if exact square roots and expressions of the form $\sin \frac{\pi}{n}$ were available. To make the reduction suitable for the Turing machine model, the data must be perturbed using fixed-precision roundings. This can be done with polynomially many bits in a way similar to the rounding procedure followed in [7,6]. We omit the details here. The construction in Section 5 uses small integral data.

2.2   Notation

Let $[n] = \{1, \ldots, n\}$ and $G([n]), E)$ be an undirected graph.

In sections 3, 4, it will be convenient to view $\mathbb{R}^{2k}$ as the product of k orthogonal planes $E_1, \ldots, E_k$, where each E_i has coordinate axes X_i, Y_i. The origin is denoted by o. The coordinates of a point $p \in \mathbb{R}^{2k}$ are denoted by $(x_1(p), y_1(p), \ldots, x_k(p), y_k(p))$. The notions of a point and vector will be used interchangeably. We denote by C_i the unit circle on E_i centered at o.

3   Minimum Enclosing Cylinder (or Stabbing Balls with One Line)

Given an undirected graph $G([n], E)$ we construct a set $\mathcal{B}$ of balls of equal radius r in $\mathbb{R}^{2k}$ such that $\mathcal{B}$ can be stabbed by a line if and only if G has an independent set of size k.

For every ball $B \in \mathcal{B}$ we will also have $-B \in \mathcal{B}$. This allows us to restrict our attention to lines through the origin: a line that stabs $\mathcal{B}$ can be translated so that it goes through the origin and still stabs $\mathcal{B}$. In this section, by a line we always mean a line through the origin. For a line l, let $\boldsymbol{l}$ be its unit direction vector.

For each plane E_i, we define $2n$ $2k$-dimensional balls, whose centers $c_{i1}, \dots,$ c_{i2n} are regularly spaced on the unit circle C_i. Let $c_{iu} \in E_i$ be the center of the ball B_{iu}, $u \in [2n]$, with

$$x_i(c_{iu}) = \cos(u-1)\tfrac{\pi}{n}, \ y_i(c_{iu}) = \sin(u-1)\tfrac{\pi}{n}.$$

We define the scaffolding ball set $\mathcal{B}^0 = \{B_{iu}, i = 1, \dots, k \text{ and } u = 1, \dots, 2n\}$. We have $|\mathcal{B}^0| = 2nk$. All balls in $\mathcal{B}^0$ will have the same radius $r < 1$, to be defined later.

Two antipodal balls $B, -B$ are stabbed by the same set of lines. A line l stabs a ball B of radius r and center c if and only if $(c \cdot \boldsymbol{l})^2 \geq \|c\|^2 - r^2$. Thus, l stabs $\mathcal{B}^0$ if and only if it satisfies the following system of nk inequalities:

$$(c_{iu} \cdot \boldsymbol{l})^2 \geq \|c_{iu}\|^2 - r^2 = 1 - r^2, \quad \text{for } i = 1, \dots, k \text{ and } u = 1, \dots, n.$$

Consider the inequality asserting that l stabs B_{iu}. Geometrically, it amounts to saying that the projection $\boldsymbol{l}_i$ of $\boldsymbol{l}$ on the plane E_i lies in one of the half-planes

$$H_{iu}^+ = \{p \in E_i | c_{iu} \cdot p \geq \sqrt{\|c_{iu}\|^2 - r^2}\}, \ H_{iu}^- = \{p \in E_i | c_{iu} \cdot p \leq -\sqrt{\|c_{iu}\|^2 - r^2}\}.$$

Consider the situation on a plane E_i. Looking at all half-planes $H_{i1}^+, H_{i1}^-, \dots, H_{in}^+,$ H_{in}^-, we see that l stabs all balls B_{iu} (centered on E_i) if and and only if $\boldsymbol{l}_i$ lies in one of the $2n$ wedges $\pm(H_{i1}^- \cap H_{i2}^+), \dots, \pm(H_{i(n-1)}^- \cap H_{in}^+), \pm(H_{i1}^- \cap H_{in}^-)$; see Fig. 1. The apices of the wedges are regularly spaced on a circle of radius $\lambda = \sqrt{2(1-r^2)/(1-\cos\frac{\pi}{n})}$, and define the set

$$A_i = \{\pm\left(\lambda\cos(2u-1)\tfrac{\pi}{2n}, \lambda\sin(2u-1)\tfrac{\pi}{2n}\right) \in E_i, u = 1, \dots, n\}.$$

For l to stab all balls B_{iu}, we must have that $\|\boldsymbol{l}_i\| \geq \lambda$. We choose $r = \sqrt{1 - (1 - \cos\frac{\pi}{n})/(2k)}$ in order to obtain $\lambda = 1/\sqrt{k}$.

Since the above hold for every plane E_i, and since $\boldsymbol{l} \in \mathbb{R}^{2k}$ is a unit vector, we have

$$1 = \|\boldsymbol{l}\|^2 = \|\boldsymbol{l}_1\|^2 + \dots + \|\boldsymbol{l}_k\|^2 \geq k\lambda^2 = 1.$$

Hence, equality holds throughout, which implies that $\|\boldsymbol{l}_i\| = 1/\sqrt{k}$, for every $i \in \{1, \dots, k\}$. Hence, for line l to stab all balls in $\mathcal{B}^0$, every projection $\boldsymbol{l}_i$ must be one of the $2n$ apices in A_i. Each projection $\boldsymbol{l}_i$ can be chosen independently. There are $2n$ choices, but since $\boldsymbol{l}$ and $-\boldsymbol{l}$ correspond to the same line, the total number of lines that stab $\mathcal{B}^0$ is $n^k 2^{k-1}$.

For a tuple $(u_1, \dots, u_k) \in [2n]^k$, we will denote by $l(u_1, \dots, u_k)$ the stabbing line with direction vector

$$\frac{1}{\sqrt{k}}\left(\cos(2u_1-1)\tfrac{\pi}{2n}, \sin(2u_1-1)\tfrac{\pi}{2n}, \dots, \cos(2u_k-1)\tfrac{\pi}{2n}, \sin(2u_k-1)\tfrac{\pi}{2n}\right).$$

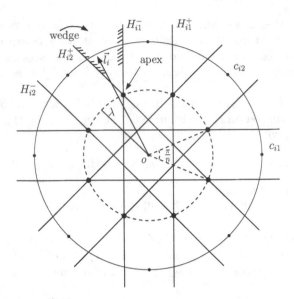

Fig. 1. Centers of the balls and their respective half-planes and wedges on a plane E_i, for $n = 4$

Two lines $l(u_1, u_2, ..., u_k)$ and $l(v_1, v_2, ..., v_k)$ are said to be equivalent if $u_i \equiv v_i$ (mod n), for all i. This relation defines n^k equivalence classes $L(u_1, \ldots, u_k)$, with $(u_1, \ldots, u_k) \in [n]^k$, where each class consists of 2^{k-1} lines.

From the discussion above, it is clear that there is a bijection between the possible equivalence classes of lines that stab $\mathcal{B}^0$ and $[n]^k$.

3.1   Constraint Balls

We continue the construction of the ball set $\mathcal{B}$ by showing how to encode the structure of G. For each pair of distinct indices $i \neq j$ ($1 \leq i, j \leq k$) and for each pair of (possibly equal) vertices $u, v \in [n]$, we define a *constraint set* $\mathcal{B}_{ij}^{uv}$ of balls with the property that (all lines in) all classes $L(u_1, \ldots, u_k)$ stab $\mathcal{B}_{ij}^{uv}$ except those with $u_i = u$ and $u_j = v$. The centers of the balls in $\mathcal{B}_{ij}^{uv}$ lie in the 4-space $E_i \times E_j$. Observe that all lines in a particular class $L(u_1, \ldots, u_k)$ project onto only two lines on $E_i \times E_j$. We use a ball B_{ij}^{uv} (to be defined shortly) of radius r that is stabbed by *all* lines $l(u_1, \ldots, u_k)$ except those with $u_i = u$ and $u_j = v$. Similarly, we use a ball $B_{ij}^{u\bar{v}}$ that is stabbed by *all* lines $l(u_1, \ldots, u_k)$ except those with $u_i = u$ and $u_j = \bar{v}$, where $\bar{v} = v + n$. Our constraint set consists then of the four balls

$$\mathcal{B}_{ij}^{uv} = \{\pm B_{ij}^{uv}, \pm B_{ij}^{u\bar{v}}\}.$$

We describe now the placement of a ball B_{ij}^{uv}. Consider a line $l = l(u_1, \ldots, u_k)$ with $u_i = u$ and $u_j = v$. The center c_{ij}^{uv} of B_{ij}^{uv} will lie on a line $z \in E_i \times E_j$ that is orthogonal to l, but not orthogonal to any line $l(u_1, \ldots, u_k)$ with $u_i \neq u$ or $u_j \neq v$. We choose the direction z of z as follows:

$$x_i(z) = \mu(\cos\theta_i - 3n\sin\theta_i), \ y_i(z) = \mu(\sin\theta_i + 3n\cos\theta_i),$$

$$x_j(z) = \mu(-\cos\theta_j - 6n^2\sin\theta_j), \ y_j(z) = \mu(-\sin\theta_j + 6n^2\cos\theta_j),$$

where $\theta_i = (2u-1)\frac{\pi}{2n}$, $\theta_j = (2u-1)\frac{\pi}{2n}$, and $\mu = 1/(9n^2 + 36n^4 + 2)$. It is straightforward to check that $l \cdot z = 0$.

Let ω be the angle between l' and z. We have the following lemma, whose proof can be found in the appendix:

Lemma 1. *For any line $l' = l(u_1, \ldots, u_k)$, with $u_i \neq u$ or $u_j \neq v$ the angle ω between l' and z satisfies $|\cos\omega| > \frac{\mu}{\sqrt{k}}$.*

This lower bound on $|\cos\omega|$ helps us place B_{ij}^{uv} sufficiently close to the origin so that it is still intersected by l', i.e., l' lies in one of the half-spaces $c_{ij}^{uv} \cdot p \geq \sqrt{\|c_{ij}^{uv}\|^2 - r^2}$ or $c_{ij}^{uv} \cdot p \leq -\sqrt{\|c_{ij}^{uv}\|^2 - r^2}$, $p \in \mathbb{R}^{2k}$.

We claim that any point c_{ij}^{uv} on z with $r < \|c_{ij}^{uv}\| < \sqrt{\frac{k}{k-\mu^2}}r$ will do. For any position of c_{ij}^{uv} on z with $\|c_{ij}^{uv}\| > r$, we have $(c_{ij}^{uv} \cdot l)^2 = 0 < \|c_{ij}^{uv}\|^2 - r^2$, i.e., l does not stab B_{ij}^{uv}. On the other hand, as argued above we need that $|c_{ij}^{uv} \cdot l'| \geq \sqrt{\|c_{ij}^{uv}\|^2 - r^2}$. Since $c_{ij}^{uv} \cdot l' = \cos\omega \cdot \|c_{ij}^{uv}\|$, we have the condition $|\cos\omega| \geq \sqrt{1 - \frac{r^2}{\|c_{ij}^{uv}\|^2}}$. By Lemma 1 we know that $|\cos\omega| > \frac{\mu}{\sqrt{k}}$, hence by choosing $\|c_{ij}^{uv}\|$ so that $\frac{\mu}{\sqrt{k}} > \sqrt{1 - \frac{r^2}{\|c_{ij}^{uv}\|^2}}$ we are done.

Reduction. Similarly to [7], the structure of the input graph $G([n], E)$ can now be represented as follows. We add to $\mathcal{B}^0$ the $4n\binom{k}{2}$ balls in $\mathcal{B}_V = \bigcup \mathcal{B}_{ij}^{uu}$, $1 \leq u \leq n$, $1 \leq i < j \leq k$, to ensure that all components u_i in a solution (class of lines $L(u_1, \ldots, u_k)$) are distinct. For each edge $uv \in E$ we also add the balls in $k(k-1)$ sets $\mathcal{B}_{ij}^{uv}$, with $i \neq j$. This ensures that the remaining classes of lines $L(u_1, \ldots, u_k)$ represent independent sets of size k. In total, the edges are represented by the $4k(k-1)|E|$ balls in $\mathcal{B}_E = \bigcup \mathcal{B}_{ij}^{uv}$, $uv \in E$, $1 \leq i, j \leq k$, $i \neq j$. The final set $\mathcal{B} = \mathcal{B}^0 \cup \mathcal{B}_V \cup \mathcal{B}_E$ has $2nk + 4\binom{k}{2}(n + 2|E|)$ balls.

As noted in above, there is a bijection between the possible equivalence classes of lines $L(u_1, \ldots, u_k)$ that stab $\mathcal{B}$ and the tuples $(u_1, \ldots, u_k) \in [n]^k$. The constraint sets of balls exclude tuples with two equal indices $u_i = u_j$ or with indices u_i, u_j when $u_i u_j \in E$, thus, the classes of lines that stab $\mathcal{B}$ represent exactly the independent sets of G. Thus, we have the following:

Lemma 2. *Set $\mathcal{B}$ can be stabbed by a line if an only if G has an independent set of size k.*

From this lemma and since this is an fpt-reduction, we conclude:

Theorem 1. *Deciding whether n unit balls in $\mathbb{R}^d$ can be stabbed with one line is W[1]-hard with respect to d.*

4   Separating Two Point Sets by Two Hyperplanes

Let P and Q be two point sets in $\mathbb{R}^d$. Two hyperplanes split space generically into four "quarters". There are three different versions of what it means to separate P and Q by two hyperplanes:

(a) Each quarter contains only points of one set.
(b) The set Q is contained in one quarter only, and set P can populate the remaining three quarters.
(c) Same as (b), but the roles of P and Q are not fixed in advance.

In the following we work only with version (a), which is the most general. For the point sets that we construct, it will turn out that if a separation according to (a) exists, it will also be valid by (b) and (c). Thus, our reduction works for all three versions of the problem.

Separation according to (a) is equivalent to requiring that every segment pq between a point $p \in P$ and a point $q \in Q$ is intersected by one of the two hyperplanes. Note that we restrict our attention to *strict* separation, i.e., no hyperplane can go through a point of P or Q. (The result extends to weak separation; see the end of this section.)

Given an undirected graph $G_0([n_0], E_0)$ with $n_0 \geq 2$ and an integer k, we construct two point sets P and Q in $\mathbb{R}^{2k}$ with the property that they can be separated by two hyperplanes if and only if G_0 has an independent set of size k. For technical reasons, we duplicate the vertices of the graph: we build a new graph with $n = 2n_0$ vertices. Every vertex $u \in [n_0]$ of the original graph gets a second copy $u' := u + n_0$, and for every original edge uv, there are now four edges uv, uv', $u'v$, $u'v'$. The new graph $G([n], E)$ has an independent set of size k if and only if the original graph has such a set.

On each plane E_i, $i = 1, \ldots, k$, we define a set P_i of n points regularly spaced on the circle C_i:

$$P_i = \{\, p_{iu} \in E_i \mid x_i(p_{iu}) = \cos(u-1)\tfrac{2\pi}{n},\ y_i(p_{iu}) = \sin(u-1)\tfrac{2\pi}{n},\ u = 1, \ldots, n \,\}.$$

For an index $u \in [n]$, it will be convenient to define its *antipodal* and *almost antipodal* partner $u' = u + \frac{n}{2}$ and $\bar{u} = u + \frac{n}{2} + 1$ respectively. (All indices are modulo n). Thus we are extending the notation u' to all (original and new) vertices u, with $(u')' = u$.

The scaffolding is defined by two sets $P = \bigcup P_i$ and $Q^0 = \{o\}$. We have $|P| = nk$.

Since the points in each P_i are regularly spaced on C_i, a hyperplane that does not contain the origin can intersect at most $n/2$ segments op_{iu} on each plane E_i. Hence, at least two hyperplanes are needed to separate P and Q^0. Actually, two suffice. One hyperplane can intersect the $n/2$ consecutive (in a counter-clockwise order) segments $op_{i\bar{u}_i}, \ldots, op_{iu_i}$ on each E_i, for a choice of $u_i \in [n]$ (see Fig. 2). There is an infinite number of such hyperplanes, forming an equivalence class $\mathcal{H}(u_1, \ldots, u_k)$. Since the planes $E_1, \ldots, E_k$ are orthogonal, each u_i independently defines which of the $n/2$ consecutive segments on

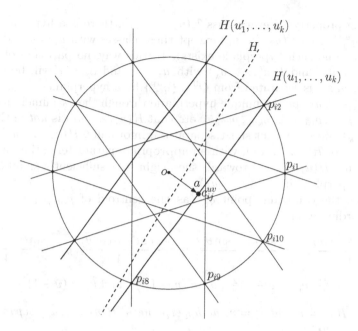

Fig. 2. Point set P_i, for $n = 10$, a hyperplane H in the class $\mathcal{H}(u_1, \ldots, u_k)$ and the corresponding boundary hyperplane $H(u_1, \ldots, u_k)$ for $u_i = 2$. The placement of q_{ij}^{uv} is shown in a two-dimensional analog.

E_i are intersected by a hyperplane in $\mathcal{H}(u_1, \ldots, u_k)$. The remaining $n/2$ segments $-op_{i\bar{u}_i}, \ldots, -op_{iu_i}$ on each E_i can then be intersected by any hyperplane in the 'complementary' class $\mathcal{H}(u_1', \ldots, u_k') = \{-H \mid H \in \mathcal{H}(u_1, \ldots, u_k)\}$. Effectively, every hyperplane in $\mathcal{H}(u_1, \ldots, u_k)$ separates Q^0 from the $\frac{kn}{2}$-point set $P(u_1, \ldots, u_k) = \{p_{1\bar{u}_1}, \ldots, p_{1u_1}\} \cup \cdots \cup \{p_{k\bar{u}_k}, \ldots, p_{ku_k}\}$. Concluding, there are n^k possible partitions of P into two groups, each separated from Q^0 by one hyperplane, in correspondence to the n^k possible tuples $(u_1, \ldots, u_k) \in [n]^k$:

Lemma 3. *The possible pairs of hyperplanes that separate P from Q^0 are of the form h, h' with $h \in \mathcal{H}(u_1, \ldots, u_k)$ and $h' \in \mathcal{H}(u_1', \ldots, u_k')$, for some $(u_1, \ldots, u_k) \in [n]^k$.*

Since by construction, the graph G has the property that $uv \in E$ iff $u'v' \in E$, the separating pairs of hyperplanes h, h' can be used to encode the potential independent sets $\{u_1, \ldots, u_k\}$: it does not matter which of h and h' we choose, the corresponding vertex set will be an independent set in both cases, or a dependent set in both cases.

4.1 Constraint Points

For each pair of indices $i \neq j$ ($1 \leq i, j \leq k$) and for each pair of (possibly equal) vertices $u, v \in [n]$, we will define a constraint point $q_{ij}^{uv} \in E_i \times E_j$ with the following property: in every class $\mathcal{H}(u_1, \ldots, u_k)$, there is a hyperplane that separates $\{q_{ij}^{uv}\}$ from $P(u_1, \ldots, u_k)$ except those classes with $u_i = u$ and $u_j = v$ (in which case no such hyperplane exists). In this way, no partition of P into sets $P(u_1, \ldots, u_k)$ and $P(u_1', \ldots, u_k')$ with $u_i = u$ and $u_j = v$ will be possible such that each set is separated from $Q^0 \cup \{q_{ij}^{uv}\}$ by a hyperplane.

Let $H(u_1, \ldots, u_k)$ be the unique hyperplane through the $2k$ affinely independent points $p_{1u_1}, p_{1\bar{u}_1}, \ldots, p_{ku_k}, p_{k\bar{u}_k}$. Note that $H(u_1, \ldots, u_k)$ is *not* in the class $\mathcal{H}(u_1, \ldots, u_k)$, since we want strict separation; informally, $H(u_1, \ldots, u_k)$ lies at the boundary of $\mathcal{H}(u_1, \ldots, u_k)$, with an appropriate parameterization of hyperplanes: moving $H(u_1, \ldots, u_k)$ towards the origin by a sufficiently small amount leads to a hyperplane in $\mathcal{H}(u_1, \ldots, u_k)$.

We define the constraint point q_{ij}^{uv} as the centroid of $p_{iu}, p_{i\bar{u}}, p_{jv}, p_{j\bar{v}}$. Its nonzero coordinates are

$$x_i = \frac{\cos\theta_i + \cos\bar{\theta}_i}{4}, y_i = \frac{\sin\theta_i + \sin\bar{\theta}_i}{4}, x_j = \frac{\cos\theta_j + \cos\bar{\theta}_j}{4}, y_j = \frac{\sin\theta_j + \sin\bar{\theta}_j}{4},$$

for $\theta_i = (u-1)\frac{2\pi}{n}$, $\bar{\theta}_i = (\bar{u}-1)\frac{2\pi}{n}$, $\theta_j = (v-1)\frac{2\pi}{n}$, and $\bar{\theta}_j = (\bar{v}-1)\frac{2\pi}{n}$.

Lemma 4. *If $u_i = u$ and $u_j = v$, no hyperplane in $\mathcal{H}(u_1, \ldots, u_k)$ separates q_{ij}^{uv} from $P(u_1, \ldots, u_k)$.*

Proof. Such a hyperplane would in particular have to separate point q_{ij}^{uv} from $p_{iu}, p_{i\bar{u}}, p_{jv}, p_{j\bar{v}}$, which is impossible.

To see that q_{ij}^{uv} does not "destroy" the classes $\mathcal{H}(u_1, \ldots, u_k)$ with $u_i \neq u$ or $u_j \neq v$, let us consider a fixed pair of indices $i \neq j$. All points q_{ij}^{uv}, $(u, v \in [n])$ lie on a sphere S_{ij} around the origin in $E_i \times E_j$ (of radius $\sqrt{1/2} \cdot \sin\frac{\pi}{n}$). The intersection $H(u_1, \ldots, u_k) \cap (E_i \times E_j)$ is a 3-dimensional hyperplane $F_{ij}^{u_i u_j}$ uniquely defined by u_i and u_j: $F_{ij}^{u_i u_j}$ goes through the four points $p_{iu_i}, p_{i\bar{u}_i}, p_{ju_j}, p_{j\bar{u}_j}$. Moreover, $q_{ij}^{u_i u_j}$ is the point where $F_{ij}^{u_i u_j}$ touches the sphere S_{ij}. (This follows from symmetry considerations, and it can also be checked by a straightforward calculation that the vector $q_{ij}^{u_i u_j}$ is perpendicular to the hyperplane $F_{ij}^{u_i u_j}$.) This allows us to conclude:

Lemma 5. *If $u_i \neq u$ or $u_j \neq v$, then q_{ij}^{uv} lies on the same side of the hyperplane $H(u_1, \ldots, u_k)$ as the origin o.*

Proof. The point q_{ij}^{uv} lies on the sphere $S_{ij} \in E_i \times E_j$ centered at the origin. This sphere lies on the same side of $H(u_1, \ldots, u_k)$ as the origin, except for the point where it touches $H(u_1, \ldots, u_k)$. But this touching point $q_{ij}^{u_i u_j}$ is different from q_{ij}^{uv}.

This means that q_{ij}^{uv} and the points in $P(u_1, \ldots, u_k)$ are on different sides of the hyperplane $H(u_1, \ldots, u_k)$ (except for the points $p_{1u_1}, p_{1\bar{u}_1}, \ldots, p_{ku_k}, p_{k\bar{u}_k}$, which lie on it). Since $q_{ij}^{uv} \notin H(u_1, \ldots, u_k)$, every sufficiently close translate of $H(u_1, \ldots, u_k)$ in $\mathcal{H}(u_1, \ldots, u_k)$ with $u_i \neq u$ or $u_j \neq v$ separates $P(u_1, \ldots, u_k)$ and $\{q_{ij}^{uv}\}$.

Reduction. Similarly to the reduction in Section 3, we encode the structure of G by adding to Q^0 the $n\binom{k}{2}$ constraint points q_{ij}^{uu} ($1 \leq u \leq n$, $1 \leq i < j \leq k$) and $2|E|\binom{k}{2}$ constraint points q_{ij}^{uv} ($uv \in E$ and $i \neq j$). Let Q be the resulting point set. Then the possible partitions of P into two sets, each separated from Q by one hyperplane, represent the independent sets of G.

Lemma 6. *Sets P and Q can be separated by two hyperplanes if and only if G has an independent set of size k.*

From this lemma, and since this is an fpt-reduction, we conclude with the following:

Theorem 2. *Deciding whether two point sets P, Q in $\mathbb{R}^d$ can be separated by two hyperplanes is W[1]-hard with respect to d.*

Remark. The construction above depends on requiring strict separation, i.e., the separating hyperplanes are not allowed to go through the given points. For the fixed-precision approximation that is necessary to make the reduction suitable for a Turing machine, we have to move the constraint points q_{ij}^{uv} a little bit further away from the center before rounding them to rational coordinates. The statement of Lemma 4 is refined and excludes the possibility of separating $P(u_1, \ldots, u_k)$ from the set $\{o, q_{ij}^{uv}\}$ rather than from the point q_{ij}^{uv} alone.

These modifications are also suitable for the version of the problem where *weak separation* is allowed, i.e., points on the separation boundary can be from P or Q arbitrarily. In this case $\binom{2k}{2}$ additional points on the coordinate planes close to the origin must be added to Q^0, in order to eliminate the coordinate hyperplanes as potential separating hyperplanes.

5   Maximum-Size Feasible Subsystem

We first consider the special problem: Given a system of linear equations find a solution that satisfies as many equations as possible. (Note that this problem is dual to the problem of covering as many points as possible by a hyperplane through the origin.) The decision version of this problem is as follows: Given a set of n hyperplanes in $\mathbb{R}^d$ and an integer l, decide whether there exists a point in $\mathbb{R}^d$ that is covered by at least l of the hyperplanes.

In the following, $\boldsymbol{x} = (x_1, \ldots, x_k) \in \mathbb{R}^k$ denotes a k-dimensional vector (a notation that is slightly different from the one used in the previous sections). We identify the grid $[n]^k$ with the set of vectors in $\mathbb{R}^k$ with integer coordinates in $[n]$.

For a set $\mathcal{H}$ of hyperplanes in $\mathbb{R}^k$ and a point $\boldsymbol{x} \in \mathbb{R}^k$ we define

$$\mathrm{depth}(\boldsymbol{x}, \mathcal{H}) = |\{h \in \mathcal{H} \mid \boldsymbol{x} \in h\}|.$$

Given an undirected graph $G([n], E)$ and $k \in \mathbb{N}$, we will now construct a set $\mathcal{H}_{G,k}$ of $nk + 2|E|\binom{k}{2}$ hyperplanes in $\mathbb{R}^k$ such that G has a clique of size k if and only if there is a point $\boldsymbol{x} \in \mathbb{R}^k$ with $\mathrm{depth}(\boldsymbol{x}, \mathcal{H}_{G,k}) = k + \binom{k}{2}$.

For $1 \leq i \leq k$ and $1 \leq v \leq n$ we define the hyperplane $h_i^v = \{ \boldsymbol{x} \mid x_i = v \}$. The scaffolding hyperplane set $\mathcal{H}^0 = \{ h_i^v \mid 1 \leq i \leq k,\ 1 \leq v \leq n \}$ consists of nk hyperplanes. Any point $\boldsymbol{x}$ is contained in at most k hyperplanes in $\mathcal{H}^0$; equality is realized for the points in $[n]^k$:

Lemma 7. depth$(\boldsymbol{x}, \mathcal{H}^0) \leq k$ for any $\boldsymbol{x} \in \mathbb{R}^k$, and depth$(\boldsymbol{x}, \mathcal{H}^0) = k$ if and only if $\boldsymbol{x} \in [n]^k$.

For $1 \leq i < j \leq k$ and $1 \leq u, v \leq n$ we define the hyperplane $h_{ij}^{uv} = \{ \boldsymbol{x} \mid (x_i - u) + n(x_j - v) = 0 \}$. This hyperplane contains only those points $\boldsymbol{x}$ of the grid for which $x_i = u$ and $x_j = v$:

Lemma 8. $\boldsymbol{x} \in h_{ij}^{uv} \cap [n]^k$ if and only if $x_i = u$ and $x_j = v$.

Proof. Assume $\boldsymbol{x} \in h_{ij}^{uv} \cap [n]^k$, i.e. $(x_i - u) + n(x_j - v) = 0$ and $x_i, x_j \in [n]$. If $x_i \neq u$, the left-hand side of the equation is not divisible by n and thus cannot be 0. Therefore, $x_i = u$ and thus, $x_j = v$. The other direction is obvious.

For $1 \leq i < j \leq k$ we define the set $\mathcal{H}_{ij}^E = \{ h_{ij}^{uv} \mid uv \in E \text{ or } vu \in E \}$ of $2|E|$ hyperplanes. All these hyperplanes are parallel; thus a point is contained in at most one hyperplane of $\mathcal{H}_{ij}^E$. By Lemma 8, a point $\boldsymbol{x} \in [n]^k$ is contained in a hyperplane of $\mathcal{H}_{ij}^E$ if and only if $x_i x_j$ is an edge of E.

We define the set $\mathcal{H}^E = \bigcup_{1 \leq i < j \leq k} \mathcal{H}_{ij}^E$ consisting of $2|E|\binom{k}{2}$ hyperplanes. From the above, we have the following facts:

Lemma 9. (a) depth$(\boldsymbol{x}, \mathcal{H}^E) \leq \binom{k}{2}$ for any $\boldsymbol{x} \in \mathbb{R}^k$.
(b) Let $\boldsymbol{x} \in [n]^k$. Then depth$(\boldsymbol{x}, \mathcal{H}^E) = |\{ (i,j) \mid 1 \leq i < j \leq k,\ x_i x_j \in E \}|$
(c) Let $\boldsymbol{x} \in [n]^k$. Then depth$(\boldsymbol{x}, \mathcal{H}^E) = \binom{k}{2}$ iff $\{x_1, \ldots, x_k\}$ is a k-clique in G.

For the set $\mathcal{H}_{G,k} = \mathcal{H}^0 \cup \mathcal{H}^E$, Lemmas 7 and 9 immediately imply:

Lemma 10. depth$(\boldsymbol{x}, \mathcal{H}_{G,k}) = k + \binom{k}{2}$ if and only if $\boldsymbol{x} \in [n]^k$ and $\{x_1, \ldots, x_k\}$ is a k-clique in G.

Note that the above construction of the set $\mathcal{H}_{G,k}$ is an fpt-reduction with respect to both the depth of the set of hyperplanes, i.e., the maximum number of hyperplanes covering any point, and the dimension. Hence, we have the following:

Theorem 3. *Given a set of n of linear equations on d variables and an integer l, deciding whether there exists a solution that satisfies l of the equations is W[1]-hard with respect to both l and d.*

Replacing each equation by 2 inequalities, an instance of the above problem is transformed into an instance with linear inequalities such that there exists a solution satisfying l out of the n equations of the original instance if and only if there exists a solution satisfying $n + l$ out of the $2n$ inequalities of the final instance; the number of variables stays the same. Hence, we have the following:

Theorem 4. *Given a set of n linear inequalities on d variables and an integer l, deciding whether there exists a solution that satisfies l of the inequalities is W[1]-hard with respect to d.*

References

1. Armaldi, E., Kann, V.: The complexity and approximability of finding maximum feasible subsystems of linear relations. Theoretical Computer Science 147, 181–210 (1995)
2. Aronov, B., Har-Peled, S.: On approximating the depth and related problems. SIAM J. Comput. 38(3), 899–921 (2008)
3. Arora, S., Babai, L., Stern, J., Sweedyk, Z.: The hardness of approximate optima in lattices, codes, and systems of linear equations. J. Comput. Syst. Sci. 54(2), 317–331 (1997)
4. Bremner, D., Chen, D., Iacono, J., Langerman, S., Morin, P.: Output-sensitive algorithms for tukey depth and related problems. Statistics and Computing 18(3), 259–266 (2008)
5. Bǎdoiu, M., Har-Peled, S., Indyk, P.: Approximate clustering via core-sets. In: Proc. 34th Annual ACM Symposium on Theory of Computing, pp. 250–257 (2002)
6. Cabello, S., Giannopoulos, P., Knauer, C., Marx, D., Rote, G.: Geometric clustering: fixed-parameter tractability and lower bounds with respect to the dimension. ACM Transactions on Algorithms (to appear, 2009)
7. Cabello, S., Giannopoulos, P., Knauer, C., Rote, G.: Geometric clustering: fixed-parameter tractability and lower bounds with respect to the dimension. In: Proc. 19th Ann. ACM-SIAM Sympos. Discrete Algorithms, pp. 836–843 (2008)
8. Chan, T.M.: A (slightly) faster algorithm for Klee's measure problem. In: Proc. 24th Annual Symposium on Computational Geometry, pp. 94–100 (2008)
9. Downey, R.G., Fellows, M.R.: Parameterized Complexity. Monographs in Computer Science. Springer, Heidelberg (1999)
10. Giannopoulos, P., Knauer, C., Whitesides, S.: Parameterized complexity of geometric problems. Computer Journal 51(3), 372–384 (2008)
11. Impagliazzo, R., Paturi, R.: On the complexity of k-SAT. J. Comput. Syst. Sci. 62(2), 367–375 (2001)
12. Langerman, S., Morin, P.: Covering things with things. Discrete & Computational Geometry 33(4), 717–729 (2005)
13. Megiddo, N.: On the complexity of polyhedral separability. Discrete & Computational Geometry 3, 325–337 (1988)
14. Megiddo, N.: On the complexity of some geometric problems in unbounded dimension. J. Symb. Comput. 10, 327–334 (1990)
15. Varadarajan, K., Venkatesh, S., Ye, Y., Zhang, J.: Approximating the radii of point sets. SIAM J. Comput. 36(6), 1764–1776 (2007)

Paths of Bounded Length and Their Cuts: Parameterized Complexity and Algorithms

Petr A. Golovach[1] and Dimitrios M. Thilikos[2,*]

[1] Department of Informatics, University of Bergen, PB 7803, 5020 Bergen, Norway
[2] Department of Mathematics, National and Kapodistrian University of Athens,
Panepistimioupolis, GR15784 Athens, Greece

Abstract. We study the parameterized complexity of two families of problems: the *bounded length disjoint paths* problem and the *bounded length cut* problem. From Menger's theorem both problems are equivalent (and computationally easy) in the unbounded case for single source, single target paths. However, in the bounded case, they are combinatorially distinct and are both NP-hard, even to approximate. Our results indicate that a more refined landscape appears when we study these problems with respect to their parameterized complexity. For this, we consider several parameterizations (with respect to the maximum length l of paths, the number k of paths or the size of a cut, and the treewidth of the input graph) of all variants of both problems (edge/vertex-disjoint paths or cuts, directed/undirected). We provide FPT-algorithms (for all variants) when parameterized by both k and l and hardness results when the parameter is only one of k and l. Our results indicate that the bounded length disjoint-path variants are structurally harder than their bounded length cut counterparts. Also, it appears that the edge variants are harder than their vertex-disjoint counterparts when parameterized by the treewidth of the input graph.

Keywords: Bounded length disjoint paths, Bounded length cuts, Parameterized Complexity, Parameterized Algorithms.

1 Introduction

We consider finite (directed and undirected) graphs without loops or multiple edges. The vertex set of a graph G is denoted by $V(G)$ and its edge set by $E(G)$. We denote undirected edges by $\{u, v\}$, and directed edges by (u, v). Given a graph G and a set $F \subseteq E(G)$ (resp. $X \subseteq V(G)$), we denote by $G \setminus F$ (resp. $G \setminus X$) the graph obtained by G if we remove from it all edges in F (resp. vertices in X).

One of the most celebrated problems in discrete algorithms and combinatorial optimization is the disjoint paths problem. Its algorithmic study dates back to Menger's Theorem [25] (see also [9]), was extended by the work of Ford and

---

* Supported by the project "Kapodistrias" (AΠ 02839/28.07.2008) of the National and Kapodistrian University of Athens (project code: 70/4/8757).

J. Chen and F.V. Fomin (Eds.): IWPEC 2009, LNCS 5917, pp. 210–221, 2009.
© Springer-Verlag Berlin Heidelberg 2009

Fulkerson [15] on network flows, and now constitutes (along with its variants) a central algorithmic problem in algorithm design.

According to Menger's theorem, given a graph G and two terminals $s, t \in V(G)$, the maximum number of vertex-disjoint (s, t)-paths in G is *equal* to the minimum cardinality of a set of vertices in $V(G) \setminus \{s, t\}$ meeting all (s, t)-paths of G. Interestingly, it appears that such a min-max equality does not hold if we restrict paths to be of bounded length. This was observed for the first time by Adámek and Koubek in [1]. Lovász, Neumann Lara, and Plummer proved in [24] that a similar min-max relation holds only for path lengths equal to 2,3, or 4. Analogous results were provided for the case where the paths are edge-disjoint in [11] and [27]. We present below the main decision versions of the problems generated by the bounded length restriction.

We need some definitions. Let G be a graph, $s, t \in V(G)$, and let l be a positive integer. We call set $F \subseteq E(G)$ (resp. $X \subseteq V(G) \setminus \{s, t\}$) (s, t)-*edge* (resp. *vertex*) l-*bounded cut* if $G \setminus F$ (resp. $G \setminus X$) contains no (s, t)-path of length at most l.

BOUNDED EDGE DIRECTED (s, t)-DISJOINT PATHS (BEDP)

Input: A directed graph G, two positive integers k, l and two distinct vertices s, t of G.

Question: Are there k edge-disjoint (s, t)-paths each of length at most l in G?

BOUNDED EDGE DIRECTED (s, t)-CUT (BEDC)

Input: A directed graph G, two positive integers k, l and two distinct vertices s, t of G.

Question: Is there a (s, t)-edge l-bounded cut $F \subseteq E(G)$ of size at most k?

Also, the first of the above problems has been extended to its multi-terminal version as follows.

BOUNDED EDGE DIRECTED MULTI-TERMINAL DISJOINT PATHS (BEDMP)

Input: A directed graph $G = (V, E)$, two positive integers k, l, and two sequences $S = (s_1, \ldots, s_k)$ (sources), $T = (t_1, \ldots, t_k)$ (targets) of vertices in G.

Question: Are there k edge-disjoint (s_i, t_i)-paths of length at most l in G for $i = 1, \ldots, k$?

Similarly to the above, we can define numerous variants depending whether the graph is directed or undirected, and whether the paths are edge-disjoint or internally vertex-disjoint. All variants and the corresponding notation are depicted in Table 1.

For all multi-terminal disjoint paths problems we can assume that all terminals are pair-wise distinct, since otherwise we can apply the following rule:

Rule (1): for every vertex v that corresponds to r terminals we first subdivide all its incident edges and then replace v by r vertices (each with one of the terminals corresponding to v) that have the same neighborhood as v (in the directed case, replacement edges maintain their original directions). The new

Table 1. Bounded length variants of the edge-disjoint paths problem and the edge cut problem

	MULTI-TERMINAL DISJOINT PATHS		(s,t)-DISJOINT PATHS		(s,t)-CUT	
	DIRECTED	UNDIRECTED	DIRECTED	UNDIRECTED	DIRECTED	UNDIRECTED
EDGE	BEDMP	BEUMP	BEDP	BEUP	BEDC	BEUC
VERTEX	BVDMP	BVUMP	BVDP	BVUP	BVDC	BVUC

graph contains k edge-disjoint paths of length at most $l + 2$ iff the original one contains k vertex-disjoint paths of length at most l.

The first algorithmic results for the above problems appeared by Itai, Perl and Shiloach in [19] where they proved that BVUP and BEUP are polynomially solvable for path lengths 2, 3 or 4, while they become NP-complete for length values bigger than 4. In the same paper they proved that if, instead of fixing the length l, we fix the number k of paths, the problem is still NP-complete even for 2 paths. Results on the fractional versions of these problems were given in [13,21,22]. The approximability of these problems was studied in [6,17] and [4,5]. Finally, for some applications of the above problems, see [29,31] and [18].

Some results for the multi-terminal variants of the bounded-length disjoint paths problem were given in [17]. We just stress that, when there is no restriction to the length of the paths, BVUMP is NP-complete in general [20] and polynomially solvable for fixed k [28], while BVDMP is NP-complete even when $k = 2$ [16].

In this paper, we provide a detailed study of the parameterized complexity of all the bounded length variants of the problems in Table 1. In a parameterized problem we distinguish some part of the input to be its parameter. Typically, a parameter is an integer, k, related to the problem input and the question is whether the problem can be solved by an algorithm (called FPT-algorithm) of time complexity $f(k) \cdot n^{O(1)}$ where n is the size of the input and f is a (super-polynomial) function depending only on the parameter (instead of worst time complexities such as $O(n^{f(k)})$ or $O(k^{f(n)})$). When a parameterized problem admits an FPT-algorithm, then it belongs in the parameterized complexity class FPT. Not all parameterized problems belong in FPT. There are several parameterized complexity classes, such as W[1], W[2], para-NP and analogous notions of hardness with respect to parameter-preserving reductions, able to prove that membership in FPT is rather non-possible (for more details, see the monographs [10,14,26]). Briefly, if a parameterized problem is W[1]-hard, this means that a complexity of type $O(n^{f(k)})$ is the best we may expected, while if a parameterized problem is para-NP-hard, then we cannot even hope for something better than a $k^{f(n)}$-algorithm.

Table 2 indicates the existence of reductions between all considered problems. Here $\Pi_1 \leq^{(i)} \Pi_2$ means that the problem Π_1 can be reduced to the problem Π_2 by the reduction rule i. The edge undirected versions are reduced to the vertex undirected ones by the following rule:

Rule (2): Take the line graph L_G of G and for every clique K of L_G, corresponding to the edges incident to a terminal v of G, add a new terminal vertex v' and connect it with all the vertices of the clique. Vertex-disjoint paths of length $l + 1$ in the new graph correspond to edge-disjoint paths of length l in the original graph, while it trivially follows that edge cuts become vertex cuts.

Certainly, vertex undirected versions are reduced to vertex directed ones by the following obvious rule:

Rule (3): Replace every edge by two opposite direction edges.

The following rule reduces all vertex directed versions to their edge directed counterparts:

Rule (4): Replace every non-terminal vertex v by a directed edge (v_t, v_h) (we call such edges *new edges*) and make v_t the head of all previous edges whose head was v and v_h the tail of all previous edges whose tail was v. Notice that every path of length at most $2l - 1$ in the new graph corresponds to a path of length at most l in the original graph and edge-disjoint paths in the new graph correspond to vertex-disjoint paths in the original graph and vice versa. This proves the correctness of Rule (4) for disjoint path problems. For cut problems we additionally observe that every vertex cut of the original graph correspond to an edge cut in the new graph. For the inverse direction, take an edge cut of the new graph and replace each non-new edges e in it with some new edge that has a common endpoint with e. This makes every edge cut in the new graph to correspond to a vertex cut in the original graph.

Notice that all rules are parameter-preserving when the parameter is k, l, or both.

Table 2. Reductions between problems

$$\text{BEUMP} \leq^{(2)} \text{BVUMP} \leq^{(3)} \text{BVDMP} \leq^{(4)} \text{BEDMP}$$
$$\text{VI}^{(1)} \qquad \text{VI}^{(1)} \qquad \text{VI}^{(1)} \qquad \text{VI}^{(1)}$$
$$\text{BEUP} \leq^{(2)} \text{BVUP} \leq^{(3)} \text{BVDP} \leq^{(4)} \text{BEDP}$$

$$\text{BEUC} \leq^{(2)} \text{BVUC} \leq^{(3)} \text{BVDC} \leq^{(4)} \text{BEDC}$$

All problems in Table 1 have two possible parameters k and l in their inputs. Therefore, we consider parameterizations of them with respect to l, k, or both, indicating which parameterization we pick in each problem. For example, the BEUP problem is denote as BEUP(k) when parameterized by the number of paths k, BEUP(l) when parameterized by the maximum length l of a path and BEUP(k, l) when parameterized by both these quantities. We follow the same notation for all problems in Table 1.

We prove that all variants of our problems are in FPT when parameterizing on both k and l.

All problems we consider are NP-hard for fixed values of l, bigger than some constant [4,19]. Using standard terminology from [14], this means that all of them, parameterized by l are para-NP-complete (i.e. they are NP-hard even for fixed values of the parameter). Moreover, the problem asking for the existence of two edge-disjoint paths between two terminals of an undirected graph is also NP-complete even for two paths, because of the results [30,23]. This implies the para-NP-completeness of all the disjoint paths variants when parameterized by k. However, no similar result can be expected (unless P=NP) for bounded cut problems, as they trivially admit an $n^{O(k)}$-step algorithm (just check all possible cuts of size at most k). It appears that this running time substantially cannot become better: we prove that these four variants are W[1]-hard (Theorem 4) and that for the directed graph variants, this holds even for DAGs (Theorem 3). This indicates that, apart from the combinatorial discrepancy between problems on paths and problems on cuts, there is also a discrepancy on the parameterized complexities of the corresponding problems. We stress that this distinction cannot be made clear by studying the classic complexity of the two families of problems (they are all NP-complete in general). Our results are depicted in Table 3.

Table 3. Summary of our results when parameterizing by l, k,l and k

	l	k,l	k
BEDMP BVDMP BVUMP BEUMP	para-NP-c [19]	FPT $O(2^{O(kl)} \cdot m \cdot \log n)$ (Th. 1)	para-NP-c [30,23]
BEDP BVDP BVUP BEUP	para-NP-c [19]	FPT $O(2^{O(kl)} \cdot m \cdot \log n)$ (Th. 1)	para-NP-c [30,23]
BEDC BVDC BVUC BEUC	para-NP-c [4]	FPT $O(l^k \cdot m)$ (Th. 2)	W[1]-h for DAGs (Th. 3) W[1]-h for DAGs (Th. 3) W[1]-h (Th. 4) W[1]-h (Th. 4)

Our next step is to study the (in general para-NP-complete) parameterized problems BVDP(l) and BVUP(l) for the special case where their input graphs are sparse. We prove (Theorem 5) that both problems admit FPT-algorithms for classes of graphs that have bounded local treewidth (typical graph class with bounded local treewidth are planar graphs or bounded-degree graphs). Moreover, this result can be extended for classes of graphs where the removal of at most one vertex includes them in some bounded local treewidth class. On the other side, we prove that this sparsity criterion cannot be relaxed: BVDP(l) (BVUP(l)) remains para-NP-complete (Theorem 6) for $l \geq 6$, on undirected (directed acyclic) graphs that can be made planar after removing 2 vertices (we

Table 4. Summary of our results for sparse graph families

	l, [bounded ltw]	l, [two-apex]	tw, l
BVUP	FPT (Th. 5)	para-NP-c (Th. 6), $l \geq 6$	FPT (Th. 5)
BVDP	FPT (Th. 5)	para-NP-c for DAGs (Th. 6), $l \geq 6$	FPT (Th. 5)
BEUP	open	para-NP-c (Th. 7), $l \geq 7$	W[1]-h for fixed $l \geq 10$ (Th. 8)
BEDP	open	para-NP-c for DAGs (Th. 7), $l \geq 7$	W[1]-h for DAGs for fixed $l \geq 10$ (Th. 8)

call these graphs *2-apex-graphs*). We also prove that the same holds for the edge variants of the same problems (Theorem 7) for $l \geq 7$. Our results suggest a rapid change on the problem complexity with respect to the minor-exclusion sparsity criterion.

Our last result concerns the case where BEUP and BEDP are parameterized by the treewidth of their input graphs. We prove that BEUP is W[1]-hard when parameterized by the treewidth of the input graph and that BEDP is W[1]-hard when parameterized by the treewidth of the underlying graph of its input graph even when the input graph is acyclic (Theorem 8). This last result indicates that the edge-disjoint variants are harder than the vertex-disjoint ones (the same parameterization leads to an FPT-algorithm for BVUP and BVDP – Theorem 5). Our results on sparse graph classes are summarized in Table 4.

2   Parameterized Algorithms

2.1   An FPT-Algorithm for BEDMP(k, l)

Our algorithm for the BEDMP(k, l) is based on the color-coding technique introduced by Alon, Yuster and Swick in [2]. In particular, we consider a family $\mathcal{F}$ of hash functions, each mapping $\{1, \ldots, m\}$ to a set of colors $\{1, \ldots, k \cdot l\}$, such that for every $S \subseteq \{1, \ldots, m\}$, where $|S| \leq k \cdot l$, there is a $f \in \mathcal{F}$ such that its restriction to S is a bijection. As mentioned in [2], such a family where $|\mathcal{F}| = 2^{O(k \cdot l)} \cdot \log m$ can be constructed in $2^{O(k \cdot l)} \cdot m \cdot \log m$ steps.

Let $\mathcal{F}$ be a family of hash functions as above where $\{1, \ldots, m\}$ represent the edges of G. Let also $\chi \in \mathcal{F}$. Given an integer $i \in \{1, \ldots, k\}$, we define a Boolean function B_i^χ such that, for every set of colors $X \subseteq \{1, \ldots, k \cdot l\}$, $B_i^\chi(X)$ is true if and only if there exists a collection of i paths $P_1, \ldots, P_i$ of length at most l where, for $j \in \{1, \ldots, i\}$, the endpoints of P_j are s_j and t_j and such that the set of the colors assigned to the edges of these paths is a subset of X (i.e. $\chi^{-1}(\cup_{j \in \{1,\ldots,i\}} E(P_j)) \subseteq X$). Notice that an instance of BEDMP(k, l) is a YES-instance if and only if there is a $\chi \in \mathcal{F}$ such that $B_k^\chi(\{1, \ldots, k \cdot l\}) = \mathsf{true}$. In general, to compute $B_i^\chi(X)$ for some $X \subseteq \{1, \ldots, k \cdot l\}$, we observe that

$$B_i^\chi(X) = \bigvee_{Y \subseteq X} (B_{i-1}^\chi(Y) \wedge C_i^\chi(X \setminus Y))$$

where C_i^χ is a Boolean function such that if $S \subseteq \{1, \ldots, k \cdot l\}$ the value of $C_i^\chi(S)$ is true if and only if the subgraph of G induced by the edges colored by colors in S contains a path between s_i and t_i of length at most l. Notice that C_i^χ can be computed in $O(m)$ steps. Moreover, computing $B_i^\chi(X)$ for all $X \subseteq \{1, \ldots, l \cdot k\}$ requires $O(3^{k \cdot l} \cdot m)$ steps. Therefore, the above dynamic programming requires in total $O(3^{k \cdot l} \cdot m \cdot k)$ steps to compute $B_k^\chi(\{1, \ldots, k \cdot l\})$. Concluding BEDMP$(k, l)$ can be solved in $O(2^{O(k \cdot l)} \log m \cdot m \cdot k)$ steps.

It is easy to observe that the above algorithm can be modified so that it also would return the requested paths when exist. We conclude to the following.

Theorem 1. *The* BEDMP(k, l) *problem (as well as* BVDMP(k, l), BVUMP (k, l), *and* BEUMP(k, l)*) can be solved by an* FPT-*algorithm that runs in* $O(2^{O(k \cdot l)} \log n \cdot m \cdot k)$ *steps where* $n = |V(G)|$ *and* $m = |E(G)|$.

2.2   An FPT-Algorithm for BEDC(k, l)

The proof of the following theorem is based on the simple observation that for any (s, t)-path of length at most l, at least one edge of it has to be included to any (s, t)-edge l-bounded cut.

Theorem 2. *The* BEDC(k, l) *problem can be solved by an* FPT-*algorithm that runs in* $O(l^k \cdot m)$ *time where* $n = |V(G)|$ *and* $m = |E(G)|$.

3   Hardness Results for (s, t)-Cuts

In this section we prove W[1]-hardness of BVDC(k) and BEUC(k). It can be noted that by the reduction rules (see Table 2) W[1]-hardness of BVDC(k) follows from a similar result for BEUC(k), but we prove here a stronger result.

Theorem 3. BVDC(k) *problem is* W[1]-*hard even for acyclic digraphs.*

Proof. We present a reduction from the MULTICOLORED CLIQUE problem:

MULTICOLORED CLIQUE

Input: A graph G with a proper k-coloring of G.
Question: Is there a clique of size k in G containing exactly one vertex from
    each color class?

The MULTICOLORED CLIQUE problem, parameterized by k, was proved to be W[1]-hard by Fellows et al. [12].

Let G be an n-vertex undirected graph. Denote by X_i the i-th color class in the given k-coloring of G. Assume without loss of a generality that $k \geq 4$. We assume also that for any pair of sets X_i, X_j, $i \neq j$, vertices of these sets are connected by the same number of edges denoted by m, and $m > 0$ (otherwise it is possible to add pairs of adjacent vertices to the graph to ensure this condition). Denote by $e_1^{(i,j)}, e_2^{(i,j)}, \ldots, e_m^{(i,j)}$ the edges which join sets X_i and X_j. Let $l = 5m + 4$.

Now we consider auxiliary constructions. For every $i, j \in \{1, 2, \ldots, k\}$, $i \neq j$, a directed graph $F_{i,j}$ is constructed as follows.

1. Two vertices s and t are created.
2. For every $r \in \{1, 2, \ldots, m\}$, vertices u_r, $a_r^{(1)}, a_r^{(2)}, a_r^{(3)}$ and $b_r^{(1)}, b_r^{(2)}, b_r^{(3)}$ are constructed, and for every $r \in \{0, 1, \ldots, m\}$, vertices v_r are introduced. It is assumed, for convenience, that $s = a_0^{(1)} = a_0^{(2)} = a_0^{(3)}$, $b_0^{(1)} = a_m^{(1)}$, $b_0^{(2)} = a_m^{(2)}$, $b_0^{(3)} = a_m^{(3)}$ and $t = b_{m+1}^{(3)} = b_{m+1}^{(3)} = b_{m+1}^{(3)}$.
3. For each vertex u_r, edges $(a_{r-1}^{(1)}, u_r)$, $(a_{r-1}^{(2)}, u_r)$, $(a_{r-1}^{(3)}, u_r)$ and $(u_r, a_r^{(1)})$, $(u_r, a_r^{(2)})$, $(u_r, a_r^{(3)})$ are added.
4. For each vertex v_r, edges $(b_r^{(1)}, v_r)$, $(b_r^{(2)}, v_r)$, $(b_r^{(3)}, v_r)$ and $(v_r, b_{r+1}^{(1)})$, $(v_r, b_{r+1}^{(2)})$, $(v_r, b_{r+1}^{(3)})$ are added.
5. Pairs of vertices $a_{r-1}^{(f)}, a_r^{(f)}$ are joined by paths of length $r + 2$ for $f = 1, 2, 3$ and $r \in \{1, 2, \ldots, m\}$.
6. Pairs of vertices $a_{r-1}^{(f)}, b_r^{(f)}$, $f = 1, 2, 3$, are joined by paths of length $3m + 4$ for $r \in \{1, 2, \ldots, m+1\}$.
7. Add vertices $w_1^{(i,j)}, w_2^{(i,j)}, \ldots, w_m^{(i,j)}$, and join every vertex v_{r-1} with $w_r^{(i,j)}$ by a path of length $3(m + 1 - r)$.

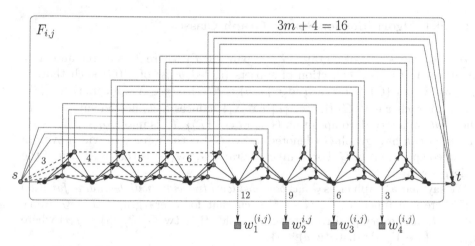

Fig. 1. Construction of $F_{i,j}$ for $m = 4$. Paths are shown by dash lines.

The graph $F_{i,j}$ is shown in Figure 1 for $m = 4$. Using these gadgets $F_{i,j}$ we construct a directed graph H from G as follows.

8. For all pairs $\{i, j\}$, $i, j \in \{1, \ldots, k\}$, $i \neq j$, graphs $F_{i,j}$ with common vertices s and t are constructed.

9. Every edge $e_f^{(i,j)} = \{x, y\}$ of G is replaced by two directed edges $(w_f^{(i,j)}, x)$ and $(w_f^{(i,j)}, y)$.

10. For each vertex $x \in V(G)$, an edge (x, t) is added.

It is easy to see that H is a directed acyclic graph. The next claim concludes the proof of the theorem.

Claim. *Graph G has a clique of size k which contains exactly one vertex from any color class if and only if there is (s,t)-vertex l-bounded cut in H with at most $k' = k^2$ vertices.*

Notice that the reduction 4 from BVDC(k) to BEDC(k) transforms a directed acyclic graph into another directed acyclic graph. So, W[1]-hardness of BEDC(k) for DAGs follows immediately. What remains to prove is the W[1]-hardness for the undirected case. The proof of the following theorem is a similar reduction from the MULTICOLORED CLIQUE problem and is omitted here.

Theorem 4. BEUC(k) *is* W[1]-*hard.*

4 (s,t)-Paths of Bounded Length for Sparse Graphs

4.1   FPT-Algorithms for Sparse Graph Classes

A *tree decomposition* of a graph G is a pair (X,T) where T is a tree and $X = \{X_i \mid i \in V(T)\}$ is a collection of subsets (called *bags*) of $V(G)$ such that: 1. $\bigcup_{i \in V(T)} X_i = V(G)$, 2. for each edge $\{x,y\} \in E(G)$, $x,y \in X_i$ for some $i \in V(T)$, and 3. for each $x \in V(G)$ the set $\{i \mid x \in X_i\}$ induces a connected subtree of T. The *width* of a tree decomposition $(\{X_i \mid i \in V(T)\}, T)$ is $\max_{i \in V(T)} \{|X_i| - 1\}$. The *treewidth* of a graph G (denoted as $\mathbf{tw}(G)$) is the minimum width over all tree decompositions of G. For a directed graph G, $\mathbf{tw}(G)$ is the treewidth of the underlying graph.

We say that a graph class $\mathcal{G}$ has *bounded local treewidth with bounding function* f if there is a function $f : \mathbb{N} \to \mathbb{N}$ such that for every graph $G \in \mathcal{G}$, every $v \in V(G)$, and every positive integer i it holds that $\mathbf{tw}(G[N_G^i[v]]) \leq f(i)$ where $N_G^i[v] = \{u \in V(G) : \mathrm{dist}_G(u,v) \leq i\}$.

It appears that many sparse classes have bounded local treewidth. Examples are planar graphs and graphs of bounded genus, bounded max-degree graphs, and graphs excluding an apex graph as a minor (an apex graph is a graph that can become planar after the removal of one vertex). The purpose of this subsection is to construct an FPT algorithm for the BVDP(l) and the BVUP(l) problems when their inputs are restricted to directed graphs whose underlying graphs belong to some (almost) bounded local treewidth graph class.

Theorem 5. *The* BVDP(l) *problems (and therefore, also* BVUP(l)*) can be solved by an* FPT-*algorithm for graph classes that have bounded local treewidth. Moreover, let $\mathcal{G}$ be a bounded local treewidth graph class, and let $\mathcal{G}'$ be a set of all graphs G such that there is a set $X \subseteq V(G)$, $|X| \leq 1$, for which $G \setminus X \in \mathcal{G}$. Then the BVDP$(l)$ can be solved by an* FPT-*algorithm in $\mathcal{G}'$.*

4.2   Vertex-Disjoint (s, t)-Paths of Bounded Length for H-Minor-Free Graphs

In this section we show that the restrictions of Theorem 5 are somehow tight. We call a graph G a *two-apex graph* if there is a set X of at most two vertices such that $G \setminus X$ is a planar graph. Due the space restrictions the proof of the following theorem is omitted here.

Theorem 6. BVUP(l) *is* NP-*complete and* BVDP(l) *is* NP-*complete for directed acyclic graphs for any fixed $l \geq 6$ for two-apex graphs.*

Consider the class of K_k-minor free graphs (i.e. none of the graphs in this class contains a subgraph that can be contracted to K_k). Notice that K_5-free graphs have bounded local treewidth. However this is not correct for K_r-minor-free graphs for $r \geq 6$. Since two-apex graphs are K_7-minor-free, Theorem 5 provides a nearly optimal estimation on the tractability of BVUP(l) and BVDP(l) on K_r-minor-free graphs. Actually, the same Theorem argues that not even a $n^{f(k)}$ step algorithm can be found for $r \geq 7$.

We can also prove the following (the proof is similar to the one of Theorem 6 and is omitted here). We consider these theorem separately instead applying reduction rules since these rules do not preserve exact value of the parameter l.

Theorem 7. *The* BEUP(l) *is* NP-*complete and* BEDP(l) *is* NP-*complete for directed acyclic graphs for any fixed $l \geq 7$ for two-apex graphs.*

It is interesting to note that BEUP(l) and BEDP(l) are more difficult than their vertex disjoint counterparts for graphs of bounded treewidth (see Theorem 5). The fact the edge-disjoint variants are harder is also indicated by the results of the following section.

4.3   Edge-Disjoint (s, t)-Paths of Bounded Length for Graphs of Bounded Treewidth

By reduction rules (see Table 2) BEUP can be reduced to BVUP, but the reduction 2 does not preserve the treewidth of the graph. The following theorem (the proof of which is omitted here due the space restrictions) shows that vertex-disjoint and edge-disjoint path problems behave very differently when parameterized by the treewidth.

Theorem 8. *For every fixed $l \geq 10$,* BEUP *is* W[1]-*hard, when parameterized by treewidth and* BEDP *is* W[1]-*hard for directed acyclic graphs when parameterized by treewidth of the underlying graph.*

5   Conclusions

A natural question about the parameterized complexity of the variants of the bounded length disjoint path and the bounded length cut problems parameterized by k and l is whether they admit polynomial kernels. In fact, using techniques from [7], we can prove that this is not the case for all the disjoint path variants. We believe that the existence of polynomial kernels for the edge cut variants as well as the planar restrictions of the disjoint path variants is an interesting open problem.

References

1. Adámek, J., Koubek, V.: Remarks on flows in network with short paths. Commentationes Mathematicae Universitatis Carolinae 12(4), 661–667 (1971)
2. Alon, N., Yuster, R., Zwick, U.: Color-coding. J. Assoc. Comput. Mach. 42, 844–856 (1995)
3. Arnborg, S., Lagergren, J., Seese, D.: Easy problems for tree-decomposable graphs. Journal of Algorithms 12, 308–340 (1991)
4. Baier, G., Erlebach, T., Hall, A., Köhler, E., Schilling, H., Skutella, M.: Length-bounded cuts and flows. In: Bugliesi, M., Preneel, B., Sassone, V., Wegener, I. (eds.) ICALP 2006. LNCS, vol. 4051, pp. 679–690. Springer, Heidelberg (2006)
5. Baier, G., Erlebach, T., Hall, A., Köhler, E., Kolman, P., Pangrác, O., Schilling, H., Skutella, M.: Length-bounded cuts and flows. ACM Transactions in Algorithms (to appear)
6. Bley, A.: On the complexity of vertex-disjoint length-restricted path problems. Comput. Complexity 12, 131–149 (2003)
7. Bodlaender, H.L., Downey, R.G., Fellows, M.R., Hermelin, D.: On problems without polynomial kernels (extended abstract). In: Aceto, L., Damgård, I., Goldberg, L.A., Halldórsson, M.M., Ingólfsdóttir, A., Walukiewicz, I. (eds.) ICALP 2008, Part I. LNCS, vol. 5125, pp. 563–574. Springer, Heidelberg (2008)
8. Borie, R.B.: Generation of polynomial-time algorithms for some optimization problems on tree-decomposable graphs. Algorithmica 14, 123–137 (1995)
9. Dantzig, G.B., Fulkerson, D.R.: On the max-flow min-cut theorem of networks, in Linear inequalities and related systems. Annals of Mathematics Studies, vol. 38, pp. 215–221. Princeton University Press, Princeton (1956)
10. Downey, R.G., Fellows, M.R.: Parameterized complexity. Monographs in Computer Science. Springer, New York (1999)
11. Exoo, G.: On line disjoint paths of bounded length. Discrete Math. 44, 317–318 (1983)
12. Fellows, M., Hermelin, D., Rosamond, F., Vialette, S.: On the parameterized complexity of multiple-interval graph problems. Theor. Comput. Sci. 410, 53–61 (2009)
13. Fleischer, L.K., Skutella, M.: The quickest multicommodity flow problem. In: Cook, W.J., Schulz, A.S. (eds.) IPCO 2002. LNCS, vol. 2337, pp. 36–53. Springer, Heidelberg (2002)
14. Flum, J., Grohe, M.: Parameterized complexity theory, Texts in Theoretical Computer Science. An EATCS Series. Springer, Berlin (2006)
15. Ford Jr., L.R., Fulkerson, D.R.: Maximal flow through a network. Canad. J. Math. 8, 399–404 (1956)
16. Fortune, S., Hopcroft, J., Wyllie, J.: The directed subgraph homeomorphism problem. Theoret. Comput. Sci. 10, 111–121 (1980)
17. Guruswami, V., Khanna, S., Rajaraman, R., Shepherd, B., Yannakakis, M.: Near-optimal hardness results and approximation algorithms for edge-disjoint paths and related problems. J. Comput. System Sci. 67, 473–496 (2003)

18. Hsu, D.: On container width and length in graphs, groups, and networks. IEICE transactions on fundamentals of electronics, communications and computer sciences 77, 668–680 (1994); Dedicated to Professor Paul Erdős on the occasion of his 80th birthday (Special Section on Discrete Mathematics and Its Applications)
19. Itai, A., Perl, Y., Shiloach, Y.: The complexity of finding maximum disjoint paths with length constraints. Networks 12, 277–286 (1982)
20. Karp, R.M.: On the computational complexity of combinatorial problems. Networks 5(1), 45–68 (1975)
21. Kolman, P., Scheideler, C.: Improved bounds for the unsplittable flow problem. In: Proceedings of the Symposium on Discrete Algorithms, pp. 184–193. ACM, New York (2002)
22. Kolman, P., Scheideler, C.: Improved bounds for the unsplittable flow problem. Journal of Algorithms 61(1), 20–44 (2006)
23. Li, C.-L., McCormick, T., Simchi-Levi, D.: The complexity of finding two disjoint paths with min-max objective function. Discrete Appl. Math. 26, 105–115 (1990)
24. Lovász, L., Neumann Lara, V., Plummer, M.: Mengerian theorems for paths of bounded length. Period. Math. Hungar. 9, 269–276 (1978)
25. Menger, K.: Über reguläre Baumkurven. Math. Ann. 96, 572–582 (1927)
26. Niedermeier, R.: Invitation to fixed-parameter algorithms. Oxford Lecture Series in Mathematics and its Applications, vol. 31. Oxford University Press, Oxford (2006)
27. Niepel, L., Šafaříková, D.: On a generalization of Menger's theorem. Acta Math. Univ. Comenian. 42, 43, 275–284 (1983)
28. Robertson, N., Seymour, P.D.: Graph minors. XIII. The disjoint paths problem. Journal of Combinatorial Theory. Series B 63, 65–110 (1995)
29. Ronen, D., Perl, Y.: Heuristics for finding a maximum number of disjoint bounded paths. Networks 14, 531–544 (1984)
30. Tragoudas, S., Varol, Y.L.: Computing disjoint paths with length constraints. In: D'Amore, F., Marchetti-Spaccamela, A., Franciosa, P.G. (eds.) WG 1996. LNCS, vol. 1197, pp. 375–389. Springer, Heidelberg (1997)
31. Wagner, D., Weihe, K.: A linear-time algorithm for edge-disjoint paths in planar graphs. Combinatorica 15, 135–150 (1995)

Fixed-Parameter Algorithms in Analysis of Heuristics for Extracting Networks in Linear Programs

Gregory Gutin[1], Daniel Karapetyan[1], and Igor Razgon[2]

[1] Department of Computer Science
Royal Holloway, University of London, UK
gutin@cs.rhul.ac.uk, daniel.karapetyan@gmail.com
[2] Department of Computer Science
University College Cork, Ireland
i.razgon@cs.ucc.ie

Abstract. A parameterized problem Π can be considered as a set of pairs (I, k) where I is the main part and k (usually an integer) is the parameter. Π is called fixed-parameter tractable (FPT) if membership of (I, k) in Π can be decided in time $O(f(k)|I|^c)$, where $|I|$ denotes the size of I, $f(k)$ is a computable function, and c is a constant independent of k and I. An algorithm of complexity $O(f(k)|I|^c)$ is called a fixed-parameter algorithm.

It often happens that although a problem is FPT, the practitioners prefer to use imprecise heuristic methods to solve the problem in the real-world situation simply because of the fact that the heuristic methods are faster. In this paper we argue that in this situation a fixed-parameter algorithm for the given problem may be still of a considerable practical use. In particular, the fixed-parameter algorithm can be used to evaluate the approximation quality of heuristic approaches.

To demonstrate this way of application of fixed-parameter algorithms, we consider the problem of extracting a maximum-size reflected network in a linear program. We evaluate a state-of-the-art heuristic SGA and two variations of it with a new heuristic and with an exact algorithm. The new heuristic and algorithm use fixed-parameter tractable procedures. The new heuristic turned out to be of little practical interest, but the exact algorithm is of interest when the network size is close to that of the linear program especially if the exact algorithm is used in conjunction with SGA. Another conclusion which has a large practical interest is that some variant of SGA can be the best choice because in most cases it returns optimal solutions; previously it was disregarded because comparing to the other heuristics it improved the solution insignificantly at the cost of much larger running times.

1 Introduction, Terminology and Notation

When a hard optimization problem is to be solved heuristically, it is often difficult to choose which heuristic to use as it rarely happens that one heuristic is both faster and more precise than another one. Often there is a tradeoff: a heuristic providing a more precise solution takes more time than a heuristic of lesser quality. In this case, the slower heuristic may be preferred if it turns out that the solution it returns is usually much closer to the optimal solution. However, to evaluate the quality of the given solution we need

J. Chen and F.V. Fomin (Eds.): IWPEC 2009, LNCS 5917, pp. 222–233, 2009.

a method that can find an optimal solution (even if finding a *provably* optimal solution takes much more time than the heuristic being analyzed). If the considered problem involves a small parameter, solving the problem to optimality can be done by a fixed-parameter algorithm. (We give a short introduction into fixed-parameter algorithmics in Section 4.) Thus, even if a fixed-parameter algorithm is not *directly* used within a problem solving software, it can still be very useful for *testing* different versions of this software. In this paper we consider a problem occurring in large-scale linear programming (LP) to demonstrate the use of fixed-parameter algorithms in this novel way.

Large-scale LP models which arise in applications usually have sparse coefficient matrices with special structure. If a special structure can be recognized, it can often be used to considerably speed up the process of solving the LP problem and/or to help in understanding the nature of the LP model. A well-known family of such special structures is networks; a number of heuristics to extract (reflected) networks in LP problems have been developed and analyzed, see, e.g., [3,6,7,8,13,14,19] (a formal definition of a reflected network is given below). From the computational point of view, it is worthwhile extracting a reflected network only if the LP problem under consideration contains a relatively large reflected network.

We consider an LP problem in the standard form stated as

$$\text{Minimize } \{p^T x; \text{ subject to } Ax = b, \ x \geq 0\}.$$

LP problems have a number of equivalent, in a sense, forms that can be obtained from each other by various operations. Often scaling operations, that is multiplications of rows and columns of the matrix A of constraints by non-zero constants, are applied, see, e.g., [3,6,8,13]. In the sequel unless stated otherwise, we assume that certain scaling operations on A have been carried out and will not be applied again apart from row reflections defined below. A matrix B is a *network (matrix)* if B is a $(0, \pm 1)$-*matrix* (that is, entries of B belong to the set $\{1, 0, -1\}$) and every column of B has at most one entry equal to 1 and at most one entry equal to -1. The operation of *reflection* of a row of a matrix B changes the signs of all non-zero entries of this row. A matrix B is a *reflected network (matrix)* if there is a sequence of row reflections that transforms B into a network matrix. The *problem of detecting a maximum embedded reflected network* (DMERN) is to find the maximum number of rows that form a submatrix B of A such that B is a reflected network. This number is denoted by $\nu(A)$. The DMERN problem is known to be NP-hard [4].

Gülpınar et al. [14] showed that the maximum size of an embedded reflected network equals the maximum order of a balanced induced subgraph of a special signed graph associated with matrix A (for details, see Section 2). This result led Gülpınar et al. [14] to a heuristic named SGA for detection of reflected networks. Computational experiments in [14] with SGA and three other heuristics demonstrated that SGA and another heuristic, RSD, were of very similar quality and clearly outperformed the two other heuristics in this respect. However, SGA was about 20 times faster, on average, than RSD. Moreover, SGA has an important theoretical property that RSD does not have: SGA always solves the DMERN problem to optimality when the whole matrix A is a reflected network [14]. Since SGA appeared to be the best choice for a heuristic for detection of reflected networks, Gutin and Zverovitch [15] investigated 'repetition' versions of SGA and found out that three times repetition of SGA (SGA3) gives about 1%

improvement, while 80 times repetition of SGA (SGA80) leads to 2% improvement. Thus, at the first glance it might seem that SGA3 and, of course, SGA80 heuristics are not of any practical interest because, taking more time, they produce a very little improvement of the solution quality.

In this paper we argue that in fact SGA80 can be viewed as *the best choice* because, being reasonably fast, in most cases it produces an *optimal solution* to the problem under consideration. To solve the DMERN problem to optimality we design a fixed-parameter algorithm for the maximum balanced subgraph problem and we compare the output of the heuristics being analyzed against the output of the algorithm. To design the FPT algorithm we reduce the maximum balanced subgraph problem to the bipartization problem and then use a fixed-parameter algorithm for the latter problem [20,25]. Thus it turns out that although the fixed-parameter algorithm we use is usually much slower than the heuristic methods, it *helps* to select the best heuristic for the DMERN problem.

As an additional contribution, we investigated another modification of SGA where the use of a greedy-type independent set extracting heuristic (which is part of SGA) is replaced by a fixed-parameter algorithm for finding a minimum vertex cover. Here we used the well-known fact that the complement of an independent set in a graph is a vertex cover. Our experiments with this modification of SGA showed very little improvement and, thus, this modification of SGA appears to be of little practical interest. However, this demonstrated that the independent set extracting heuristic need not be replaced by a more powerful heuristic or exact algorithm.

The rest of the paper is organized as follows. In Section 2 we introduce necessary notation, Section 3 presents the SGA heuristic and its variants, and Section 4 introduces the fixed-parameter algorithms. Section 5 describes a fixed-parameter algorithm for the maximum balanced subgraph problem. In Section 6 we report empirical results and analyze them. Concluding remarks are made in Section 7.

2   Embedded Networks and Signed Graphs

In this section, we assume, for simplicity, that A is a $(0, \pm 1)$-matrix itself (since all rows containing entries not from the set $\{-1, 0, +1\}$ cannot be part of a reflected network). Here we allow graphs to have parallel edges, but no loops. A graph $G = (V, E)$ along with a function $s : E \rightarrow \{-, +\}$ is called a *signed graph*. Signed graphs have been studied by many researchers, see, e.g., [16,17,18,26].

We assume that signed graphs have no parallel edges of the same sign, but may have parallel edges of opposite signs. An edge is *positive* (*negative*) if it is assigned plus (minus). For a $(0, \pm 1)$-matrix $A = [a_{ik}]$ with n rows, we construct a signed graph $G(A)$ as follows: the vertex set of $G(A)$ is $\{1, 2, \ldots, n\}$; $G(A)$ has a positive (negative) edge ij if and only if $a_{ik} = -a_{jk} \neq 0$ $(a_{ik} = a_{jk} \neq 0)$ for some k. Let $G = (V, E, s)$ be a signed graph. For a non-empty subset W of V, the W-*switch* of G is the signed graph G^W obtained from G by changing the signs of the edges between W and $V(G) \setminus W$. A signed graph $G = (V, E, s)$ is *balanced* if there exists a subset W of V (W may coincide with V) such that G^W has no negative edges. Let $\eta(G)$ be the largest order of a balanced induced subgraph of G.

The following important result was proved in [14]. This result allows us to search for a largest balanced induced subgraph of $G(A)$ instead of a largest reflected network in A.

Theorem 1. *[14] Let A be a $(0, \pm 1)$-matrix. A set R of rows in A forms a reflected network if and only if the vertices of $G(A)$ corresponding to R induce a balanced subgraph of $G(A)$. In particular, $\nu(A) = \eta(G(A))$.*

3   SGA and Its Variations

The heuristic SGA introduced in [14] is based on the following:

Lemma 1. *[14] Every signed tree T is a balanced graph.*

Proof. We prove the lemma by induction on the number of edges in T. The lemma is true when the number of edges is one. Let x be a vertex of T of degree one. By the induction hypothesis, there is a set $W \subseteq V(T) - x$ such that $(T - x)^W$ has no negative edges. In T^W the edge e incident to x is positive or negative. In the first case, let $W' = W$ and the second case, let $W' = W \cup \{x\}$. Then, $T^{W'}$ has no negative edges.

Heuristic SGA:
Step 1: Construct signed graph $G = G(A) = (V, E, s)$.
Step 2: Find a spanning forest T in G.
Step 3: Using a recursive algorithm based on the proof of Lemma 1, compute $W \subseteq V$ such that T^W has no negative edges.
Step 4: Let N be the subgraph of G^W induced by the negative edges. Apply the following greedy-degree algorithm [23] to find a maximal independent set I in N: starting from empty I, append to I a vertex of N of minimum degree, delete this vertex together with its neighbors from N, and repeat the above procedure till N has no vertex.
Step 5: Output I.

Proposition 1. *[14] If G is balanced, then $I = V$.*

Proof. It is well-known (see, e.g., Theorem 2.8 in [14]) that a signed graph is balanced if and only if it does not contain cycles with odd number of negative edges. Let T be

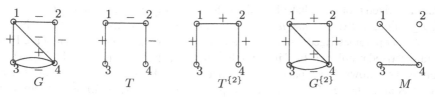

Fig. 1. Illustration for SGA; M is the subgraph $G^{\{2\}}$ induced by the negative edges of $G^{\{2\}}$

a a spanning forest in G. Since T^W has no negative edges, G^W cannot have negative edges. Indeed, if xy was a negative edge in G^W, it would be the unique negative edge in a cycle formed by xy and the (x, y)-path of T^W, a contradiction.

Gutin and Zverovitch [15] investigated a repetition version of SGA where Steps 2-4 were repeated several time (each time the vertices of G were pseudo-randomly permuted and a new spanning forest of G was built). They found out that three times repetition of SGA gives about 1% improvement, while 80 times repetition of SGA leads to 2% improvement, on average. In our experiments we used a larger text bed and better scaling procedure than in [15] and, thus, we run SGA and its 3 and 80 times repetitions on the new set of instances of the DMERN problem (see Section 6). We will denote these repetition versions of SGA by SGA3 and SGA80, respectively.

In Section 6 we also report results on another modification of SGA, SGA+VC, where we replace Step 4 with finding a vertex cover C of G^W and setting $I = V(G^W) \setminus C$. Since the vertex cover problem is well studied in the area of parameterized complexity [1,9,22], to find C we can use a fixed-parameter algorithm for the problem.

4   Fixed-Parameter Algorithmics

We recall some most basic notions of fixed-parameter algorithmics (FPA) here, for a more in-depth treatment of the topic we refer the reader to the monographs [11,12,22].

FPA is a relatively new approach for dealing with intractable computational problems. In the framework of FPA we introduce a parameter k, which is often a positive integer (but may be a vector, graph, or any other object for some problems) such that the problem at hand can be solved in time $O(f(k)n^c)$, where n is the size of the problem instance, c is a constant not dependent on n or k, and $f(k)$ is an arbitrary computable function not dependent on n. The ultimate goal is to obtain $f(k)$ and c such that for small or even moderate values of k the problem under consideration can be completely solved in a reasonable amount of time.

As an example, consider the *Vertex Cover problem (VC)*: given an undirected graph G (with n vertices and m edges), find a minimum number of vertices such that every edge is incident to at least one of these vertices. In the (naturally) parameterized version of VC, k-VC, given a graph G, we are to check whether G has a vertex cover with at most k vertices. k-VC admits an algorithm of running time $O(1.2738^k + kn)$ obtained in [9] that allows us to solve VC with k up to several hundreds. Without using FPA, we would be likely to end up with the obvious algorithm of complexity $O(mn^k)$. The last algorithm is far too slow even for small values of k such as $k = 10$.

Parameterized problems that admit algorithms of complexity $O(f(k)n^c)$ (we refer to such algorithms as *fixed-parameter*) are called *fixed-parameter tractable (FPT)*. Notice that not every parameterized problem is FPT, but there are many problems that are FPT. A parameterized problem is FPT if and only if it admits kernelization [11,12,22], which is defined as follows. For a parameterized decision problem Π given by pairs (I, k), where I is an instance of Π and k is the parameter, a *kernelization* is a polynomial time (in the size of I and k) reduction $(I, k) \mapsto (I', k')$ such that I is a Yes-instance if and only if I' is a Yes-instance, the size of I' is bounded (from above) by a function

$g(k)$ depending on k only and $k' \leq k$. The instances (I', k') comprise a *kernel* of Π of size $g(k)$.

5 Minimum Balanced Deletion Problem

By our discussions above, we are interested in the following parameterized problem.

> **The minimum balanced deletion problem (MBD)**
> *Input:* A signed graph $G = (V, E, s)$, an integer k.
> *Parameter:* k.
> *Output:* A set of at most k vertices whose removal makes G balanced or 'NO' if no such set exists.

We show that the MBD problem is FPT by transforming it into the Bipartization problem defined as follows.

> **The Bipartization problem**
> *Input:* A graph G, an integer k
> *Parameter:* k
> *Output:* A set of at most k vertices whose removal makes G bipartite or 'NO' if no such set exists.

The transformation is described in the following theorem.

Theorem 2. *The MBD problem is FPT and can be solved in time* $O^*(3^k)$.

Proof. It is well-known (see, e.g., Theorem 2.8 in [14]) that a signed graph is balanced if and only if it does not contain cycles involving odd number of negative edges. Hence, the MBD problem in fact asks for at most k vertices whose removal breaks all cycles containing an odd number of negative edges.

Let G' be the (unsigned) graph obtained from G by *subdividing* each positive edge. In other words, for each positive edge $\{u, v\}$, we introduce a new vertex w and replace $\{u, v\}$ by $\{u, w\}$ and $\{w, v\}$. We claim that G has a set of at most k vertices breaking all cycles with an odd number of negative edges if and only if G' can be made bipartite by removal of at most k vertices.

Assume the former and let K be a set of at most k vertices whose removal breaks all cycles with an odd number of negative edges. It follows that $G' - K$ is bipartite. Indeed, each cycle C' of $G' - K$ can be obtained from a cycle C of $G - K$ by subdivision of its positive edges. Hence, C' can be of an odd length only if C has an odd number of negative edges which is impossible according to our assumption about K.

Conversely, let K be a set of at most k vertices such that $G' - K$ is bipartite. We may safely assume that K does not contain the new vertices subdividing positive edges: otherwise each such vertex can be replaced by one of its neighbors. Thus, $K \subseteq V(G)$. Observe that $G - K$ does not have cycles with odd number of negative edges. Indeed, by subdividing positive edges, any such cycle translates into an odd cycle of $G' - K$ in contradiction to our assumption about K.

It follows from the above argumentation that the MBD problem can be solved as follows. Transform G into G' and run on G' the $O^*(3^k)$ algorithm solving the bipartization problem [20]. If the algorithm returns 'NO' then return 'NO'. Otherwise, replace each subdividing vertex by one of its neighbors and return the resulting set of vertices. Clearly, the complexity of the resulting algorithm is $O^*(3^k)$.

Remarks. Unfortunately, it is not known yet whether the Bipartization problem has a polynomial-size problem kernel [20]. Thus, it is not known yet whether the MBD problem has a polynomial-size problem kernel. (If one was known, we could try to use it to speed up our fixed-parameter algorithm.)

Note that a version of the MBD problem, where edge-deletions rather than vertex-deletions are used was considered in [5,10].

6   Experimental Evaluation

In this section we provide and discuss our experiment results for the heuristics SGA, SGA3, SGA80, SGA+VC descried in Section 3 and the exact algorithm given in Section 5. Note that in our experiments we use a larger test bed and better scaling procedure than in [15].

Recall that we consider an LP problem in the standard form stated as

$$\text{Minimize } \{p^T x; \text{ subject to } Ax = b, \ x \geq 0\}.$$

In Section 2, to simplify our notation we assumed that A is a $(0, \pm 1)$-matrix. However, in general, in real LP problems A is not a $(0, \pm 1)$-matrix. Therefore, in reality, the first phase in solving the DMERN problem is applying a scaling procedure whose aim is to increase the number of $(0, \pm 1)$-rows by scaling rows and columns. Here we describe a scaling procedure that we have used. Our computational experiments indicate that this scaling is often better than the scaling procedures we found in the literature. Let us describe our scaling procedure. Let $A = [a_{ij}]_{n \times m}$.

First we apply simple row scaling, i.e., scale all the rows which contain only zeros and $\pm x$, where $x > 0$ is some constant: for every $i \in \{1, 2, \ldots, n\}$ set $a_{ij} = a_{ij}/x$ for $j = 1, 2, \ldots, m$ if $a_{ij} \in \{0, -x, +x\}$ for every $j \in \{1, 2, \ldots, m\}$.

Then we apply a more sophisticated procedure. Let $[r_i]_n$ be an array of boolean values, where r_i indicates whether the ith row is a $(0, \pm 1)$-row. Let $[b_j]_m$ be an array of boolean values, where b_j indicates whether the jth column is bounded, i.e., whether it has at least one nonzero value in a $(0, \pm 1)$-row: for some $j \in \{1, 2, \ldots, m\}$ the value $b_j = true$ if and only if there exists some i such that $r_i = true$ and $a_{ij} \neq 0$.

Next we do the following for every non $(0, \pm 1)$-row (note that at this stage any non $(0, \pm 1)$-row contains at least two nonzero elements). Let J be the set of indices of bounded columns with nonzero elements in the current row c: $J = \{j : a_{cj} \neq 0 \text{ and } b_j = true\}$. If $J = \emptyset$, i.e., all the columns corresponding to nonzero elements in the current row are unbounded, then we simply scale every of these columns: $a_{ij} = a_{ij}/a_{cj}$ for every $i = 1, 2, \ldots, n$ and for every j such that $a_{cj} \neq 0$. If $J \neq \emptyset$ and $a_{cj} \in \{+x, -x\}$ for every $j \in J$, where x is some constant, then we scale accordingly the current row ($a_{cj} = a_{cj}/x$ for every $j \in \{1, 2, \ldots, m\}$) and scale the unbounded columns: $a_{ij} = a_{ij}/a_{cj}$ for every $j \notin J$ if $a_{cj} \neq 0$. Otherwise we do nothing for the current row.

Every time when we scale rows or columns we update the arrays r and b.

Since the matrices processed by this heuristic are usually sparse, we use a special data structure to store them. In particular, we store only nonzero elements providing the row and column indices for each of them. We also store a list of references to the corresponding nonzero elements for every row and for every column of the matrix.

The computational results for all heuristics apart from SGA+VC as well as for the exact algorithm are provided in Table 1. As a test bed we use all the instances provided in Netlib (http://netlib.org/lp/data/). In the table, n denotes the number of $(0, \pm 1)$-rows in the instance, i.e., the number of vertices in the corresponding signed graph G. Also $k_{min}, k_1, k_3, k_{80}$ denote the values of the difference between n and the number of vertices in a maximum induced balanced subgraph of G found by the exact algorithm and SGA, SGA3 and SGA80, respectively, and t, t_1, t_3, t_{80} stand for the running time (in seconds) of the exact algorithm and SGA, SGA3 and SGA80, respectively. When the exact algorithm could not produce a solution after 1 hour, it was terminated. The average values of k are given over all the instances for SGA, SGA3 and SGA80. We also provide the averages for the instances solved by the exact algorithm (see the optimal set average row).

All algorithms were implemented in C++ and the evaluation platform is based on an AMD Athlon 64 X2 3.0 GHz processor. For the exact algorithm we used a code of Hüffner http://theinf1.informatik.uni-jena.de/~hueffner/. In SGA+VC we used a vertex cover code based on [2].

The results with SGA+VC are not provided since SGA+VC managed to improve SGA only for four instances: D6CUBE ($k_{SGA+VC} = 59$), DEGEN2 ($k_{SGA+VC} = 230$),

Table 1. Experiment results for the SGA, SGA3 and SGA80 heuristics and for the exact algorithm

Instance	n	k_{min}	k_1	k_3	k_{80}	t	t_1	t_3	t_{80}
25FV47	283	15	25	25	22	4.40	0.02	0.03	0.39
80BAU3B	1629	—	42	40	40	> 1h	0.08	0.25	9.75
ADLITTLE	31	1	1	1	1	0.02	0.00	0.00	0.00
AFIRO	16	0	0	0	0	0.00	0.00	0.00	0.00
AGG	159	—	107	104	104	> 1h	0.02	0.00	0.09
AGG2	153	—	85	85	83	> 1h	0.00	0.00	0.03
AGG3	153	—	85	85	83	> 1h	0.02	0.00	0.08
BANDM	143	23	24	24	23	1493.12	0.00	0.00	0.08
BEACONFD	118	3	3	3	3	0.00	0.00	0.00	0.02
BLEND	24	1	1	1	1	0.00	0.00	0.00	0.00
BNL1	315	14	17	17	14	1.83	0.00	0.02	0.19
BNL2	1549	—	127	110	99	> 1h	0.05	0.17	4.96
BOEING1	145	—	49	49	48	> 1h	0.00	0.00	0.03
BOEING2	79	15	17	17	15	0.05	0.00	0.02	0.02
BORE3D	131	12	14	13	12	0.14	0.00	0.00	0.03
BRANDY	122	6	7	6	6	0.00	0.00	0.00	0.05
CAPRI	126	—	40	37	34	> 1h	0.00	0.00	0.05

Table 1. (*continued*)

Instance	n	k_{min}	k_1	k_3	k_{80}	t	t_1	t_3	t_{80}
CYCLE	700	—	34	34	34	> 1h	0.02	0.06	1.64
CZPROB	912	1	1	1	1	0.27	0.02	0.03	1.73
D2Q06C	980	—	67	67	67	> 1h	0.02	0.11	3.56
D6CUBE	122	—	61	52	46	> 1h	0.02	0.00	0.16
DEGEN2	444	—	234	233	226	> 1h	0.02	0.03	0.83
DEGEN3	1503	—	822	819	813	> 1h	0.17	0.53	16.91
DFL001	6022	—	2818	2818	2802	> 1h	1.53	5.87	166.05
E226	100	15	18	17	16	1.09	0.00	0.00	0.03
ETAMACRO	145	12	20	20	20	0.47	0.00	0.00	0.09
FFFFF800	178	—	50	41	41	> 1h	0.00	0.02	0.14
FINNIS	325	—	121	120	119	> 1h	0.00	0.02	0.31
FIT1D	10	6	6	6	6	0.00	0.00	0.00	0.00
FIT1P	1	0	0	0	0	0.00	0.00	0.00	0.03
FIT2D	10	6	7	6	6	0.00	0.00	0.02	0.33
FIT2P	4	2	2	2	2	0.00	0.03	0.02	0.76
FORPLAN	61	1	1	1	1	0.00	0.00	0.00	0.05
GANGES	822	—	83	83	77	> 1h	0.03	0.05	1.89
GFRD-PNC	616	—	68	68	68	> 1h	0.03	0.05	0.78
GREENBEA	970	—	48	48	45	> 1h	0.06	0.12	3.46
GREENBEB	970	—	48	48	45	> 1h	0.03	0.17	3.42
GROW15	15	0	0	0	0	0.00	0.00	0.00	0.02
GROW22	22	0	0	0	0	0.00	0.00	0.00	0.02
GROW7	7	0	0	0	0	0.00	0.00	0.02	0.02
ISRAEL	30	8	9	9	8	0.02	0.00	0.00	0.00
KB2	15	1	1	1	1	0.00	0.00	0.00	0.00
LOTFI	105	18	24	22	19	11.23	0.00	0.00	0.02
MAROS-R7	50	0	0	0	0	0.00	0.05	0.05	0.83
MAROS	340	11	17	15	11	0.23	0.00	0.03	0.33
MODSZK1	374	—	237	237	237	> 1h	0.02	0.02	0.30
NESM	232	10	13	11	10	0.03	0.02	0.03	0.19
PEROLD	235	—	28	25	24	> 1h	0.00	0.00	0.12
PILOT.JA	318	16	18	16	16	11.72	0.00	0.02	0.31
PILOT	337	—	45	42	41	> 1h	0.00	0.03	0.70
PILOT.WE	295	—	34	29	28	> 1h	0.00	0.02	0.34
PILOT4	151	3	3	3	3	0.00	0.02	0.02	0.08
PILOT87	479	—	77	76	70	> 1h	0.03	0.02	1.25
PILOTNOV	329	19	21	21	19	201.29	0.03	0.00	0.59
RECIPE	61	0	0	0	0	0.00	0.00	0.00	0.02
SC105	75	16	17	17	17	12.56	0.00	0.00	0.02
SC205	148	—	36	36	36	> 1h	0.00	0.02	0.03
SC50A	35	8	8	8	8	0.02	0.00	0.02	0.00
SC50B	33	6	6	6	6	0.02	0.00	0.00	0.00

Table 1. (*continued*)

Instance	n	k_{min}	k_1	k_3	k_{80}	t	t_1	t_3	t_{80}
SCAGR25	299	0	0	0	0	0.03	0.00	0.02	0.19
SCAGR7	83	0	0	0	0	0.02	0.00	0.00	0.02
SCFXM1	154	12	13	12	12	0.30	0.00	0.00	0.08
SCFXM2	308	—	26	26	24	> 1h	0.00	0.00	0.22
SCFXM3	462	—	39	38	36	> 1h	0.02	0.02	0.56
SCORPION	214	1	1	1	1	0.02	0.00	0.00	0.14
SCRS8	281	9	9	9	9	0.06	0.00	0.02	0.19
SCSD1	39	0	0	0	0	0.03	0.00	0.02	0.02
SCSD6	74	0	0	0	0	0.14	0.00	0.02	0.02
SCSD8	199	0	0	0	0	0.59	0.00	0.00	0.09
SCTAP1	120	0	0	0	0	0.00	0.00	0.00	0.03
SCTAP2	470	0	0	0	0	0.00	0.00	0.03	0.53
SCTAP3	620	0	0	0	0	0.00	0.00	0.06	0.92
SEBA	408	—	274	271	269	> 1h	0.00	0.05	1.95
SHARE1B	49	4	5	4	4	0.00	0.00	0.00	0.02
SHARE2B	36	6	6	6	6	0.00	0.02	0.00	0.00
SHELL	536	2	2	2	2	1.51	0.00	0.02	0.51
SHIP04L	394	—	36	36	36	> 1h	0.02	0.05	0.53
SHIP04S	394	—	36	36	36	> 1h	0.00	0.02	0.41
SHIP08L	762	—	64	64	64	> 1h	0.05	0.08	1.87
SHIP08S	762	—	64	64	64	> 1h	0.02	0.03	1.31
SHIP12L	1141	—	96	96	96	> 1h	0.03	0.14	3.71
SHIP12S	1141	—	96	96	96	> 1h	0.03	0.12	2.76
SIERRA	1161	—	400	399	387	> 1h	0.02	0.05	2.71
STAIR	82	8	11	10	8	0.02	0.00	0.02	0.03
STANDATA	245	—	53	53	53	> 1h	0.00	0.00	0.17
STANDGUB	247	—	53	53	53	> 1h	0.02	0.02	0.19
STANDMPS	353	—	54	54	54	> 1h	0.00	0.02	0.31
STOCFOR1	56	0	0	0	0	0.00	0.00	0.00	0.00
STOCFOR2	1306	—	258	258	243	> 1h	0.05	0.12	3.88
TUFF	175	16	26	17	16	0.58	0.02	0.00	0.11
VTP.BASE	30	4	6	4	4	0.00	0.00	0.00	0.00
WOOD1P	74	0	0	0	0	0.02	0.00	0.00	0.11
WOODW	329	0	0	0	0	0.00	0.00	0.05	0.78
Total average		—	79.3	78.3	76.9	—	0.03	0.10	2.66
Optimal set average		5.8	7.0	6.6	6.1	32.26	0.00	0.01	0.19

DEGEN3 ($k_{SGA+VC} = 806$), and DFL001 ($k_{SGA+VC} = 2809$). Note that in three of these instances $k_{80} < k_{SGA+VC}$. Since the running time of SGA+VC usually exceeds that of SGA80 and the quality of SGA+VC is not much different even from that of SGA, SGA+VC appears to be of little practical interest. However, SGA+VC demonstrates that there is no need to replace Step 4 of SGA by a more powerful heuristic or exact algorithm.

Observe that the exact algorithm completed its computations for 54 instances out of the total of 93, and for 52 instances the running time was at most 1 minute. Note that SGA achieved the optimal solution in 33 out of 54 cases, SGA3 in 39 cases and SGA80 in 49 cases. Observe that in almost all the cases feasible for the exact algorithm, SGA80, being much faster than the exact algorithm, managed to compute an optimal solution! Although SGA80 is slower than SGA and SGA3, it is much more precise and its running time is still reasonable. It follows that SGA80 is the best choice with respect to the tradeoff between running time and precision. Note that this conclusion can be made only *given the knowledge* about the optimal solution and without the considered fixed-parameter algorithm such knowledge would be very hard to obtain (for example, the instance PILOTNOV with $n = 329$ and $k_{min} = 19$ would hardly be feasible to a brute-force exploration of all $\binom{329}{19}$ possibilities).

7 Conclusions

In this paper we have demonstrated a novel way of use of fixed-parameter algorithms where they do not substitute heuristic methods but are used to evaluate them. As a case study, we considered heuristics for the problem of extracting a maximum-size reflected network in an LP problem. The main conclusion of our empirical study is that the slowest heuristic, which provided only a minor improvement over the other ones and was basically disregarded due to this fact, has turned out to be the best among the heuristics under consideration because, being reasonably fast, it often produces an optimal solution. The fixed-parameter algorithm in this case has helped us to check whether a solution returned by the heuristic being analyzed is indeed optimal.

We believe that this way of applying fixed-parameter algorithms can be useful for other problems as well. One candidate might be the problem of finding whether the given CNF formula has at most k variables so that their removal makes the resulting formula Renameable Horn. This is called the Renameable Horn deletion backdoor problem and was recently shown FPT [24]. Heuristics for this problem are widely used in modern SAT solvers for identifying a small subset of variables on which an exponential-time branching is to be performed [21]. Currently it is unclear whether substituting a heuristic approach by the exact fixed-parameter algorithm would result in a better SAT solver. But even if it is not the case, the exact algorithm can be still of a considerable use for ranking the heuristic techniques, especially as producing small Renameable Horn backdoors is vitally important for reducing the exponential-time impact on the runtime of SAT solvers.

Acknowledgements We are grateful to Michael Langston and his group for providing us with a vertex cover code. The work of G. Gutin was supported in part by a grant from EPSRC. The work of I. Razgon was supported by Science Foundation Ireland grant 05/IN/I886.

References

1. Abu-Khzam, F.N., Fellows, M.R., Langston, M., Suters, W.H.: Crown Structures for Vertex Cover Kernelization. Theory of Computing Systems 41, 411–430 (2007)

2. Abu-Khzam, F.N., Langston, M., Shanbhag, P., Symons, C.T.: Scalable Parallel Algorithms for FPT Problems. Algorithmica 45, 269–284 (2006)
3. Baker, B.M., Maye, P.J.: A Heuristic for Finding Embedded Network Structure in Mathematical Programmes. Europ. Jour. Oper. Res. 67, 52–63 (1993)
4. Bartholdi, J.J.: A Good Submatrix is Hard to Find. Oper. Res. Letters 1, 190–193 (1982)
5. Betzler, N., Hüffner, F., Niedermeier, R.: Optimal edge deletions for signed graph balancing. In: Demetrescu, C. (ed.) WEA 2007. LNCS, vol. 4525, pp. 297–310. Springer, Heidelberg (2007)
6. Bixby, R.E., Fourer, R.: Finding Embedded Network Rows in Linear Programs I. Extraction Heuristics. Manag. Science 34, 342–376 (1988)
7. Bixby, R.E., Cunningham, W.H.: Converting Linear Programs to Network Problems. Math. Oper. Res. 5, 321–356 (1980)
8. Brown, G.G., Wright, W.G.: Automatic Identification of Embedded Network Rows in Large-Scale Optimization Models. Math. Prog. 29, 41–56 (1984)
9. Chen, J., Kanj, I.A., Xia, G.: Simplicity is beauty: Improved upper bounds for vertex cover. Tech. Report TR05-008, DePaul University, Chicago IL (2005)
10. DasGupta, B., Enciso, G.A., Sontag, E.D., Zhang, Y.: Algorithmic and complexity results for decompositions of biological networks into monotone subsystems. In: Àlvarez, C., Serna, M. (eds.) WEA 2006. LNCS, vol. 4007, pp. 253–264. Springer, Heidelberg (2006)
11. Downey, R.G., Fellows, M.R.: Parameterized Complexity. Springer, Heidelberg (1999)
12. Flum, J., Grohe, M.: Parameterized Complexity Theory. Springer, Heidelberg (2006)
13. Gülpınar, N., Gutin, G., Mitra, G., Maros, I.: Detecting Embedded Pure Network Structures by Using GUB and Independent Set Algorithms. Comput. Optim. Applic. 15, 235–247 (2000)
14. Gülpınar, N., Gutin, G., Mitra, G., Zverovitch, A.: Extracting Pure Network Submatrices in Linear Programs Using Signed Graphs. Discrete Applied Mathematics 137, 359–372 (2004)
15. Gutin, G., Zverovitch, A.: Extracting pure network submatrices in linear programs using signed graphs, Part 2. Communications of DQM 6, 58–65 (2003)
16. Hansen, P.: Labelling Algorithms for Balance in Signed Graphs. In: Problémes Combinatoires et Theorie des Graphes, Colloq. Internat., Orsay, pp. 215–217 (1976); Colloques Internat. du CNRS 260 Paris (1978)
17. Harary, F., Kabell, J.A.: A Simple Algorithm to Detect Balance in Signed Graphs. Math. Social Science 1, 131–136 (1980-1981)
18. Heller, I., Tompkins, C.B.: An Extension of a Theorem of Dantzig's. In: Kuhn, H.W., Tucker, A.W. (eds.) Linear Inequalities and Related Systems. Annals Math. Studies, vol. 38, pp. 247–252. Princeton Univ. Press, Princeton (1956)
19. Hsu, A.C., Fourer, R.: Identification of Embedded Network Structure in Linear Programming Models. GSIA Working Paper, 1997-58
20. Hüffner, F.: Algorithm Engineering for Optimal Graph Bipartization. Journal of Graph Algorithms and Applications 13, 77–98 (2009)
21. Kottler, S., Kaufmann, M., Sinz, C.: Computation of Renameable Horn Backdoors. In: Kleine Büning, H., Zhao, X. (eds.) SAT 2008. LNCS, vol. 4996, pp. 154–160. Springer, Heidelberg (2008)
22. Niedermeier, R.: Invitation to Fixed-Parameter Algorithms. Oxford University Press, Oxford (2006)
23. Paschos, V.T.: A $\Delta/2$-Approximation for the Maximum Independent Set Problem. Inform. Proc. Let. 44, 11–13 (1992)
24. Razgon, I., O'Sullivan, B.: Almost 2-SAT is fixed-parameter tractable. Journal of Computer and System Sciences (in press)
25. Reed, B., Smith, K., Vetta, A.: Finding odd cycle transversals. Operations Research Letters 32, 299–301 (2004)
26. Zaslavsky, T.: Signed Graphs. Discete Applied Math. 4, 47–74 (1982)

A Probabilistic Approach to Problems Parameterized above or below Tight Bounds

Gregory Gutin[1], Eun Jung Kim[1], Stefan Szeider[2], and Anders Yeo[1]

[1] Department of Computer Science
Royal Holloway, University of London
Egham, Surrey TW20 0EX, UK
{gutin,eunjung,anders}@cs.rhul.ac.uk
[2] Department of Computer Science, Durham University,
Durham DH1 3LE, England, UK
stefan@szeider.net

Abstract. We introduce a new approach for establishing fixed-parameter tractability of problems parameterized above tight lower bounds or below tight upper bounds. To illustrate the approach we consider two problems of this type of unknown complexity that were introduced by Mahajan, Raman and Sikdar (J. Comput. Syst. Sci. 75, 2009). We show that a generalization of one of the problems and three nontrivial special cases of the other problem admit kernels of quadratic size.

1 Introduction

A parameterized problem Π can be considered as a set of pairs (I, k) where I is the *main part* and k (usually an integer) is the *parameter*. Π is called *fixed-parameter tractable* (FPT) if membership of (I, k) in Π can be decided in time $O(f(k)|I|^c)$, where $|I|$ denotes the size of I, $f(k)$ is a computable function, and c is a constant independent of k and I (for further background and terminology on parameterized complexity we refer the reader to the monographs [7,8,16]). If the nonparameterized version of Π (where k is just a part of the input) is NP-hard, then the function $f(k)$ must be superpolynomial provided P $\neq$ NP. Often $f(k)$ is "moderately exponential," which makes the problem practically feasible for small values of k. Thus, it is important to parameterize a problem in such a way that the instances with small values of k are of real interest.

Consider the following well-known problem: given a digraph $D = (V, A)$, find an acyclic subdigraph of D with the maximum number of arcs. We can parameterize this problem "naturally" by asking whether D contains an acyclic subdigraph with at least k arcs. It is easy to prove that this parameterized problem is fixed-parameter tractable by observing that D always has an acyclic subdigraph with at least $|A|/2$ arcs. (Indeed, consider a bijection $\alpha : V \to \{1, \ldots, |V|\}$ and the following subdigraphs of D: $(V, \{ xy \in A : \alpha(x) < \alpha(y) \})$ and $(V, \{ xy \in A : \alpha(x) > \alpha(y) \})$. Both subdigraphs are acyclic and at least one of them has at least $|A|/2$ arcs.) However, $k \leq |A|/2$ for every small value of k

J. Chen and F.V. Fomin (Eds.): IWPEC 2009, LNCS 5917, pp. 234–245, 2009.

and almost every practical value of $|A|$ and, thus, our "natural" parameterization is of almost no practical or theoretical interest.

Instead, one should consider the following parameterized problem: decide whether $D = (V, A)$ contains an acyclic subdigraph with at least $|A|/2 + k$ arcs. We choose $|A|/2 + k$ because $|A|/2$ is a *tight lower bound* on the size of a largest acyclic subdigraph. Indeed, the size of a largest acyclic subdigraph of a symmetric digraph $D = (V, A)$ is precisely $|A|/2$. (A digraph $D = (V, A)$ is *symmetric* if $xy \in A$ implies $yx \in A$.)

In a recent paper [15] Mahajan, Raman and Sikdar provided several examples of problems of this type and argued that a natural parameterization is one above a tight lower bound for maximization problems, and below a tight upper bound for minimization problems. Furthermore, they observed that only a few non-trivial results are known for problems parameterized above a tight lower bound [10,11,13,14], and they listed several problems parameterized above a tight lower bound whose complexity is unknown. The difficulty in showing whether such a problem is fixed-parameter tractable can be illustrated by the fact that often we even do not know whether the problem is in XP, i.e., can be solved in time $O(|I|^{g(k)})$ for a computable function $g(k)$. For example, it is non-trivial to see that the above-mentioned digraph problem is in XP when parameterized above the $|A|/2$ bound.

In this paper we introduce the *Strictly Above/Below Expectation Method (SABEM)*, a novel approach for establishing the fixed-parameter tractability of maximization problems parameterized above tight lower bounds and minimization problems parameterized below tight upper bounds. The new method is based on probabilistic arguments and utilizes certain probabilistic inequalities. We will state the equalities in the next section, and in the subsequent sections we will apply SABEM to two open problems posed in [15].

Now we give a very brief description of the new method with respect to a given problem Π parameterized above a tight lower bound or below a tight upper bound. We first apply some reductions rules to reduce Π to its special case Π'. Then we introduce a random variable X such that the answer to Π is YES if and only if X takes, with positive probability, a value greater or equal to the parameter k. Now using some probabilistic inequalities on X, we derive upper bounds on the size of NO-instances of Π' in terms of a function of the parameter k. If the size of a given instance exceeds this bound, then we know the answer is YES; otherwise, we produce a *problem kernel* [7]. In many cases, we obtain problem kernels of polynomial size.

In Section 3, we consider the LINEAR ORDERING problem, a generalization of the problem discussed above: Given a digraph $D = (V, A)$ in which each arc ij has a positive integral weight w_{ij}, find an acyclic subdigraph of D of maximum weight. Observe that $W/2$, where W is the sum of all arc weights, is a tight lower bound for LINEAR ORDERING. We prove that the problem parameterized above $W/2$ is fixed-parameter tractable and admits a quadratic kernel. Note that this parameterized problem generalizes the parameterized maximum acyclic subdigraph problem considered in [15]; thus, our result answers the corresponding open question of [15].

In Section 4, we consider the problem MAX LIN-2: Given a system of m linear equations $e_1, \ldots, e_m$ in n variables over GF(2), and for each equation e_j a positive integral weight w_j; find an assignment of values to the n variables that maximizes the total weight of the satisfied equations. We will see that $W/2$, where $W = w_1 + \cdots + w_m$, is a tight lower bound for MAX LIN-2. The complexity of the problem parameterized above $W/2$ is open [15]. We prove that the following three special cases of the parameterized problem are fixed-parameter tractable: (1) there is a set U of variables such that each equation has an odd number of variables from U, (2) there is a constant r such that each equation involves at most r variables, (3) there is a constant ρ such that any variable appears in at most ρ equations. For all three cases we obtain kernels with $O(k^2)$ variables and equations. We also show that if we allow the weights w_j to be positive reals, the problem is NP-hard already if $k = 1$ and each equation involves two variables.

In Section 5, we briefly mention minimization problems parameterized below tight upper bounds, provide further discussions of problems considered in this paper and point out to a very recent result obtained using our new method.

2   Probabilistic Inequalities

In our approach we introduce a random variable X such that the answer to the problem parameterized above a tight lower bound or below a tight upper bound is YES if and only if X takes with positive probability a value greater or equal to the parameter k.

In this paper all random variables are real. A random variable is *discrete* if its distribution function has a finite or countable number of positive increases. A random variable X is a *symmetric* if $-X$ has the same distribution function as X. If X is discrete, then X is symmetric if and only if $\text{Prob}(X = a) = \text{Prob}(X = -a)$ for each real a. Let X be a symmetric variable for which the first moment $\mathbb{E}(X)$ exists. Then $\mathbb{E}(X) = \mathbb{E}(-X) = -\mathbb{E}(X)$ and, thus, $\mathbb{E}(X) = 0$. The following is easy to prove [17].

Lemma 1. *If X is a symmetric random variable and $\mathbb{E}(X^2) < \infty$, then*

$$\text{Prob}(\ X \geq \sqrt{\mathbb{E}(X^2)}\ ) > 0.$$

See Sections 3 and 4 for applications of Lemma 1. Unfortunately, often X is not symmetric, but Lemma 2 provides an inequality that can be used in many such cases. This lemma was proved by Alon et al. [2]; a weaker version was obtained by Håstad and Venkatesh [12].

Lemma 2. *Let X be a random variable and suppose that its first, second and forth moments satisfy $\mathbb{E}(X) = 0$, $\mathbb{E}(X^2) = \sigma^2 > 0$ and $\mathbb{E}(X^4) \leq b\sigma^4$, respectively. Then $\text{Prob}(\ X > \frac{\sigma}{4\sqrt{b}}\ ) \geq \frac{1}{4^{4/3}b}.$*

Since it is often rather nontrivial to evaluate $\mathbb{E}(X^4)$ in order to check whether $\mathbb{E}(X^4) \leq b\sigma^4$ holds, one can sometimes use the following result by Bourgain [5].

Lemma 3. *Let* $f = f(x_1, \ldots, x_n)$ *be a polynomial of degree* r *in* n *variables* $x_1, \ldots, x_n$ *with domain* $\{-1, 1\}$. *Define a random variable* X *by choosing a vector* $(\epsilon_1, \ldots, \epsilon_n) \in \{-1, 1\}^n$ *uniformly at random and setting* $X = f(\epsilon_1, \ldots, \epsilon_n)$. *Then* $\mathbb{E}(X^4) \leq 2^{6r}(\mathbb{E}(X^2))^2$.

3   Linear Ordering

Let $D = (V, A)$ be a digraph with no loops or parallel arcs in which every arc ij has a positive weight w_{ij}. The problem of finding an acyclic subdigraph of D of maximum weight, known as LINEAR ORDERING, has applications in economics [3]. Let $n = |V|$ and consider a bijection $\alpha : V \to \{1, \ldots, n\}$. Observe that the subdigraphs $(V, \{ij \in A : \alpha(i) < \alpha(j)\})$ and $(V, \{ij \in A : \alpha(i) > \alpha(j)\})$ are acyclic. Since the two subdigraphs contain all arcs of D, at least one of them has weight at least $W/2$, where $W = \sum_{ij \in A} w_{ij}$, the *weight* of D. Thus, $W/2$ is a lower bound on the maximum weight of an acyclic subdigraph of D. Consider a digraph D where for every arc ij of D there is also an arc ji of the same weight. Each maximum weight subdigraph of D has weight exactly $W/2$. Hence the lower bound $W/2$ is tight.

LINEAR ORDERING ABOVE TIGHT LOWER BOUND (LOALB)
Instance: A digraph $D = (V, A)$, each arc ij has an integral positive weight w_{ij}, and a positive integer k.
Parameter: The integer k.
Question: Is there an acyclic subdigraph of D of weight at least $W/2 + k$, where $W = \sum_{ij \in A} w_{ij}$?

Mahajan, Raman, and Sikdar [15] asked whether LOALB is fixed-parameter tractable for the special case when all arcs are of weight 1 (i.e., D is unweighted). In this section we will prove that LOALB admits a kernel with $O(k^2)$ arcs; consequently the problem is fixed-parameter tractable. Note that if we allow weights to be positive reals, then we can show, similarly to the NP-completeness proof given in the next section, that LOALB is NP-complete already for $k = 1$.
    Consider the following reduction rule:

Reduction Rule 1. *Assume* D *has a directed 2-cycle* iji; *if* $w_{ij} = w_{ji}$ *delete the cycle, if* $w_{ij} > w_{ji}$ *delete the arc* ji *and replace* w_{ij} *by* $w_{ij} - w_{ji}$, *and if* $w_{ji} > w_{ij}$ *delete the arc* ij *and replace* w_{ji} *by* $w_{ji} - w_{ij}$.

It is easy to check that the answer to LOALB for a digraph D is YES if and only if the answer to LOALB is YES for a digraph obtained from D using the reduction rule as long as possible.
    Let $D = (V, A)$ be an oriented graph, let $n = |V|$ and $W = \sum_{ij \in A} w_{ij}$. Consider a random bijection: $\alpha : V \to \{1, \ldots, n\}$ and a random variable

$X(\alpha) = \frac{1}{2} \sum_{ij \in A} \epsilon_{ij}(\alpha)$, where $\epsilon_{ij}(\alpha) = w_{ij}$ if $\alpha(i) < \alpha(j)$ and $\epsilon_{ij}(\alpha) = -w_{ij}$, otherwise. It is easy to see that $X(\alpha) = \sum \{ w_{ij} : ij \in A, \alpha(i) < \alpha(j) \} - W/2$. Thus, the answer to LOALB is YES if and only if there is a bijection $\alpha : V \to \{1, \ldots, n\}$ such that $X(\alpha) \geq k$. Since $\mathbb{E}(\epsilon_{ij}) = 0$, we have $\mathbb{E}(X) = 0$.

Let $W^{(2)} = \sum_{ij \in A} w_{ij}^2$. We will prove the following:

Lemma 4. $\mathbb{E}(X^2) \geq W^{(2)}/12$.

Proof. Let $N^+(i)$ and $N^-(i)$ denote the sets of out-neighbors and in-neighbors of a vertex i in D. By the definition of X,

$$4 \cdot \mathbb{E}(X^2) = \sum_{ij \in A} \mathbb{E}(\epsilon_{ij}^2) + \sum_{ij, pq \in A} \mathbb{E}(\epsilon_{ij} \epsilon_{pq}), \qquad (1)$$

where the second sum is taken over ordered pairs of distinct arcs. Clearly, $\sum_{ij \in A} \mathbb{E}(\epsilon_{ij}^2) = W^{(2)}$. To compute $\sum_{ij, pq \in A} \mathbb{E}(\epsilon_{ij} \epsilon_{pq})$ we consider the following cases:

Case 1: $\{i, j\} \cap \{p, q\} = \emptyset$. Then ϵ_{ij} and ϵ_{pq} are independent and $\mathbb{E}(\epsilon_{ij} \epsilon_{pq}) = \mathbb{E}(\epsilon_{ij}) \mathbb{E}(\epsilon_{pq}) = 0$.

Case 2a: $|\{i, j\} \cap \{p, q\}| = 1$ and $i = p$. Since the probability that $i < \min\{j, q\}$ or $i > \max\{j, q\}$ is $2/3$, $\epsilon_{ij} \epsilon_{iq} = w_{ij} w_{iq}$ with probability $\frac{2}{3}$ and $\epsilon_{ij} \epsilon_{iq} = -w_{ij} w_{iq}$ with probability $\frac{1}{3}$. Thus, for every $i \in V$ we have $\sum_{ij, iq \in A} \mathbb{E}(\epsilon_{ij} \epsilon_{iq}) = \frac{1}{3} \sum \{ w_{ij} w_{iq} : j \neq q \in N^+(i) \} = \frac{1}{3} (\sum_{j \in N^+(i)} w_{ij})^2 - \frac{1}{3} \sum_{j \in N^+(i)} w_{ij}^2$.

Case 2b: $|\{i, j\} \cap \{p, q\}| = 1$ and $j = q$. Similarly to Case 2a, we obtain $\sum_{ij, pj \in A} \mathbb{E}(\epsilon_{ij} \epsilon_{pj}) = \frac{1}{3} (\sum_{i \in N^-(j)} w_{ij})^2 - \frac{1}{3} \sum_{i \in N^-(j)} w_{ij}^2$.

Case 3a: $|\{i, j\} \cap \{p, q\}| = 1$ and $i = q$. Since $\epsilon_{ij} \epsilon_{pi} = w_{ij} w_{pi}$ with probability $\frac{1}{3}$ and $\epsilon_{ij} \epsilon_{pi} = -w_{ij} w_{pi}$ with probability $\frac{2}{3}$, we obtain $\sum_{ij, pi \in A} \mathbb{E}(\epsilon_{ij} \epsilon_{pi}) = -\frac{1}{3} \sum \{ w_{ij} w_{pi} : j \in N^+(i), p \in N^-(i) \} = -\frac{1}{3} \sum_{j \in N^+(i)} w_{ij} \sum_{p \in N^-(i)} w_{pi}$.

Case 3b: $|\{i, j\} \cap \{p, q\}| = 1$ and $j = p$. Similarly to Case 3a, we obtain $\sum_{ij, jq \in A} \mathbb{E}(\epsilon_{ij} \epsilon_{jq}) = -\frac{1}{3} \sum_{i \in N^-(j)} w_{ij} \sum_{q \in N^+(j)} w_{jq}$.

Equation (1) and the subsequent computations imply that $4 \cdot \mathbb{E}(X^2) = W^{(2)} + \frac{1}{3}(Q - R)$, where $Q = \sum_{i \in V} \left(\left(\sum_{j \in N^+(i)} w_{ij} \right)^2 - \sum_{j \in N^+(i)} w_{ij}^2 + \left(\sum_{j \in N^-(i)} w_{ji} \right)^2 - \sum_{j \in N^-(i)} w_{ji}^2 \right)$, and $R = 2 \cdot \sum_{i \in V} \left(\sum_{j \in N^+(i)} w_{ij} \right) \left(\sum_{j \in N^-(i)} w_{ji} \right)$. By the inequality of arithmetic and geometric means, for each $i \in V$, we have

$$\left(\sum_{j \in N^+(i)} w_{ij} \right)^2 + \left(\sum_{j \in N^-(i)} w_{ji} \right)^2 - 2 \left(\sum_{j \in N^+(i)} w_{ij} \right) \left(\sum_{j \in N^-(i)} w_{ji} \right) \geq 0.$$

Therefore,

$$Q - R \geq -\sum_{i \in V} \sum_{j \in N^+(i)} w_{ij}^2 - \sum_{i \in V} \sum_{j \in N^-(i)} w_{ji}^2 = -2W^{(2)},$$

and $4 \cdot \mathbb{E}(X^2) \geq W^{(2)} - 2W^{(2)}/3 = W^{(2)}/3$, implying $\mathbb{E}(X^2) \geq W^{(2)}/12$. $\qquad \square$

Now we can prove the main result of this section.

Theorem 1. *The problem* LOALB *admits a kernel with* $O(k^2)$ *arcs.*

Proof. Let H be a digraph. We know that the answer to LOALB for H is YES if and only if the answer to LOALB is YES for a digraph D obtained from H using Reduction Rule 1 as long as possible. Observe that D is an oriented graph. Let $\mathcal{B}$ be the set of bijections from V to $\{1, \ldots, n\}$. Observe that $f : \mathcal{B} \to \mathcal{B}$ such that $f(\alpha(v)) = |V| + 1 - \alpha(v)$ for each $\alpha \in \mathcal{B}$ is a bijection. Note that $X(f(\alpha)) = -X(\alpha)$ for each $\alpha \in \mathcal{B}$. Therefore, $\mathrm{Prob}(X = a) = \mathrm{Prob}(X = -a)$ for each real a and, thus, X is symmetric. Thus, by Lemmas 1 and 4, we have $\mathrm{Prob}(\ X \geq \sqrt{W^{(2)}/12}\ ) > 0$. Hence, if $\sqrt{W^{(2)}/12} \geq k$, there is a bijection $\alpha : V \to \{1, \ldots, n\}$ such that $X(\alpha) \geq k$ and, thus, the answer to LOALB (for both D and H) is YES. Otherwise, $|A| \leq W^{(2)} < 12 \cdot k^2$. $\qquad\square$

We close this section by outlining how Theorem 1 can be used to actually find a solution to LOALB if one exists. Let (D, k) be an instance of LOALB where $D = (V, A)$ is a directed graph with integral positive arc-weights and $k \geq 1$ is an integer. Let W be the total weight of D. As discussed above, we may assume that D is an oriented graph. If $|A| < 12k^2$ then we can find a solution, if one exists, by trying all subsets $A' \subseteq A$, and testing whether (V, A') is acyclic and has weight at least $W/2 + k$; this search can be carried out in time $2^{O(k^2)}$. Next we assume $|A| \geq 12k^2$. We know by Theorem 1 that (D, k) is a YES-instance; it remains to find a solution.

For a vertex $i \in V$ let $d_D(i)$ denote its unweighted degree in D, i.e., the number of arcs (incoming or outgoing) that are incident with i. Consider the following reduction rule:

Reduction Rule 2. *If there is a vertex $i \in V$ with $|A| - 12k^2 \geq d_D(i)$, then delete i from D.*

Observe that by applying the rule we obtain again a YES-instance $(D - i, k)$ of LOALB since $D - i$ has still at least $12k^2$ arcs. Moreover, if we know a solution D'_i of $(D - i, k)$, then we can efficiently obtain a solution D' of (D, k): if $\sum_{j \in N^+(i)} w_{ij} \geq \sum_{j \in N^-(i)} w_{ij}$ then we add i and all outgoing arcs $ij \in A$ to D'_i; otherwise, we add i and all incoming arcs $ji \in A$ to D'_i. After multiple applications of Rule 2 we are left with an instance (D_0, k) to which Rule 2 cannot be applied. Let $D_0 = (V_0, A_0)$. We pick a vertex $i \in V_0$. If i has a neighbor j with $d_{D_0}(j) = 1$, then $|A_0| \leq 12k^2$, since $|A_0| - d_{D_0}(j) < 12k^2$. On the other hand, if $d_{D_0}(j) \geq 2$ for all neighbors j of i, then i has less than $2 \cdot 12k^2$ neighbors, since $D_0 - i$ has less than $12k^2$ arcs; thus $|A_0| < 3 \cdot 12k^2$. Therefore, as above, time $2^{O(k^2)}$ is sufficient to try all subsets $A'_0 \subseteq A_0$ to find a solution to the instance (D_0, k). Let n denote the input size of instance (D, k). Rule 2 can certainly be applied in polynomial time $n^{O(1)}$, and we apply it less than n times. Hence, we can find a solution to (D, k), if one exists, in time $n^{O(1)} + 2^{O(k^2)}$.

Recall that a kernelization reduces in polynomial time an instance (I, k) of a parameterized problem to a *decision-equivalent* instance (I', k'), its *problem kernel*, where $k' \leq k$ and the size of I' is bounded by a function of k. Solutions for (I, k) and solutions for (I', k') are possibly unrelated to each other. We call (I', k') a *faithful problem kernel* if from a solution for (I', k') we can construct a solution for (I, k) in time polynomial in $|I|$ and k. Clearly the above (D_0, k) is a faithful kernel.

4   Max Lin-2

Consider a system of m linear equations $e_1, \ldots, e_m$ in n variables $z_1, \ldots, z_n$ over GF(2), and suppose that each equation e_j has a positive integral weight w_j, $j = 1, \ldots, m$. The problem MAX LIN-2 asks for an assignment of values to the variables that maximizes the total weight of the satisfied equations. Let $W = w_1 + \cdots + w_m$.

To see that the total weight of the equations that can be satisfied is at least $W/2$, we describe a simple procedure suggested in [12]. We assign values to the variables $z_1, \ldots, z_n$ one by one and simplify the system after each assignment. When we wish to assign 0 or 1 to z_i, we consider all equations reduced to the form $z_i = b$, for a constant b. Let W' be the total weight of all such equations. We set $z_i := 0$, if the total weight of such equations is at least $W'/2$, and set $z_i := 1$, otherwise. If there are no equations of the form $z_i = b$, we set $z_i := 0$. To see that the lower bound $W/2$ is tight, consider a system consisting of pairs of equations of the form $\sum_{i \in I} z_i = 1$ and $\sum_{i \in I} z_i = 0$ where both equations have the same weight.

The parameterized complexity of MAX LIN-2 parameterized above the tight lower bound $W/2$ was stated by Mahajan, Raman and Sikdar [15] as an open question:

> MAX LIN-2 PARAMETERIZED ABOVE TIGHT LOWER BOUND (LINALB)
> *Instance:* A system S of m linear equations $e_1, \ldots, e_m$ in n variables $z_1, \ldots, z_n$ over GF(2), each equation e_i with a positive integral weight w_i, $i = 1, 2, \ldots, m$, and a positive integer k. Each equation e_j can be written as $\sum_{i \in I_j} z_i = b_j$, where $\emptyset \neq I_j \subseteq \{1, \ldots, n\}$.
> *Parameter:* The integer k.
> *Question:* Is there an assignment of values to the variables $z_1, \ldots, z_n$ such that the total weight of the satisfied equations is at least $W/2 + k$, where $W = \sum_{i=1}^{m} w_i$?

Let r_j be the number of variables in equation e_j, and let $r(S) = \max_{i=1}^{m} r_j$. We are not able to determine whether LINALB is fixed-parameter tractable or not, but we can prove that the following three special cases are fixed-parameter tractable: (1) there is a set U of variables such that each equation contains an odd number of variables from U, (2) there is a constant r such that $r(S) \leq r$, (3) there is a constant ρ such that any variable appears in at most ρ equations.

Notice that in our formulation of LINALB it is required that each equation has a positive integral weight. In a relaxed setting in which an equation may have any positive real number as its weight, the problem is NP-complete even for $k = 1$ and each $r_j = 2$. Indeed, let each linear equation be of the form

$z_u + z_v = 1$. Then the problem is equivalent to MAXCUT, the problem of finding a cut of total weight at least L in an undirected graph G, where $V(G)$ is the set of variables, $E(G)$ contains (z_u, z_v) if and only if there is a linear equation $z_u + z_v = 1$, and the weight of an edge (z_u, z_v) equals the weight of the corresponding linear equation. The problem MAXCUT is a well-known NP-complete problem. Let us transform an instance I of MAXCUT into an instance I' of the "relaxed" LINALB by replacing the weight w_i by $w_i' := w_i/(L - W/2)$. We may assume that $L - W/2 > 0$ since otherwise the instance is immediately seen as a YES-instance. Observe that the new instance I' has an assignment of values with total weight at least $W'/2 + 1$ if and only if I has a cut with total weight at least L. We are done.

Let A be the matrix of the coefficients of the variables in S. It is well-known that the maximum number of linearly independent columns of A equals rank$A$, and such a collection of columns can be found in time polynomial in n and m, using, e.g., the Gaussian elimination on columns [4]. We have the following reduction rule and supporting lemma.

Reduction Rule 3. *Let A be the matrix of the coefficients of the variables in S, let $t = $ rank$A$ and let columns $a^{i_1}, \ldots, a^{i_t}$ of A be linearly independent. Then delete all variables not in $\{z_{i_1}, \ldots, z_{i_t}\}$ from the equations of S.*

Lemma 5. *Let T be obtained from S by Rule 3. Then T is a YES-instance if and only if S is a YES-instance. Moreover, T can be obtained from S in time polynomial in n and m.*

Proof. If $t = n$, set $T := S$, so assume that $t < n$. The remark before the lemma immediately implies that T can be obtained from S in time polynomial in n and m. Let S' be a system of equations from S and let T' be the corresponding system of equations from T. It is sufficient to prove the following claim:

There is an assignment of values to $z_1, \ldots, z_n$ satisfying all equations in S' and falsifying the rest of equations in S if and only if there is an assignment of values to $z_{i_1}, \ldots, z_{i_t}$ satisfying all equations in T' and falsifying the rest of equations in T.

Let an assignment z^0 of values to $z = (z_1, \ldots, z_n)$ satisfy all equations of S' and falsify the equations of S'', where $S'' = S \setminus S'$. This assignment satisfies all equations of R, the system obtained from S by replacing the right hand side b_j of each equation in S'' by $1 - b_j$. Note that R has the same matrix A of coefficients with columns $a^1, \ldots, a^n$. Let a column $a^i \notin \{a^{i_1}, \ldots, a^{i_t}\}$. Then, by definition of $a^{i_1}, \ldots, a^{i_t}$, $a^i = \lambda_1 a^{i_1} + \cdots + \lambda_t a^{i_t}$ for some numbers $\lambda_j \in \{0, 1\}$. Knowing the numbers λ_j, we may eliminate a variable z_i from R by replacing a^i with the sum of all columns from $\{a^{i_1}, \ldots, a^{i_t}\}$ for which $\lambda_j = 1$ and carrying out the obvious simplification of the system. Thus, we may eliminate from R all variables $z_i \notin \{z_{i_1}, \ldots, z_{i_t}\}$ and get $y_{i_1} a^{i_1} + \cdots + y_{i_t} a^{i_t} = b'$, where b' is the right hand side of R and each $y_j \in \{0, 1\}$. Now replace, in the modified R, the right hand side b_j' of each equation corresponding to an equation in S'' by $1 - b_j'$ obtaining T. Clearly, $(y_{i_1}, \ldots, y_{i_t})$ satisfies all equations of T' and falsifies all equations in $T'' = T \setminus T'$.

Suppose now that $(y_{i_1}, \ldots, y_{i_t})$ satisfies all equations of T' and falsifies all equations in T''. Then $(y_1, \ldots, y_n)$, where $y_j = 0$ if $j \notin \{i_1, \ldots, i_t\}$, satisfies all equations of S' and falsifies all equations in S''. Thus, the claim has been proved.   □

Consider the following reduction rule for LINALB used in [12].

Reduction Rule 4. *If we have, for a subset I of $\{1, 2, \ldots, n\}$, the equation $\sum_{i \in I} z_i = b'$ with weight w', and the equation $\sum_{i \in I} z_i = b''$ with weight w'', then we replace this pair by one of these equations with weight $w' + w''$ if $b' = b''$ and, otherwise, by the equation whose weight is bigger, modifying its new weight to be the difference of the two old ones. If the resulting weight is 0, we omit the equation from the system.*

If Rule 4 is not applicable to a system we call the system *reduced under Rule 4*. Note that the problem LINALB for S and the system obtained from S by applying Rule 4 as long as possible have the same answer.

Let $I_j \subseteq \{1, \ldots, n\}$ be the set of indices of the variables participating in equation e_j, and let $b_j \in \{0, 1\}$ be the right hand side of e_j. Define a random variable $X = \sum_{j=1}^{m} X_j$, where $X_j = (-1)^{b_j} w_j \prod_{i \in I_j} \epsilon_i$ and all the ϵ_i are independent uniform random variables on $\{-1, 1\}$ (X was first introduced in [12]). We set $z_i = 0$ if $\epsilon_i = 1$ and $z_i = 1$, otherwise, for each i. Observe that $X_j = w_j$ if e_j is satisfied and $X_j = -w_j$, otherwise.

Lemma 6. *Let S be reduced under Rule 4. The weight of the satisfied equations is at least $W/2 + k$ if and only if $X \geq 2k$. We have $\mathbb{E}(X) = 0$ and $\mathbb{E}(X^2) = \sum_{j=1}^{m} w_j^2$.*

Proof. Observe that X is the difference between the weights of satisfied and falsified equations. Therefore, the weight of the satisfied equations equals $(X + W)/2$, and it is at least $W/2 + k$ if and only if $X \geq 2k$. Since ϵ_i are independent, $\mathbb{E}(\prod_{i \in I_j} \epsilon_i) = \prod_{i \in I_j} \mathbb{E}(\epsilon_i) = 0$. Thus, $\mathbb{E}(X_j) = 0$ and $\mathbb{E}(X) = 0$ by linearity of expectation. Moreover,

$$\mathbb{E}(X^2) = \sum_{j=1}^{m} \mathbb{E}(X_j^2) + \sum_{1 \leq j \neq q \leq m} \mathbb{E}(X_j X_q) = \sum_{j=1}^{m} w_j^2 > 0$$

as $\mathbb{E}(\prod_{i \in I_j} \epsilon_i \cdot \prod_{i \in I_q} \epsilon_i) = \mathbb{E}(\prod_{i \in I_j \Delta I_q} \epsilon_i) = 0$ implies $\mathbb{E}(X_j X_q) = 0$, where $I_j \Delta I_q$ is the symmetric difference between I_j and I_q ($I_j \Delta I_q \neq \emptyset$ due to Rule 4).   □

Lemma 7. *Let S be reduced under Rule 4 and suppose that no variable appears in more than $\rho \geq 2$ equations of S. Then $\mathbb{E}(X^4) \leq 2\rho^2 (\mathbb{E}(X^2))^2$.*

Proof. Observe that

$$\mathbb{E}(X^4) = \sum_{(p,q,s,t) \in [m]^4} \mathbb{E}(X_p X_q X_s X_t), \tag{2}$$

where $[m] = \{1, \ldots, m\}$. Note that if the product $X_p X_q X_s X_t$ contains a variable ϵ_i in only one or three of the factors, then $\mathbb{E}(X_p X_q X_s X_t) = A \cdot \mathbb{E}(\epsilon_i) = 0$, where

A is a polynomial in random variables ϵ_l, $l \in \{1, \dots, n\} \setminus \{i\}$. Thus, the only nonzero terms in (2) are those for which either (1) $p = q = s = t$, or (2) there are two distinct integers j, l such that each of them coincides with two elements in the sequence p, q, s, t, or (3) $|\{p, q, s, t\}| = 4$, but each variable ϵ_i appears in an even number of the factors in $X_p X_q X_s X_t$. In Cases 1 and 2, we have $\mathbb{E}(X_p X_q X_s X_t) = w_p^4$ and $\mathbb{E}(X_p X_q X_s X_t) = w_j^2 w_l^2$, respectively. In Case 3,

$$\mathbb{E}(X_p X_q X_s X_t) \leq w_p w_q w_s w_t \leq (w_p^2 w_q^2 + w_s^2 w_t^2)/2.$$

Let $1 \leq j < l \leq m$. Observe that $\mathbb{E}(X_p X_q X_s X_t) = w_j^2 w_l^2$ in Case 2 for $\binom{4}{2} = 6$ 4-tuples $(p, q, s, t) \in [m]^4$. In Case 3, we claim that $j, l \in \{p, q, s, t\}$ for at most $4 \cdot (\rho - 1)^2$ 4-tuples $(p, q, s, t) \in [m]^4$. To see this, first note that $w_p^2 w_q^2$ and $w_s^2 w_t^2$ appear in our upper bound on $\mathbb{E}(X_p X_q X_s X_t)$ (with coefficient $1/2$). Therefore, there are only four possible ways for $w_j^2 w_l^2$ to appear in our upper bound, namely the following: (i) $j = p, l = q$, (ii) $l = p, j = q$, (iii) $j = s, l = t$, and (iv) $l = s, j = t$. Now assume, without loss of generality, that $j = p$ and $l = q$. Since S is reduced under Rule 4, the product $X_j X_l$ must have a variable ϵ_i of degree one. Thus, ϵ_i must be in X_s or X_t, but not in both (two choices). Assume that ϵ_i is in X_s. Observe that there are at most $\rho - 1$ choices for s. Note that $X_j X_l X_s$ must contain contain a variable $\epsilon_{i'}$ of odd degree. Thus, $\epsilon_{i'}$ must be in X_t and, hence, there are at most $\rho - 1$ choices for t.

Therefore, we have

$$\mathbb{E}(X^4) \leq \sum_{j=1}^{m} w_j^4 + (6 + 4(\rho - 1)^2) \sum_{1 \leq j < l \leq m} w_j^2 w_l^2 < 2\rho^2 \left(\sum_{j=1}^{m} w_j^2 \right)^2.$$

Thus, by Lemma 6, $\mathbb{E}(X^4) \leq 2\rho^2 (\mathbb{E}(X^2))^2$.   □

Case 1 of Theorem 2 is of interest since its condition can be checked in polynomial time due to the following:

Proposition 1. *We can check, in polynomial time, whether there exists a set U of variables such that each equation of S contains an odd number of variables from U.*

Proof. Observe that such a set U exists if and only if the unweighted system S' of linear equations over $GF(2)$ obtained from S by replacing each b_j with 1 has a solution. Indeed, if U exists, set $z_j = 1$ for each $z_j \in U$ and $z_j = 0$ for each $z_j \notin U$. This assignment is a solution to S'. If a solution to S' exists, form U by including in it all variables z_j which equal 1 in the solution. We can check whether S' has a solution using the Gaussian elimination or other polynomial-time algorithms, see, e.g., [6].   □

Now we can prove the following:

Theorem 2. *Let S be reduced under Rule 4. The following three special cases of* LinALB *are fixed-parameter tractable: (1) there is a set U of variables such* *that each equation contains an odd number of variables from U, (2) there is a constant r such that $r(S) \leq r$, (3) there is a constant ρ, such that any variable appears in at most ρ equations. In each case, there exists a kernel with $O(k^2)$ equations and variables.*

Proof. **Case 1.** Let $z^0 = (z_1^0, \ldots, z_n^0) \in \{0,1\}^n$ be an assignment of values to the variables $z_1, \ldots, z_n$, and let $-z^0 = (z_1', \ldots, z_n')$, where $z_i' = 1 - z_i^0$ if $z_i \in U$ and $z_i' = z_i^0$, otherwise, $i = 1, \ldots, n$. Observe that $f : z^0 \mapsto -z^0$ is a bijection on the set of assignments and $X(-z^0) = -X(z^0)$. Thus, X is a symmetric random variable. Therefore, by Lemmas 1 and 6, $\mathrm{Prob}(X \geq \sqrt{m}) \geq \mathrm{Prob}(X \geq \sqrt{\sum_{j=1}^m w_j^2}) > 0$. Hence, if $\sqrt{m} \geq 2k$, the answer to LINALB is YES. Otherwise, $m < 4k^2$ and after applying Rule 3, we obtain a kernel with $O(k^2)$ equations and variables.

Case 2. Since X is a polynomial of degree at most r, it follows by Lemma 3 that $\mathbb{E}(X^4) \leq 2^{6r}\mathbb{E}(X^2)^2$. This inequality and the results in the previous paragraph show that the conditions of Lemma 2 are satisfied and, thus, $\mathrm{Prob}\left(X > \frac{\sqrt{\sum_{j=1}^m w_j^2}}{4 \cdot 8^r}\right) > 0$, implying $\mathrm{Prob}\left(X > \frac{\sqrt{m}}{4 \cdot 8^r}\right) > 0$. Consequently, if $2k - 1 \leq \sqrt{m}/(4 \cdot 8^r)$, then there is an assignment of values to the variables $z_1, \ldots, z_n$ which satisfies equations of total weight at least $W/2 + k$. Otherwise, $2k - 1 > \sqrt{m}/(4 \cdot 8^r)$ and $m < 16(2k - 1)^2 64^r$. After applying Rule 3, we obtain the required kernel.

Case 3. If $\rho = 1$, it is easy to find an assignment to the variables that satisfies all equations of S. Thus, we may assume that $\rho \geq 2$. To prove that there exists a kernel with $O(k^2)$ equations, we can proceed as in Case 2, but use Lemma 7 rather than Lemma 3. □

Remark 1. Note that even if S does not satisfy Case 2 of the theorem, T, the system obtained from S using Rule 3, may still satisfy Case 2. However, we have not formulated the theorem for S reduced under Rule 3 as the reduced system depends on the choice of a maximum linear independent collection of columns of A.

5   Discussions

We have showed that the new method allows us to prove that some maximization problems parameterized above tight lower bounds are fixed-parameter tractable. Our method can also be used for minimization problems parameterized below tight upper bounds. As a simple example, consider the feedback arc problem: given a digraph $D = (V, A)$ find a minimum set F of arcs such that $D - F$ is acyclic. Certainly, $|A|/2$ is a tight upper bound on a minimum feedback set and we can consider the parameterized problem which asks whether D has a feedback arc set with at most $|A|/2 - k$ arcs. Fixed-parameter tractability of this parameterized problem follows immediately from fixed-parameter tractability of LOALB, but we could prove this result directly using essentially the same approach as for LOALB.

Theorem 2.1 in [1] allows one to obtain a smaller kernel for LOALB on dense digraphs than given in our paper. The proof of Theorem 2.1 in [1] also uses a probabilistic approach, but it is more specialized than SABEM.

It would be interesting to obtain applications of our method to other problems parameterized above tight lower bounds or below tight upper bounds. One such very recent application is given in [9], where an open problem due to Benny Chor and described in [16] was solved.

Acknowledgments. Research of Gutin, Kim and Yeo was supported in part by an EPSRC grant.

References

1. Alon, N.: Voting paradoxes and digraphs realizations. Advances in Applied Math. 29, 126–135 (2002)
2. Alon, N., Gutin, G., Krivelevich, M.: Algorithms with large domination ratio. J. Algorithms 50(1), 118–131 (2004)
3. Bang-Jensen, J., Gutin, G.: Digraphs: Theory, Algorithms and Applications, 2nd edn. Springer, London (2009)
4. Blyth, T.S., Robertson, E.F.: Basic Linear Algebra. Springer, Heidelberg (2000)
5. Bourgain, J.: Walsh subspaces of L^p-product spaces. In: Seminar on Functional Analysis (1979–1980) (French)
6. Coppersmith, D.: Solving linear systems over GF(2): block Lanczos algorithm. Lin. Algebra Applic. 192, 33–60 (1993)
7. Downey, R.G., Fellows, M.R.: Parameterized Complexity. Springer, Heidelberg (1999)
8. Flum, J., Grohe, M.: arameterized Complexity Theory. Texts in Theoretical Computer Science. An EATCS Series, vol. XIV. Springer, Heidelberg (2006)
9. Gutin, G., Kim, E.J., Mnich, M., Yeo, A.: Ordinal Embedding Relaxations Parameterized Above Tight Lower Bound. Tech. Report arXiv:0907.5427
10. Gutin, G., Rafiey, A., Szeider, S., Yeo, A.: The linear arrangement problem parameterized above guaranteed value. Theory Comput. Syst. 41, 521–538 (2007)
11. Gutin, G., Szeider, S., Yeo, A.: Fixed-parameter complexity of minimum profile problems. Algorithmica 52(2), 133–152 (2008)
12. Håstad, J., Venkatesh, S.: On the advantage over a random assignment. In: Proceedings of the Thirty-Fourth Annual ACM Symposium on Theory of Computing, vol. 25(2), pp. 117–149 (2002)
13. Heggernes, P., Paul, C., Telle, J.A., Villanger, Y.: Interval completion with few edges. In: STOC 2007—Proceedings of the 39th Annual ACM Symposium on Theory of Computing, pp. 374–381. ACM, New York (2007); Full version appeared in SIAM J. Comput. 38(5) (2008-2009)
14. Mahajan, M., Raman, V.: Parameterizing above guaranteed values: MaxSat and MaxCut. J. Algorithms 31(2), 335–354 (1999)
15. Mahajan, M., Raman, V., Sikdar, S.: Parameterizing above or below guaranteed values. J. of Computer and System Sciences 75(2), 137–153 (2009)
16. Niedermeier, R.: Invitation to Fixed-Parameter Algorithms. Oxford Lecture Series in Mathematics and its Applications. Oxford University Press, Oxford (2006)
17. Vovk, V.: Private communication (August 2009)

Polynomial Kernels and Faster Algorithms for the Dominating Set Problem on Graphs with an Excluded Minor

Shai Gutner[*]

School of Computer Science, Tel-Aviv University, Tel-Aviv, 69978, Israel
gutner@tau.ac.il

Abstract. The domination number of a graph $G = (V, E)$ is the minimum size of a dominating set $U \subseteq V$, which satisfies that every vertex in $V \setminus U$ is adjacent to at least one vertex in U. The notion of a problem kernel refers to a polynomial time algorithm that achieves some provable reduction of the input size. Given a graph G whose domination number is k, the objective is to design a polynomial time algorithm that produces a graph G' whose size depends only on k, and also has domination number equal to k. Note that the graph G' is constructed without knowing the value of k. All the constructions in this paper are of **monotone kernels**, that is, the kernel G' is a subgraph of the input graph G. Problem kernels can be used to obtain efficient approximation and exact algorithms for the domination number, and are also useful in practical settings.

In this paper, we present the first nontrivial result for the general case of graphs with an excluded minor, as follows. For every fixed h, given a graph G with n vertices that does not contain K_h as a topological minor, our $O(n^{3.5} + k^{O(1)})$ time algorithm constructs a subgraph G' of G, such that if the domination number of G is k, then the domination number of G' is also k and G' has at most k^c vertices, where c is a constant that depends only on h. This result is improved for graphs that do not contain $K_{3,h}$ as a topological minor, using a simpler algorithm that constructs a subgraph with at most ck vertices, where c is a constant that depends only on h.

Our results imply that there is a problem kernel of polynomial size for graphs with an excluded minor and a linear kernel for graphs that are $K_{3,h}$-minor-free. The only previous kernel results known for the dominating set problem are the existence of a linear kernel for the planar case as well as for graphs of bounded genus. Using the polynomial kernel construction, we give an $O(n^{3.5} + 2^{O(\sqrt{k})})$ time algorithm for finding a dominating set of size at most k in an H-minor-free graph with n vertices. This improves the running time of the previously best known algorithm.

Keywords: H-minor-free graphs, degenerated graphs, dominating set problem, fixed-parameter tractable algorithms, problem kernel.

[*] Research supported in part by an ERC Advanced grant. This paper forms part of a Ph.D. thesis written by the author under the supervision of Prof. N. Alon and Prof. Y. Azar in Tel Aviv University.

J. Chen and F.V. Fomin (Eds.): IWPEC 2009, LNCS 5917, pp. 246–257, 2009.

1   Introduction

The input to a parameterized problem is a pair (x, k), where x is the problem instance, k is the parameter, and $n := |(x, k)|$ denotes the input size. A parameterized problem is fixed-parameter tractable if it can be solved in time $f(k) \cdot n^c$, for a computable function $f : \mathbb{N} \to \mathbb{N}$ and a constant c. A kernelization is a polynomial time computable function that given input (x, k) constructs an equivalent input (x', k'), such that $k' \leq k$ and $|x'| \leq g(k)$ for a computable function $g : \mathbb{N} \to \mathbb{N}$. The image x' is called the problem kernel of x. In this paper, the notion of a kernel for the dominating set problem refers to a polynomial time algorithm that given a graph G whose domination number is k, constructs a graph G' whose size depends only on k, and also has domination number equal to k.

It is easy and known that a parameterized problem is kernelizable if and only if it is fixed-parameter tractable. Thus, a fixed-parameter algorithm for the dominating set problem gives a trivial kernel whose size is some function of k, not necessarily a polynomial. Problem kernels can be used to obtain efficient approximation and exact algorithms for the domination number, and are also useful in practical settings.

Our main result is a polynomial problem kernel for the case of graphs with an excluded minor. This is the most general class of graphs for which a polynomial problem kernel has been established. To the best of our knowledge, the only previous results are a linear kernel for the planar case as well as for graphs of bounded genus. For a general introduction to the field of parameterized complexity, the reader is referred to [13],[15], and [23].

Fixed-Parameter Algorithms for the Dominating Set Problem. The dominating set problem on general graphs is known to be $W[2]$-complete [13]. This means that most likely there is no $f(k) \cdot n^c$-algorithm for finding a dominating set of size at most k in a graph of size n for any computable function $f : \mathbb{N} \to \mathbb{N}$ and constant c. This suggests the exploration of specific families of graphs for which this problem is fixed-parameter tractable.

The method of bounded search trees has been used to give an $O(8^k n)$ time algorithm for the dominating set problem in planar graphs [3] and an $O((4g + 40)^k n^2)$ time algorithm for the problem in graphs of bounded genus $g \geq 1$ [14]. The algorithms for planar graphs were improved to $O(4^{6\sqrt{34k}} n)$ [1], then to $O(2^{27\sqrt{k}} n)$ [20], and finally to $O(2^{15.13\sqrt{k}} k + n^3 + k^4)$ [17]. Fixed-parameter algorithms are now known also for map graphs [10] and for constant powers of H-minor-free graphs [11]. The running time given in [11] for finding a dominating set of size k in an H-minor-free graph G with n vertices is $2^{O(\sqrt{k})} n^c$, where c is a constant depending only on H. In a previous paper with Alon, we proved that the dominating set problem is fixed-parameter tractable for degenerated graphs, by establishing an algorithm with running time $k^{O(dk)} n$ for finding a dominating set of size k in a d-degenerated graph with n vertices [6].

Kernels for the Dominating Set Problem. The reduction rules introduced by Alber, Fellows, and Niedermeier were the first to establish a linear problem kernel for planar graphs [4]. The kernel obtained was of size $335k$, where k is the domination number of the graph. Fomin and Thilikos proved that the same rules of Alber et al. provide a linear kernel of size $O(k+g)$ for graphs of genus g [16]. Chen et al. improved the upper bound for the planar case to $67k$ [9]. They also gave the first lower bound, by proving that for any $\epsilon > 0$, there is no $(2-\epsilon)k$ kernel for the planar dominating set problem, unless $P = NP$. It is interesting to note that Alber, Dorn, and Niedermeier introduced a reduction rule that explores the joint neighborhood of l distinct vertices [2], but this general rule has been applied only for $l = 1$ and $l = 2$, in order to prove that the directed dominating set problem on planar graphs has a linear size kernel. Their reduction rule generates a constraint, which is encoded by a corresponding gadget in the graph. Thus, the kernel constructed is not necessarily a subgraph of the input graph.

Our Results. We present the first nontrivial result that provides a kernel for the dominating set problem on the general class of H-minor-free graphs. The proofs that our new reduction rules bring us to a polynomial kernel are based on deep and new combinatorial results on the structure of dominating sets in graphs. This gives an $O(n^{3.5} + 2^{O(\sqrt{k})})$ time algorithm for finding a dominating set of size at most k in an H-minor-free graph with n vertices. For graphs that are $K_{3,h}$-minor-free, the reduction rules of Alber, Fellows, and Niedermeier [4] are shown to give a linear problem kernel (the proof is postponed to the full version of the paper). All the reduction rules described in this paper have the property that the only modifications made to an input graph are the removal of vertices and edges. This implies that the graph obtained, as a result of applying the rules, is a subgraph of the input graph, and this is why we call it a **monotone kernel**. The advantages of this approach are its simplicity and the fact that it preserves monotone properties, like planarity, being H-minor-free, and degeneracy. We show that the rules of Alber et al. can also be described in such a way.

2   Preliminaries

The paper deals with undirected and simple graphs. Generally speaking, we will follow the notation used in [8] and [12]. For a graph $G = (V, E)$ and a vertex $v \in V$, $N(v)$ denotes the set of all vertices adjacent to v (not including v itself), whereas $N[v]$ denotes $N(v) \cup \{v\}$. This is generalized to the neighborhood of arbitrary sets by defining $N(A) := \left(\bigcup_{v \in A} N(v)\right) \setminus A$ and $N[A] := \bigcup_{v \in A} N[v]$. The graph obtained from G by deleting a vertex v is denoted $G - v$. The subgraph of G induced by some set $V' \subseteq V$ is denoted by $G[V']$.

A dominating set of a graph $G = (V, E)$ is a subset of vertices $U \subseteq V$, such that every vertex in $V \setminus U$ is adjacent to at least one vertex in U. The domination number of a graph G, denoted by $\gamma(G)$, is the minimum size of a dominating set. For a set of vertices A, if $U \subseteq N[A]$, then we say that A *dominates* U.

A graph G is d-*degenerated* if every induced subgraph of G has a vertex of degree at most d. A d-degenerated graph with n vertices has less than dn edges. An edge is said to be *subdivided* when it is deleted and replaced by a path of length two connecting its ends, the internal vertex of this path being a new vertex. A *subdivision* of a graph G is a graph that can be obtained from G by a sequence of edge subdivisions. If a subdivision of a graph H is the subgraph of another graph G, then H is a *topological minor* of G. A graph H is called a *minor* of a graph G if it can be obtained from a subgraph of G by a series of edge contractions.

In this paper, we consider only simple paths, that is, paths of the form $x_0 - x_1 - \cdots - x_k$, where the x_i are all distinct. The vertices $x_1, \ldots, x_{k-1}$ are the inner vertices of the path. The number of edges of a path is its length. Suppose that $G = (V, E)$ is a graph, $U \subseteq V$, and r and l are two integers. We denote by $\widehat{U}_{r,l}$ the set of all vertices $v \in V \setminus U$ for which there are r vertex disjoint paths of length at most l from v to r different vertices of U. To avoid confusion, we stress the fact that v is the starting vertex of all the paths, but any other vertex belongs to at most one of the paths. The vertices of $\widehat{U}_{r,l}$ are called *central vertices*, and when the values of r and l are clear from the context, the simpler notation $\widehat{U}$ will be used.

3   Dominating Sets in Degenerated Graphs

Graphs with either an excluded minor or with no topological minor are known to be degenerated. We will apply the following useful propositions.

Proposition 1. *[7,21] There exists a constant c such that, for every h, every graph that does not contain K_h as a topological minor is ch^2-degenerated.*

Proposition 2. *[22,25,26] There exists a constant c such that, for every h, every graph with no K_h minor is $ch\sqrt{\log h}$-degenerated.*

Some of our results for graphs with no topological K_h use the constant from Proposition 1. The results can be improved for graphs that are K_h-minor-free using Proposition 2.

A major part of Rule 2, described in Section 5, involves getting a succinct representation of all sets of some bounded size that dominate a specific set of vertices in a degenerated graph. This useful representation is achieved by applying a $k^{O(dk)}n$ time algorithm from [6] for finding a dominating set of size at most k in a d-degenerated graph with n vertices. We need the following combinatorial lemma proved in that paper.

Lemma 1. *Let $G = (V, E)$ be a d-degenerated graph, and assume that $B \subseteq V$. If $|B| > (4d + 2)k$, then there are at most $(4d + 2)k$ vertices in G that dominate at least $|B|/k$ vertices of B.*

This gives the following useful characterization of dominating sets in degenerated graphs.

Theorem 1. *Suppose that $G = (V, E)$ is a d-degenerated graph with n vertices, $B \subseteq V$, and $k \geq 1$. There is a $k^{O(dk)}n$ time algorithm for finding a family $\mathcal{F}$ of $t \leq (4d + 2)^k k!$ pairs (D_i, B_i) of subsets of V, such that $|D_i| \leq k$ and*

$|B_i| \leq (4d+2)k$ for every $1 \leq i \leq t$, for which the following holds. If $D \subseteq V$ is a subset of size at most k that dominates B, then some i, $1 \leq i \leq t$, satisfies that $D_i \subseteq D$ and $B_i = B \setminus N[D_i]$.

Proof. The algorithm uses the method of bounded search trees. In each step of the algorithm, B denotes the vertices that still need to be dominated. If $|B| > (4d+2)k$, then denote by R the set of all vertices that dominate at least $|B|/k$ vertices of B. Every set of size at most k that dominates B must contain a vertex from R. It follows from Lemma 1 that $|R| \leq (4d+2)k$, so we can build our search tree, by creating $|R|$ branches and checking all possible options of adding one of the vertices of R to the dominating set. For each such vertex $v \in R$, we add v to the dominating set, assign $B := B \setminus N(v)$, and remove v from the graph. We continue until $|B| \leq (4d+2)k$ in all the leaves of the search tree. The search tree can grow to be of size at most $(4d+2)^k k!$, and each subset $D \subseteq V$ of size at most k that dominates the original input set B will correspond to one of the leaves of this search tree, as needed. □

Though the dominating set problem has a polynomial time approximation scheme when restricted to a class of graphs with an excluded minor [18], for our purposes, a fast algorithm that achieves a constant approximation is required. The following combinatorial Theorem is proved in [6] (note that $\uplus$ denotes disjoint set union).

Theorem 2. *Let s be the constant from Proposition 1. Suppose that the graph $G = (B \uplus W, E)$ satisfies that W is an independent set, all vertices of W have degree at least 2, and $N(w_1) \neq N(w_2)$ for every two distinct vertices $w_1, w_2 \in W$ for which $|N(w_1)| < h - 1$. If G does not contain K_h as a topological minor, then there exists a vertex $b \in B$ of degree at most $(3sh)^h$.*

This gives the following constant factor approximation algorithm.

Theorem 3. *Let s be the constant from Proposition 1. Suppose that the graph $G = (B \uplus W, E)$ does not contain K_h as a topological minor, and there is a set of size k that dominates B. There is an $O(nk)$ time algorithm that finds a set of size at most $(3sh)^h k$ that dominates B.*

Proof. Start with a solution $D := \emptyset$. Given a graph $G = (B \uplus W, E)$, remove all edges whose two endpoints are in W and all vertices of W of degree 0 or 1. As long as there are two different vertices $w_1, w_2 \in W$ with $N(w_1) = N(w_2)$, $|N(w_1)| < h - 1$, remove one of them from the graph. As proved in [6], these modifications can be performed in time $O(|E|)$ and they obviously do not affect the minimum size of a set that dominates B. It follows from Theorem 2 that there is a vertex $b \in B$ of degree at most $(3sh)^h$. We assign $D := D \cup N[b]$, move the vertices of $N(N[b]) \cap B$ from B to W, and remove the vertices of $N[b]$ from the graph. The size of the optimal solution decreased by at least one, since every set that dominates b must contain at least one vertex from $N[b]$. We continue as before in the resulting graph, and after at most k steps, the algorithm will compute a dominating set as needed. □

4   Bounds on the Number of Central Vertices

For graphs with no topological K_h, the following bound applies.

Lemma 2. *Let s be the constant from Proposition 1. If the graph $G = (V, E)$ does not contain K_h as a topological minor, and $U \subseteq V$ is of size k, then for every l, $|\widehat{U}_{h-1,l}| \leq (2sh^2 l)^{hl} k$.*

Proof. Denote $d = sh^2$. To bound the size of $\widehat{U}$, we initially define the set B to be equal U, and then in a series of $1 + (h - 1)(l - 1)$ phases, vertices will be added to B, until eventually $\widehat{U} \subseteq B$. As proved later, after every phase i, $1 \leq i \leq 1 + (h - 1)(l - 1)$, the set B will be of size at most $(1 + sh^2(2l - 1))^i k$. This gives the needed bound for $\widehat{U}$, by setting $i = 1 + (h - 1)(l - 1)$.

The following is the description of a phase. At the beginning of phase i, the set B is of size at most $(1 + sh^2(2l - 1))^{i-1} k$. Consider the vertices of $V \setminus B$ in some arbitrary order. For each vertex $w \notin B$, if there exist two vertex disjoint paths of length at most l from w to two vertices $b_1, b_2 \in B$, such that b_1 and b_2 are not connected, and all the inner vertices of the two paths are not in B, then add the edge $\{b_1, b_2\}$ to G and remove the vertex w from the graph together with the two paths (the vertices b_1 and b_2 remain in the graph). Denote the resulting graph by G'. Obviously, $G'[B]$ does not contain K_h as a topological minor and therefore has at most $d|B| = sh^2|B|$ edges. The number of edges in the induced subgraph $G'[B]$ is at least the number of deleted vertices divided by $(2l - 1)$, which means that at most $sh^2(2l - 1)|B|$ vertices were deleted so far. All the vertices that were removed from the graph during this phase are added to the set B, and now we start the next phase with the original graph G and a new set B of size at most $(1 + sh^2(2l - 1))^i k$.

Consider a vertex $v \in \widehat{U}$ at the beginning of a phase. There are $h - 1$ vertex disjoint paths of length at most l from v to a set H of $h - 1$ different vertices of U. Assume that when v is considered in the arbitrary order, all the vertices of these $h - 1$ paths are still in the graph. We claim that the $h - 1$ vertices of H cannot all be adjacent to each other, since otherwise they form a topological K_h together with v. Thus, if v was not removed from the graph during the phase, then this can only happen in case there exists a vertex $u \notin B$ on one of the $h - 1$ vertex disjoint paths, which was removed from the graph before v was considered. This vertex u was later added to B at the end of the phase. There are $h - 1$ vertex disjoint paths of length at most l from v to H, and these paths contain at most $(h - 1)(l - 1)$ inner vertices. Thus, after at most $1 + (h - 1)(l - 1)$ phases, the vertex v will be added to B. $\square$

Itai, Perl, and Shiloach [19] proved that given a graph G with two distinct vertices s and t, the problem of deciding whether there exist m vertex disjoint paths of length at most K from s to t is NP-complete for $K \geq 5$ and polynomially solvable for $K \leq 4$. Thus, $\widehat{U}_{r,3}$ can be efficiently computed as follows.

Lemma 3. *There is an $O(|V|^{1.5}|E|)$ time algorithm for computing $\widehat{U}_{r,3}$ for a graph $G = (V, E)$, a subset $U \subseteq V$, and an integer r.*

Proof. Suppose that $v \in V \setminus U$, and let w be a new vertex that is connected to all the vertices of U. By definition, $v \in \widehat{U}_{r,3}$ if and only if there are r vertex disjoint paths of length at most 4 from v to w. To determine this, apply the $O(|V|^{0.5}|E|)$ time algorithm of Itai et al. [19] for finding the maximum number of vertex disjoint paths of length at most 4 from v to w. □

5 Problem Kernel for Graphs with an Excluded Minor

The reduction rules described in [4] examine the neighborhood of either a single vertex or a pair of vertices. In this section we generalize these definitions to a neighborhood of a set of arbitrary size.

Definition 1. *Consider a subset of vertices $A \subseteq V$ of a given graph $G = (V, E)$. The neighborhood of A is partitioned into four disjoint sets $N_1(A)$, $N_2(A)$, $N_3(A)$, and $N_4(A)$.*

- $N_1(A) := \{u \in N(A) : N(u) \setminus N[A] \neq \emptyset\}$
- $N_2(A) := \{u \in N(A) \setminus N_1(A) : N(u) \cap N_1(A) \neq \emptyset\}$
- $N_3(A) := \{u \in N(A) \setminus (N_1(A) \cup N_2(A)) : N(u) \cap N_2(A) \neq \emptyset\}$
- $N_4(A) := N(A) \setminus (N_1(A) \cup N_2(A) \cup N_3(A))$

In the original definitions from [4] the neighborhood is partitioned into only three parts. Here, the definition of $N_3(A)$ is modified and $N_4(A)$ takes the role of the "inner neighborhood" of A.

Proposition 3. *Let D be a dominating set of a graph G. If $v \notin N_4(A) \cup A$, then there is a path of length at most 4 from v to a vertex of D, and the path does not contain any vertices of A.*

Proof. Since $v \notin N_4(A) \cup A$, there is a path of length at most 3 from v to a vertex $w \notin N[A]$, and the path does not contain any vertices of A. Since D is a dominating set, this vertex w is adjacent to some vertex $d \in D$. Since $w \notin N[A]$, then obviously $d \notin A$ (it could be that $d \in N(A)$). This gives a path of length at most 4 from v to d, as needed. □

We now define our two reduction rules. Rule 2 applies Rule 1 as a subroutine. Rule 1 removes a vertex u from the graph in case there are two other vertices v and w such that $\{u, v, w\}$ is an independent set and $N(u) = N(v) = N(w) \neq \emptyset$. Rule 2 examines the "inner neighborhood" $N_4(A)$ of a subset A of size k. By applying a fixed-parameter algorithm for finding dominating sets in degenerated graphs, it calculates a small set W that contains all the vertices that dominate many vertices of $N_3(A) \cup N_4(A)$. More formally, for every set D of size of at most k that dominates $N_3(A) \cup N_4(A)$, there is a subset $D' \subseteq D$, such that $D' \subseteq W$ and $(N_3(A) \cup N_4(A)) \setminus N[D'] \subseteq W$. In case $N_4(A)$ is large, many of the vertices of $N_4(A) \setminus W$ can be removed from the graph. The main goal of this section will be to analyze graphs for which Rule 2 cannot be applied anymore.

Rule 1: Let $A \subseteq V$ be an independent set of the graph $G = (V, E)$ and assume that $N(v) \neq \emptyset$ for every $v \in A$.

- Partition the set A into disjoint subsets $A_1, A_2, \ldots, A_t$ according to the neighborhoods of vertices of A. That is, every two vertices $v, w \in A_i$ satisfy $N(v) = N(w)$, whereas every two vertices $v \in A_i$ and $w \in A_j$ for $i \neq j$ satisfy $N(v) \neq N(w)$.
- For every $1 \leq i \leq t$ for which $|A_i| > 2$, let $v, w \in A_i$ be two arbitrary distinct vertices. Remove all the vertices of $A_i \setminus \{v, w\}$ from the graph.

Rule 2: Suppose that $G = (V, E)$ is d-degenerated and $A \subseteq V$ is a subset of k vertices. If $|N_4(A)| > 2^{(4dk+3k)^{k+1}}$, do the following.

- Let $\mathcal{F}$ be a family of $t \leq (4d+2)^k k!$ pairs (D_i, B_i) of subsets of V, such that $|D_i| \leq k$ and $|B_i| \leq (4d+2)k$ for every $1 \leq i \leq t$ for which the following holds. If $D \subseteq V$ is a subset of size at most k that dominates $N_3(A) \cup N_4(A)$, then some i, $1 \leq i \leq t$, satisfies that $D_i \subseteq D$ and $B_i = (N_3(A) \cup N_4(A)) \setminus N[D_i]$.
- Denote $W := A \cup \bigcup_{i=1}^{t}(D_i \cup B_i)$. Remove all edges between vertices of $(N_3(A) \cup N_4(A)) \setminus W$.
- Apply Rule 1 to the resulting graph and the independent set $N_4(A) \setminus W$.

The next two Lemmas prove the correctness of these rules.

Lemma 4. *Let $A \subseteq V$ be an independent set of the graph $G = (V, E)$. Applying Rule 1 to G and A does not change the domination number.*

Proof. It is enough to prove that if $\{x, y, z\}$ is an independent set, such that $N(x) = N(y) = N(z) \neq \emptyset$, then $\gamma(G-z) = \gamma(G)$. To prove that $\gamma(G) \leq \gamma(G-z)$, let D be a dominating set of $G - z$. If $D \cap N(x) = \emptyset$, then $\{x, y\} \subseteq D$, and therefore $(D \setminus \{y\}) \cup \{u\}$ is a dominating set of G, for any $u \in N(x)$.

To prove that $\gamma(G - z) \leq \gamma(G)$, let D be a *minimum* dominating set of G. It cannot be the case that $\{x, y, z\} \subseteq D$, since adding one of the vertices of $N(x)$ to $D \setminus \{y, z\}$ results in a smaller dominating set. We can assume, without loss of generality, that $z \notin D$, and therefore D is a dominating set of $G - z$. □

Lemma 5. *Suppose that $G = (V, E)$ is d-degenerated and $A \subseteq V$ is a subset of k vertices. In case Rule 2 is applied to G and A, then at least one vertex is removed from the graph, whereas the domination number does not change.*

Proof. Using the notations of Rule 2, denote by G' the graph obtained from G by removing all edges between vertices of $(N_3(A) \cup N_4(A)) \setminus W$, just before Rule 1 is applied. It follows from Lemma 4 that in order to verify that Rule 2 does not change the domination number, it is enough to prove that $\gamma(G') = \gamma(G)$. It is obvious that $\gamma(G') \geq \gamma(G)$, since removing edges cannot decrease the domination number. We now prove that $\gamma(G') \leq \gamma(G)$. Let D be a *minimum* dominating set of G, and let $D' \subseteq D$ be a subset of *minimum* size that dominates $N_3(A) \cup N_4(A)$. This implies that $D' \subseteq A \cup N_2(A) \cup N_3(A) \cup N_4(A)$ and $N[D'] \subseteq N[A]$. Obviously $|D'| \leq k$, since otherwise $(D \setminus D') \cup A$ would be a smaller dominating set of G. Thus, from Theorem 1, some i, $1 \leq i \leq t$, satisfies that $D_i \subseteq D'$ and $B_i = (N_3(A) \cup N_4(A)) \setminus N[D_i]$. To prove that D is also a dominating set of G', we need to show that the vertices of $(N_3(A) \cup N_4(A)) \setminus W$ are dominated by D in G', since the neighborhood of all other vertices remained the same. Assume that

$v \in (N_3(A) \cup N_4(A)) \setminus W$. Since $B_i \subseteq W$, it follows that $v \notin B_i$, and therefore v is dominated in G by some vertex $d' \in D_i$. This means that v is still dominated by d' in G', since $D_i \subseteq W$. This completes the proof that Rule 2 does not change the domination number.

We now prove that when Rule 2 is applied, at least one vertex of $N_4(A) \setminus W$ is removed from the graph G'. First, note that $(N_3(A) \cup N_4(A)) \setminus W$ is an independent set, and therefore $N_4(A) \setminus W$ is also independent. Given a vertex $v \in N_4(A) \setminus W$, obviously $N(v) \subseteq A \cup N_3(A) \cup N_4(A)$ and $N(v) \neq \emptyset$, since it is adjacent to at least one vertex of A. The important property of v is that it is adjacent in G' only to vertices of W, since all other edges incident at v were removed. Since $W = A \cup \bigcup_{i=1}^{t}(D_i \cup B_i)$, it follows that $|W| \leq k + (4d+2)^k k!(k + (4d+2)k)) = (4d+3)k(4d+2)^k k! + k$. It is easy to verify that $2 \cdot 2^{|W|} + |W| \leq 2^{|W|+2} \leq 2^{(4dk+3k)^{k+1}} < N_4(A)$. Thus, $|N_4(A) \setminus W| \geq |N_4(A)| - |W| > 2 \cdot 2^{|W|}$. By the pigeonhole principle, we conclude that there are three distinct vertices $x, y, z \in N_4(A) \setminus W$, such that $N(x) = N(y) = N(z) \neq \emptyset$. One of these three vertices will be removed by Rule 1. $\square$

The following Lemma is useful for showing that given a graph with an excluded minor and a dominating set D of size k, there exists a subset of vertices U whose size is linear in k, such that all vertices not in $D \cup U$ belong to the "inner neighborhood" $N_4(A)$ of a subset $A \subseteq D \cup U$ of constant size.

Lemma 6. *Let D be a dominating set of the graph $G = (V, E)$. If $r \geq 1$ and $v \notin D \cup \widehat{D}_{r+1,4}$, then there exists a subset $A \subseteq D \cup \widehat{D}_{r+1,3}$ of size at most $40r^5$, such that $v \in N_4(A)$.*

Proof. To simplify the notation, the symbol $\widehat{D}$ will refer to $\widehat{D}_{r+1,3}$. Let q be the maximum number of disjoint paths of length at most 4 from v to q different vertices of D. Since $v \notin D \cup \widehat{D}_{r+1,4}$, it follows from the definition of $\widehat{D}_{r+1,4}$ that $q \leq r$. Construct q such paths of length at most 4, whose total length is the minimum possible. Denote by B the set of all vertices that appear in these q paths and call the inner vertices of these paths $B' := B \setminus (D \cup \{v\})$. Assign $t := 3r(r + r^2 + r^4) + 1$, and assume, by contradiction, that $v \notin N_4(A)$ for all subsets $A \subseteq D \cup \widehat{D}$ of size at most $4(r + t - 1)$. Note that $4(r + t - 1) \leq 40r^5$.

We will now construct t paths of length at most 4 and a series of t subsets $A_1 \subseteq A_2 \subseteq \cdots \subseteq A_t$ of size at most $4(r+t-1)$. Let $A_1 := B \cap (D \cup \widehat{D})$. For each i from 1 to t, do the following. According to our assumption $v \notin N_4(A_i) \cup A_i$, which means by Proposition 3 that there is a path of length at most 4 from v to a vertex of D, and this path does not contain any vertices from A_i. Denote by P_i the vertices of a minimum length path, which satisfies these properties. Define $A_{i+1} := A_i \cup (P_i \cap (D \cup \widehat{D}))$ and proceed to the next iteration to construct P_{i+1}.

Note that $|A_1| \leq 4r$ and $|A_{i+1}| \leq |A_i| + 4$. Thus, all the sets A_i are of size at most $4r + 4(t-1) = 4(r+t-1)$. After completing this process, we get t paths of length at most 4 that start at v. Note that a vertex $u \in D \cup \widehat{D}$ can participate in at most one of these paths, since once it appears in a path P_i, it is immediately added to A_{i+1}. Because of the maximality of q, each path P_i must contain a vertex of B'. From now on, we will consider the last appearance of a vertex from B' in a path P_i as the starting point of the path. This means that all the paths P_i start at a vertex of B' and are of length at most 3. Since $|B'| \leq 3q \leq 3r$ and the number of paths is $t = 3r(r + r^2 + r^4) + 1$, by the pigeonhole principle there must be a vertex $b \in B'$ that is a starting point of $r + r^2 + r^4 + 1$ paths of length at most 3. We now prove that $b \in \widehat{D}$. There are three possible cases.

Case 1: The vertex b starts at least $r + 1$ paths of length 1. This means that b is adjacent to $r + 1$ vertices of D and therefore $b \in \widehat{D}$.

Case 2: The vertex b starts at least $r^2 + 1$ paths of length 2. It follows from the construction that all these paths are from b to a different vertex of D. A vertex u cannot be the middle vertex of more than r of these paths, since this would imply that $u \in \widehat{D}$, but as mentioned before, vertices of $\widehat{D}$ can appear in at most one path. Thus, there are at least $r + 1$ middle vertices that are part of $r + 1$ vertex disjoint paths of length 2 from b to D, which implies that $b \in \widehat{D}$.

Case 3: The vertex b starts at least $r^4 + 1$ paths of length 3. The vertex b is the first vertex of these paths, whereas the fourth vertex is always a different vertex from D. Denote by U_2 and U_3 the vertices that appear as a second and third vertex on one of these paths, respectively. Recall that when creating the paths P_i, we always chose a path of minimum length that leads to a vertex of D. This implies that $U_2 \cap U_3 = \emptyset$. As before, vertices of U_2 and U_3 can belong to at most r^2 and r paths, respectively. The total number of paths is $r^4 + 1$, and therefore $|U_2| \geq r^2 + 1$. Since a vertex of U_3 belongs to at most r paths, we can find $r + 1$ vertices of U_2 that can be matched to $r + 1$ different vertices of U_3 in a way which would give $r + 1$ vertex disjoint paths of length 3 from b to $r + 1$ different vertices of D. This implies that $b \in \widehat{D}$.

In all three cases $b \in \widehat{D}$, which means that $b \in A_1$. Thus, b cannot belong to any path P_i, and we get a contradiction. $\square$

Theorem 4. *For every fixed h, given a graph G with n vertices that does not contain K_h as a topological minor, there is an $O(n^{3.5} + k^{O(1)})$ time algorithm that constructs a subgraph G' of G, such that if $\gamma(G) = k$, then $\gamma(G') = k$ and G' has at most k^c vertices, where c is a constant that depends only on h.*

Proof. Let s be the constant from Proposition 1. Suppose that the graph G contains no K_h as a topological minor and $\gamma(G) = k > 1$. To construct the kernel, we perform at most n iterations, as follows. The iteration starts by applying the $O(nk)$ time approximation algorithm described in Theorem 3 in order to compute a dominating set D of size at most $(3sh)^h k$. It followed from Lemmas 2 and 3 that the set $\widehat{D}_{h-1,3}$ is of size at most $(6sh^2)^{3h}|D|$, and can be computed in time $O(n^{2.5})$. In case there is a subset $A \subseteq D \cup \widehat{D}_{h-1,3}$ of size $40(h-2)^5$, for which the conditions of Rule 2 are satisfied, then the rule is applied. It follows from Lemma 5 that at least one vertex is removed from the graph and the domination number does not change. We continue to the next iteration with the resulting graph. Upon termination, this process computes a kernel G' with $\gamma(G') = k$, and a dominating set D of size at most $(3sh)^h k$.

As for the kernel size, Lemma 2 implies that $|\widehat{D}_{h-1,4}| = O(k)$, whereas from Lemma 6 we know that if $v \notin D \cup \widehat{D}_{h-1,4}$, then there exists a subset $A \subseteq D \cup \widehat{D}_{h-1,3}$ of size at most $40(h-2)^5$, such that $v \in N_4(A)$. The number of such subsets A is $k^{O(1)}$ and it follows from Lemma 5 that each subset A satisfied that $N_4(A) = O(1)$, since Rule 2 cannot be applied anymore. We conclude that the number of vertices not in $D \cup \widehat{D}_{h-1,4}$ is $k^{O(1)}$, and the theorem is proved. $\Box$

Theorem 5. *There is an $O(n^{3.5} + 2^{O(\sqrt{k})})$ time algorithm for finding a dominating set of size at most k in an H-minor-free graph with n vertices that contains such a set.*

Proof. Construct a problem kernel G' using Theorem 4 and apply the $2^{O(\sqrt{k})} n^c$ time algorithm of Demaine et al. [11] on the graph G'. $\Box$

6   Concluding Remarks and Open Problems

- An interesting open problem, stated in a preliminary version of this paper is to decide whether there is a polynomial size kernel for the dominating set problem on degenerated graphs [5]. This problem has been very recently resolved by Philip et al. [24], who exhibited a polynomial kernel in $K_{i,j}$-free and degenerated graphs. In their reduction, the kernel constructed is not a monotone kernel, and therefore cannot be used for obtaining Theorem 5.
- Another challenging question is to characterize the families of graphs for which the dominating set problem admits a linear kernel. We cannot rule out the possibility that a linear kernel can be obtained for graphs with any fixed excluded minor.
- A kernel is called an **induced kernel** in case the only modifications made to an input graph are the removal of vertices. The following variant of Rule 2 gives a polynomial induced kernel. When Rule 2 is applied to an input graph $G = (V, E)$, a set of edges E' and later a set of vertices V' are removed from the graph. In the modified rule, the resulting graph is replaced by the induced subgraph $G[V \setminus V']$. The details are deferred to the full version.

References

1. Alber, J., Bodlaender, H.L., Fernau, H., Kloks, T., Niedermeier, R.: Fixed parameter algorithms for DOMINATING SET and related problems on planar graphs. Algorithmica 33(4), 461–493 (2002)
2. Alber, J., Dorn, B., Niedermeier, R.: A general data reduction scheme for domination in graphs. In: Wiedermann, J., Tel, G., Pokorný, J., Bieliková, M., Štuller, J. (eds.) SOFSEM 2006. LNCS, vol. 3831, pp. 137–147. Springer, Heidelberg (2006)
3. Alber, J., Fan, H., Fellows, M.R., Fernau, H., Niedermeier, R., Rosamond, F.A., Stege, U.: A refined search tree technique for dominating set on planar graphs. J. Comput. Syst. Sci. 71(4), 385–405 (2005)
4. Alber, J., Fellows, M.R., Niedermeier, R.: Polynomial-time data reduction for dominating set. Journal of the ACM 51(3), 363–384 (2004)

5. Alon, N., Gutner, S.: Kernels for the dominating set problem on graphs with an excluded minor. Electronic Colloquium on Computational Complexity (ECCC) 15(066) (2008)
6. Alon, N., Gutner, S.: Linear time algorithms for finding a dominating set of fixed size in degenerated graphs. Algorithmica 54(4), 544–556 (2009)
7. Bollobás, B., Thomason, A.: Proof of a conjecture of Mader, Erdös and Hajnal on topological complete subgraphs. Eur. J. Comb. 19(8), 883–887 (1998)
8. Bondy, J.A., Murty, U.S.R.: Graph theory with applications. American Elsevier Publishing Co., Inc., New York (1976)
9. Chen, J., Fernau, H., Kanj, I.A., Xia, G.: Parametric duality and kernelization: lower bounds and upper bounds on kernel size. SIAM J. Comput. 37(4), 1077–1106 (2007)
10. Demaine, E.D., Fomin, F.V., Hajiaghayi, M., Thilikos, D.M.: Fixed-parameter algorithms for (k, r)-center in planar graphs and map graphs. ACM Transactions on Algorithms 1(1), 33–47 (2005)
11. Demaine, E.D., Fomin, F.V., Hajiaghayi, M., Thilikos, D.M.: Subexponential parameterized algorithms on bounded-genus graphs and H-minor-free graphs. Journal of the ACM 52(6), 866–893 (2005)
12. Diestel, R.: Graph theory, 3rd edn. Graduate Texts in Mathematics, vol. 173. Springer, Berlin (2005)
13. Downey, R.G., Fellows, M.R.: Parameterized complexity. Monographs in Computer Science. Springer, New York (1999)
14. Ellis, J.A., Fan, H., Fellows, M.R.: The dominating set problem is fixed parameter tractable for graphs of bounded genus. J. Algorithms 52(2), 152–168 (2004)
15. Flum, J., Grohe, M.: Parameterized complexity theory. Texts in Theoretical Computer Science. An EATCS Series. Springer, Berlin (2006)
16. Fomin, F.V., Thilikos, D.M.: Fast parameterized algorithms for graphs on surfaces: Linear kernel and exponential speed-up. In: Díaz, J., Karhumäki, J., Lepistö, A., Sannella, D. (eds.) ICALP 2004. LNCS, vol. 3142, pp. 581–592. Springer, Heidelberg (2004)
17. Fomin, F.V., Thilikos, D.M.: Dominating sets in planar graphs: Branch-width and exponential speed-up. SIAM J. Comput. 36(2), 281–309 (2006)
18. Grohe, M.: Local tree-width, excluded minors, and approximation algorithms. Combinatorica 23(4), 613–632 (2003)
19. Itai, A., Perl, Y., Shiloach, Y.: The complexity of finding maximum disjoint paths with length constraints. Networks 12(3), 277–286 (1982)
20. Kanj, I.A., Perkovic, L.: Improved parameterized algorithms for planar dominating set. In: Diks, K., Rytter, W. (eds.) MFCS 2002. LNCS, vol. 2420, pp. 399–410. Springer, Heidelberg (2002)
21. Komlós, J., Szemerédi, E.: Topological cliques in graphs II. Combinatorics. Probability & Computing 5, 79–90 (1996)
22. Kostochka, A.V.: Lower bound of the Hadwiger number of graphs by their average degree. Combinatorica 4(4), 307–316 (1984)
23. Niedermeier, R.: Invitation to fixed-parameter algorithms. Oxford University Press, Oxford (2006)
24. Philip, G., Raman, V., Sikdar, S.: Solving dominating set in larger classes of graphs: FPT algorithms and polynomial kernels. In: Fiat, A., Sanders, P. (eds.) ESA 2009. LNCS, vol. 5757, pp. 694–705. Springer, Heidelberg (2009)
25. Thomason, A.: An extremal function for contractions of graphs. Math. Proc. Cambridge Philos. Soc. 95(2), 261–265 (1984)
26. Thomason, A.: The extremal function for complete minors. J. Comb. Theory, Ser. B 81(2), 318–338 (2001)

Partitioning into Sets of Bounded Cardinality

Mikko Koivisto*

Helsinki Institute for Information Technology HIIT,
Department of Computer Science, University of Helsinki,
P.O.Box 68, FI-00014 University of Helsinki, Finland
`mikko.koivisto@cs.helsinki.fi`

Abstract. We show that the partitions of an n-element set into k members of a given set family can be counted in time $O((2-\epsilon)^n)$, where $\epsilon > 0$ depends only on the maximum size among the members of the family. Specifically, we give a simple combinatorial algorithm that counts the perfect matchings in a given graph on n vertices in time $O(\text{poly}(n)\varphi^n)$, where $\varphi = 1.618\ldots$ is the golden ratio; this improves a previous bound based on fast matrix multiplication.

1 Introduction

The generic set partitioning problem is as follows. Given an n-element universe N, a family $\mathcal{F}$ of subsets of N, and an integer k, decide whether there exists a partition of N into k members of $\mathcal{F}$, that is, pairwise disjoint sets $S_1, S_2, \ldots, S_k$ such that the union $S_1 \cup S_2 \cup \cdots \cup S_k$ equals N; we call the set $\{S_1, S_2, \ldots, S_k\}$ a *k-partition*, or simply a *partition*, and the tuple $(S_1, S_2, \ldots, S_k)$ an *ordered k-partition* or just an *ordered partition*.

Oftentimes, the family $\mathcal{F}$ is given implicitly by a description of size only polynomial in n. For example, in the graph coloring problem, $\mathcal{F}$ consists of the independent sets of a graph with vertex set N, while in the domatic partitioning problem, $\mathcal{F}$ consists of the dominating sets; these problems are NP-hard. In general, however, the size of the input may already be of order 2^n, and the best one can hope for is an algorithm with complexity within a polynomial factor of 2^n. Fairly recently [2], such a bound was indeed achieved via solving a somewhat harder-looking problem, namely that of *counting* all valid partitions. An intriguing question is, whether the base of the exponent can be lowered to $2 - \epsilon$ for some $\epsilon > 0$, given that the size of the set family $\mathcal{F}$ is within a polynomial factor of c^n for some $c < 2$.

In this paper, we answer the question affirmatively in the special case where the given set family consists of sets whose cardinality is bounded by a constant. Throughout the paper the O^* notation suppresses a factor polynomial in n.

Theorem 1. *Given an n-element universe N, a number k, and a family $\mathcal{F}$ of subsets of N, each of cardinality at most r, the partitions of N into k members of $\mathcal{F}$ can be counted in time $O^*\left(|\mathcal{F}|\, 2^{n\lambda_r}\right)$, where $\lambda_r = (2r-2)/\sqrt{(2r-1)^2 - 2\ln 2}$.*

---

* This research was supported in part by the Academy of Finland, Grant 125637.

J. Chen and F.V. Fomin (Eds.): IWPEC 2009, LNCS 5917, pp. 258–263, 2009.
© Springer-Verlag Berlin Heidelberg 2009

Previously, such an improved bound has been found in the special case where $\mathcal{F}$ contains only 2-sets, that is, pairs $\{u, v\} \subseteq N$. Then a valid partitioning corresponds to a perfect matching in a graph with vertex set N and edge set $\mathcal{F}$. While the existence of a perfect matching can be decided in polynomial time, the counting version is #P-complete [6]. The fastest known exact algorithm is by Björklund and Husfeldt [1], inspired by Williams's construction [7] and running in time $O^*\left(2^{n\omega/3}\right)$ where ω is the exponent of matrix multiplication. The Coppersmith–Winograd algorithm [4] shows $\omega < 2.38$ and, hence, the bound $O(1.732^n)$ [1]. The bound in Theorem 1 turns out to be slightly better, $O(1.653^n)$. In fact, the bound in Theorem 1 is somewhat crude for small r, and a specialized analysis yields yet a better bound.

Theorem 2. *The perfect matchings in a given graph on n vertices can be counted in time $O^*\left(\varphi^n\right)$, where $\varphi = (1 + \sqrt{5})/2 = 1.618\ldots$ is the golden ratio.*

Note, however, that if $\omega = 2$, as conjectured by many, then the matrix multiplication algorithm remains faster, running in time $O(1.588^n)$.

We remark that the coefficient λ_r in Theorem 1 is only slightly larger than $(2r - 2)/(2r - 1) = 1 - 1/(2r - 1)$ and amounts to a rather moderate growth of the bound with r. For example, for $r = 3, 4, 5$, and 6, Theorem 1 gives the bounds $O^*(|\mathcal{F}|\, c^n)$ with $c = 1.769, 1.827, 1.862$, and 1.885, respectively.

We will prove Theorems 1 and 2 (in Section 2) by giving a simple variant of the following folklore dynamic programming algorithm. For any $S \subseteq N$ and $j = 1, 2, \ldots, k$, let $f_j(S)$ be the number of ordered partitions of S into j members of $\mathcal{F}$. Then we have the recurrence

$$f_1(S) = [S \in \mathcal{F}] \,, \quad f_j(S) = \sum_{X \subseteq S} f_{j-1}(S \setminus X)\, [X \in \mathcal{F}] \quad \text{for } j > 1 \,, \tag{1}$$

where $[P]$ is 1 if P is true and 0 otherwise. We note that by dynamic programming, the number of k-partitions of N, given as $f_k(N)/k!$, can be computed in time $O^*(|\mathcal{F}|\, 2^n)$, or for large $|\mathcal{F}|$ better in time $O^*(3^n)$. The bound can be reduced to $O^*(2^n)$ by implementing the dynamic programming step (1) using fast subset convolution [3].[1]

To lower the base of the exponent below 2, we will apply an innocent-looking modification, stemming from the idea of counting an ordered partition $(S_1, S_2, \ldots, S_k)$ only if its members are lexicographically ordered. It turns out that this simple constraint yields a substantial exponential speedup when the family $\mathcal{F}$ contains only sets whose cardinality is at most some constant r.

Finally, we note that our dynamic programming algorithm and the runtime analysis readily generalize to arbitrary commutative semirings. Thus, the bounds in Theorems 1 and 2 extend, for example, to the following variant in the min–sum semiring. Given a family of subsets of N, each member S associated with a real-valued cost $f(S)$, find the minimum total cost $f(S_1) + f(S_2) + \cdots + f(S_k)$ over the k-partitions $(S_1, S_2, \ldots, S_k)$, each S_i from the given family.

---

[1] If dynamic programming is replaced altogether by an inclusion–exclusion algorithm, the running times $O^*(|\mathcal{F}|\, 2^n)$ and $O^*(3^n)$ are achieved in polynomial space [2,3].

2   Proof of Theorems 1 and 2

We modify the dynamic programming algorithm (1) to consider the members of a partition in a specific order. To this end, let N be an n-element set and $\mathcal{F}$ a family of subsets of N, each of size at most r. Fix a linear order $<$ on N and label the elements of N by $a_1 < a_2 < \cdots < a_n$. For any nonempty subset $S \subset N$ the minimum in S, $\min S$, is defined with respect to $<$ in the obvious way. Furthermore, define a lexicographic order, $\prec$, among the subsets of N, and hence in $\mathcal{F}$, with respect to the order $<$ on N in the usual manner; for instance, $\{a_1, a_2, a_5\} \prec \{a_1, a_3, a_4\} \prec \{a_2, a_4\}$.

While we are interested in counting the partitions of N into k members of $\mathcal{F}$, it turns out to be useful to consider ordered k-partitions $(S_1, S_2, \ldots, S_k)$ of N with the members from $\mathcal{F}$ and listed in the lexicographic order, that is, $S_i \prec S_j$ when $i < j$. We denote by $\mathcal{L}_k$ the set of such *lexicographically ordered k-partitions*, treating N and $\mathcal{F}$ as fixed. Since for any k-partition of N, the ordering of its members into the lexicographic order is unique, we have the following.

Lemma 1. *The number of partitions of N into k members of $\mathcal{F}$ equals the cardinality of $\mathcal{L}_k$.*

The lexicographic order implies certain constraints on the tuples $(S_1, S_2, \ldots, S_k) \in \mathcal{L}_k$, which amount to a reduction in the number of subsets of N that need be considered by a dynamic programming algorithm similar to (1). For example, the first set S_1 obviously must contain the smallest element of N. In general, the ith set S_i must contain the smallest element of N not contained by the preceding sets $S_1, S_2, \ldots, S_{i-1}$. Let $\mathcal{R}_j$ denote the family of sets S that can be expressed as the union of j such sets $S_1, S_2, \ldots, S_j$. Formally, we define the family of relevant sets $\mathcal{R}_j$, for $j = 1, 2, \ldots, n$, by the recurrence

$$\mathcal{R}_1 = \{X : X \in \mathcal{F}, \min N \in X\};$$
$$\mathcal{R}_j = \{Y \cup X : Y \in \mathcal{R}_{j-1}, X \in \mathcal{F}, Y \cap X = \emptyset, \min N \setminus Y \in X\}.$$

We proceed by defining, for each $j = 1, 2, \ldots, n$, a set function g_j that associates any set $S \subseteq N$ with the number of ordered partitions $(S_1, S_2, \ldots, S_j)$ of S into j members of $\mathcal{F}$ such that the following condition holds:

$$\min N \setminus (S_1 \cup S_2 \cup \cdots \cup S_{i-1}) \in S_i \quad \text{for all } i = 1, 2, \ldots, j. \tag{2}$$

We note that for $S = N$, this condition is satisfied if and only if $(S_1, S_2, \ldots, S_j)$ is a lexicographically ordered partition of N. Thus, $g_k(N)$ equals the cardinality of $\mathcal{L}_k$. Our modified dynamic programming algorithm evaluates $g_k(N)$ using the following recurrence.

Lemma 2. *Let $S \subseteq N$. Then*

$$g_1(S) = [S \in \mathcal{R}_1] = [a_1 \in S] \tag{3}$$

and

$$g_j(S) = \sum_{Y \subseteq S} g_{j-1}(Y) \, [S \setminus Y \in \mathcal{F}] \, [\min N \setminus Y \in S \setminus Y]. \tag{4}$$

Proof. The first equality (3) holds by the definition of $\mathcal{R}_1$.

We then prove the recurrence (4). For any $Y \subseteq S$, define $g_j(S; Y)$ as the number of ordered partitions $(S_1, S_2, \ldots, S_j)$ of S into j members of $\mathcal{F}$ satisfying (2) and $S_1 \cup S_2 \cup \cdots \cup S_{j-1} = Y$. We note that

$$g_j(S; Y) = g_{j-1}(Y) [S \setminus Y \in \mathcal{F}] [\min N \setminus Y \in S \setminus Y].$$

Because every $(S_1, S_2, \ldots, S_j)$ determines a unique Y, we have $g_j(S) = \sum_{Y \subseteq S} g_j(S; Y)$. □

It remains to analyze the time complexity of computing the values $g_j(S)$ for all relevant sets S via the recurrence (3–4). Straightforward induction shows that each g_j vanishes outside $\mathcal{R}_j$. Thus, the number of additions, multiplications and basic set operations of a straightforward implementation that first computes $g_1(S)$ for all $S \in \mathcal{R}_1$, then $g_2(S)$ for all $S \in \mathcal{R}_2$, and so on, is proportional to

$$\left(|\mathcal{R}_1| + |\mathcal{R}_2| + \cdots + |\mathcal{R}_k| \right) |\mathcal{F}| . \tag{5}$$

In the remainder of this section we derive upper bounds for this expression.

We begin with the special case where every member of the set family contains *exactly* 2 elements. In this case we have $|\mathcal{R}_j| \leq \binom{n-j}{j}$, because each set in $\mathcal{R}_j$ is of size $2j$ and must contain the first j elements $a_1, a_2, \ldots, a_j$ and exactly j other elements from $\{a_{j+1}, a_{j+2}, \ldots, a_n\}$. Now, we make use of the following well-known relations[2] of the diagonal sums of the binomial coefficients, the Fibonacci sequence (F_n), and the golden ratio $\varphi = (1 + \sqrt{5})/2$:

$$\sum_{j=0}^{n} \binom{n-j}{j} = F_{n+1} = \left(\varphi^{n+1} - (1-\varphi)^{n+1} \right) \Big/ \sqrt{5} < \varphi^n , \tag{6}$$

This suffices for proving the bound $O^*(\varphi^n)$ for (5), and hence Theorem 2.

It is easy to generalize the bound $O^*(\varphi^n)$ to the case where every member of the set family contains *at most* 2 elements. In this case we have $|\mathcal{R}_j| \leq \sum_{s=j}^{2j} \binom{n-j}{s-j} \leq \sum_{t=0}^{j} \binom{n-t}{t}$, because each set in $\mathcal{R}_j$ is of size at most $2j$ and must contain the first j elements $a_1, a_2, \ldots, a_j$ and at most j other elements from $\{a_{j+1}, a_{j+2}, \ldots, a_n\}$. Thus, by (6), the sum $|\mathcal{R}_1| + |\mathcal{R}_2| + \cdots + |\mathcal{R}_k|$ is at most $k\varphi^n$.

We finally turn to the case of an arbitary size bound r. In this case we have $|\mathcal{R}_j| \leq \sum_{s=j}^{rj} \binom{n-j}{s-j}$, because each set in $\mathcal{R}_j$ is of size at most rj and must contain the first j elements $a_1, a_2, \ldots, a_j$ and 0 to $rj - j$ other elements from $\{a_{j+1}, a_{j+2}, \ldots, a_n\}$. Now, the above analysis for $r = 2$ seems not to extend to $r > 2$, as it relies heavily on the special property of the diagonal sums of binomial coefficients. We therefore resort to a somewhat less accurate analysis, making use of the following specialization of the Hoeffding bounds:

---

[2] The author was pointed to these relations by two anonymous reviewers.

Theorem 3 (Hoeffding [5]). *Let $X_1, X_2, \ldots, X_n$ be independent Bernoulli trials with $\Pr\{X_i = 1\} = \mu_i$ for $i = 1, 2, \ldots, n$. Let $X = \sum_{i=1}^n X_i$, $\mu = \sum_{i=1}^n \mu_i$, and $0 < t < 1 - \mu/n$. Then*

$$\Pr\{X \leq \mu - tn\} \leq \exp[-2nt^2] .$$

Substituting $\mu_i \equiv 1/2$ and $t = 1/2 - k/n$ gives us a useful bound:

Corollary 1. *If $n > 2k$, then*

$$\sum_{j=0}^k \binom{n}{j} \leq 2^n \exp\left[-2n\left(\frac{1}{2} - \frac{k}{n}\right)^2\right] .$$

We are now ready to prove the following lemma, which completes the proof of Theorem 1.

Lemma 3. *Let n and r be natural numbers. Then*

$$\sum_{s=j}^{jr} \binom{n-j}{s-j} < 2^{n\lambda_r} , \quad \text{with } \lambda_r = \frac{r-1}{\sqrt{(r-1/2)^2 - \ln\sqrt{2}}} .$$

Proof. We consider two cases. First, suppose $jr - j \geq (n-j)/2$. Then $j \geq n/(2r-1)$, and we can bound the sum of the binomial coefficients above by $2^{n-j} \leq 2^{n(2r-2)/(2r-1)}$; the claim follows.

In the remaining case, suppose $jr - j < (n-j)/2$. Now it is handy to use $\ell = r - 1$. By Corollary 1,

$$\sum_{i=0}^{j\ell} \binom{n-j}{i} \leq 2^{n-j} \exp\left[-2(n-j)\left(\frac{1}{2} - \frac{j\ell}{n-j}\right)^2\right] .$$

Letting $n - j = xn$, with $2\ell/(2\ell+1) \leq x \leq 1$, and

$$\psi(x) = x\left[\ln 2 - 2\left(\frac{1}{2} + \ell - \frac{\ell}{x}\right)^2\right]$$

the bound becomes simply $\exp[n\psi(x)]$.

We next bound $\psi(x)$ in the relevant range. The derivative of $\psi(x)$ is

$$\psi'(x) = \ln 2 - 2\left(\frac{1}{2} + \ell - \frac{\ell}{x}\right)^2 - x4\left(\frac{1}{2} + \ell - \frac{\ell}{x}\right)\frac{\ell}{x^2} .$$

In terms of a new variable $y = \ell/x$, write

$$\psi'(\ell/y) = \ln 2 - 2\left(\frac{1}{2} + \ell - y\right)^2 - 4\left(\frac{1}{2} + \ell - y\right)y$$

$$= \ln 2 - 2\left(\frac{1}{2} + \ell - y\right)\left(\frac{1}{2} + \ell + y\right) .$$

Solving for $\psi'(\ell/y) = 0$ yields

$$(\ln 2)/2 - \left(\frac{1}{2} + \ell\right)^2 + y^2 = 0$$

$$y^2 = \left(\frac{1}{2} + \ell\right)^2 - \ln \sqrt{2} .$$

Thus, $\psi(x)$ is maximized at

$$\tilde{x} = \frac{\ell}{\sqrt{(1/2 + \ell)^2 - \ln \sqrt{2}}} > \frac{\ell}{1/2 + \ell} = \frac{2\ell}{2\ell + 1} .$$

Now we may bound $\psi(\tilde{x})$ as

$$\psi(\tilde{x}) < \tilde{x} \ln 2 = \frac{\ell \ln 2}{\sqrt{(1/2 + \ell)^2 - \ln \sqrt{2}}} .$$

Recalling $\ell = r - 1$ we arrive at the claimed bound.   □

Acknowledgements

The author is grateful to Andreas Björklund, Thore Husfeldt, and Petteri Kaski for valuable discussions, and to four anonymous reviewers for suggestions that helped to improve the presentation.

References

1. Björklund, A., Husfeldt, T.: Exact algorithms for Exact Satisfiability and Number of Perfect Matchings. Algorithmica 52, 226–249 (2008)
2. Björklund, A., Husfeldt, T., Koivisto, M.: Set partitioning via inclusion–exclusion. SIAM J. Comput., Special Issue for FOCS 2006 (to appear)
3. Björklund, A., Husfeldt, T., Kaski, P., Koivisto, M.: Fourier meets Möbius: fast subset convolution. In: 39th ACM Symposium on Theory of Computing (STOC 2007), pp. 67–74. ACM Press, New York (2007)
4. Coppersmith, D., Winograd, S.: Matrix multiplication via arithmetic progressions. J. Symb. Comput. 9, 251–280 (1990)
5. Hoeffding, W.: Probability inequalities for sums of bounded random variables. J. American Stat. Assoc. 58, 13–30 (1963)
6. Valiant, L.G.: The complexity of computing the permanent. Theor. Comput. Sci. 8, 189–201 (1979)
7. Williams, R.: A new algorithm for optimal 2-constraint satisfaction and its implications. Theor. Comput. Sci. 348, 357–365 (2005)

Two Edge Modification Problems without Polynomial Kernels

Stefan Kratsch and Magnus Wahlström

Max-Planck-Institut für Informatik, 66123 Saarbrücken, Germany

Abstract. Given a graph G and an integer k, the Π Edge Completion/Editing/Deletion problem asks whether it is possible to add, edit, or delete at most k edges in G such that one obtains a graph that fulfills the property Π. Edge modification problems have received considerable interest from a parameterized point of view. When parameterized by k, many of these problems turned out to be fixed-parameter tractable and some are known to admit polynomial kernelizations, i.e., efficient preprocessing with a size guarantee that is polynomial in k. This paper answers an open problem posed by Cai (IWPEC 2006), namely, whether the Π Edge Deletion problem, parameterized by the number of deletions, admits a polynomial kernelization when Π can be characterized by a finite set of forbidden induced subgraphs. We answer this question negatively based on recent work by Bodlaender et al. (ICALP 2008) which provided a framework for proving polynomial lower bounds for kernelizability. We present a graph H on seven vertices such that H-free Edge Deletion and H-free Edge Editing do not admit polynomial kernelizations, unless NP $\subseteq$ coNP/poly. The application of the framework is not immediate and requires a lower bound for a Not-1-in-3 SAT problem that may be of independent interest.

1   Introduction

In recent years the kernelizability of edge modification problems has received considerable attention [5,9,13,14,15]. For a graph property Π the Π Edge Completion/Editing/Deletion problem is defined as

Input: A graph $G = (V, E)$ and an integer k.
Parameter: k.
Task: Decide whether adding, editing, or deleting at most k edges in G yields a graph with property Π.

Edge modification problems have a number of applications, including machine learning, numerical algebra, and molecular biology [6,17,18,19]. In typical applications the input graphs arise from experiments and edge modification serves to correct the (hopefully) few errors. For Similarity Clustering, for example, the vertices represent entities that are linked by an edge if their similarity exceeds a certain threshold. Given a perfect similarity measure one would obtain a cluster graph, i.e., a disjoint union of cliques. In practice though, the obtained graph will

J. Chen and F.V. Fomin (Eds.): IWPEC 2009, LNCS 5917, pp. 264–275, 2009.
© Springer-Verlag Berlin Heidelberg 2009

Table 1. Kernelization bounds for some edge modification problems

Property/Class	Modification	Kernel size	Characterization
Chain	Deletion	$O(k^2)$ [13]	$(2K_2, K_3, C_5)$-free
Split	Deletion	$O(k^4)$ [13]	$(2K_2, C_4, C_5)$-free
Threshold	Deletion	$O(k^3)$ [13]	$(2K_2, C_4, P_4)$-free
Co-Trivially Perfect	Deletion	$O(k^3)$ [13]	$(2K_2, P_4)$-free
Triangle-Free	Deletion	$6k$ [5]	K_3-free
Cluster	Editing	$4k$ [14]	P_3-free
Grid	Editing	$O(k^4)$ [9]	None finite
Bi-connectivity	Completion	$O(k^2)$ [15]	None finite
Bridge-connectivity	Completion	$O(k^2)$ [15]	None finite

at best be close to a cluster graph, i.e., within few edge modifications. Clearly a large number of modifications would yield a clustering of the entities that deviates strongly from the similarity measure, allowing the conclusion that the measure is probably faulty. Thus *fpt-algorithms* that solve the problem efficiently when the number of modifications is not too large are a good way of solving this and similar problems (see Section 2 for formal definitions).

Another parameterized way of dealing with edge modification problems lies in kernelization. The notion of *kernelization* or *reduction to a problem kernel* is an important contribution of parameterized complexity, in formalizing preprocessing to allow a rigorous study. A kernelization is a polynomial-time computable mapping that transforms a given instance, say (x, k), of a parameterized problem into an equivalent instance (x', k') with size of x' and value of k' bounded by a computable function in k. It is known that a parameterized problem is fixed-parameter tractable if and only if it is decidable and kernelizable (cf. [11]). However this fact does not imply the existence of a polynomial kernelization.

Related Work: There exists rich literature on the complexity of edge modification problems; recent surveys were given by Natanzon et al. [17] and Burzyn et al. [6]. Cai [7] showed that a very general version of this problem, also allowing vertex deletions, is FPT when Π can be characterized by a finite set of forbidden induced subgraphs. A finite set of graphs $\mathcal{H}$ is a *finite forbidden subgraph characterization* of a property Π if for any graph G, G has property Π if and only if G is $\mathcal{H}$-*free*, i.e., if it has no graph from $\mathcal{H}$ as an induced subgraph.

In Table 1 we give an overview of kernelization results for a number of edge modification problems, along with their $\mathcal{H}$-free characterizations, where applicable. Six of the properties listed in the table can be characterized by such a finite set of forbidden induced subgraphs, leading to the question whether a finite characterization implies the existence of a polynomial kernelization. It is easy to see that the answer is yes if only vertex deletions are allowed since this translates directly to an instance of d-HITTING SET, where the constant d is the largest number of vertices among forbidden induced subgraphs (see [1] for a d-HITTING SET kernelization). Cai posed the question whether the same is true for the Π Edge Deletion problem, i.e., achieving property Π by at most k

edge deletions when $\Pi = \mathcal{H}$-free for some finite set $\mathcal{H}$ of graphs (see [2]). To our best knowledge the kernelizability of the $\mathcal{H}$-free Edge Editing problem is also open. We point out that $\mathcal{H}$-free Edge Completion is equivalent to $\bar{\mathcal{H}}$-free Edge Deletion, where $\bar{\mathcal{H}}$ contains the complements of the graphs in $\mathcal{H}$.

Our Work: The contribution of this paper is to present a small graph H of seven vertices such that H-free Edge Deletion and H-free Edge Editing do not admit polynomial kernelizations unless the polynomial hierarchy collapses (more specifically, unless NP $\subseteq$ coNP/poly, which is known to imply PH= Σ_3^P [20]). This result builds upon recent seminal work by Bodlaender et al. [3] showing that a polynomial kernelization for any so-called compositional problem whose unparameterized version is NP-complete would imply such a collapse.

The application of [3] on the H-free modification problems is not immediate. We first define a problem Not-1-in-3 SAT and prove that it does not admit a polynomial kernelization, then reduce this problem to H-free Edge Deletion and H-free Edge Editing, respectively, by polynomial-time reductions with polynomial bounds on the new parameter values. Such reductions were proposed by Bodlaender et al. [4] as a way to extend kernelization results to further problems.

Structure of the paper: In Section 2 we introduce some notation as well as giving the necessary definitions from parameterized complexity and reviewing briefly the framework for lower bounds for kernelization. In Section 3 we prove that Not-1-in-3 SAT does not admit a polynomial kernelization, unless NP $\subseteq$ coNP/poly, and extend this result to the version without repeated variables. The latter problem is then reduced to H-free Edge Deletion and H-free Edge Editing, proving the claimed lower bounds, in Section 4. We conclude in Section 5.

2   Preliminaries

2.1   Notation

Graphs: We consider undirected, simple graphs $G = (V, E)$. An *induced subgraph* of G is a graph $G' = (V', E')$ with $V' \subseteq V$ and with $E' = \{\{u, v\} \in E \mid u, v \in V'\}$. We denote this subgraph by $G[V']$. A graph G is *H-free* if it does not contain H as an induced subgraph (i.e., a subgraph G' which is isomorphic to H). For a set $\mathcal{H}$ of graphs, G is *$\mathcal{H}$-free* if it is H-free for every graph $H \in \mathcal{H}$. We denote by K_i, C_i, and P_i the clique, cycle, and path of i vertices respectively. The graph $2K_2$ is the disjoint union of two K_2.

Not-1-in-3 SAT: We define Not-1-in-3 SAT as a satisfiability problem asking for a feasible assignment with at most k ones to a conjunction of not-1-in-3-constraints $R(x, y, z)$, and constraints $(x \neq 0)$. Inputs to the problem consist of such a conjunction F and an integer k. An assignment ϕ to the variables of F is feasible if $\phi(x) = 1$ for every variable x with a constraint $(x \neq 0)$ and

$$(\phi(x), \phi(y), \phi(z)) \in \{(0,0,0), (0,1,1), (1,0,1), (1,1,0), (1,1,1)\}$$

for every not-1-in-3-constraint $R(x, y, z)$. The *weight* of an assignment is the number of ones it assigns.

Parameterized Complexity: A *parameterized problem* Q over a finite alphabet Σ is a subset of $\Sigma^* \times \mathbb{N}$. The second component is called the *parameter*. The problem Q is *fixed-parameter tractable* if there is an algorithm that decides whether $(x, k) \in Q$ in time $f(k) \cdot |x|^c$, where f is a computable function and c is a constant independent of k. Such an algorithm is called an *fpt-algorithm* for Q.

A *kernelization* of Q is a polynomial-time computable mapping $K : \Sigma^* \times \mathbb{N} \to \Sigma^* \times \mathbb{N} : (x, k) \mapsto (x', k')$ such that

$$\forall (x, k) \in \Sigma^* \times \mathbb{N} : ((x, k) \in Q \Leftrightarrow (x', k') \in Q) \text{ and } |x'|, k' \leq h(k),$$

for some computable function $h : \mathbb{N} \to \mathbb{N}$. The kernelization K is *polynomial* if h is a polynomial. We say that Q *admits a (polynomial) kernelization* if there exists a (polynomial) kernelization of Q. We refer the reader to [10] for an introduction to parameterized complexity.

2.2 Polynomial Lower Bounds for Kernelization

Bodlaender et al. [3] provided the first polynomial lower bounds for the kernelizability of some parameterized problems. They introduced the notions of *or-* respectively *and-compositionality* for parameterized problems and showed that such an algorithm together with a polynomial kernelization yields an *or-* respectively *and-distillation algorithm* for the unparameterized version of the problem. Based on the hypothesis that NP-complete problems do not admit either variant of distillation algorithm, this proves non-existence of polynomial kernelizations for some problems. A related article by Fortnow and Santhanam [12] showed that an or-distillation algorithm for any NP-complete problem would imply $NP \subseteq coNP/poly$, causing a collapse of the polynomial hierarchy [20]. In this paper we refer only to the results for or-compositional problems and hence omit the prefix for readability.

Let Q be a parameterized problem. A *composition algorithm* for Q is an algorithm that on input $(x_1, k), \ldots, (x_t, k) \subseteq \Sigma^* \times \mathbb{N}$ uses time polynomial in $\sum_{i=1}^{t} |x_i| + k$ and outputs (y, k') such that

1. k' is bounded by a polynomial in k and
2. $(y, k') \in Q$ if and only if $(x_i, k) \in Q$ for at least one $i \in \{1, \ldots, t\}$.

The *unparameterized version* or *derived classical problem* $\tilde{Q}$ of Q is defined by $\tilde{Q} = \{x\#1^k \mid (x, k) \in Q\}$, where $\# \notin \Sigma$ is the blank letter and 1 is any letter from Σ.

Theorem 1 ([3,12]). *Let L be a compositional parameterized problem whose unparameterized version $\tilde{L}$ is NP-complete. The problem L does not admit a polynomial kernelization unless $NP \subseteq coNP/poly$.*

In [4] Bodlaender et al. proposed the use of polynomial-time reductions, where the new parameter value is polynomially bounded in the old one, as a tool to extend the lower bounds to further problems.

A *polynomial time and parameter transformation* from Q to Q' is a polynomial-time mapping $H : \Sigma^* \times \mathbb{N} \to \Sigma^* \times \mathbb{N} : (x, k) \mapsto (x', k')$ such that

$$\forall (x, k) \in \Sigma^* \times \mathbb{N} : ((x, k) \in Q \Leftrightarrow (x', k') \in Q') \text{ and } k' \leq p(k),$$

for some polynomial p. We use the name *polynomial parameter transformation*.

Theorem 2 ([4]). *Let P and Q be parameterized problems, and let $\tilde{P}$ and $\tilde{Q}$ be their derived classical problems. Suppose that $\tilde{P}$ is NP-complete and $\tilde{Q} \in NP$. If there is a polynomial parameter transformation from P to Q and Q admits a polynomial kernelization, then P also admits a polynomial kernelization.*

3   Hardness of Not-One-In-Three SAT

In this section, we show that the Not-1-in-3 SAT problem does not admit a polynomial kernelization, unless $NP \subseteq coNP/poly$. A polynomial parameter transformation to an H-free Edge Deletion problem and an H-free Edge Editing problem in Section 4 will then yield the desired lower bounds.

To apply Theorem 1 we need to show that Not-1-in-3 SAT is compositional and that the unparameterized version is NP-hard. For showing this, two basic observations are useful. Firstly, we have the ability to force $x = 0$ for a variable x of any instance of Not-1-in-3 SAT, by the next lemma. Secondly, using such a variable, we are able to force $x = y$ by a not-1-in-3-constraint $R(x, y, 0)$ where 0 is a variable that has been forced to be false.

Lemma 1. *Let (F, k) be an instance of Not-1-in-3 SAT containing a variable x. Adding only not-1-in-3-constraints, we can create an instance (F', k) such that there is a feasible assignment for F' with at most k ones if and only if there is a feasible assignment ϕ for F with at most k ones such that $\phi(x) = 0$.*

Proof. For each $1 \leq i \leq k$, add a not-1-in-3-constraint $R(x, y_{2i-1}, y_{2i})$, where y_i are new variables. Now for any feasible assignment ϕ for F', if $\phi(x) = 1$, then k further variables must be assigned 1, implying that ϕ assigns more than k ones. Thus feasible assignments ϕ for F' of weight at most k have $\phi(x) = 0$, and can be restricted to feasible assignments for F.

Feasible assignments ϕ for F with $\phi(x) = 0$ can be extended to feasible assignments for F', setting $y_i = 0$ for all added variables, at no extra cost.   □

Lemma 2. *Not-1-in-3 SAT is NP-complete and compositional.*

Proof. The NP-completeness follows from Khanna et al. [16]. This extends also to the variant where k is encoded in unary since values of k greater than the number of variables are meaningless (i.e., can be replaced by the number of variables). We will now show that Not-1-in-3 SAT is compositional.

Let $(F_1, k), \ldots, (F_t, k)$ be instances of Not-1-in-3 SAT; we will create a new instance (F', k') with $k' = O(k)$ that is a yes-instance if and only if (F_i, k) is a yes-instance for some $1 \leq i \leq t$. If $t \geq 3^k$, then we have time to solve every

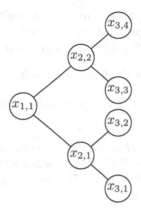

Fig. 1. First three levels of a composition tree. Every parent-node is in a not-1-in-3-constraint with its two sons, e.g. $R(x_{1,1}, x_{2,1}, x_{2,2})$. If $x_{1,1}$ is true, then on every level i some variable $x_{i,j}$ must be true.

instance exactly by branching, and output some dummy yes- or no-instance. We assume for the rest of the proof that $t < 3^k$. We also assume that t is a power of two (or else duplicate some instance (F_i, k)): say $t = 2^l$.

Start building F' by creating variable-disjoint copies of all formulas F_i. For every formula F_i, additionally create an activity variable s_i, and for every variable x in F_i such that F_i contains a constraint $x \neq 0$, replace this constraint by $x = s_i$ (i.e., a not-1-in-3-constraint $R(x, s_i, 0)$ where 0 is a variable forced to be false as in Lemma 1). The variable s_i now decides whether F_i is active: if $s_i = 0$ then every variable of F_i can be set to zero, and if $s_i = 1$, then all constraints of the formula F_i are active. We will complete the construction by creating a formula that forces some s_i to be true, in a form of *composition tree*.

Figure 1 illustrates the construction used for this part. We use variables $x_{i,j}$ where i is the *level* of the variable, and $1 \leq j \leq 2^{i-1}$. The first-level variable $x_{1,1}$ is forced to be true by a constraint $(x_{1,1} \neq 0)$, and for every internal variable in the composition tree, we use a not-1-in-3-constraint $R(x_{i,j}, x_{i+1,2j-1}, x_{i+1,2j})$. Finally, for every variable $x_{l+1,j}$ on the last level, we add a constraint such that $s_j = x_{l+1,j}$. Since a not-1-in-3-constraint $R(x, y, z)$ implies $x \rightarrow (y \vee z)$, we see inductively that at least one variable per level must be true, and furthermore for every variable $x_{l+1,j}$ on the last level, there exists an assignment where $x_{l+1,j} = 1$ and exactly one variable per level is true. Thus, assuming a minimal assignment to the composed instance F', we can treat the composition tree as adding weight exactly $l + 1$ and encoding a disjunction over the activity variables s_j. We obtain the composed instance $(F', k + l + 2)$.

Now if F' has an assignment ϕ' of weight at most $k + l + 2$ then $\phi'(s_j) = 1$ for some $j \in \{1, \ldots, t\}$. Thus the constraints of F_j are active and ϕ' can be restricted to a feasible assignment for F_j (excluding s_j) of weight at most k. If some input formula F_j has an assignment of weight at most k then this can be extended to an assignment for F' of weight at most $k + l + 2$ by assigning 1 to each $x_{i,l}$ on the path from s_j to $x_{1,1}$ in the composition tree and 0 to all other variables of F'. Finally, the parameter $k' = k + l + 2$ is $O(k)$ since $t < 3^k$, and the work performed is polynomial in the length of all input instances.   □

Now by Theorem 1 we obtain the following result.

Theorem 3. *Not-1-in-3 SAT does not admit a polynomial kernelization unless $NP \subseteq coNP/poly$.*

To simplify the reduction to H-free Edge Deletion in the following section, we consider a variant of Not-1-in-3 SAT that does not allow repeated variables in the not-1-in-3-constraints, e.g., $R(x, x, y)$. The following lemma implicitly extends our lower bound to that variant.

Lemma 3. *Not-1-in-3 SAT reduces to Not-1-in-3 SAT without repeated variables by a polynomial parameter transformation.*

Proof. Observe that both the construction of Lemma 1 to force a variable to take value 0, and the usage of a not-1-in-3-constraint $R(x, y, 0)$ to force $x = y$ work without repeating variables. Therefore, from an instance (F, k) of Not-1-in-3 SAT, we can create an instance (F', k') of Not-1-in-3 SAT without repeated variables where $k' = 2k$ in the following manner:

1. Place constraints in F' forcing $z = 0$ for some unique variable z according to Lemma 1.
2. For every variable x in F, create two variables x, x' in F'. Force $x = x'$ using a not-1-in-3-constraint $R(x, x', z)$.
3. For every not-1-in-3-constraint $R(x, x, y)$ in F, with $x \neq y$, place $R(x, x', y)$ in F'. Copy not-1-in-3-constraints $R(x, y, z)$ for distinct variables x, y, z and constraints $x \neq 0$ into F', and ignore any not-1-in-3-constraints $R(x, x, x)$.

The formula F' has a solution with at most $2k$ true variables iff F has a solution with at most k true variables.   □

4   Lower Bounds for Two H-Free Modification Problems

Throughout this section let $H = (V_H, E_H)$ with vertex set $V_H = \{a, b, c, d, e, f, g\}$ and edge set $E_H = \{\{a, b\}, \{a, c\}, \{a, d\}, \{a, e\}, \{a, f\}, \{a, g\}, \{b, c\}, \{d, e\}\}$, as depicted in Figure 2. We will show that Not-1-in-3 SAT without repeated variables reduces to H-free Edge Deletion by a polynomial parameter transformation, thereby proving that the latter problem does not admit a polynomial kernelization unless NP $\subseteq$ coNP/poly, according to Theorem 2. Consecutively we use a slightly more involved reduction from Not-1-in-3 SAT without repeated variables to H-free Edge Editing.

Lemma 4. *Not-1-in-3 SAT without repeated variables reduces to H-free Edge Deletion by a polynomial parameter transformation.*

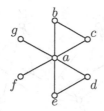

Fig. 2. The graph H

Proof. Let (F, k) be an instance of Not-1-in-3 SAT without repeated variables. We will construct an equivalent instance (G, k') of H-free Edge Deletion.

Let α be the number of variables x of F that have a constraint $(x \neq 0)$. Clearly, if $k < \alpha$ then (F, k) is a no-instance and we map it to a dummy no-instance of H-free Edge Deletion, e.g. $(H, 0)$. Henceforth we assume that $k \geq \alpha$ and we let $k' = k - \alpha$. Starting from the empty graph we use the following two steps to construct G:

1. For each variable x of F we add two vertices p_x and q_x to G. We add the edge $\{p_x, q_x\}$ to G if there is no constraint $(x \neq 0)$ in the formula F.
2. For each not-1-in-3-constraint $R(x, y, z)$ add $k' + 1$ new vertices $r_1, \ldots, r_{k'+1}$ to G. Connect each r_i to p_x, q_x, p_y, q_y, p_z, and q_z. Recall that x, y, and z are different by our restriction on the source problem.

See Figure 3 for an example of a formula F and the resulting graph. It is easy to see that this construction can be accomplished in polynomial time and that $k' = k - \alpha \in O(k)$.

We denote by X the set of vertices created in Step 1 and by C the set of vertices created in Step 2. Thus $G = (X \cup C, E)$. We make two simple observations:

(1) Each vertex of X has at most one neighbor in X.
(2) C is an independent set in G.

We will now show that (F, k) is a yes-instance of Not-1-in-3 SAT without repeated variables if and only if (G, k') is a yes-instance of H-free Edge Deletion.

Suppose that (F, k) is a yes-instance of Not-1-in-3 SAT without repeated variables and let ϕ be a feasible assignment for F with at most k ones. We define a subset $D \subseteq E$ as those edges $\{p_x, q_x\}$ of G with $\phi(x) = 1$. Observe that $\{p_x, q_x\} \in E$ if and only if F contains no constraint $(x \neq 0)$. Therefore $|D| \leq k - \alpha = k'$ since at most k variables x have $\phi(x) = 1$ but α of those variables have a constraint $(x \neq 0)$.

We assume for contradiction that $G - D$ is not H-free. Let $a, \ldots, g$ be vertices of G that induce a subgraph H. For simplicity let the adjacencies be the same as in Figure 2.

It is easy to see that $a \in C$: Otherwise, if $a \in X$, then by Observation (1) at most one of its neighbors can be in X. This would imply that $b, c \in C$ or $d, e \in C$ contradicting Observation (2), namely, that C is an independent set.

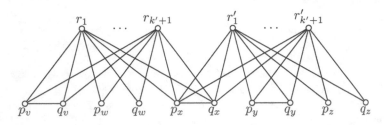

Fig. 3. The resulting graph for $F = R(v, w, x) \land R(x, y, z) \land (w \neq 0) \land (z \neq 0)$

Since $a \in C$ the other vertices $b, \ldots, g$ must be in X by Observation (2). Recall that in $G[X]$, i.e., among vertices of X, there are only edges of the form $\{p_x, q_x\}$ where x is a variable of F. Since $\{b, c\}$ and $\{d, e\}$ are edges of $G[X]$ there must be variables x and y of F such that w.l.o.g. $b = p_x$, $c = q_x$, $d = p_y$, and $e = q_y$. By Step 2 of the construction there must be a third variable z of F such that w.l.o.g. $f = p_z$ and $g = q_z$.

From $\{p_x, q_x\}, \{p_y, q_y\} \in E \setminus D$ (i.e., the edge set of $G - D$) it follows that $\phi(x) = \phi(y) = 0$. Since $\{p_z, q_z\} \notin E \setminus D$ it follows that $\{p_z, q_z\} \in D$ or $\{p_z, q_z\} \notin E$. In the first case we have that $\phi(z) = 1$. In the latter case there must be a constraint $(z \neq 0)$ in F, which also implies that $\phi(z) = 1$ since ϕ is feasible for F. However, $\phi(x) = \phi(y) = 0$ and $\phi(z) = 1$ contradict the feasibility of ϕ since $R(x, y, z)$ is a not-1-in-3-constraint of F. Hence $G - D$ is an H-free graph, implying that (G, k') is a yes-instance of H-free Edge Deletion, since $|D| \leq k'$.

Now, suppose that (G, k') is a yes-instance of H-free Edge Deletion. We will construct a feasible assignment ϕ with at most k ones for F. Let $D \subseteq E$ with $|D| \leq k'$ such that $G - D$ is H-free. We define ϕ by

$$\phi(x) = \begin{cases} 1 & \text{if } \{p_x, q_x\} \in D, \\ 1 & \text{if } (x \neq 0) \text{ is a constraint of } F, \\ 0 & \text{otherwise.} \end{cases}$$

Thus the number of ones of ϕ is at most the cardinality of D plus the number of variables x of F with an $(x \neq 0)$-constraint, i.e., $|D| + \alpha \leq k$.

We assume for contradiction that ϕ is not feasible for F. This implies that there is a not-1-in-3-constraint, say $R(x, y, z)$, in F that is not satisfied by ϕ. The reason is that, by definition, ϕ satisfies all $(x \neq 0)$-constraints. Not satisfying $R(x, y, z)$ implies that ϕ sets exactly one of the three variables to 1. Recall that x, y, and z must be pairwise different by problem definition. W.l.o.g. we assume that $\phi(x) = 1$ and $\phi(y) = \phi(z) = 0$. This immediately implies that there are no constraints $(y \neq 0)$ or $(z \neq 0)$ in F and that $\{p_y, q_y\}$ and $\{p_z, q_z\}$ are not contained in D, by definition of ϕ. Hence $\{p_y, q_y\}$ and $\{p_z, q_z\}$ are contained in the edge set $E \setminus D$ of $G - D$.

Since $\phi(x) = 1$ we conclude that $\{p_x, q_x\} \in D$ or that $(x \neq 0)$ is a constraint of F. In the latter case, by Step 1 in the construction of G, the edge $\{p_x, q_x\}$ does not occur in G. Thus, in both cases, $\{p_x, q_x\}$ is not an edge of $G - D$, i.e., $\{p_x, q_x\} \notin E \setminus D$.

We will now consider the vertices that were added for $R(x, y, z)$ in Step 2 of the construction, say $r_1, \ldots, r_{k'+1}$. Since $|D| \leq k'$ at least one r_i is adjacent to all six vertices p_x, q_x, p_y, q_y, p_z, and q_z. Together with r_i those vertices induce a subgraph H. This contradicts the choice of D, implying that ϕ is feasible for F.

Thus (F, k) is a yes-instance of Not-1-in-3 SAT without repeated variables if and only if (G, k') is a yes-instance of H-free Edge Deletion. □

By Theorem 2 we obtain the desired result.

Theorem 4. *For the graph $H = (V_H, E_H)$ with $V_H = \{a, b, c, d, e, f, g\}$ and with $E_H = \{\{a, b\}, \{a, c\}, \{a, d\}, \{a, e\}, \{a, f\}, \{a, g\}, \{b, c\}, \{d, e\}\})$ the H-free Edge Deletion problem does not admit a polynomial kernelization unless $NP \subseteq coNP/poly$.*

The following corollary proves that the H-free Edge Editing problem does not admit a polynomial kernelization either.

Corollary 1. *For the same graph H as in Theorem 4, the H-free Edge Editing problem does not admit a polynomial kernelization unless $NP \subseteq coNP/poly$.*

Proof. We extend the construction used in Lemma 4 to obtain a polynomial parameter transformation from Not-1-in-3 SAT without repeated variables to H-free Edge Editing. Let (F, k) be an instance of Not-1-in-3 SAT without repeated variables. We create an instance (G, k') according to the construction in Lemma 4, except for using $3k' + 1$ vertices r_i per not-1-in-3-constraint in Step 2.

Essentially we only need one additional gadget to ensure that no edges are added between vertices of X in G (recall that X contains vertices p_x and q_x for every variable of F). Adding this gadget for every non-edge of $G[X]$ will yield a graph G' and we will show that (G', k') and (F, k) are equivalent.

Firstly, let us recall that the only edges in $G[X]$ are of the form $\{p_x, q_x\}$ for some variable x of F that has no constraint ($x \neq 0$). For every other pair of vertices, say (p_x, p_y) with $x \neq y$, we add $k' + 1$ vertices s_i, t_i, u_i, v_i, and w_i, and for each i connect s_i to p_x, p_y, t_i, u_i, v_i, and w_i, and connect t_i to u_i. That is, create constructions which will induce $k' + 1$ subgraphs H if the edge $\{p_x, p_y\}$ is added. Note that these subgraphs have only one pair of vertices in common. The same construction applies also for pairs (p_x, q_y) and (q_x, q_y) as well as for pairs (p_x, q_x) where x has a constraint ($x \neq 0$) in F. Let G' denote the graph that is obtained by adding these vertices and edges to G.

Now, for proving equivalence of (G', k') and (F, k), let us assume that (G', k') is a yes-instance of H-free Edge Editing and let D be a set of edges, with $|D| \leq k'$ such that $G' \triangle D := (V(G'), E(G') \triangle D)$ is H-free.

Firstly, we claim that D adds no edge between vertices of X. Otherwise, if for example $\{p_x, p_y\} \in D$ then, referring to the new gadget, there would be $k' + 1$ induced subgraphs H that share only p_x and p_y. Since removal of these $k' + 1$ subgraphs would demand at least $k' + 1$ modifications, this would imply that $G' \triangle D$ is not H-free. Thus D restricted to $G'[X]$ contains only deletions; let D' be these deletions. Clearly $|D'| \leq k'$.

Let us consider the r-vertices that are added in Step 2. Clearly, since $|D| \leq k'$, for every not-1-in-3-constraint of F at least $k'+1$ of its corresponding vertices r_i are not incident with edges of D (recall that we started with $3k'+1$ vertices r_i per constraint). We select a set R of vertices, picking, for every not-1-in-3-constraint of F, $k'+1$ of its corresponding r-vertices which are not incident with edges of D. Let G'' be the graph induced by $R \cup X$.

Now let (G_0, k') be the H-free Edge Deletion-instance created from (F, k) in Lemma 4. It is easy to see that G_0 and G'' are isomorphic, and that the only modifications made to G'' by D are the deletions D'. Since $G' \triangle D$ is H-free, so is $G'' \triangle D' = (G' \triangle D)[R \cup X]$. Thus D' constitutes a solution to the H-free Edge Deletion instance (G'', k), which by Lemma 4 maps to a solution to (F, k).

Let us now assume that (F, k) is a yes-instance and let ϕ be a feasible assignment of weight at most k. Let D be obtained as in Lemma 4, i.e., D will only delete some $\{p_x, q_x\}$-edges from G'. We show that $G' \triangle D$ is H-free.

Consider the vertex which would form the vertex a of H, i.e., the vertex of degree six in H. It cannot be a p-vertex, since the neighborhood of a p-vertex contains only edges incident to the corresponding q-vertex (all r- and s-vertices are independent of one another, and a p-vertex has only such neighbors, in addition to its q-vertex). By symmetry, this excludes q-vertices as well. The r-vertices have only six neighbors, so any induced H would correspond to a contradicted constraint R which by assumption does not exist (same argument as in Lemma 4). Furthermore, in the solution constructed, the neighborhood of any s-vertex contains only one edge, as D only deletes edges. The remaining types of vertices, i.e., the t, u, v, and w vertices of the gadget, have degree less than six.   □

5   Conclusion

We have presented a small graph H such that the H-free Edge Deletion problem and the H-free Edge Editing problem do not admit polynomial kernelizations under a reasonable complexity hypothesis. This answers an open question by Cai, namely, whether the $\mathcal{H}$-free Edge Deletion problem admits a polynomial kernelization for every finite set $\mathcal{H}$ of graphs.

It is an interesting open problem to further characterize kernelizability of $\mathcal{H}$-free edge modification problems. Considering the structure of the known positive examples (see introduction), one might first ask whether the $\mathcal{H}$-free Edge Deletion problem always admits a polynomial kernelization when $\mathcal{H}$ consists only of paths, cycles, and cliques (or sums thereof, such as $2K_2$). One particular interesting open case here is Cograph Deletion, i.e., P_4-free Edge Deletion.

References

1. Abu-Khzam, F.N.: Kernelization algorithms for d-hitting set problems. In: Dehne, et al. (eds.) [8], pp. 434–445
2. Bodlaender, H.L., Cai, L., Chen, J., Fellows, M.R., Telle, J.A., Marx, D.: IWPEC 2006. LNCS, vol. 4169. Springer, Heidelberg (2006)

3. Bodlaender, H.L., Downey, R.G., Fellows, M.R., Hermelin, D.: On problems without polynomial kernels (extended abstract). In: Aceto, L., Damgård, I., Goldberg, L.A., Halldórsson, M.M., Ingólfsdóttir, A., Walukiewicz, I. (eds.) ICALP 2008, Part I. LNCS, vol. 5125, pp. 563–574. Springer, Heidelberg (2008)
4. Bodlaender, H.L., Thomassé, S., Yeo, A.: Kernel bounds for disjoint cycles and disjoint paths. In: Fiat, A., Sanders, P. (eds.) ESA 2009. LNCS, vol. 5757, pp. 635–646. Springer, Heidelberg (2009)
5. Brügmann, D., Komusiewicz, C., Moser, H.: On generating triangle-free graphs. Electronic Notes in Discrete Mathematics 32, 51–58 (2009)
6. Burzyn, P., Bonomo, F., Durán, G.: NP-completeness results for edge modification problems. Discrete Applied Mathematics 154(13), 1824–1844 (2006)
7. Cai, L.: Fixed-parameter tractability of graph modification problems for hereditary properties. Information Processing Letters 58(4), 171–176 (1996)
8. Dehne, F., Sack, J.-R., Zeh, N. (eds.): WADS 2007. LNCS, vol. 4619. Springer, Heidelberg (2007)
9. Díaz, J., Thilikos, D.M.: Fast fpt-algorithms for cleaning grids. In: Durand, B., Thomas, W. (eds.) STACS 2006. LNCS, vol. 3884, pp. 361–371. Springer, Heidelberg (2006)
10. Downey, R.G., Fellows, M.R.: Parameterized Complexity. Monographs in Computer Science. Springer, Heidelberg (1998)
11. Flum, J., Grohe, M.: Parameterized Complexity Theory. Texts in Theoretical Computer Science. An EATCS Series. Springer, Heidelberg (2006)
12. Fortnow, L., Santhanam, R.: Infeasibility of instance compression and succinct PCPs for NP. In: STOC, pp. 133–142. ACM, New York (2008)
13. Guo, J.: Problem kernels for NP-complete edge deletion problems: Split and related graphs. In: Tokuyama, T. (ed.) ISAAC 2007. LNCS, vol. 4835, pp. 915–926. Springer, Heidelberg (2007)
14. Guo, J.: A more effective linear kernelization for cluster editing. Theoretical Computer Science 410(8-10), 718–726 (2009)
15. Guo, J., Uhlmann, J.: Kernelization and complexity results for connectivity augmentation problems. In: Dehne, et al [8], pp. 483–494
16. Khanna, S., Sudan, M., Trevisan, L., Williamson, D.P.: The approximability of constraint satisfaction problems. SIAM Journal on Computing. 30(6), 1863–1920 (2000)
17. Natanzon, A., Shamir, R., Sharan, R.: Complexity classification of some edge modification problems. Discrete Applied Mathematics 113(1), 109–128 (2001)
18. Rose, D.J.: A graph-theoretic study of the numerical solution of sparse positive definite systems of linear equations. Graph Theory and Computing, pp. 183–217 (1972)
19. Sharan, R., Maron-Katz, A., Shamir, R.: CLICK and EXPANDER: a system for clustering and visualizing gene expression data. Bioinformatics 19(14), 1787–1799 (2003)
20. Yap, C.-K.: Some consequences of non-uniform conditions on uniform classes. Theoretical Computer Science 26, 287–300 (1983)

On the Directed Degree-Preserving Spanning Tree Problem

Daniel Lokshtanov[1], Venkatesh Raman[2], Saket Saurabh[1], and Somnath Sikdar[2]

[1] The University of Bergen, Norway
{daniello,saket.saurabh}@ii.uib.no
[2] The Institute of Mathematical Sciences, India
{vraman,somnath}@imsc.res.in

Abstract. In this paper we initiate a systematic study of the REDUCED DEGREE SPANNING TREE (d-RDST) problem, where given a digraph D and a nonnegative integer k, the goal is to construct a spanning out-tree with at most k vertices of reduced out-degree. We show that this problem is fixed-parameter tractable and admits a problem kernel with at most $8k$ vertices on strongly connected digraphs and $O(k^2)$ vertices on general digraphs. We also give an algorithm for this problem on general digraphs with run-time $O^*(5.942^k)$. We also consider the dual of d-RDST: given a digraph D and a nonnegative integer k, construct a spanning out-tree of D with at least k vertices of full out-degree. We show that this problem is W[1]-hard on two important digraph classes: directed acyclic graphs and strongly connected digraphs.

1 Introduction

Given a directed graph $D = (V, A)$, we say that a subdigraph T of D is an *out-tree* if it is an oriented tree with exactly one vertex s of in-degree zero (called the *root*). An out-tree that contains all vertices of D is an *out-branching* of D. Given a digraph $D = (V, A)$ and an out-tree T of D, we say that a vertex $v \in V$ is of *full-degree* if its out-degree in T is the same as that in D; otherwise, v is said to be of *reduced-degree*. We define the DIRECTED REDUCED DEGREE SPANNING TREE (d-RDST) problem as follows.

Input: Given a directed graph $D = (V, A)$ and a positive integer k.
Parameter: The integer k.
Question: Does there exist an out-branching of D in which at most k vertices are of reduced degree?

We call the dual of this problem the DIRECTED FULL-DEGREE SPANNING TREE (d-FDST) problem and it is defined as follows.

Input: Given a directed graph $D = (V, A)$ and a positive integer k.
Parameter: The integer k.
Question: Does there exist an out-branching of D in which at least k vertices are of full degree?

J. Chen and F.V. Fomin (Eds.): IWPEC 2009, LNCS 5917, pp. 276–287, 2009.

d-RDST and d-FDST are natural generalizations of their undirected counterparts, namely, VERTEX FEEDBACK EDGE SET and FULL-DEGREE SPANNING TREE, respectively. In the VERTEX FEEDBACK EDGE SET problem, given a connected undirected graph $G = (V, E)$ and a non-negative integer k, the goal is to find an edge subset E' incident on at most k vertices such that the deletion of the edges in E' leaves a tree. Note that the resulting graph will span the entire vertex set. Bhatia et al. [3] show that this problem is MAX SNP-hard and describe a $(2 + \epsilon)$-approximation algorithm for it for any fixed $\epsilon > 0$. Guo et al. [9] show that this problem is fixed-parameter tractable by demonstrating a problem kernel with at most $4k$ vertices.

The FULL-DEGREE SPANNING TREE problem asks, given a connected undirected graph $G = (V, E)$ and a non-negative integer k as inputs, whether G has a spanning tree T in which at least k vertices have the same degree in T as in G. This was first studied by Pothof and Schut [12] in the context of water distribution networks and has attracted a lot of attention of late [4,5,3,9,8]. Bhatia et al. [4] studied this problem from the point-of-view of approximation algorithms and gave a factor-$\Theta(\sqrt{|V|})$ algorithm for it. They also showed that this problem admits no factor $O(|V|^{1/2-\epsilon})$ approximation algorithm unless NP $=$ co-R. For planar graphs, a polynomial-time approximation scheme (PTAS) was presented. Independently, Broersma et al. [5] developed a PTAS for planar graphs and showed that this problem can be solved in polynomial time in special classes of graphs such as bounded treewidth graphs and co-comparability graphs. Guo et al. [9] studied the parameterized complexity of this problem and showed it to be W[1]-hard. Gaspers et al. [8] give an $O^*(1.9465^{|V|})$ algorithm for this problem.

Organization of the Paper and Our Contribution. We start in Section 2 by defining relevant notions related to digraphs. In Section 3, we show that d-RDST is fixed-parameter tractable (FPT) by exhibiting a problem kernel with at most $O(k^2)$ vertices We first demonstrate that for strongly connected digraphs, d-RDST admits a kernel with at most $8k$ vertices and use the ideas therein to develop the $O(k^2)$ kernel for general digraphs. In Section 4, we describe a branching algorithm for the d-RDST problem with run-time $O^*(5.942^k)$. Our fixed parameter tractable and kernelization algorithms are sufficiently nontrivial and exploit structures provided by the problem in an elegant way. Finally in Section 5, we show that d-FDST is W[1]-hard on two classes of digraphs: directed acyclic graphs (DAGs) and strongly connected digraphs. This also proves that both d-RDST and d-FDST are NP-complete on these graph classes.

2 Preliminaries

The notation and terminology for digraphs that we follow are from [2]. Given a digraph D we let $V(D)$ and $A(D)$ denote the vertex set and arc set, respectively, of D. If $u, v \in V(D)$, we say that u is an *in-neighbour* (*out-neighbour*) of v if $(u, v) \in A(D)$ $((v, u) \in A(D))$. The in-degree $d^-(u)$ (out-degree $d^+(u)$) of u is the number of in-neighbours (out-neighbours) of u. Given a subset $V' \subseteq V(D)$, we let $D[V']$ denote the digraph induced on V'.

A digraph with no dicycles is called a *directed acyclic graph (DAG)*. A digraph D is *strongly connected* if for every pair of distinct vertices $u, v \in V(D)$, there exists a (u, v)-walk and a (v, u)-walk. A *strong component* of a digraph is a maximal induced subdigraph that is strongly connected. A strong component S of a digraph D is a *source strong component* if no vertex in S has an in-neighbour in $V(D) \backslash V(S)$. The following is a necessary and sufficient condition for a digraph to have an out-branching.

Proposition 1. [2] *A digraph D has an out-branching if and only if D has a unique source strong component.*

Therefore one can in time $O(|V(D)| + |A(D)|)$ check whether a digraph admits an out-branching or not. We assume that our graphs have no self-loops.

3   The d-RDST Problem: An $O(k^2)$-Vertex Kernel

In this section we show that d-RDST admits a problem-kernel with $O(k^2)$ vertices. We first consider the special case when the input digraph is strongly connected and establish a kernel with $8k$ vertices for this case. This will give some insight as to how to tackle the general case. The fact that d-RDST is NP-hard on the class of strongly connected digraphs follows from the fact that d-FDST is NP-hard on this class of graph (see Theorem 4).

3.1   A Linear Kernel for Strongly Connected Digraphs

We actually establish the $8k$-vertex kernel for a more general class of digraphs, those in which every vertex has out-degree at least one. It is easy to see that this class includes strongly connected digraphs (SCDs) as a proper subclass. Call this class of digraphs *out-degree at least one digraphs*.

A common technique to establish a problem-kernel is to devise a set of *reduction rules* which when applied to the input instance (in some specified sequence) produces the kernel. Recall that a *reduction rule* for a parameterized problem Q is a polynomial-time algorithm that takes an input (I, k) of Q and outputs either

1. YES or NO, in which case the input instance is a YES- or NO-instance, respectively, or
2. an "equivalent" instance (I', k') of Q such that $k' \leq k$.

Two instances are *equivalent* if they are both YES-instances or both NO-instances. An instance (I, k) of a parameterized problem Q is *reduced with respect to (wrt)* a set $\mathcal{R}$ of *reduction rules* if the instance (D', k') output by any reduction rule in $\mathcal{R}$ is the original instance (D, k) itself.

There are three simple reduction rules for the case where the input is an out-degree at least one digraph. We assume that the input is (D, k).

Rule 1. If there exists $u \in V(D)$ such that $d^-(u) \geq k + 2$ then return NO; else return (D, k).

Rule 2. If there are $k + 1$ vertices of out-degree at least $k + 1$ then return NO; else return (D, k).

Rule 3. (The Path Rule.) Let $x_0, x_1 \ldots, x_{p-1}, x_p$ be a sequence of vertices in D such that for $0 \leq i \leq p - 1$ we have $d^+(x_i) = 1$ and $(x_i, x_{i+1}) \in A(D)$. Let Y_0 be the set of in-neighbours of $x_1, \ldots, x_{p-1}$ and let $Y := Y_0 \setminus \{x_0, x_1, \ldots, x_{p-2}\}$. Delete the vertices $x_1, \ldots, x_{p-1}$ and add two new vertices z_1, z_2 and the arcs $(x_0, z_1), (z_1, z_2), (z_2, x_p)$. If $y \in Y$ has at least two out-neighbors in $\{x_1, \ldots, x_{p-1}\}$ then add arcs $(y, z_1), (y, z_2)$. If $y \in Y$ has exactly one out-neighbor in $\{x_1, \ldots, x_{p-1}\}$ then add the arc (y, z_1). Return (D, k).

If a vertex v has in-degree at least $k + 2$ then at least $k + 1$ in-neighbors of u must be of reduced degree in any out-branching. This shows that Rule 1 is a reduction rule. If a vertex u has out-degree $k + 1$ and is of full degree in some out-branching T then T has at least $k + 1$ leaves. Since the input digraph is such that every vertex has out-degree at least one, this means that in T there are at least $k + 1$ vertices of reduced degree. This shows that any vertex of out-degree $k+1$ must necessarily be of reduced degree in any solution out-branching. Therefore if there are $k + 1$ such vertices the given instance is a NO-instance. This proves that Rule 2 is a reduction rule.

Lemma 1. [⋆][1] *Rules 3 is a reduction rule for the d-RDST problem.*

It is easy to see that Rules 1 and 2 can be applied in $O(n)$ time and that Rule 3 can be applied in $O(n + m)$ time. Note that Rule 3 is *parameter independent*, that is, an application of the rule does not affect the parameter. Consequently, it also makes sense to talk about a digraph being reduced wrt Rule 3 as distinct from an instance (D, k) of d-RDST being reduced wrt Rule 3. Our kernelization algorithm consists in applying Rules 1 to 3 repeatedly, in that order, until the given instance is reduced.

We next describe a lemma that we repeatedly make use of in the sequel. Given a directed graph D, we let $V_i(D) \subseteq V(D)$ denote the set of vertices of out-degree i; $V_{\geq i}(D) \subseteq V(D)$ denotes the set of vertices of out-degree at least i.

Lemma 2. *Let D be a directed graph reduced wrt the Path Rule (Rule 3) and let T be an out-branching of D with root r such that X is the set of vertices of reduced out-degree. Then*

$$|V(T)| \leq 4|V_0(T) \cup V_{\geq 2}(T) \cup X| \leq 4(|V_0(T)| + |X \cup V_0(T)|).$$

Proof. If we view the out-branching T as an undirected graph, $V_0(T)$ is the set of leaves and $V_{\geq 2}(T)$ is the set of vertices of degree at least three along the root r, if $d_T^+(r) \geq 2$. Thus $V_{\geq 2}(T)$ has at most one vertex of total degree two and all other vertices are of total degree three. It is a well-known fact that a tree with l leaves has at most $l - 1$ internal vertices of degree at least three. Since $V_{\geq 2}(T)$ has at most one vertex of total degree two, we have $|V_{\geq 2}(T)| \leq |V_0(T)|$.

---

[1] Proofs labeled with a ⋆ have been omitted due to space constraints. For full proofs see [10].

Now consider the vertices of the out-branching T which have out-degree exactly one. Define $W := X \cup V_0(T) \cup V_{\geq 2}(T)$ and let $\mathcal{P}$ be the set of maximal dipaths in T such that for any dipath $P = x_0 \rightarrow x_1 \rightarrow \cdots \rightarrow x_p$ in $\mathcal{P}$ we have that (1) $d_D^+(x_i) = 1$ for $0 \leq i \leq p-1$, and (2) $x_p \in W$. Observe that every vertex with out-degree exactly one in T is contained in exactly one path in $\mathcal{P}$. Also observe that the set of vertices of out-degree exactly one in T *not* contained in W is precisely the set $V_1(T) \setminus X$. Therefore $|V_1(T) \setminus X| \leq \sum_{P \in \mathcal{P}}(|P| - 1)$, where $|P|$ denotes the number of *vertices* in the path P. By Rule 3, any dipath $P \in \mathcal{P}$ has at most four vertices and since the number of dipaths in $\mathcal{P}$ is at most $|W|$, we have $|V_1(T) \setminus X| \leq 3 \cdot |\mathcal{P}| \leq 3 \cdot |W| \leq 3|X \cup V_0(T) \cup V_{\geq 2}(T)|$. Since $|V(T)| \leq |V_1(T) \setminus X| + |X \cup V_0(T) \cup V_{\geq 2}(T)|$, we have $|V(T)| \leq 4|V_0(T) \cup V_{\geq 2}(T) \cup X| \leq 4(|V_0(T)| + |V_0(T) \cup X|)$. This completes the proof of the lemma. □

We can now bound the size of a yes-instance of the d-RDST problem on out-degree at least one digraphs that have been reduced wrt Rules 1 to 3.

Theorem 1. *Let (D, k) be a yes-instance of the d-RDST problem on out-degree at least one digraphs reduced wrt Rules 1 to 3. Then $|V(D)| \leq 8k$.*

Proof. Since (D, k) is a YES-instance of the problem, let T be an out-branching of D and let X be the set of vertices of reduced degree in T, where $|X| \leq k$. Every vertex of D is of out-degree at least one and hence $V_0 \subseteq X$, where V_0 is the set of leaves in T. Consequently $|X \cup V_0| \leq k$ and $|V_0| \leq k$ and by Lemma 2, we have $|V(T)| \leq 8k$, as claimed. □

Observe that the crucial step in the proof above was to bound the number of leaves in the solution out-branching. For out-degree at least one digraphs this is easy since every leaf is a vertex of reduced degree. This is not the case with general digraphs which may have an arbitrary number of vertices of out-degree zero, all of which are of full degree in any out-branching. In the next subsection we present a set of reduction rules for the d-RDST problem in general digraphs which help us bound the number of vertices of out-degree zero in terms of the parameter k.

3.2   An $O(k^2)$-Vertex Kernel in General Digraphs

For general digraphs, the reduction rules that we will describe work with an annotated instance of the problem. We assume that we are given (D, k) and a set $X \subseteq V(D)$ of vertices which will be of reduced degree in *any* out-branching with at most k vertices of reduced degree. The question in this case is to decide whether D admits an out-branching where the set of vertices of reduced degree is $X \cup S$, where $S \subseteq V(D) \setminus X$ and $|S| \leq k$. The parameter in this case is k. Call such an out-branching a *solution out-branching*. To obtain a kernel for d-RDST, we apply the reduction rules to an instance (D, k) after setting $X = \emptyset$.

Given an instance (D, X, k), we define the *conflict set* of a vertex $u \in V(D) \setminus X$ as $C(u) := \{v \in V(D) \setminus X : N^+(u) \cap N^+(v) \neq \emptyset\}$. Clearly vertices of out-degree zero have an empty conflict set. If a vertex v has a non-empty conflict set then in any out-branching either v loses its degree or *every* vertex in $C(v)$ loses its degree. Moreover if $u \in C(v)$ then $v \in C(u)$ and in this case we say that u and v are in conflict. The *conflict number* of D is defined as $c(D) := \sum_{v \in V(D) \setminus X} |C(v)|$.

We assume that the input instance is (D, X, k) and the kernelization algorithm consists in applying each reduction rule repeatedly, in the order given below, until no longer possible. Therefore when we say that Rule i is indeed a reduction rule we assume that the input instance is reduced wrt the rules preceding it.

Rule 1. If there exists $u \in V(D)$ such that the number of in-neighbors of u in $V(D) \setminus X$ is at least $k + 2$ then return NO; else return (D, X, k).

In the last subsection, we already showed that this rule is indeed a reduction rule.

Rule 2. If $u \in V(D) \setminus X$ such that $|C(u)| > k$, set $X \leftarrow X \cup \{u\}$ and $k \leftarrow k - 1$. Furthermore if $d^+(u) = 1$ then delete the out-arc from u and return (D, X, k).

If the conflict set $C(u)$ of $u \in V(D) \setminus X$ is of size at least $k + 1$ and if u is of full degree in some out-branching T, then every vertex in $C(u)$ must be of reduced degree in T. Therefore if (D, X, k) is a YES-instance then u must lose its degree in *any* solution out-branching. Moreover if u has out-degree exactly one, then the out-arc from u must be deleted. This shows that Rule 2 is a reduction rule.

Rule 3. If $c(D) > 2k^2$ then return NO, else return (D, X, k).

Lemma 3. *Rule 3 is a reduction rule for the d-RDST problem.*

Proof. To see why Rule 3 qualifies to be a reduction rule, construct the *conflict graph* $\mathcal{C}_{D,X}$ of the instance (D, X, k) which is defined as follows. The vertex set $V(\mathcal{C}_{D,X}) := V(D) \setminus X$ and two vertices in $V(\mathcal{C}_{D,X})$ have an edge between them if and only if they are in conflict. Since the size of the conflict set of any vertex is at most k, the degree of any vertex in $\mathcal{C}_{D,X}$ is at most k. The key observation is that if T is any solution out-branching of (D, X, k) in which the set of vertices of reduced degree is $X \cup S$ with $S \subseteq V(D) \setminus X$, then S forms a vertex cover of $\mathcal{C}_{D,X}$. Since we require that $|S| \leq k$, the number of edges in $\mathcal{C}_{D,X}$ is at most k^2. For a vertex $v \in V(D) \setminus X$, let $d'(v)$ be the number of neighbors of vertex v in the conflict graph $\mathcal{C}_{D,X}$. Observe that $c(D) := \sum_{v \in V(D) \setminus X} |C(v)| = \sum_{v \in V(D) \setminus X} d'(v) \leq 2k^2$. The last inequality follows from the fact that sum of degrees of vertices in a graph is equal to twice the number of edges. $\square$

Rule 4. If $u \in V(D)$ such that $d^+(u) = 0$ and $d^-(u) = 1$ then delete u from D and return (D, X, k).

It is easy to see that Rule 4 is a reduction rule: vertex u does not determine whether its parent is of full or reduced degree in a solution out-branching and therefore can be safely deleted.

Rule 5. Let $u \in V(D)$ be of out-degree zero and let $v_1, \ldots, v_r$ be its in-neighbors, where $r > 2$. Delete u and add $\binom{r}{2}$ new vertices $u_{12}, u_{13}, \ldots,$ $u_{r-1,r}$; for a newly added vertex u_{ij} add the arcs (v_i, u_{ij}) and (v_j, u_{ij}). Return (D, X, k).

Note that vertex u forces at least $r - 1$ vertices from $\{v_1, \ldots, v_r\}$ to be of reduced degree in any out-branching. This situation is captured by deleting u and introducing $\binom{r}{2}$ vertices as described in the rule. The upshot is that each vertex of out-degree zero has in-degree exactly two.

Rule 6. If $u, v \in V(D) \setminus X$ have $p > 1$ common out-neighbors of out-degree zero, delete all but one of them. Return (D, X, k).

Rule 7. If $u \in V(D)$ is of out-degree zero such that at least one in-neighbor of u is in X, delete u. Return (D, X, k).

By Rule 5, it is clear that if $u, v \in V(D) \setminus X$ have at least two common out-neighbors of out-degree zero then these out-neighbors have in-degree exactly two. It is intuitively clear that these out-neighbors are equivalent in some sense and it suffices to preserve just one of them. It is easy to show that the original instance has a solution out-branching if and only if the instance obtained by one application of Rule 6 has a solution out-branching. As for Rule 7, if u has in-neighbors v and w and if $v \in X$, we can delete the arc (v, u) without altering the solution structure. But then v is a private neighbor of w of out-degree zero and hence can be deleted by Rule 4.

Rule 8. This is the Path Rule (Rule 3) from Section 3.1.

In the previous subsection, this was shown to be a reduction rule for the d-RDST problem (note that the proof of Lemma 1 did not use the fact that the input was an out-degree at least one digraph). By Rule 2, no vertex on the path $x_0, x_1, \ldots, x_{p-1}$ is in X and therefore the proof of Lemma 1 continues to hold for the annotated case as well.

Note that a single application of Rule 5 or 6 takes time $O(n^2)$; all other rules take time $O(n + m)$. We are now ready to bound the number of vertices of out-degree zero in a reduced instance of the annotated problem.

Lemma 4. *Let (D, X, k) be a yes-instance of the annotated d-RDST problem that is reduced wrt Rules 1 through 8 mentioned above. Then the number of vertices of out-degree zero in D is at most k^2.*

Proof. Let u be a vertex of out-degree zero. By Rules 4 and 5, it must have exactly two in-neighbors, say, x and y. By Rule 7, neither x nor y is in X and are therefore still in conflict in the reduced graph. Hence, either x or y must be of reduced degree in any solution out-branching. Furthermore any vertex not in X can have at most k out-neighbors of out-degree zero since, by Rule 2, any vertex not in X is in conflict with at most k other vertices and, by Rule 6, two vertices in conflict have at most one out-neighbor of out-degree zero. Since (D, X, k) is assumed to be a YES-instance, at most k vertices can lose their out-degree in any solution out-branching. Moreover any vertex of out-degree zero is an out-neighbor of at least one vertex of reduced degree. Therefore the total number of vertices of out-degree zero is at most k^2.   □

Lemma 5. *Let (D, k) be a* YES-*instance of the d-RDST problem and suppose that (D_1, X, k_1) is an instance of the annotated d-RDST problem reduced wrt Rules 1 through 8 by repeatedly applying them on (D, k), by initially setting $X = \emptyset$. Then $|V(D_1)| \leq 8(k^2 + k)$.*

Proof. If T_1 is a solution out-branching of (D_1, X, k_1), then a leaf of T_1 is either a vertex of out-degree zero in D_1 or a vertex of reduced degree. By Lemma 4, the total number of vertices of out-degree zero in D_1 is at most $k_1^2 \leq k^2$ and the total number of vertices of reduced degree in T_1 is at most $k_1 + |X| \leq k$. Thus the number of leaves of T_1 is at most $k^2 + k$ and by Lemma 2 we have $|V(T_1)| \leq 4(k^2 + k + k^2 + k) = 8(k^2 + k)$.   □

We now show how to obtain a kernel for the original (unannotated) version of the problem. Let (D, k) be an instance of the d-RDST problem and let (D', X, k') be the instance obtained by applying reduction rules 1 through 8 on (D, k) until no longer possible, by initially setting $X = \emptyset$. By Lemma 5, we know that $|V(D')| \leq 5k^2 + 9k$ if (D, k) is a YES-instance, and that $k' + |X| = k$. To get back an instance of the unannotated version, apply the following transformation on (D', X, k'). Let $X = \{x_1, \ldots, x_r\}$. For each $x_i \in X$ add $k + 1$ new vertices $z_{i1}, \ldots, z_{i,k+1}$ and out-arcs (x_i, z_{ij}) for all $1 \leq j \leq k + 1$. Then add $k + 1$ new vertices $u_1, \ldots, u_{k+1}$ and out-arcs (u_j, z_{ij}) for all $1 \leq i \leq r$ and $1 \leq j \leq k + 1$. Finally add a vertex u and out-arcs (u, u_j) for $1 \leq j \leq k + 1$ and (x_i, u) for $1 \leq i \leq r$. Call the resulting digraph D'' and set $k'' = k' + |X|$. We show that (D', X, k') is a YES-instance of the annotated version of d-RDST if and only if (D'', k'') is a YES-instance of the (unannotated) d-RDST problem.

If D' has an out-branching T' with at most $k' + |X|$ vertices of reduced degree with all vertices in X of reduced degree, then modify T' into an out-branching T'' for D'' as follows. Add arcs (x_1, u), (u, u_j) for all $1 \leq j \leq k + 1$ and (u_j, z_{ij}) for all i, j. Clearly T'' is an out-branching of D'' and has the same number of vertices of reduced degree as T'. Conversely if D'' admits an out-branching T'' with $k' + |X|$ vertices of reduced degree, then it must be the case that all vertices in X are of reduced degree in T''. For if $x_i \in X$ is of full-degree then the vertices $u_1, \ldots, u_{k+1}$ are of reduced degree, contradicting the fact that T'' has at most $k' + |X| = k$ vertices of reduced degree. Therefore in T'', we may assume that the vertices $u, u_1, \ldots, u_{k+1}$ and the z_{ij}'s are of full-degree. This implies that in T'' there are at most k' vertices from $V(D') \setminus X$ of reduced degree. Furthermore in T'', the vertex u has as in-neighbor a vertex $x_i \in X$. Therefore by deleting $u, u_1, \ldots, u_{k+1}$ and the z_{ij}'s, we obtain an out-branching T' with at most $k' + |X|$ vertices of reduced degree with all vertices of X of reduced degree. This completes the proof of the reduction from the annotated to the unannotated case.

Since we add at most $k(k + 1) + k + 2$ vertices in the process, we have

Theorem 2. *The d-RDST problem admits a problem kernel with at most $6k^2 + 11k + 2$ vertices, where the parameter k is the number of vertices of reduced degree.*

4 An Algorithm for the d-RDST Problem

In this section we describe a branching algorithm for the d-RDST problem with run-time $O^*(5.942^k)$. We first observe that in order to construct a solution out-branching of a given digraph, it is sufficient to know which vertices will be of reduced degree.

Lemma 6. [⋆] *Let $D = (V, A)$ be a digraph and let X be the set of vertices of reduced degree in some out-branching of D. Given D and X, one can in polynomial time construct an out-branching of D in which the set of vertices of reduced degree is a subset of X.*

Proof. (Sketch) One can construct the desired out-branching as follows: Run through all possible choices of the root. For each choice of root, perform a breadth-first search starting at the root and greedily add out-arcs from the vertices of X to the out-branching. □

By Lemma 6 and Theorem 2, there exists an $O^*(k^{O(k)})$ algorithm for the d-RDST problem. In the rest of this section, we give an improved algorithm with run-time $O^*(c^k)$, for a constant c. Our algorithm (see Figure 1) is based on the simple observation that if two vertices u and v of the input digraph D have a common out-neighbor then one of them must be of reduced degree in *any* out-branching

---

RDST (D, X, k)

Input: A digraph $D = (V, A)$; $X \subseteq V$, such that the vertices in X will be of reduced degree in the out-branching that is being constructed; an integer parameter k. The algorithm is initially called after setting $X = \emptyset$.

Output: An out-branching of D in which every vertex of X is of reduced degree and with at most k vertices of reduced degree in total, if one exists, or NO, signifying that no such out-branching exists.

1. If $k < 0$ or $|X| > k$ return NO.
2. If no two vertices in $V(D) \setminus X$ have a common out-neighbor then
   (a) Reduce (D, X, k) wrt Rules $1'$ through $5'$.
   (b) For each $(k - |X|)$-sized subset Y of $V(D) \setminus X$, check if there exists an out-branching of D in which the vertex set of reduced degree is $X \cup Y$. If yes, then "expand" this out-branching to an out-branching for the original instance and return the solution; else return NO.
3. Let $u, v \in V(D) \setminus X$ be two vertices with a common out-neighbor then
   (a) $X \leftarrow X \cup \{u\}$; $Z = $ Call **RDST**$(D, X, k - 1)$.
   (b) If $Z \neq$ NO then return Z.
   (c) $X \leftarrow X \cup \{v\}$; Return **RDST**$(D, X, k - 1)$.

Fig. 1. Algorithm **RDST**

of D. The algorithm recurses on vertex-pairs that have a common out-neighbor and, along each branch of the recursion tree, builds a set X of vertices which would be the candidate vertices of reduced degree in the out-branching that it attempts to construct. When there are no vertices to branch on, it reduces the instance (D, X, k) wrt the following rules.

Rule 1'. If $u \in X$ and $d^+(u) = 1$, delete the out-arc from u and return (D, X, k).

Rule 2'. Let $u \in V(D)$ be of out-degree zero and let $v_1, \ldots, v_r$ be its in-neighbors. If $v_i \in X$ for all $1 \leq i \leq r$, assign v_1 as the parent of u and delete u. If there exists $1 \leq i \leq r$ such that $v_i \notin X$ then assign v_i as the parent of u and delete u. Return (D, X, k).

Rule 3'. This is Rule 8 from Section 3.2.

Rule 1' is a reduction rule because a vertex of out-degree exactly one that is of reduced degree must necessarily lose its only out-arc. As for Rule 2', we know that in the instance (D, X, k) obtained after the algorithm finishes branching, no two vertices of $V(D) \setminus X$ have a common out-neighbor and therefore at least $r - 1$ in-neighbors of u must be of reduced degree. If all in-neighbors of u are of reduced degree, we arbitrarily fix one of them as parent of u (so that we can construct an out-branching of the original instance later on) and delete u. If exactly $r - 1$ in-neighbors of u are already of reduced degree, we choose that in-neighbor not in X as the parent of u and delete u. Also note that when applying Rule 3' to a path $x_0, x_1, \ldots, x_{p-1}, x_p$, the vertices $x_0, x_1, \ldots, x_{p-1}$ are not in X, by Rule 1'. Therefore if Y is the set of in-neighbors of $x_1, \ldots, x_{p-1}$, excluding $\{x_0, x_1, \ldots, x_{p-2}\}$, then $Y \subseteq X$.

Observe the following:

1. By Rule 2', no vertex in the reduced instance (D, X, k) has out-degree zero.
2. Every vertex in the subdigraph induced by $V(D) \setminus X$ has in-degree at most one and hence each connectivity component (a connected component in the undirected sense) is either a dicycle, or an out-tree or a dicycle which has out-trees rooted at its vertices. Thus each connectivity component has at most one dicycle and if a component does have a dicycle then it can be transformed into an out-branching by deleting an arc from the cycle.

We now reduce the instance (D, X, k) wrt the following two rules:

Rule 4'. If at least $k + 1 - |X|$ connectivity components of $D[V \setminus X]$ contain dicycles, then return NO; else return (D, X, k).

Rule 5'. If a connectivity component of $D[V \setminus X]$ is a dicycle C such that no vertex in $V(C)$ has an out-neighbor in X, pick a vertex $u \in X$ with an arc to C and fix it as the "entry point" to C; delete C and set $k \leftarrow k - 1$; return (D, X, k).

Rule 4' is a reduction rule as every connectivity component that has a dicycle contains at least one vertex that will be of reduced degree. If the number of such components if at least $k + 1 - |X|$, one cannot construct an out-branching with at most k vertices of reduced degree. To see that Rule 5' is a reduction rule, first note that since C has no out-arcs, it cannot contain the root of the proposed out-branching. Any path from the root to C must necessarily include a vertex from X and it does not matter which arc out of X we use to get to C, since every vertex in X is of reduced degree anyway. Moreover in any out-branching, exactly one vertex of C must be of reduced degree. Therefore if (D', X', k') is the instance obtained by one application of Rule 5' to the instance (D, X, k), then it is easy to see that these instances must be equivalent. Also note that each application of Rule 1' through 5' takes time $O(n + m)$.

Lemma 7. *Let (D, X, k) be an instance of the d-RDST problem in which no two vertices of $V(D) \setminus X$ have a common out-neighbor, and reduced wrt Rules 1' through 5'. Then $|V(D) \setminus X| \leq 7|X|$.*

Proof. Let D' be a digraph obtained from D by deleting all out-arcs from the vertices in X. Therefore in D', every vertex of X has out-degree zero and in-degree at most one. We show that a connectivity component of D' that has p vertices of X has at most $7p$ vertices of $V(D') \setminus X$. This will prove the lemma.

If a connectivity component of D' is an out-tree T', then every leaf of this out-tree is a vertex of X. If T' has p leaves, then applying Lemma 2 to T', we have that $|V(T')| \leq 8p$. Since exactly p of these vertices are from X, the number of vertices of $V(D') \setminus X$ in the out-tree is at most $7p$. Therefore let R be a connectivity component of D' containing a dicycle such that $|V(R) \cap X| = p$. Then R has exactly one dicycle, say C. By Rule 5', C has a vertex x with an out-neighbor in $V(R) \setminus V(C)$, and therefore $d_R^+(x) \geq 2$. Let y be the out-neighbor of x in C. Delete the arc (x, y) to obtain an out-branching T with root x. Note that the number of leaves in T is the same as that in R. Moreover in transforming R to T, only one vertex loses its out-degree. By Lemma 2,

$$|V(T)| \leq 4|x \cup V_0(T) \cup V_{\geq 2}(T)| \leq 4|V_0(T)| + 4|x \cup V_{\geq 2}(T)|,$$

and since $|x \cup V_{\geq 2}(T)| \leq 1 + (|V_0(T)| - 1) = p$, we have $|V(T)| \leq 8p$. Consequently $|V(R) \setminus X| \leq 7p$. $\qquad\square$

To construct an out-branching, it is sufficient to choose the remaining $k - |X|$ vertices of reduced degree from the vertices in $V(D) \setminus X$. Setting $|X| = c$, the exponential term in the run-time of the algorithm is bounded above by the function

$$\sum_{c=0}^{k} 2^c \cdot \binom{7c}{k-c} \leq k \cdot \max_{0 \leq c \leq k} 2^c \cdot \binom{7c}{k-c}.$$

In the latter function, we must have $k - c \leq 7c$ which implies that $k/8 \leq c$, and one can show that this function attains a maximum at $c = k/2$ where its value is $k \cdot 2^{k/2} \cdot \binom{7k/2}{k/2}$. Using the inequality $\binom{n}{r} \leq n^n / (r^r \cdot (n-r)^{n-r})$, we can bound this by $k \cdot 5.942^k$.

Theorem 3. *Given a digraph D and a nonnegative integer k, one can decide whether D has an out-branching with at most k vertices of reduced degree, and if so, construct such an out-branching in time $O^*(5.942^k)$.*

5   The *d*-FDST Problem

We now show that *d*-FDST is W[1]-hard even on DAGs. This is a modification of the reduction presented in [3] (Lemma 3.2).

Theorem 4. [⋆] *The d-FDST problem parameterized by the solution size is W[1]-hard on directed acyclic graphs and strongly connected digraphs. Also the d-RDST problem is NP-hard on the class of strongly connected digraphs.*

References

1. Arnborg, S., Lagergren, J., Seese, D.: Easy Problems for Tree-Decomposable Graphs. Journal of Algorithms 12, 308–340 (1991)
2. Bang-Jensen, J., Gutin, G.: Digraphs: Theory, Algorithms and Applications. Springer, Heidelberg (2000)
3. Bhatia, R., Khuller, S., Pless, R.: On Local Search and Placement of Meters in Networks. SIAM Journal on Computing 32(2), 470–487 (2003)
4. Bhatia, R., Khuller, S., Pless, R., Sussmann, Y.: The Full-Degree Spanning Tree Problem. Networks 36, 203–209 (2000)
5. Broersma, H.J., Huck, A., Kloks, T., Koppios, O., Kratsch, D., Müller, H., Tuinstra, H.: Degree-Preserving Forests. Networks 35, 26–39 (2000)
6. Downey, R.G., Fellows, M.R.: Parameterized Complexity. Springer, Heidelberg (1999)
7. Flum, J., Grohe, M.: Parameterized Complexity Theory. Springer, Heidelberg (2006)
8. Gaspers, S., Saurabh, S., Stepanov, A.A.: Moderately Exponential-Time Algorithm for Full-Degree Spanning Tree. In: Agrawal, M., Du, D.-Z., Duan, Z., Li, A. (eds.) TAMC 2008. LNCS, vol. 4978, pp. 478–489. Springer, Heidelberg (2008)
9. Kára, J., Kratochvíl, J.: Fixed Parameter Tractability of Independent Set in Segment Intersection Graphs. In: Bodlaender, H.L., Langston, M.A. (eds.) IWPEC 2006. LNCS, vol. 4169, pp. 166–174. Springer, Heidelberg (2006)
10. Lokshtanov, D., Raman, V., Saurabh, S., Sikdar, S.: On the Directed Degree-Preserving Spanning Tree Problem. Tech. Report
11. Niedermeier, R.: An Invitation to Fixed-Parameter Algorithms. Oxford University Press, Oxford (2006)
12. Pothof, I.W.M., Schut, J.: Graph-Theoretic Approach to Identifiability in a Water Distribution Network. Memorandum 1283, Universiteit Twente, Twente, The Netherlands (1995)

Even Faster Algorithm for SET SPLITTING!

Daniel Lokshtanov[1] and Saket Saurabh[2]

[1] Department of Informatics, University of Bergen, N-5020 Bergen, Norway
daniello@ii.uib.no
[2] The Institute of Mathematical Sciences, C.I.T. Campus, Chennai 600 113
saket@imsc.res.in

Abstract. In p-SET SPLITTING we are given a universe U, a family $\mathcal{F}$ of subsets of U and a positive integer k and the objective is to find a partition of U into W and B such that there are at least k sets in $\mathcal{F}$ that have non-empty intersection with both B and W. In this paper we study p-SET SPLITTING from kernelization and algorithmic view points. Given an instance $(U, \mathcal{F}, k)$ of p-SET SPLITTING, our kernelization algorithm obtains an equivalent instance with at most $2k$ sets and k elements in polynomial time. Finally, we give a fixed parameter tractable algorithm for p-SET SPLITTING running in time $O(1.9630^k + N)$, where N is the size of the instance. Both our kernel and our algorithm improve over the best previously known results. Our kernelization algorithm utilizes a classical duality theorem for a connectivity notion in hypergraphs. We believe that the duality theorem we make use of, will turn out to be an important tool from combinatorial optimization in obtaining kernelization algorithms.

1   Introduction

In the MAX CUT problem we are given a graph G with vertex set $V(G)$ and edge set $E(G)$ and asked to find a partitioning of $V(G)$ into W (*white*) and B (*black*) such that the number of edges with one endpoint in W and one in B is maximized. The MAX CUT problem is one of Karp's 21 NP-hard problems [13] and also the first problem for which an approximation algorithm using semi-definite programming was obtained [12]. The problem has also been studied from the viewpoint of parameterized algorithms [16,18].

A natural generalization of MAX CUT to *hypergraphs* is the SET SPLITTING problem, also known as MAX HYPERGRAPH 2-COLORING. A *hypergraph* $H = (\mathcal{V}, \mathcal{E})$ consists of a vertex set $\mathcal{V}$ and a set $\mathcal{E}$ of hyperedges. A *hyperedge* $e \in \mathcal{E}$ is a subset of the vertex set $\mathcal{V}$. By $V(e)$ we denote the subset of vertices corresponding to the edge e. In the SET SPLITTING problem we are given a family $\mathcal{F}$ of sets over a universe U. We say that a partitioning (W, B) of U *splits* a set $S \in \mathcal{F}$ if $S \cap W \neq \emptyset$ and $S \cap B \neq \emptyset$. The objective is to partition U into W and B such that the number of sets in $\mathcal{F}$ that are split is maximized. If the SET SPLITTING instance $(U, \mathcal{F})$ is viewed as a hypergraph $H = (U, \mathcal{F})$ the objective is to color the vertices of H black or white, maximizing the number of hyperedges containing at least one white and at least one black vertex. It

J. Chen and F.V. Fomin (Eds.): IWPEC 2009, LNCS 5917, pp. 288–299, 2009.

Table 1. List of known results about p-SET SPLITTING in chronological order. The row marked with $\star$ represents result in the current article.

History of p-Set Splitting			
Dehne, Fellows and Rosamond	WG 2003	$O(72^k N^{O(1)})$	Deterministic
Dehne, Fellows, Rosamond and Shaw	IWPEC 2004	$O(8^k N^{O(1)})$	Deterministic
Lokshtanov and Sloper	ACiD 2005	$O(2.6499^k N^{O(1)})$	Deterministic
Chen and Lu	COCOON 2007	$O(2^k + N)$	Randomized
Lokshtanov and Saurabh*	2009	$O(1.96^k + N)$	Deterministic

should be noted that MAX CUT is the special case of SET SPLITTING when all sets in $\mathcal{F}$ have cardinality 2. The SET SPLITTING and (MAX) HYPERGRAPH 2-COLORING problems have been studied intensively from a combinatorial as well as an algorithmic viewpoint [1,2,5,6,7,15,17,21,22].

We study SET SPLITTING from the parameterized algorithms perspective. In parameterized algorithms every instance x comes with a *parameter* k and an algorithm for the problem with running time $f(k)n^{O(1)}$ is said to be *fixed parameter tractable*. Formally a parameterized problem Π is a subset of $\Gamma^* \times \mathbb{N}$ for some finite alphabet Γ and an instance of the problem consists of (x, k), where k is the parameter. The problem Π is said to admit a $g(k)$ *kernel* if there is a polynomial time algorithm that transforms any instance (x, k) to an equivalent instance (x', k') such that $|x'| \leq g(k)$ and $k' \leq g(k)$. If $g(k) = k^{O(1)}$ or $g(k) = O(k)$ we say that Π admits a polynomial kernel and linear kernel respectively. We remark that for most kernels, and in particular all kernels mentioned in this article k' is in fact bounded by k. In the parameterized version of MAX CUT, called p-MAX CUT the input is a graph G and an integer k and the objective is to partition $V(G)$ into W and B such that at least k edges have one white and one black endpoint. Similarly, in p-SET SPLITTING an input instance is a family $\mathcal{F}$ of sets over a universe U and an integer k. The objective is to find a partitioning (W, B) of U that splits at least k sets. Throughout this paper we denote the size of an instance $(U, \mathcal{F}, k)$ of p-SET SPLITTING by N.

Related Work. The fastest known parameterized algorithm for the p-MAX CUT problem has running time $O(1.2418^k + |V(G)| + |E(G)|)$ [18] and the smallest kernel has $2k$ vertices and k edges [16,18]. In fact, bounding the number of vertices by k is easy - any connected graph G has a spanning tree with $|V(G)|-1$ edges. Since trees are bipartite we can partition $V(G)$ into (W, B) such that all edges in the spanning tree have one endpoint in W and one in B. Hence, if $|V(G)| - 1 \geq k$ we can immediately answer yes. This immediately yields a $O(2^k + |V(G)| + |E(G)|)$ time algorithm for the problem.

On the other hand, until this work, no deterministic algorithm with running time $O(2^k + N)$ was known for p-SET SPLITTING, even though the problem is quite well-studied in parameterized algorithms. Dehne, Fellows and Rosamond [4] initiated the study of p-SET SPLITTING and gave an algorithm running in time $O(72^k N^{O(1)})$. They also provided kernel for the problem with at most

$2k$ sets in the family. Later Dehne, Fellows, Rossmand and Shaw [5] obtained an algorithm with running time $O(8^k N^{O(1)})$. Continuing this chain of improvement Lokshtanov and Sloper [14] gave an algorithm with running time $O(2.65^k N^{O(1)})$ and obtained a kernel with both universe size and family size at most $2k$. Finally, Chen and Lu [2] provided a randomized algorithm with running time $O(2^k + N)$ for a weighted version of problem. We refer to Table 1 for a quick reference on the history of the p-SET SPLITTING problem.

Our Results. The first part of this article is devoted to generalizing the simple kernelization algorithm for p-MAX CUT to hypergraphs and giving a kernel with at most $2k$ sets and k elements for the p-SET SPLITTING problem. To this end, we make a detour and introduce notions of *spanning trees* and *strong cut-sets* in a hypergraph. The purpose of these notions is to be able to generalize the statement "every connected graph has a spanning tree" to "every hypergraph without a strong cut-set has a spanning tree". Making this generalization turned out to be non-trivial and required using an interesting duality theorem for a connectivity notion in hypergraphs.

Theorem 1. *p-SET SPLITTING admits a kernel with $2k$ sets and k elements.*

On the face of it Theorem 1 could look like a simple improvement over the previous known kernel with $2k$ sets and $2k$ elements but it is not. Observe that Theorem 1 yields as a corollary the *fastest* known deterministic algorithm for p-SET SPLITTING running in time $O(2^k + N)$. In the last section of this article we break the "2^k barrier" and give a $O(1.9630^k + N)$ time algorithm for problem using memoization and the *Measure & Conquer* paradigm.

Theorem 2. *There is an $O(1.9630^k + N)$ time algorithm for the p-SET SPLITTING problem*

The *Measure & Conquer* paradigm has been extensively applied to obtain faster exact exponential time algorithms. We refer to [9,10] for a reference on *Measure & Conquer*. Even though *Measure & Conquer* has been applied to several problems to obtain exact exponential time algorithms, its applicability in obtaining parameterized algorithm has been limited to an algorithm for 3-HITTING SET by Wahlström [20]. Our fixed parameter algorithm for p-SET SPLITTING provides another example of application of *Measure & Conquer* in parameterized algorithms.

Throughout this paper for an undirected graph G by $V(G)$ we denote its vertex set and by $E(G)$ we denote its edge set. For a subset $V' \subseteq V(G)$, by $G[V']$ we mean the subgraph of G induced on V'.

2   Kernelization Algorithm

In this section we first give an algorithmic version of a classical duality theorem for a connectivity notion in hypergraphs. Next, we use this duality result to get a kernel for p-SET SPLITTING with at most $2k$ sets and k elements.

2.1   A Duality Theorem for Hypergraph Connectivity

We begin with a few definitions related to hypergraphs. With every hypergraph $H = (\mathcal{V}, \mathcal{E})$ we can associate the following graph: The *primal graph*, also called the *Gaifmann graph*, $P(H)$ has the same vertices $\mathcal{V}$ as H and, two vertices $u, v \in \mathcal{V}$ are connected by an edge in $P(H)$ if there is a hyperedge $e \in \mathcal{E}$, such that $\{u, v\} \subseteq V(e)$. We say that H is *connected* or has r components if the corresponding primal graph $P(H)$ is connected or has r components. Now we define the notions of *strong cut-sets* and forests in hypergraphs.

Definition 1 (Strong Cut-Set). *A subset $X \subseteq \mathcal{E}$ is called a strong cut-set if the hypergraph $H' = (\mathcal{V}, \mathcal{E} \setminus X)$ has at least $|X| + 2$ connected components.*

Definition 2 (Hypergraph Forest). *A forest $\mathcal{F}$ of a hypergraph H is a pair (F, g) where F is a forest on the vertex set $\mathcal{V}$ with edge set $E(F)$ where F is a forest in normal graph sense and $g : E(F) \to \mathcal{E}$ is an injective map such that for every $uv \in E(F)$ we have $\{u, v\} \subseteq V(g(uv))$. The number of edges in $\mathcal{F}$ is $|E(F)|$.*

Observe that if a forest $\mathcal{F}$ has $|\mathcal{V}| - 1$ edges then F is a spanning tree on $\mathcal{V}$. In this case we say that $\mathcal{F}$ is a *spanning tree* of H. Frank, Király, and Kriesell proved the following duality result relating spanning trees and strong cut-sets in hypergraphs [11, Corollary 2.6].

Proposition 1 ([11]). *A hypergraph H contains a hypertree if and only if H does not have a strong cut-set.*

We give an algorithmic version of Proposition 1 in Theorem 3 which is central to our kernelization algorithm. We start with a few observations about forest in hypergraphs and a definition useful for the proof of Theorem 3. Given a forest $\mathcal{F} = (F, g)$ we classify the edges of $\mathcal{E}$ as follows. An edge $e \in \mathcal{E}$ is

- a *forest edge* if there exists an edge f in $E(F)$ such that $g(f) = e$;
- a *cut edge* if there exist two connected components C_1 and C_2 of F such that $V(e) \cap V(C_1) \neq \emptyset$ and $V(e) \cap V(C_2) \neq \emptyset$.
- an *unused edge* if there does not exist an edge f in $E(F)$ such that $g(f) = e$; that is e is not in the image of the map g.

We remark that an edge e can be a forest edge as well as a cut edge at the same time. Similarly an edge can be a cut edge as well as an unused edge at the same time.

Definition 3. *For a hypergraph $H = (\mathcal{V}, \mathcal{E})$, a forest $\mathcal{F} = (F, g)$ and $e_1, e_2 \in \mathcal{E}$, we say that an edge e_2 follows e_1 if e_1 is a forest edge of F and e_2 is a cut edge with respect to $\mathcal{F}' = (F', g')$ where $F' = (\mathcal{V}, E(F) \setminus \{g^{-1}(e_1)\})$ and $g'(f) = g(f)$ for $f \in E(F')$.*

We are now in position to state an algorithmic version of Proposition 1 which will be used in our kernelization algorithm.

Theorem 3. [1] *There is a polynomial time algorithm that given a connected hypergraph $H = (\mathcal{V}, \mathcal{E})$ and a forest $\mathcal{F} = (F, g)$ of H such that $|E(F)| < |\mathcal{V}| - 1$ finds either a forest $\mathcal{F}' = (F', g')$ of H with $|E(F')| \geq |E(F)| + 1$ or a strong cut-set X of H.*

2.2   Kernel for Set Splitting

In this section we show how to utilize Theorem 3 to give a kernel with $2k$ sets and k elements for the p-SET SPLITTING problem. Since Theorem 3 is phrased in terms of hypergraphs, it is useful to view the p-SET SPLITTING instance $(U, \mathcal{F}, k)$ as a hypergraph $H = (U, \mathcal{F})$ and integer k. We start by showing that if H contains a strong cut-set, then the instance $(U, \mathcal{F}, k)$ can be reduced.

Definition 4. *Let $f : \mathcal{V} \to \{0, 1\}$ be a function from set of vertices of the hypergraph H to the set $\{0, 1\}$. Then $Split(f)$ is the set of hyperedges such that for every hyperedge $e \in Split(f)$ there exist vertices $u, v \in V(e)$ such that $f(u) = 0$ and $f(v) = 1$.*

Lemma 1. *There is a polynomial time algorithm that given a strong cut-set X of a connected hypergraph $H = (\mathcal{V}, \mathcal{E})$ finds a cut-set $X' \subseteq X$ such that $X' \neq \emptyset$ and $(H = (\mathcal{V}, \mathcal{E}), k)$ is a yes instance of p-SET SPLITTING if and only if $(H' = (\mathcal{V}, \mathcal{E} \setminus X'), k - |X'|)$ is a yes instance of p-SET SPLITTING.*

Proof. Let $H^* = (\mathcal{V}, \mathcal{E} \setminus X)$ and let $|X| = t$. By assumption, X is a strong cut-set and hence the primal graph $P(H^*)$ has at least $t + 2$ connected components. Let the connected components of $P(H^*)$ be $\mathcal{C} = \{C_1, \ldots, C_q\}$ where $q \geq t + 2$ and $X = \{e_1, \ldots, e_t\}$. We construct an auxiliary bipartite graph $\mathcal{B}$ with vertex set $A \cup B$ with a vertex $a_i \in A$ for every edge $e_i \in X$ and a vertex $b_i \in B$ for every connected component $C_i \in \mathcal{C}$. There is an edge $a_i b_j$ if $V(e_i) \cap V(C_j) \neq \emptyset$.

We prove the statement of the lemma by induction on $|X|$. For the base case we assume that $|X| = 1$ and $X = \{e_1\}$. In particular, we show that given any $f : \mathcal{V} \to \{0, 1\}$ there exists a function $g : \mathcal{V} \to \{0, 1\}$ such that $Split(g) = Split(f) \cup \{e_1\}$ which will prove the desired assertion. If $e_1 \in Split(f)$ the statement follows, so assume that $e_1 \notin Split(f)$. Since $P(H)$ is connected we have that $a_1 b_j$, $j \in \{1, \ldots, q\}$ are edges in $\mathcal{B}$. Let $g : \mathcal{V} \to \{0, 1\}$ be such that $g(v) = f(v)$ if and only if $v \notin C_1$. That is, for all vertices in C_1, g flips the assignment given by f. Observe that $e_1 \in Split(g)$ since $V(e_1)$ contains a vertex $u \in C_1$ and a vertex $v \in C_2$. Since $f(u) = f(v)$, $g(u) \neq g(v)$ and hence $e_1 \in Split(g)$. For every edge in $Split(f)$ we have that $V(e)$ is completely contained in one of the components and hence, $e \in Split(f)$ implies $e \in Split(g)$. This completes the proof for the base case. So we assume that $|X| \geq 2$ and that the statement of the lemma holds for all X' satisfying the conditions of the lemma and $|X'| < |X|$. In inductive step we consider two cases:

(a) there does not exist a matching in $\mathcal{B}$ which saturates A; or
(b) there is a matching saturating A in $\mathcal{B}$.

---

[1] Due to space restrictions, the proof of this theorem has been omitted.

In Case (a) by Hall's theorem we know that there exists a subset $A' \subseteq A$, $A' \neq \emptyset$ such that $|A'| > |N(A')|$ and such a set can be found in polynomial time. We claim that $X' = X \setminus \{e_j \mid a_j \in A'\}$ is a strong cut-set and is of smaller size than X. It is clear that $|X'| < |X|$ as $A' \neq \emptyset$. We now show that X' is indeed a strong cut-set. Let $\mathcal{C}' = \mathcal{C} \setminus \{C_j \mid b_j \in N(A')\}$. Observe that in $H' = (\mathcal{V}, \mathcal{E} \setminus X')$, every $C_i \in \mathcal{C}'$ is a connected component. The size of $\mathcal{C}'$ is bounded as follows

$$|\mathcal{C}'| = |\mathcal{C}| - |N(A')| \geq (t+2) - |N(A')| > (t+2) - |A'| = t - |A'| + 2 = |X'| + 2,$$

and hence X' is indeed a strong cut-set. In this case the statement of the lemma follows from the induction hypothesis as $|X'| < |X|$.

For Case (b) we assume that we have a matching M saturating A. Without loss of generality let M be $a_1 b_1, \ldots, a_t b_t$. Let $U = \{b_{t+1}, \ldots, b_q\}$ be the set of vertices in B that are unsaturated by M. Iteratively we construct a set U' containing U as follows. Initially we set $U' := U$ and $\tilde{A} = A$.

- Check whether there exists a neighbor of a vertex in U' in $\tilde{A}$; if yes go to the next step. Otherwise, output U'.
- Let a_j be a vertex in $\tilde{A}$ having a neighbor in U'. Set $U' := U' \cup \{b_j\}$ (b_j is the matching end point of a_j in B), $\tilde{A} := \tilde{A} \setminus \{a_j\}$ and go to the first step.

Let U' be the set returned by the iterative process above. Observe that $U \subsetneq U'$ and $\tilde{A} \subsetneq A$ as $P(H)$ is connected and hence there exists at least one vertex a_j having a neighbor in U and hence the above iteration does not stop in the first round. Let $A' = A \setminus \tilde{A}$ and let $X' = \{e_j \mid a_j \in A'\}$. In what follows we prove that X' is the desired subset of X mentioned in the statement of the lemma.

We first show that X' is a strong cut-set. Let $\mathcal{C}' = \{C_j \mid b_j \in U'\}$. From the construction it follows that every $C_i \in \mathcal{C}'$ is a connected component of $H' = (\mathcal{V}, \mathcal{E} \setminus X')$. The size of $\mathcal{C}'$ is bounded as follows

$$|\mathcal{C}'| = |U'| = |U| + |A'| \geq |A'| + 2 = |X'| + 2,$$

and hence X' is a strong cut set.

We show that given any $f : \mathcal{V} \to \{0,1\}$ there exists a function $g : \mathcal{V} \to \{0,1\}$ such that $Split(g) = Split(f) \cup X'$. This will complete the proof of the lemma. Let $U' \setminus U = \{b_{1'}, b_{2'}, \ldots, b_{r'}\}$ and without loss of generality assume that $b_{1'}, b_{2'}, \ldots, b_{r'}$ is the order in which these elements are included in the set U'. Let $\mathcal{B}_i = \mathcal{B}[U \cup \{b_{1'}, \ldots, b_{i'}\} \cup \{a_{1'}, \ldots, a_{i'}\}]$. Iteratively we construct the function $g : \mathcal{V} \to \{0,1\}$ as follows. Initially we set $g := f$ and $i := 1$ and repeat the following until $i = r$:

Check whether the restriction of g to $\{C_1, \ldots, C_{i'}\}$ splits $e_{i'}$. If yes then set $i := i+1$ and repeat. Otherwise let $C_{i'}$ be the connected component corresponding to $b_{i'}$ having vertex set $V(C_{i'})$. Now for every vertex $u \in V(C_{i'})$ change $g(u)$ to $1 - f(u)$. Basically, we flip the assignment of 0 and 1 in the vertex set $V(C_{i'})$. Set $i := i+1$ and repeat the procedure.

Now we show that $Split(g) = Split(f) \cup X'$. Observe that when we flip the assignment of the vertex set $V(C_{i'})$ the only hyperedges which could go out of the set $Split(g)$ are $\{e_{i'}, e_{(i+1)'}, \ldots, e_{r'}\}$. The reason we flip the assignment is because $e_{i'} \notin Split(g)$ at that point. Also notice that by construction there exists a $b_{j'} \in \{b_{1'}, b_{2'} \ldots, b_{(i-1)'}\}$ such that $V(e) \cap V(C_{j'}) \neq \emptyset$. Hence after we flip the assignment of the vertex set $V(C_{i'})$ we have that $e_{i'} \in Split(g)$. Hence after the rth step of the procedure we have that $Split(g) = Split(f) \cup X'$. This concludes the proof.   □

Lemma 1 naturally gives rise to a reduction rule for the p-SET SPLITTING problem. Given a strong cut set X, a strong cut set X' obtained by the Lemma 1 is called *reducible strong cut-set*. This brings us to the following reduction rule.

REDUCTION RULE 1 : *Let $(H = (\mathcal{V}, \mathcal{E}), k)$ be an instance of p-SET SPLITTING and X' be a reducible strong cut-set of H. Remove X' from the set of hyperedges and reduce k to $k - |X'|$, that is, obtain an instance $(H' = (\mathcal{V}, \mathcal{E} \setminus X), k - |X|)$.*

When the hypergraph H is disconnected we can give a simple reduction rule.

REDUCTION RULE 2 : *Let $(H = (\mathcal{V}, \mathcal{E}), k)$ be an instance of p-SET SPLITTING such that $P(H)$ has connected components $P(H)[C_1], \ldots, P(H)[C_t]$. Let $v_1, \ldots, v_t$ be vertices such that $v_i \in C_i$. Construct a hypergraph $H' = (\mathcal{V}', \mathcal{E}')$ from H by unifying the vertices $v_1, \ldots, v_t$. In particular $\mathcal{V}' = \mathcal{V} \setminus \{v_2, \ldots, v_t\}$ and for every hyperedge $e \in \mathcal{E}$ make the edge $e' \in \mathcal{E}'$ where $e' = e$ if $v_i \notin e$ for every i and $e' = (V(e) \setminus \{v_2, \ldots, v_t\}) \cup \{v_1\}$ otherwise. We obtain the instance (H', k).*

The correctness proof for this reduction rule is simple, and given for example in [19] for the case of p-MAX CUT.

Proof **(Proof of Theorem 1).** Given an instance (H, k) of p-SET SPLITTING we first obtain an equivalent instance with at most $2k$ sets and at most $2k^2$ elements by applying the kernelization algorithm of Chen and Lu [2], given in Theorem 1 of their paper. We then apply Reduction Rules 1 and 2 exhaustively. Let $(H' = (\mathcal{V}', \mathcal{E}'), k')$ be the reduced instance. Since both our rules and the rules of Chen and Lu [2] only reduce k we have that $k' \leq k$. Let H' have n elements and $m \leq 2k$ sets. We show that if $n > k'$ then (H', k') is a yes-instance. In particular, since Reduction Rule 2 does not apply, H' is connected. Since Reduction Rule 1 does not apply, H' does not have a strong cut-set. By Theorem 3 we can find in polynomial time a forest $\mathcal{F} = (F, g)$ of H' with $n - 1$ edges. Since F is a forest, F is bipartite. Let $W \uplus B$ be bipartitions of $\mathcal{V}'$. By the definition of a forest in a hypergraph, the bipartitions (W, B) splits all sets corresponding to hyperedges in $\mathcal{F}$. Since F has $n - 1$ edges, at least $n - 1 \geq k$ hyperedges are split and hence (H', k') is a yes-instance. Thus if $n > k$ for a reduced instance, the kernelization algorithm outputs that (H', k') is a yes-instance. Hence any unresolved reduced instance has at most $k' \leq k$ elements. This concludes the proof.   □

3   Faster Parameterized Algorithm for p-SET SPLITTING

Theorem 1 yields a simple $O(2^k k^2 + N)$ time algorithm for the p-SET SPLITTING problem by looping over all possible bipartitions of set of elements into (W, B)

and for each checking whether they split at least k edges. Previously, only a randomized $O(2^k k^2 + N)$ time algorithm [2] and a deterministic $O(2.65^k + N)$ time algorithm [14] was known. In this section we give an algorithm for for the p-SET SPLITTING problem running in $O(1.9630^k + N)$ time. Our algorithm first obtains a kernel with $2k$ sets and at most k elements using Theorem 1. Then the algorithm proceeds to solve the small instance recursively.

The subcases generated by the algorithm are naturally phrased as a colored version of the p-SET SPLITTING. In this version of the problem the sets in $\mathcal{F}$ are either *uncolored* or colored *white* or *black*. A black set S is split by a partitioning of U into W and B if $S \cap W \neq \emptyset$. Similarly a white set S is split if $S \cap B \neq \emptyset$. Hence, an instance to the COLORED p-SET SPLITTING (p-CSS) problem is a universe U, a family $\mathcal{F} = \mathcal{F}_u \uplus \mathcal{F}_w \uplus \mathcal{F}_b$ over U and an integer k. The families $\mathcal{F}_u$, $\mathcal{F}_w$, and $\mathcal{F}_b$ denote the families of uncolored, white and black sets respectively.

Our algorithm is based on a single branching step. For a particular element v of U we try putting v in W or in B. If v is inserted into W, all sets in $\mathcal{F}_b$ containing v are split and all sets in $\mathcal{F}_u$ containing v are put into $\mathcal{F}_w$ instead. The sets that are split are removed from $\mathcal{F}_b$ and k is decreased by the number of newly split sets. Finally v is removed from the universe U and from all sets containing v. Similarly, if v is inserted into B then all sets in $\mathcal{F}_w$ containing v are split and all sets in $\mathcal{F}_u$ containing v are put into $\mathcal{F}_b$ instead. For a vertex v let $\mathcal{N}_u(v)$, $\mathcal{N}_b(v)$ and $\mathcal{N}_w(v)$ be the set of uncolored, black and white sets containing v respectively. We call $d_u(v) = |\mathcal{N}_u(v)|$, $d_b(v) = |\mathcal{N}_b(v)|$ and $d_w(v) = |\mathcal{N}_w(v)|$ the uncolored, black and white degree of v. The degree of v is $d(v) = d_u(v) + d_w(v) + d_b(v)$. Formalizing the discussion above we obtain the following recurrence.

$$(U, \mathcal{F}_u, \mathcal{F}_w, \mathcal{F}_b, k) \in p\text{-CSS}$$

$$\Longleftrightarrow$$

$$(U \setminus \{v\}, \mathcal{F}_u \setminus \mathcal{N}_u(v), \mathcal{F}_w \cup \mathcal{N}_u(v), \mathcal{F}_b \setminus \mathcal{N}_b(v), k - d_b(v)) \in p\text{-CSS} \qquad (1)$$

$$\vee$$

$$(U \setminus \{v\}, \mathcal{F}_u \setminus \mathcal{N}_u(v), \mathcal{F}_w \setminus \mathcal{N}_w(v), \mathcal{F}_b \cup \mathcal{N}_u(v), k - d_w(v)) \in p\text{-CSS}$$

We now describe the algorithm for p-SET SPLITTING using Recurrence 1. We first formulate the p-SET SPLITTING instance $(U\mathcal{F}, k)$ as a p-CSS instance $(U, \mathcal{F}_u, \emptyset, \emptyset, k)$ where $\mathcal{F}_u = \mathcal{F}$. We fix $K = k$, and fix $\alpha = 0.027$, $\beta = 0.31$, and $\gamma = 0.13$.

Preprocessing. The algorithm computes a table for all subproblems $(U', \mathcal{F}'_u, \mathcal{F}'_w, \mathcal{F}'_b, k')$ where

- $U' \subseteq U$ and $|U'| \leq \alpha k$;
- $\mathcal{F}'_u \uplus \mathcal{F}'_w \uplus \mathcal{F}'_b \subseteq \mathcal{F}_u$ and $|\mathcal{F}'_u \uplus \mathcal{F}'_w \uplus \mathcal{F}'_b| \leq 4\alpha k$;
- $k' \leq k$.

An entry of the table contains true if the corresponding instance $(U', \mathcal{F}'_u, \mathcal{F}'_w, \mathcal{F}'_b, k')$ is in p-CSS. The table can be filled using Recurrence 1 and bottom up dynamic programming in time linear in the number of table entries. Thus the total time required to perform the preprocessing step is

$$O\left(\binom{k}{\alpha k} \cdot \binom{2k}{4\alpha k} \cdot 3^{4\alpha k} \cdot k\right).$$

Rewriting $\binom{a}{b}$ as $\frac{a!}{b!(a-b)!}$, using Stirling's approximation for $n!$ and plugging in the value of $\alpha = 0.027$ yields that the preprocessing step is done in $O(1.9630^k)$ time.

Branching. The algorithm selects an element $v \in U$ of highest degree and branches on this vertex using Recurrence 1. If the algorithm generates a subcase for which the answer is already stored in the preprocessing table, the algorithm returns this answer. If k reaches 0 or a negative number the algorithm returns "yes" and if k is positive and $U = \emptyset$ the algorithm returns "no". While k might not be the same in different recursive calls, the value of K fixed initially remains the same throughout the algorithm. Correctness of the algorithm follows directly from Recurrence 1. We now proceed to analyze the running time of the algorithm.

Running Time Analysis. We use the *Measure & Conquer* paradigm to analyze the running time of the algorithm. For two constants β and γ we define

$$\mu = \mu(U, \mathcal{F}_u, \mathcal{F}_w, \mathcal{F}_b, k) = |U| + \beta k + \gamma |\mathcal{F}_u|.$$

The running time of the algorithm is bounded from above by the number of leaves in the search tree, modulo a polynomial in k. Let $T(\mu, |U|)$ be an upper bound on the number of leaves in the search tree of the algorithm on an instance with measure μ and universe size $|U|$. We first need an auxiliary claim about the size of the search tree when the degree of any element is at most 4.

Claim. Let K and α be fixed as in the discussion above and $(U, \mathcal{F}_u, \mathcal{F}_w, \mathcal{F}_b, k)$ be an instance of p-CSS such that $|U| \geq \alpha K$ generated during a recursive call such that the degree of any element is at most 4. Then $T(\mu, |U|) \leq 2^{|U|-\alpha K}$.

Proof. We prove the claim by induction on $|U|$. If $|U| = \alpha K$ then the algorithm solves the instance by looking up in the preprocessing table as $|\mathcal{F}| \leq 4\alpha K$ and hence $T(\mu, |U|) = 1$. Assume now that the statement holds whenever $|U| = t$ for some fixed $t \geq \alpha K$ and consider an instance with $|U| = t + 1$. The algorithm makes two recursive calls applying Recurrence 1. In each recursive call all elements have degree at most 4 and the size of $|U|$ is exactly t. By the induction hypothesis the number of leaves in the search tree of the two subinstances is at most $2^{t-\alpha K}$. Hence the total number of leaves in the search tree is bounded from above by $2 \cdot 2^{t-\alpha K} = 2^{|U|-\alpha K}$.   □

We now extend the analysis in Claim 3 to instances with no constraints on element degree.

Claim. Let K and α be fixed as in the discussion above and $(U, \mathcal{F}_u, \mathcal{F}_w, \mathcal{F}_b, k)$ be an instance of p-CSS generated during a recursive call. Then $T(\mu, |U|) \leq 2^{|U|-\alpha K} + 1.5222^{\mu}$.

Proof. We prove the claim by induction on $|U|$. If there are no elements of degree at least 5 and $|U| \geq \alpha K$ then the statement of the claim holds by Claim 3. If there are no elements of degree at least 5 and $|U| < \alpha K$ then $T(\mu, |U|) = 1 \leq 1.5222^{\mu}$. Assume now that there are elements of degree at least 5 and let v be the element on which the algorithm branches. Since the algorithm picks an element of largest degree, v has degree at least 5. If the uncolored, white and black degrees of v are $d_u(v), d_w(v), d_b(v)$ we let d'_u, d'_w and d'_b be non-negative integers such that $d'_u + d'_w + d'_b = 5$ and $d'_u \leq d_u(v)$, $d'_w \leq d_w(v)$ and $d'_b \leq d_b(v)$. When we apply Recurrence 1 to branch on an element v we get the following recurrence for T

$$T(\mu, |U|) \leq T(\mu - 1 - \beta d'_b - \gamma d'_u, |U| - 1) + T(\mu - 1 - \beta d'_w - \gamma d'_u, |U| - 1).$$

One can verify that if we pick $c = 1.5222$ then for any choice of (d'_u, d'_w, d'_b) such that $d'_u + d'_w + d'_b = 5$ we have

$$\begin{aligned}
T(\mu, |U|) &\leq T(\mu - 1 - \beta d'_b - \gamma d'_u, |U| - 1) + T(\mu - 1 - \beta d'_w - \gamma d'_u, |U| - 1) \\
&\leq c^{\mu-1-\beta d'_b - \gamma d'_u} + 2^{|U|-1-\alpha K} + c^{\mu-1-\beta d'_w - \gamma d'_u} + 2^{|U|-1-\alpha K} \\
&= c^{\mu} \cdot (c^{-1-\beta d'_b - \gamma d'_u} + c^{-1-\beta d'_w - \gamma d'_u}) + 2 \cdot 2^{|U|-1-\alpha K} \\
&\leq c^{\mu} + 2^{|U|-\alpha K}.
\end{aligned}$$

Hence $T(\mu, |U|) \leq 2^{|U|-\alpha K} + 1.5222^{\mu}$, concluding the proof.   □

Summing up the above analysis, noticing that $\mu \leq K + \beta K + \gamma 2K$ in a reduced instance, and inserting this into the bound for $T(\mu, |U|)$ from Claim 3 yields an upper bound of $O(1.9630^k)$ for the running time of the branching part of the algorithm. Since both parts of the algorithm take $O(1.9630^k)$ time, this completes the proof of Theorem 2.

4   Conclusion and Discussions

In this paper we gave a smaller kernel and a faster algorithm for the p-SET SPLITTING problem improving over the previosuly known results. The number of elements and sets in our kernel matches the number of vertices and edges in the best known kernel for p-MAX CUT. It should be noted that both the kernel and the algorithm for p-SET SPLITTING presented here also work for the p-NOT ALL EQUAL SAT problem. The reduction rule we use to handle instances with strong cut-sets has similarities with reduction rules based on *crown decompositions* [3,8,19], and it seems that crown decompositions and strong cut-sets are closely related. This similarity also makes us believe that the duality theorem we made us of in our kenrelization algorithm will be a useful tool in the field of kernelization.

Acknowledgments

We would like to thank Stéphan Thomassé, Fedor V. Fomin, Magnus Wahlström and Gregory B. Sorkin for valuable suggestions and insightful discussions. We especially thank Stéphan Thomassé for pointing us to the Proposition 1 and the reference [11].

References

1. Andersson, G., Engebretsen, L.: Better approximation algorithms for SET SPLIT-TING and NOT-ALL-EQUAL SAT. Inf. Process. Lett. 65(6), 305–311 (1998)
2. Chen, J., Lu, S.: Improved algorithms for weighted and unweighted set splitting problems. In: Lin, G. (ed.) COCOON 2007. LNCS, vol. 4598, pp. 537–547. Springer, Heidelberg (2007)
3. Chlebík, M., Chlebíková, J.: Crown reductions for the minimum weighted vertex cover problem. Discrete Applied Mathematics 156(3), 292–312 (2008)
4. Dehne, F.K.H.A., Fellows, M.R., Rosamond, F.A.: An FPT algorithm for set split-ting. In: Bodlaender, H.L. (ed.) WG 2003. LNCS, vol. 2880, pp. 180–191. Springer, Heidelberg (2003)
5. Dehne, F.K.H.A., Fellows, M.R., Rosamond, F.A., Shaw, P.: Greedy localization, iterative compression, modeled crown reductions: New fpt techniques, an improved algorithm for set splitting, and a novel $2k$ kernelization for vertex cover. In: Downey, R.G., Fellows, M.R., Dehne, F. (eds.) IWPEC 2004. LNCS, vol. 3162, pp. 271–280. Springer, Heidelberg (2004)
6. Erdős, P.: On a combinatorial problem, I. Nordisk Mat. Tidskrift 11, 5–10 (1963)
7. Erdős, P.: On a combinatorial problem, II. Acta Math, Hungary 15, 445–447 (1964)
8. Fellows, M.R.: Blow-ups, win/win's, and crown rules: Some new directions in FPT. In: Bodlaender, H.L. (ed.) WG 2003. LNCS, vol. 2880, pp. 1–12. Springer, Heidel-berg (2003)
9. Fomin, F.V., Grandoni, F., Kratsch, D.: Measure and conquer: Domination - a case study. In: Caires, L., Italiano, G.F., Monteiro, L., Palamidessi, C., Yung, M. (eds.) ICALP 2005. LNCS, vol. 3580, pp. 191–203. Springer, Heidelberg (2005)
10. Fomin, F.V., Grandoni, F., Kratsch, D.: Measure and conquer: a simple o($2^{0.288}$) independent set algorithm. In: SODA, pp. 18–25 (2006)
11. Frank, A., Király, T., Kriesell, M.: On decomposing a hypergraph into k connected sub-hypergraphs. Discrete Applied Mathematics 131(2), 373–383 (2003)
12. Goemans, M.X., Williamson, D.P.: Improved approximation algorithms for maximum cut and satisfiability problems using semidefinite programming. J. ACM 42(6), 1115–1145 (1995)
13. Karp, R.M.: Reducibility among combinatorial problems. In: Complexity of Computer Computations, pp. 85–103 (1972)
14. Lokshtanov, D., Sloper, C.: Fixed parameter set splitting, linear kernel and improved running time. In: ACiD. Texts in Algorithmics, vol. 4, pp. 105–113 (2005)
15. Lovász, L.: Covering and coloring of hypergraphs. In: Proceedings of the 4th South-eastern Conference on Combinatorics, Graph Theory and Computing. Utilitas Mathematica Publishing, pp. 3–12 (1973)
16. Prieto, E.: The method of extremal structure on the k-maximum cut problem. In: CATS, pp. 119–126 (2005)

17. Radhakrishnan, J., Srinivasan, A.: Improved bounds and algorithms for hypergraph 2-coloring. Random Struct. Algorithms 16(1), 4–32 (2000)
18. Raman, V., Saurabh, S.: Improved fixed parameter tractable algorithms for two "edge" problems: MAXCUT and MAXDAG. Inf. Process. Lett. 104(2), 65–72 (2007)
19. Sloper, C.: Techniques in parameterized algorithm design. PhD thesis, University of Bergen (2005)
20. Wahlström, M.: Algorithms, measures, and upper bounds for satisfiability and related problems. PhD thesis, Linkþing University (2007)
21. Zhang, J., Ye, Y., Han, Q.: Improved approximations for max set splitting and max NAE SAT. Discrete Applied Mathematics 142(1-3), 133–149 (2004)
22. Zwick, U.: Outward rotations: A tool for rounding solutions of semidefinite programming relaxations, with applications to max cut and other problems. In: STOC, pp. 679–687 (1999)

Stable Assignment with Couples: Parameterized Complexity and Local Search

Dániel Marx and Ildikó Schlotter

Department of Computer Science and Information Theory,
Budapest University of Technology and Economics,
H-1521 Budapest, Hungary
{dmarx,ildi}@cs.bme.hu

Abstract. We study the Hospitals/Residents with Couples problem, a variant of the classical Stable Marriage problem. This is the extension of the Hospitals/Residents problem where residents are allowed to form pairs and submit joint rankings over hospitals. We use the framework of parameterized complexity, considering the number of couples as a parameter. We also apply a local search approach, and examine the possibilities for giving FPT algorithms applicable in this context. Furthermore, we also investigate the matching problem containing couples that is the simplified version of the Hospitals/Residents problem modeling the case when no preferences are given.

1 Introduction

The classical Hospitals/Residents problem (which is a generalization of the well-known Stable Marriage problem) was introduced by Gale and Shapley [6] to model the following situation. We are given a set of hospitals, each having a number of open positions, and a set of residents applying for jobs in the hospitals. Each resident has a ranking over the hospitals, and conversely, each hospital has a ranking over the residents. Our aim is to assign as many residents to a hospital as possible, with the restrictions that the capacities of the hospitals are not exceeded and the resulting assignment is stable (no hospital-resident pair would benefit from rejecting the assignment and contracting each other).

The original version of the Hospitals/Residents problem is well understood: a stable assignment always exists, and every stable assignment has the same size. (The size of an assignment is the number of residents that have a job.) Moreover, the classical Gale-Shapley algorithm [6] can find a stable assignment in linear time. However, several practical applications motivate some kind of extension or modification of the problem (see e.g. the NRMP program for assigning medical residents in the USA [17,18]), and in the recent decade various versions have been investigated. We study an extension of this problem, called Hospitals/Residents with Couples (or HRC), where residents may form couples, and thus have joint rankings over the hospitals. This extension models a situation that arises in many real world applications [18], and was introduced by Roth [17] who also discovered that a stable assignment need not exist when couples are involved.

J. Chen and F.V. Fomin (Eds.): IWPEC 2009, LNCS 5917, pp. 300–311, 2009.

Later, Ronn [16] proved that it is NP-hard to decide whether a stable assignment exists in such a setting. There have been investigations of different assumptions on the preferences of couples that guarantee some kind of tractability [10,14].

Algorithmic approaches. In the Hospitals/Residents problem, practical scenarios usually involve much fewer couples than singles, e.g. the ratio of couples to singles participating in the NRMP program is around 2.5 percent[1]. Thus, the number of couples in the HRC problem is a natural parameter. Investigating the parameterized complexity of HRC with this parameter is our first goal.

Local search is a basic technique that has been widely applied in heuristics for practical optimization problems for several decades [1]. However, investigations considering the connection of local search and parameterized algorithms have only been started a few years ago, and research in this area has been gaining increasing attention lately [12]. The basic idea of local search is to find an optimal solution by an iteration in which we improve the current solution step by step through local modifications. Local search can become more efficient if we can decide whether there exists a better solution S' that is ℓ modification steps away from a given solution S. Typically, the ℓ-neighborhood of a solution S can be explored in $n^{O(\ell)}$ time by examining all possibilities to find those parts of S that should be modified. (Here n is the input size.) The question whether an FPT algorithm with parameter ℓ can be found for the neighborhood exploration problem has already been studied in connection with different optimization problems ([9,13,4]). Our second goal is to investigate this approach for assignment problems.

We also contribute to the framework of parameterized local search algorithms by distinguishing between "strict" algorithms that perform the local search step in some neighborhood of a solution as described above, and "permissive" algorithms whose task is the following: given some problem with an initial solution S, find *any* better solution, provided that a better solution exists in the local neighborhood of S. Our motivation for this distinction is that finding an improved solution in the neighborhood of a given solution may be hard, even for problems where an optimal solution is easily found.

Most of the questions examined here are also worth studying in a simplified model that does not involve preferences. In the Maximum Matching with Couples problem, or shortly MMC, no stability requirement is given, and we aim for an assignment of maximum size.

Results. For lack of space, we stated some of our results without proof, see the full version of the paper for these proofs. Our main results are outlined below (see Table 1). We denote by C the set of couples in a problem instance, and we denote by ℓ the neighborhood size in a given local search problem.

- Theorem 1 gives a randomized FPT algorithm with parameter $|C|$ for Maximum Matchig with Couples. The presented algorithm uses an FPT result from matroid theory.

---

[1] http://www.nrmp.org/data/resultsanddata2008.pdf

Table 1. Summary of our results (assuming $W[1] \neq FPT$)

Task:	Existence problem	Maximum problem	Local search algorithm with FPT running time	
Parameter:	$|C|$		ℓ	$(|C|, \ell)$
MMC (no pref's)	P (trivial)	randomized FPT (Theorem 1)	No permissive alg. (Theorem 3)	Permissive alg. (Theorem 1)
HRC (with pref's)	W[1]-hard (Theorem 4)	W[1]-hard (Theorem 4)	No permissive alg. (Theorem 5)	Strict alg. (Theorem 7)

- Theorem 3 shows that no permissive local search FPT algorithm exists for MMC, where the parameter is ℓ, unless $W[1] = FPT$.
- Theorem 4 proves that the existence version of the HRC problem is W[1]-hard with parameter $|C|$.
- Theorem 5 shows that no permissive local search FPT algorithm exists for the maximization version of HRC with parameter ℓ, unless $W[1] = FPT$.
- Theorem 7 presents a strict local search FPT algorithm for the maximization version of HRC, with combined parameters $|C|$ and ℓ. The algorithm uses color coding and a set of non-trivial reduction rules.

2   Preliminaries

For some integer k, we use $[k] = \{1, 2, \ldots, k\}$, and $\binom{[k]}{2} = \{(i, j) | 1 \le i < j \le k\}$. If a matching M in a graph contains an edge xy, then we write $M(x) = y$ and $M(y) = x$. For other graph theoretic concepts, we use standard notation. We assume basic knowledge of matroid theory in Sect. 4. We also assume that the reader is familiar with the framework of parameterized complexity. For an introduction, see [15] or [5].

To formalize the task of a local search algorithm, let Q be an optimization problem with an objective function T which we want to maximize. To define the concept of neighborhoods, we suppose there is some *distance* $d(x, y)$ defined for each pair (x, y) of solutions for some instance I of Q. A *strict local search* algorithm for Q has the following task:

> Strict local search for Q
> Input: (I, S_0, ℓ) where I is an instance of Q, S_0 is a solution for I, and $\ell \in \mathbb{N}$.
> Task: If there exists a solution S for I such that $d(S, S_0) \le \ell$ and $T(S) > T(S_0)$, then output such an S.

In contrast, a *permissive local search* algorithm for Q is allowed to output a solution that is not close to S_0, provided that it is better than S_0. In local search methods, such an algorithm is as useful as its strict version.

Permissive local search for Q
Input: (I, S_0, ℓ) where I is an instance of Q, S_0 is a solution for I, and $\ell \in \mathbb{N}$.
Task: If there exists a solution S for I such that $d(S, S_0) \le \ell$ and $T(S) >$
$T(S_0)$, then output *any* solution S' for I with $T(S') > T(S)$.

Note that if an optimal solution can be found by some algorithm, then this yields a permissive local search algorithm for the given problem. On the other hand, finding a strict local search algorithm might be hard even if an optimal solution is easily found. An example for such a case is the MINIMUM VERTEX COVER problem for bipartite graphs [9]. Besides, proving that no permissive local search algorithm exists for some problem is clearly harder than it is for strict local search algorithms. We also present results of this kind.

3   Hospitals/Resident with Couples

A *couples' market with preference*, or shortly *cmp*, consists of the sets S, C and H representing singles, couples and hospitals, respectively, a *capacity* $f(h)$ for each $h \in H$, and a *preference list* $L(a)$ for each $a \in S \cup C \cup H$. The set $A = S \cup C \cup H$ is called the set of *agents*. Each couple c is a pair $(c(1), c(2))$, and we call the elements of the set $R = \bigcup_{c \in C}\{c(1), c(2)\} \cup S$ *residents*. For a hospital h, $L(h)$ is a list of residents, for a single s, $L(s)$ is a list of hospitals, and for a couple c, $L(c)$ is a list containing pairs of hospitals, or more precisely, a list containing elements from $(H \cup \{u\}) \times (H \cup \{u\}) \setminus \{(u, u)\}$ where u is a special symbol indicating that someone is unemployed. The preference lists can be incomplete, but cannot involve ties, so these lists are strictly ordered.

The set of elements appearing in the list $L(a)$ is $A^L(a)$, and some x is considered *acceptable* for a if $x \in A^L(a)$. Clearly, we may assume that acceptance is mutual, so $h \in A^L(s)$ holds if and only if $s \in A^L(h)$ for each $s \in S$ and $h \in H$, and $(h_1, h_2) \in A^L(c)$ implies $c(i) \in A^L(h_i)$ or $h_i = u$ for both $i \in \{1, 2\}$, for each $c \in C$. For some $x \in A^L(a)$, the *rank* of x w.r.t. a, denoted by $\rho(a, x)$, is $r \in \mathbb{N}$ if x is the r-th element in $L(a)$. If $x \notin A^L(a)$, then we let $\rho(a, x) = \infty$ for all meaningful x. We say that the cmp is f_0-*uniform* if $f \equiv f_0$ for some $f_0 \in \mathbb{N}$.

An *assignment* is a function $M : R \to H \cup \{u\}$ such that $M(s) \in A^L(s) \cup \{u\}$ for each $s \in S$, $M(c) \in A^L(c) \cup \{(u, u)\}$ for each $c \in C$, and $|M(h)| \le f(h)$ holds for each hospital h. Here, $M(c)$ denotes the pair $(M(c(1)), M(c(2)))$, and $M(h) = \{r | r \in R, M(r) = h\}$ is the set of residents assigned to h in M. We say that an assignment M *covers* a resident r if $M(r) \ne u$, and M covers a couple c, if it covers $c(1)$ or $c(2)$. We define the *size* of M, denoted by $|M|$, to be the number of residents covered by M. The *distance* $d(M, M')$ of two assignments M and M' is the number of residents r for which $M(r) \ne M'(r)$.

We say that x is *beneficial* for the agent a with respect to an assignment M if $x \in A^L(a)$ and one of the following cases holds: (1) $a \in S \cup C$ and either a is not covered by M or $\rho(a, x) < \rho(a, M(a))$, (2) $a \in H$ and either $|M(a)| < f(a)$ or there exists a resident $r' \in M(a)$ such that $\rho(a, x) < \rho(a, r')$. A *blocking pair* for M can be of three types:

- it is either a pair formed by a single s and a hospital h such that both s and h are beneficial for each other w.r.t. M,
- or a pair formed by a couple c and a pair (h_1, h_2) with $h_1 \neq h_2$ such that (h_1, h_2) is beneficial for c w.r.t. M, and for both $i \in \{1, 2\}$ it holds that if $h_i \neq u$ then either $c(i)$ is beneficial for h_i w.r.t. M or $c(i) \in M(h_i)$,
- or a pair formed by a couple c and a hospital h such that (h, h) is beneficial for c w.r.t. M, and *the couple c is beneficial for h*. If h prefers $c(1)$ to $c(2)$, this latter means that either $|M(h)| \leq f(h) - 2$, or $|M(h)| \leq f(h) - 1$ and $\rho(h, c(1)) < \rho(h, r)$ for some $r \in M(h)$, or $\rho(h, c(1)) < \rho(h, r_1)$ and $\rho(h, c(2)) < \rho(h, r_2)$ for some $r_1 \neq r_2$ in $M(h)$. [2]

An assignment M for I is *stable* if there is no blocking pair for M.

The input of the HOSPITALS/RESIDENTS WITH COUPLES problem is a cmp I, and the task is to determine a stable assignment for I, if such an assignment exists. If no couples are involved, then a stable assignment can always be found in linear time with the Gale-Shapley algorithm [6]. In the case when couples are present, a stable assignment may not exist, as first proved by Roth [17]. Ronn proved that deciding whether a stable assignment exists for a cmp is NP-complete [16]. Moreover, an instance of the HOSPITALS/RESIDENTS WITH COUPLES problem may admit stable assignments of different sizes, see the full paper for an example. In the optimization problem MAXIMUM HOSPITALS/RESIDENTS WITH COUPLES, the task is to determine a stable assignment of maximum size for a given cmp. This problem is trivially NP-hard, as it contains the HOSPITALS/RESIDENTS WITH COUPLES problem. We study these problems in Sect. 5.

We also study a version of the HOSPITALS/RESIDENTS WITH COUPLES problem that does not contain preferences and only deals with the notion of acceptability. To describe the input of this problem, we define a *couples' market with acceptance*, or shortly *cma*, as a quintuple (S, C, H, f, A) where S, C, H and f are defined analogously as in a cmp, but $A(a)$ defines only the set of acceptable elements for an agent a, without any ordering. Each concept described above that does not rely on the preference lists (and thus on stability) is inherited also for cmas in the straightforward way. In Sect. 4, we investigate the optimization problem MAXIMUM MATCHING WITH COUPLES, where given a cma I, the task is to find an assignment for I of maximum size.

4   Matching without Preferences

First, we investigate a slightly modified version of MAXIMUM MATCHING WITH COUPLES, denoted as (k, n)-MATCHING WITH COUPLES: given a cma I and two integers k and n, find an assignment for I that covers at least k couples and n singles, if possible. Such an assignment is called a (k, n)-*assignment*. Clearly, if there are no couples in a given instance, then the problem is equivalent to finding a maximum matching in a bipartite graph, and can be solved by standard techniques. If couples are involved, the problem becomes hard. More precisely,

---

[2] We thank David Manlove for pointing out this case.

the decision version of this problem is NP-complete [8,3], even in the following special case: each hospital has capacity 2, and the acceptable hospital pairs for a couple are always of the form (h, h) for some $h \in H$. However, if the number of couples is small, which is a reasonable assumption in many practical applications, (k, n)-MATCHING WITH COUPLES becomes tractable, as shown by Theorem 1.

Theorem 1. (k, n)-MATCHING WITH COUPLES *can be solved in randomized FPT time with parameter* $|C|$.

To prove Theorem 1, we need a variant of a result from [11] concerning matroids.

Theorem 2. *Let* $\mathcal{M}(U, \mathcal{I})$ *be a linear matroid and let* $\mathcal{X} = \{X_1, X_2, \ldots X_n\}$ *be a collection of subsets of* U, *each of size* b. *Given a linear representation* A *of* $\mathcal{M}$, *it can be determined in* $f(k, b) \cdot ||A||^{O(1)}$ *randomized time whether there is an independent set that is the union of* k *disjoint sets in* $\mathcal{X}$.

Proof (of Theorem 1). Let (S, C, H, f, A) be the cma for which we have to find a (k, n)-assignment. W.l.o.g. we can assume that each hospital has capacity 1 as otherwise we can "clone" the hospitals, i.e. for each $h \in H$ we can substitute h with the newly introduced hospitals $h^1, \ldots, h^{f(h)}$, also modifying $A(p)$ for each $p \in S \cup C$ appropriately. (As $f(h) \leq |S| + 2|C|$ can be assumed, this increases the input size only polynomially.) Note that the case $k < |C|$ can be solved by finding a (k, n)-assignment for (S, C', H, f, A') for every $C' \subseteq C$ where $|C'| = k$ and A' is the restriction of A on $S \cup C'$. As this increases the running time only with a factor of at most $2^{|C|}$, it is sufficient to give an FPT algorithm for the case $|C| = k$. Moreover, we can assume $A(c) \subseteq H \times H$, since for each $c \in C$ we can eliminate each pair of the form (h, u) or (u, h) ($h \in H$) in $A(c)$ by adding a new hospital u_c to H with capacity 1 and substituting u with u_c.

Now, let $G(H, S; E)$ be the bipartite graph where a single $s \in S$ is connected with a hospital $h \in H$ if and only if $h \in A(s)$. We can assume w.l.o.g. that G has a matching of size at least n as otherwise no solution may exist, and this case can be detected easily in polynomial time. We define $\mathcal{M}(H, \mathcal{I})$ to be the matroid where a set $X \subseteq H$ is independent if and only if there is a matching in G that covers at least n singles but covers no hospitals from X. Observe that $\mathcal{M}$ is exactly the dual of the n-truncation of the transversal matroid of G, and thus it is indeed a matroid. By a lemma in [11], we can find a linear representation A of $\mathcal{M}$ in randomized polynomial time.

We define the matroid $\mathcal{M}'(U, \mathcal{I}')$ with ground set $U = H \cup C$ such that $X \subseteq U$ is independent in $\mathcal{M}'$ if $X \cap H$ is independent in $\mathcal{M}$. A representation of $\mathcal{M}'$ can be obtained by taking the direct sum of the matrices A and E_k where E_k is the unit matrix of size $k \times k$. Let $\mathcal{X}$ be the collection of the sets that are of the form $\{c, h_1, h_2\}$ where $c \in C$ and $(h_1, h_2) \in A(c)$.

Observe that if $X_1, X_2, \ldots, X_k$ are k disjoint sets in $\mathcal{X}$ whose union is independent in $\mathcal{M}'$, then we can construct a (k, n)-assignment as follows. For each $\{c, h_1, h_2\} \in \{X_1, \ldots, X_k\}$ we choose $M(c)$ from $\{(h_1, h_2), (h_2, h_1)\} \cap A(c)$ arbitrarily. The disjointness of the sets $X_1, \ldots, X_k$ guarantees that this way we assign exactly one resident to each hospital in $X = \bigcup_{i \in [k]} X_i \cap H$. Now, let N

be a matching in G that covers at least n singles, but no hospitals from X. Such a matching exists, as X is independent in $\mathcal{M}$. Thus letting $M(s)$ to be $N(s)$ if s is covered by N and u otherwise for each $s \in S$ yields that M is a (k, n)-assignment. Conversely, if M is a (k, n)-assignment then the sets $\{c, h_1, h_2\}$ for each $c \in C$ and $M(c) = (h_1, h_2)$ form a collection of k disjoint sets in $\mathcal{X}$ whose union is independent in $\mathcal{M}'$. By Theorem 2, such a collection can be found in randomized FPT time when k is the parameter, yielding a solution if exists.   □

We remark that Theorem 1 also applies to the following cases.

- Markets containing groups of fixed size instead of couples.
- Maximization (or minimization) of an arbitrary function $f(k, n)$, where k and n are the number of covered couples and singles, respectively.
- Minimizing the makespan in the scheduling problem containing parallel machines and independent jobs with job assignment restrictions, if the processing time is $p \in \mathbb{N}$ for k jobs, and 1 for the others, and k is the parameter.

Considering the parameterized complexity of the local search approach for the MMC problem with parameter ℓ denoting the neighborhood size, Theorem 3 shows that no FPT local search algorithm is likely to exists. We omit the proof.

Theorem 3. *No permissive local search algorithm for 2-uniform* MAXIMUM MATCHING WITH COUPLES *runs in FPT time with parameter ℓ, if $W[1] \neq FPT$.*

5   Matching with Preferences

In this section, we investigate several versions of the Hospitals/Residents problem, where couples are involved and preferences play an important role.

After presenting some hardness results, Theorem 7 gives an FPT time strict local search algorithm for the MAXIMUM HOSPITALS/RESIDENTS WITH COUPLES problem, where $|C|$ and ℓ are parameters. In contrast, Theorem 4 shows the W[1]-hardness of the HOSPITALS/RESIDENTS WITH COUPLES problem with parameter $|C|$, which clearly implies that MAXIMUM HOSPITALS/RESIDENTS WITH COUPLES is also W[1]-hard with parameter $|C|$.

However, supposing that a stable assignment has already been determined by some method, it is a valid question whether we can increase its size. We will denote this problem INCREASE HOSPITALS/RESIDENTS WITH COUPLES. Formally, its input is a cmp I and a stable assignment M_0 for I, and the task is to find a stable assignment with size at least $|M_0|+1$. If no couples are involved, then all stable assignments for the instance have the same size, so this problem is trivially polynomial-time solvable. Theorem 4 shows that INCREASE HOSPITALS/RESIDENTS WITH COUPLES is also W[1]-hard with parameter $|C|$.

Theorem 4. *(1) The decision version of* HOSPITALS/RESIDENTS WITH COUPLES *is W[1]-hard with parameter $|C|$, even in the 1-uniform case.*
(2) The decision version of INCREASE HOSPITALS/RESIDENTS WITH COUPLES *is W[1]-hard with parameter $|C|$, even in the 1-uniform case.*

Considering the applicability of the local search approach for the MAXIMUM HOSPITALS/RESIDENTS WITH COUPLES problem, Theorem 5 shows that no permissive local search algorithm is likely to run in FPT time with parameter ℓ. However, if we regard $|C|$ as a parameter as well, then even a strict local search algorithm with FPT running time can be given, as presented in Theorem 7.

Theorem 5. *No permissive local search algorithm for the 1-uniform* MAXIMUM HOSPITALS/RESIDENTS WITH COUPLES *runs in FPT time with parameter ℓ, if $W[1] \neq FPT$.*

To prove Theorems 4 and 5, we give FPT-reductions from the parameterized CLIQUE problem, both reductions relying on the same idea. Although we omit the proofs, we describe the key structure used, whose main properties are stated in Lemma 6. For a graph G and some $k \in \mathbb{N}$, we introduce a cmp $I^{G,k} = (S, C, H, f, L)$ as follows (see Fig. 1).

Let $V(G) = \{v_i | i \in [n]\}$, $|E(G)| = m$ and let ν be a bijection from $[m]$ into the set $\{(x, y) | v_x v_y \in E(G), x < y\}$. First, we construct a *node-gadget* $\mathcal{G}^i$ for each $i \in [k]$ and an *edge-gadget* $\mathcal{G}^{i,j}$ for each pair $(i, j) \in \binom{[k]}{2}$. The node-gadget $\mathcal{G}^i$ contains hospitals $H^i \cup G^i \cup \{f^i\}$, singles $S^i \cup T^i$ and a couple a^i. Analogously, the edge-gadget $\mathcal{G}^{i,j}$ contains hospitals $H^{i,j} \cup G^{i,j} \cup \{f^{i,j}\}$, singles $S^{i,j} \cup T^{i,j}$ and a couple $a^{i,j}$. Here $T^i = \{t^i_j | j \in [n-1]\}$ and $T^{i,j} = \{t^{i,j}_e | e \in [m-1]\}$, $H^i = \{h^i_j | j \in [n]\}$ and $H^{i,j} = \{h^{i,j}_e | e \in [m]\}$, and we define G^i, S^i and $G^{i,j}, S^{i,j}$ similarly to H^i and $H^{i,j}$. Observe that $|C| = k + \binom{k}{2}$.

We let $f \equiv 1$, so $I^{G,k}$ is 1-uniform. The precedence lists for each agent in $I^{G,k}$ are defined below. The notation $[X]$ for some set X in a preference list denotes an arbitrary ordering of the elements of X. We write Q^i_x for the set $\{s^{i,j}_e | i < j \leq k, \exists y : \nu(e) = (x, y)\} \cup \{s^{j,i}_e | 1 \leq j < i, \exists y : \nu(e) = (y, x)\}$ and $Q^{i,j}_e$ for $\{h^i_x, h^j_y\}$ where $\nu(e) = (x, y)$. The indices in the precedence lists take all possible values if not stated otherwise, and the symbol α can be any index in $[k]$ or a pair of indices in $\binom{[k]}{2}$. If α takes a value in $[k]$ then $N(\alpha) = n$, otherwise $N(\alpha) = m$.

$$L(g^\alpha_x) : t^\alpha_{x-1}, a^\alpha(2), t^\alpha_x \quad \text{if } 1 < x < N(\alpha)$$
$$L(g^\alpha_1) : a^\alpha(2), t^\alpha_1$$
$$L(g^\alpha_{N(\alpha)}) : t^\alpha_{N(\alpha)-1}, a^\alpha(2), a^\alpha(1)$$
$$L(t^\alpha_x) : g^\alpha_x, g^\alpha_{x+1}$$
$$L(f^\alpha) : s^\alpha_1, s^\alpha_2, \ldots, s^\alpha_{N(\alpha)}, a^\alpha(2)$$
$$L(a^\alpha) : (g^\alpha_{N(\alpha)}, f^\alpha), (h^\alpha_1, g^\alpha_{N(\alpha)}), (h^\alpha_2, g^\alpha_{N(\alpha)-1}), \ldots, (h^\alpha_{N(\alpha)}, g^\alpha_1)$$

$$L(h^i_x) : a^i(1), [Q^i_x], s^i_x$$
$$L(h^{i,j}_e) : a^{i,j}(1), s^{i,j}_e$$
$$L(s^i_x) : h^i_x, f^i$$
$$L(s^{i,j}_e) : h^{i,j}_e, [Q^{i,j}_e], f^{i,j}$$

Lemma 6. *For a graph G and $k \in \mathbb{N}$, $I^{G,k}$ has a stable assignment $M^{G,k}_0$ that covers each resident. Moreover, statements (1), (2), and (3) are equivalent:*

(1) *There is a clique in G of size k.*
(2) *There is a stable assignment M for $I^{G,k}$ with the following property, which we will call property π: $M(f^{i,j}) \subseteq S^{i,j}$ for each $(i, j) \in \binom{[k]}{2}$.*
(3) *There is a stable assignment for $I^{G,k}$ with property π covering each resident.*

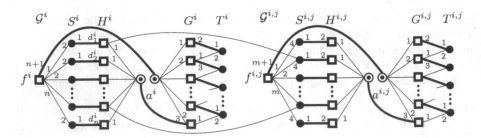

Fig. 1. A node- and an edge-gadget of $I^{G,k}$. Hospitals, singles, and couples are represented by rectangles, black, and double circles, resp. We connect $h \in H$ and $r \in R$ if $r \in A^L(h)$. Numbers show ranks, bold edges represent $M_0^{G,k}$, and $d_x^i = |Q_x^i| + 2$.

Proof. To see the first claim, we define an assignment M_0 by letting $M_0(a^\alpha) = (g_{N(\alpha)}^\alpha, f^\alpha)$, $M_0(t_x^\alpha) = g_x^\alpha$, and $M_0(s_x^\alpha) = h_x^\alpha$ for all possible α and x. As each resident is assigned to his best choice, M_0 is stable and covers each resident.

To prove (2) $\Rightarrow$ (1), suppose that $I^{G,k}$ has a stable assignment M with property π. Let us define $\sigma(i,j)$ for each $(i,j) \in \binom{[k]}{2}$ such that $M(f^{i,j}) = \{s_{\sigma(i,j)}^{i,j}\}$. Since $s_{\sigma(i,j)}^{i,j}$ prefers $h_{\sigma(i,j)}^{i,j}$ to $f^{i,j}$, the stability of M implies $M(h_{\sigma(i,j)}^{i,j}) = \{a^{i,j}(1)\}$. From this, we get that $M(s_e^{i,j}) = h_e^{i,j}$ must hold for each $e \in [m] \setminus \{\sigma(i,j)\}$ as otherwise $(s_e^{i,j}, h_e^{i,j})$ would be a blocking pair. Note that each single in $S^{i,j}$ is assigned to a hospital in $H^{i,j} \cup \{f^{i,j}\}$. As this holds for each $(i,j) \in \binom{[k]}{2}$, we get that $M(h_x^i) \subseteq S^i \cup \{a^i(1)\}$ holds for each $i \in [k]$, $x \in [n]$.

Let $\nu(\sigma(i,j)) = (x,y)$ for some $(i,j) \in \binom{[k]}{2}$. Since $s_{\sigma(i,j)}^{i,j}$ prefers the hospitals in $Q_{\sigma(i,j)}^{i,j} = \{h_x^i, h_y^j\}$ to $f^{i,j}$, M can only be stable if both h_x^i and h_y^j prefer their partner in M to $s_{\sigma(i,j)}^{i,j}$. This implies $M(h_x^i) = \{a^i(1)\}$ and $M(h_y^j) = \{a^j(1)\}$. Thus, by defining $\sigma(i)$ to be x if $M(a^i) = (h_x^i, g_{n+1-x}^i)$ for each $i \in [k]$, we obtain $\nu(\sigma(i,j)) = (\sigma(i), \sigma(j))$. From the definition of ν, this implies that $v_{\sigma(i)}$ and $v_{\sigma(j)}$ are adjacent in G. As this holds for every $(i,j) \in \binom{[k]}{2}$, we get that $\{v_{\sigma(i)} | i \in [k]\}$ is a clique in G.

Now we prove (1) $\Rightarrow$ (3). If $v_{\sigma(1)}, v_{\sigma(2)}, \ldots, v_{\sigma(k)}$ form a clique in G, then define $\sigma(i,j)$ such that $\sigma(i,j) = \nu^{-1}(\sigma(i), \sigma(j))$. We define a stable assignment M fulfilling property π and covering every resident as follows:

$$M(a^\alpha) = (h_{\sigma(\alpha)}^\alpha, g_{N(\alpha)+1-\sigma(\alpha)}^\alpha) \qquad M(t_x^\alpha) = g_x^\alpha \quad \text{if } x \in [N(\alpha) - \sigma(\alpha)]$$
$$M(s_{\sigma(\alpha)}^\alpha) = f^\alpha \qquad\qquad\qquad M(t_x^\alpha) = g_{x+1}^\alpha \quad \text{otherwise}$$
$$M(s_x^\alpha) = h_x^\alpha \quad \text{if } x \in [N(\alpha)] \setminus \{\sigma(\alpha)\}$$

The stability of M can be verified by simply checking all possibilities to find a blocking pair. (We note that many agents are only contained in $I^{G,k}$ to assure that M is indeed stable.) As (3) $\Rightarrow$ (2) is trivial, this finishes the proof. $\square$

Theorem 7. *There is an FPT time strict local search algorithm for* MAXIMUM HOSPITALS/RESIDENTS *with* COUPLES *with combined parameter* $(\ell, |C|)$.

Fig. 2. A possible component of G^δ. Winners and losers are marked by '+' and '−' signs, respectively. Bold edges represent M_0, normal edges represent M.

Proof. Let $I = (S, C, H, f, L)$ be given with the stable assignment M_0 and the integer ℓ. Although the case $f \equiv 1$ is different from the general case in many aspects, the trick of cloning the hospitals is applicable in our case (see the full version). Therefore, w.l.o.g. we may assume $f \equiv 1$. Thus, if $M(r) = h$ for some $r \in R$, then we will write $M(h) = r$ instead of $M(h) = \{r\}$.

Before describing the strict local search algorithm for MAXIMUM HOSPITALS/ RESIDENTS WITH COUPLES, we introduce some notation to capture the structure of the solution. The bipartite graph G underlying I has vertex set $H \cup R$ and edge set $E = \{hr | h \in H, r \in A^L(h)\}$. Clearly, an assignment M for I determines a matching $E(M)$ in G in the natural way: $hr \in E(M)$ if and only if $M(r) = h$ for some resident r and hospital h. Suppose that M is a closest solution, i.e. a stable assignment for I with $|M| > |M_0|$ and $d(M, M_0) \leq \ell$ that is the closest to M_0 among all such assignments. Let $A^\delta = \{a \in R \cup H | M(a) \neq M_0(a)\}$, and E^δ be the symmetric difference of $E(M_0)$ and $E(M)$. Note that E^δ covers exactly the vertices of A^δ, and $G^\delta = (A^\delta, E^\delta)$ is the union of paths and cycles which contain edges from M_0 and M in an alternating manner. It is well-known that for a cmp not containing couples every stable assignment covers exactly the same agents [7]. Thus, it is easy to see that the stability of M and M_0 imply that if a component of G^δ contains only single residents, then it must be a cycle. Let $\mathcal{K}_0$ denote the set of such cycles, and $\mathcal{K}_1$ the set of the remaining components of G^δ. We write C^δ for $(R \setminus S) \cap A^\delta$, and we define $B(a) = \{b | a \text{ is beneficial for } b \text{ w.r.t. } M_0\}$ for every $a \in S \cup H$. We also let $S^+ = \{s \in S | M(s) \text{ is beneficial for } s \text{ w.r.t. } M_0\}$, and $S^- = \{s \in S | M_0(s) \text{ is beneficial for } s \text{ w.r.t. } M\}$. Note that $S^+ \cup S^- = S \cap A^\delta$. We define H^+ and H^- analogously. We call agents in $A^+ = S^+ \cup H^+$ *winners* and agents in $A^- = S^- \cup H^-$ *losers*. For a simple illustration see Fig. 2.

Now, we describe an algorithm that finds vertices of A^δ. The algorithm first branches on guessing $|A^\delta|$ and a copy $\bar{G}$ of the graph G^δ. Let φ denote an isomorphism from $\bar{G}$ to G^δ. The algorithm also guesses the vertex sets $\varphi^{-1}(C^\delta)$, $\varphi^{-1}(H^+), \varphi^{-1}(H^-), \varphi^{-1}(S^+), \varphi^{-1}(S^-)$, and edge sets $\bar{E}_{M_0}$ and $\bar{E}_M$ denoting $\varphi^{-1}(E(M_0) \cap E^\delta)$ and $\varphi^{-1}(E(M) \cap E^\delta)$, respectively. Since $|A^\delta| \leq 2\ell$, it can be achieved by careful implementation that the algorithm branches into at most $(2\ell) \cdot 6^{2\ell}$ directions in this phase. Now, let Γ be an ordering of $V(\bar{G})$, i.e. a bijection from $V(\bar{G})$ to $[|A^\delta|]$. The algorithm colors the vertices of G with $|A^\delta| \leq 2\ell$ colors randomly with uniform and independent distribution, $\gamma(a)$ denotes the color of a. The coloring γ is *nice*, if $\gamma(\varphi(a)) = \Gamma(a)$ for each $a \in V(\bar{G})$. We suppose that γ is nice, which clearly holds with probability $|A^\delta|^{-|A^\delta|} \geq (2\ell)^{-2\ell}$.

Given a coloring, the algorithm grows a subset $X \subseteq V(\bar{G})$ on which φ is already known. It applies the following extension rules repeatedly (see Fig. 3), until none of them applies. When Rule 1 is applied, the algorithm branches into at most $2|C|$ branches, but no other branchings happen. We write $\bar{X} = V(\bar{G}) \setminus X$.

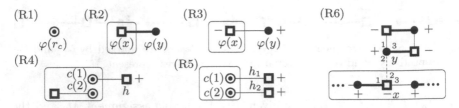

Fig. 3. Subgraphs of G^δ illustrating the rules of Theorem 7. Agents of $\varphi(X)$ are shown in a rectangular box. Bold edges represent $\bar{E}_{M_0}$, normal edges represent $\bar{E}_M$.

Rule 1 [guessing a member of a couple]: applicable if $r_c \in \bar{X} \cap \varphi^{-1}(C^\delta)$. In this case we simply branch on the vertices of $(R \setminus S) \cap \{a | \gamma(a) = \Gamma(c)\}$ to choose $\varphi(r_c)$. Note that this means at most $2|C|$ branches.

Rule 2 [finding pairs by M_0]: applicable if $x \in X, y \in \bar{X}$ and $xy \in \bar{E}_{M_0}$ for some x and y. By $\varphi(y) = M_0(\varphi(x))$, we can extend φ by adding y to X.

Rule 3 [finding pairs by M for losers]: applicable if $x \in X \cap \varphi^{-1}(A^-)$, $y \in \bar{X} \cap \varphi^{-1}(A^+)$ and $xy \in \bar{E}_M$ for some x and y. Let y^* be the first element in the list $L(\varphi(x))$ contained in $B(\varphi(x))$ having color $\Gamma(y)$. We claim $y^* = \varphi(y)$. Clearly, $\varphi(y) \in B(\varphi(x))$ holds because $\varphi(y)$ is a winner, and its color must be $\Gamma(y)$ as γ is nice. Now, suppose for contradiction that y^* precedes $\varphi(y)$ in $L(\varphi(x))$. Since the only vertex in A^δ having color $\Gamma(y)$ is $\varphi(y)$, we get $M(y^*) = M_0(y^*)$ implying that y^* and $\varphi(x)$ form a blocking pair for M. Thus, $\varphi(y) = y^*$ can be found in linear time, so we can extend φ by adding y to X.

Rule 4 [finding pairs by M for couples with one winner hospital]: applicable if $c(i) \in C^\delta \cap \varphi(X)$, $y \in \varphi^{-1}(H^+) \cap \bar{X}$, $\varphi^{-1}(c(i))y \in \bar{E}_M$, and $M(c(i'))$ is already known for some $c \in C$, $i \neq i'$ and y. W.l.o.g. we assume $i = 1$. Let h be defined such that $(h, M(c(2)))$ is the first element in $L(c)$ for which $h \in B(c(1))$ and h has color $\Gamma(y)$. We claim $\varphi(y) = h$. Observe that $\varphi(y) \in B(c(1))$ must hold because $\varphi(y)$ is a winner. As γ is nice, $\varphi(y)$ indeed has color $\Gamma(y)$. Thus, if $h \neq \varphi(y)$ then $(h, M(c(2)))$ precedes $(\varphi(y), M(c(2)))$ in $L(c)$, but this implies that the couple c and $(h, M(c(2)))$ form a blocking pair for M. Therefore, we get $\varphi(y) = h$, and we can extend φ in linear time by adding y to X.

Rule 5 [finding pairs by M for couples with two winner hospitals]: applicable if $c(i) \in C^\delta \cap \varphi(X)$, $y_i \in \varphi^{-1}(H^+) \cap \bar{X}$, and $\varphi^{-1}(c(i))y_i \in \bar{E}_M$ holds for both $i \in \{1, 2\}$, for some $c \in C$, y_1 and y_2. We let (h_1, h_2) be the first element in $L(c)$ such that $h_i \in B(c(i))$ and $\gamma(h_i) = \Gamma(y_i)$ for both $i \in \{1, 2\}$. Using the same arguments as in the previous case, we can show $\varphi(y_1) = h_1$ and $\varphi(y_2) = h_2$. Thus, we can extend φ in linear time by adding both y_1 and y_2 to X.

Rule 6 [dissolving a blocking pair]: applicable if $M(a) \in \varphi(X)$ if and only if $a \in \varphi(X)$ for all $a \in A^\delta$, and xy is a blocking pair for the *actual assignment* M_X. We define M_X by setting $M_X(a) = M_0(a)$ if $a \notin \varphi(X)$ and $M_X(a) = M(a)$ if $a \in \varphi(X)$, for each agent a. Note that by our first condition, M_X is indeed an assignment. Now, as xy cannot be a blocking pair for M or M_0, either $x \in \varphi(X)$

and $y \in A^\delta \setminus \varphi(X)$, or vice versa. W.l.o.g. we suppose the former. By defining $\bar{y} \in V(\bar{G})$ such that $\Gamma(\bar{y}) = \gamma(y)$, it can be seen that $\varphi(\bar{y}) = y$ must hold because γ is nice. Thus, φ can be extended by adding $\bar{y}$ to X.

Lemma 8. *If none of the rules is applicable, then $\varphi(X) = A^\delta$.*

If no extension rule is applicable, then we can obtain the solution M by Lemma 8. Each step takes linear time, the number of steps is at most 2ℓ, and the algorithm branches into at most $(2\ell)6^{2\ell}(2|C|)^\ell$ branches in total, thus the overall running time is $O(\ell(72|C|)^\ell|I|)$. The output is correct if the coloring γ is nice, which holds with probability at least $(2\ell)^{-2\ell}$. [3] $\square$

References

1. Aarts, E.H.L., Lenstra, J.K. (eds.): Local Search in Combinatorial Optimization. Wiley, New York (1997)
2. Alon, N., Yuster, R., Zwick, U.: Color-coding. J. ACM 42, 844–856 (1995)
3. Biró, P., McDermid, E.J.: Matching with couples is hard, but extra beds help (manuscript) (2009)
4. Fellows, M.R., Fomin, F.V., Lokshtanov, D., Rosamond, F.A., Saurabh, S., Villanger, Y.: Local search: Is brute-force avoidable? In: IJCAI 2009 (2009)
5. Flum, J., Grohe, M.: Parameterized Complexity Theory. Springer, Heidelberg (2006)
6. Gale, D., Shapley, L.S.: College admissions and the stability of marriage. American Mathematical Monthly 69, 9–15 (1962)
7. Gale, D., Sotomayor, M.: Some remarks on the stable matching problem. Discrete Appl. Math. 11, 223–232 (1985)
8. Glass, C.A., Kellerer, H.: Parallel machine scheduling with job assignment restrictions. Naval Research Logistics 54(3), 250–257 (2007)
9. Krokhin, A., Marx, D.: On the hardness of losing weight. In: Aceto, L., Damgård, I., Goldberg, L.A., Halldórsson, M.M., Ingólfsdóttir, A., Walukiewicz, I. (eds.) ICALP 2008, Part I. LNCS, vol. 5125, pp. 662–673. Springer, Heidelberg (2008)
10. Klaus, B., Klijn, F.: Stable matchings and preferences of couples. J. Econ. Theory 121, 75–106 (2005)
11. Marx, D.: A parameterized view on matroid optimization problems. In: Bugliesi, M., Preneel, B., Sassone, V., Wegener, I. (eds.) ICALP 2006. LNCS, vol. 4051, pp. 656–667. Springer, Heidelberg (2006)
12. Marx, D.: Local Search. Parameterized Complexity News 3, 7–8 (2008)
13. Marx, D.: Searching the k-change neighborhood for TSP is W[1]-hard. Oper. Res. Lett. 36(1), 31–36 (2008)
14. McDermid, E.J., Manlove, D.F.: Keeping partners together: Algorithmic results for the Hospitals/Residents problem with couples. To appear in J. of Comb. Opt.
15. Niedermeier, R.: Invitation to Fixed-Parameter Algorithms. Oxford University Press, Oxford (2006)
16. Ronn, E.: NP-Complete stable matching problems. J. Algorithms 11, 285–304 (1990)
17. Roth, A.E.: The evolution of the labor market for medical interns and residents: a case study in game theory. J. Polit. Econ. 92, 991–1016 (1984)
18. Roth, A.E., Sotomayor, M.: Two Sided Matching: A Study in Game-Theoretic Modelling and Analysis. Cambridge University Press, Cambridge (1990)

---

[3] To derandomize the algorithm, we can use the standard method of k-perfect hash functions [2], yielding a running time of $O(\ell^{O(\ell)}|C|^\ell|I| \log |I|)$.

Improved Parameterized Algorithms for the Kemeny Aggregation Problem

Narges Simjour

David Cheriton School of Computer Science
University of Waterloo
Waterloo, Ontario, Canada N2L 3G1
nsimjour@cs.uwaterloo.ca

Abstract. We give improvements over fixed parameter tractable (FPT) algorithms to solve the Kemeny aggregation problem, where the task is to summarize a multi-set of preference lists, called votes, over a set of alternatives, called candidates, into a single preference list that has the minimum total τ-distance from the votes. The τ-distance between two preference lists is the number of pairs of candidates that are ordered differently in the two lists. We study the problem for preference lists that are total orders. We develop algorithms of running times $O^*(1.403^{k_t})$, $O^*(5.823^{k_t/m}) \leq O^*(5.823^{k_{avg}})$ and $O^*(4.829^{k_{max}})$ for the problem, ignoring the polynomial factors in the O^* notation, where k_t is the optimum total τ-distance, m is the number of votes, and k_{avg} (resp. k_{max}) is the average (resp. maximum) over pairwise τ-distances of votes. Our algorithms improve the best previously known running times of $O^*(1.53^{k_t})$ and $O^*(16^{k_{avg}}) \leq O^*(16^{k_{max}})$ [3,4], which also implies an $O^*(16^{2k_t/m})$ running time. We also show how to enumerate all optimal solutions in $O^*(36^{k_t/m}) \leq O^*(36^{k_{avg}})$ time.

Keywords: Kemeny score, parameterized, exact algorithms, enumeration.

1 Introduction

Preference lists are typical elements of psychology questionnaires and social science surveys. In many cases, we wish to combine the gathered preference lists into a single list that reflects the opinion of the surveyed group as much as possible. The Kemeny aggregation problem, introduced by Kemeny in 1959, is a famous abstract form of this problem [9]. Given a set of m total orders, called *votes*, over a set of n alternatives, called *candidates*, the Kemeny-optimal aggregation problem asks for a total order over candidates, called an *optimal aggregation*, that minimizes the sum of τ-distances from the votes, where the τ-distance between total orders π_1, π_2 is the number of pairs of candidates that are ordered differently in the two total orders.

Bartholdi et al. [2] showed that the problem is NP-hard. Later, Dwork et al. [8] proved that the problem remains NP-hard for constant even m's as small

J. Chen and F.V. Fomin (Eds.): IWPEC 2009, LNCS 5917, pp. 312–323, 2009.

as $m = 4$. Their proof had a small error, which was fixed by Biedl et al. [5]. Dwork et al. used Kemeny's formalization [9] in their search for an effective spam filtering method that combined the results of multiple search engines. Their article [8] initiated a series of papers studying algorithmic aspects of the Kemeny aggregation problem. The problem was shown to have an $O(n^{2.5} + mn^2)$ 2-approximation [8]. Ailon et al. developed randomized approximation algorithms of ratios $11/7$ and $4/3$ [1]. Later, Kenyon-Mathieu and Schudy developed a PTAS for the feedback arc set problem for special weighted tournaments, which solves the Kemeny aggregation problem as a special case [10]. Despite being a theoretical breakthrough, this algorithm could not be used in practice. Recently, in an attempt to develop practical approximation algorithms, Williamson and van Zuylen derived a deterministic $8/5$-approximation algorithm for the problem [13]. The reader is referred to a survey by Charon and Hudry [6] for a detailed list of results.

Computational experiments of Davenport and Kalagnanam [7] suggest that the Kemeny aggregation problem might be easier to solve when an optimal aggregation is close to the input votes. In this direction, Betzler et al. [3] parameterized the problem with the sum of the τ-distances of the optimal aggregation from the input votes, denoted by k_t, and the maximum pairwise τ-distances of the input votes, denoted by k_{max}. They developed $O^*(1.53^{k_t})$ and $O^*((3k_{max}+1)!)$ time algorithms, giving the first FPT algorithms for the Kemeny aggregation problem [3]. Later, they parameterized the problem by the average pairwise τ-distances of the input votes, denoted by k_{avg}, and developed an algorithm that ran in time $O^*(16^{k_{avg}}) \leq O^*(16^{k_{max}})$ [4].

Our results. We develop parameterized algorithms of running times $O^*(1.403^{k_t})$, $O^*(5.823^{k_t/m}) \leq O^*(5.823^{k_{avg}})$, and $O^*(4.829^{k_{max}})$ for the problem, improving the previous best running times of $O^*(1.53^{k_t})$ [3] and $O^*(16^{k_{avg}}) \leq O^*(16^{k_{max}})$ [4]. We also give an algorithm to enumerate all optimal solutions in $O^*(36^{k_t/m})$ time. We are the first to parameterize the problem in terms of k_t/m, although, as $k_{avg} \leq 2k_t/m$, any FPT algorithm in terms of k_{avg} implies an FPT algorithm in terms of k_t/m. It is worth mentioning that k_t/m is smaller than k_{avg} and k_{max}, and therefore, parameterizing the problem with k_t/m, instead of k_{avg} or k_{max}, leads to a potentially tighter analysis of FPT algorithms.

m	Our Results			Previous Running Times		
	k_t	k_{avg}	k_{max}	k_t	k_{avg}	k_{max}
3	1.403^{k_t}	$1.968^{k_{avg}}$	$1.968^{k_{max}}$			
4	1.403^{k_t}	$2.760^{k_{avg}}$	$2.760^{k_{max}}$			
5	1.342^{k_t}	$3.241^{k_{avg}}$	$3.241^{k_{max}}$			
6	1.342^{k_t}	$4.348^{k_{avg}}$	$4.348^{k_{max}}$			
...	...	...	...			
m	$(2.415^{1/\lceil m/2 \rceil})^{k_t}$	$5.833^{k_{avg}}$	$4.829^{k_{max}}$	1.53^{k_t} [3]	$16^{k_{avg}}$ [4]	$16^{k_{max}}$ [4]

Fig. 1. A summary of the running times proved in this paper and the best previous running times. Only the exponential terms are listed.

Figure 1 summarizes the running times proved in this paper and the best previous running times, in terms of the three parameters k_t, k_{avg} and k_{max}.

We fix pertinent notation in Section 2 and explain the parameterized algorithms in Section 3.

2 Preliminaries

We use U to denote the set of candidates. A *binary relation* on U is a subset of $U \times U$. A binary relation R is *irreflexive* if no (x, x), $x \in U$, is in R, and is *asymmetric* if $(x, y) \in R$, $x, y \in U$, and $x \neq y$ implies $(y, x) \notin R$. In this article, we only work with irreflexive asymmetric binary relations. We may use $x <_R y$ to denote $(x, y) \in R$, and describe it as "R orders x before y".

A binary relation R is called *complete* if for any $x, y \in U$, $x \neq y$, either $(x, y) \in R$ or $(y, x) \in R$. A binary relation R is *transitive* if $(w, x) \in R$ and $(x, y) \in R$ imply $(w, y) \in R$. A *total order* is an irreflexive asymmetric binary relation that is complete and transitive. We use $\mathcal{T}_U$ to denote the set of total orders on U. We use R^+ for a transitive binary relation R to denote the transitive closure of R.

For any set $R \subseteq U \times U$, we use $rev(R)$ to denote $\{(b, a) : (a, b) \in R\}$; we may abuse the notation a little bit and use $rev((a, b))$ instead of $rev(\{(a, b)\})$. We say that $R_1 \subseteq U \times U$ is *consistent* with $R_2 \subseteq U \times U$ if $R_1 \cap rev(R_2) = \emptyset$.

Definition 1. *The τ-distance between $\pi_1, \pi_2 \in \mathcal{T}_U$, denoted by $\tau(\pi_1, \pi_2)$, is the cardinality of $\pi_1 - \pi_2$. For a multi-set $\mathcal{I}$ over $\mathcal{T}_U$, $\tau(\pi_1, \mathcal{I})$ is defined as the sum of $\tau(\pi_1, \pi_2)$ over all total orders π_2 in $\mathcal{I}$.*

An *optimal aggregation* of a multi-set $\mathcal{I}$ on $\mathcal{T}_U$ is a total order $\sigma \in \mathcal{T}_U$ that minimizes $\tau(\sigma, \mathcal{I})$. We use $OPT(\mathcal{I})$ to denote the set of all optimal aggregations. The *Kemeny aggregation problem* is the problem of finding an optimal aggregation for any given multi-set $\mathcal{I}$ on $\mathcal{T}_U$. For the case $|\mathcal{I}| = 1$ or 2, any $\sigma \in \mathcal{I}$ is an optimal aggregation [8]. Therefore, we are only interested in input instances that include more than two total orders.

We let *unanimity*$(\mathcal{I})$ denote the binary relation $\bigcap_{\pi \in \mathcal{I}} \pi$.

Observation 1. *[11] For any $\sigma \in OPT(\mathcal{I})$, unanimity$(\mathcal{I}) \subseteq \sigma$.*

Therefore, the Kemeny aggregation problem reduces to determining the order of dirty pairs, defined below:

Definition 2. *The set of* dirty pairs *of $\mathcal{I}$, denoted by dirty$(\mathcal{I})$, is $\{\{a, b\} : (a, b) \in \bigcup_{\pi \in \mathcal{I}}(\pi - unanimity(\mathcal{I}))\}$.*

We use $num_{(a,b)}(\mathcal{I})$ to denote the cardinality of $\{\pi \in \mathcal{I} : a <_\pi b\}$.

Definition 3. *The* majority graph *of $\mathcal{I}$, denoted by $M(\mathcal{I})$, is a weighted directed graph constructed as follows: for each $a \in U$, we put a vertex in $M(\mathcal{I})$ labeled as a. For each pair of vertices a and b, we put an edge from a to b if $num_{(a,b)}(\mathcal{I}) > num_{(b,a)}(\mathcal{I})$, and set its weight to $num_{(a,b)}(\mathcal{I}) - num_{(b,a)}(\mathcal{I})$.*

We refer to the edge set of a graph G as $E(G)$. We use weight(e) to refer to the weight of an edge $e \in E(G)$, and use weight(S) to refer to $\sum_{e \in S}$ weight(e) for any subset S of edges.

Definition 4. *A tournament majority graph of $\mathcal{I}$ is a supergraph TM of $M(\mathcal{I})$ whose set of vertices is U, which is a tournament, and in which the weight of any edge in $E(TM) - E(M(\mathcal{I}))$ is zero.*

Dwork et al. observed that σ is in $OPT(\mathcal{I})$ if and only if $E(M(\mathcal{I})) - \sigma$ is a minimum-weight feedback arc set for $M(\mathcal{I})$ [8]; a subset F of edges of a graph G is called a *feedback arc set* for G if $(E(G) - F) \cup rev(F)$ is transitive. The same observation is true for any tournament majority graph of $\mathcal{I}$:

Observation 2. *A total order σ is in $OPT(\mathcal{I})$ if and only if $E(TM) - \sigma$ is a minimum-weight feedback arc set for a tournament majority graph of $\mathcal{I}$.*

For weighted tournament graphs, the search tree algorithm of Raman and Saurabh [12] can be used to find a minimum-weight feedback arc set of size at most k edges in $O^*(2.415^k)$ time. We should mention that the original algorithm is designed for weighted tournaments with edge weights greater than or equal to one; however, the algorithm can be used for general weights if the search is confined to feedback arc sets that have no more than k edges. This variant is especially useful for finding a minimum-weight feedback arc set in tournament majority graphs, since although these graphs can have zero-weight edges, we will show that they have a minimum-weight feedback arc set with small number of edges.

We use MINFAS(G, k) to refer to this version of the algorithm, which is shown in Algorithm 1 for the sake of completeness. We use C_t to denote a cycle of length t.

Algorithm 1. MINFAS

> **Require** : G, k
> 1 $O \leftarrow$ BOUNDEDSEARCHTREE1$(G, \emptyset, k)$;
> 2 **return** $F \in \{E(G) - \sigma : \sigma \in O\}$ that has the minimum weight(F);

Lemma 1. *[12] Suppose that G is a weighted tournament graph and k is a positive integer. Then, MINFAS(G, k) returns a minimum-weight feedback arc set of G with at most k edges, if one exists, in time $O^*((1+\sqrt{2})^k) \approx O^*(2.415^k)$.*

Definition 5. *For any multi-set $\mathcal{I}$ with an optimal aggregation σ, $k_t = \tau(\sigma, \mathcal{I})$, $k_{avg} = avg\{\tau(\pi_1, \pi_2) : \pi_1, \pi_2 \in \mathcal{I}\}$, and $k_{max} = max\{\tau(\pi_1, \pi_2) : \pi_1, \pi_2 \in \mathcal{I}\}$.*

Observation 3. $k_t/(m-1) \leq k_{avg} \leq 2k_t/m$.

Proof. Since $\tau(\sigma, \mathcal{I}) \leq \tau(\pi_i, \mathcal{I})$, for any $\pi_i \in \mathcal{I}$, the inequality $\sum_{1 \leq i \leq m} \tau(\pi_i, \sigma) \leq \sum_{1 \leq i \leq m} \tau(\pi_i, \pi_j)$ holds for any $\pi_j \in \mathcal{I}$. Summing over all j's proves the first part of the observation. The second part follows from applying the triangular inequality $\tau(\pi_i, \pi_j) \leq \tau(\pi_i, \sigma) + \tau(\sigma, \pi_j)$ for every i and j. $\square$

Algorithm 2. BOUNDEDSEARCHTREE1

Require : G, L, k
1 **if** G does not have a C_3 **then**   /* no cycles remain */
2 | **return** $\{E(G)\}$;
3 **else if** $|L| = k$ **then**   /* cannot afford more edges */
4 | **return** $\emptyset$;
5 **else if** G has a C_3, $C = (V_c, E_c)$, with $E_c \cap L \neq \emptyset$ **then**
6 | **if** $E_c \subseteq L$ **then return** $\emptyset$;   /* L has a cycle */
7 | **else**
| $P \leftarrow \{\pi \in \mathcal{T}_{V_c} : (E_c - \pi)$ is a minimal FAS for C, π is consistent with $L\}$;
8 **else if** G has a C_4, $C = (V_c, E_c)$, **then**
9 | $P \leftarrow \{\pi \in \mathcal{T}_{V_c} : (E_c - \pi)$ is a minimal FAS for C, π is consistent with $L\}$;
10 **else**   /* C_3's in G do not have common edges */
11 | let (V_c, E_c) be a C_3 in G;
12 | let e be a minimum-weight edge in E_c;
13 | $P \leftarrow \{E_c - \{e\} \cup rev(\{e\})\}$;
14 **return** $\bigcup_{\pi \in P}$ BOUNDEDSEARCHTREE1$((G - rev(\pi)) + \pi, L \cup rev(E(G) - \pi))$;

We use $O^*(f(k, |\mathcal{I}|))$ to denote $O(f(k, |\mathcal{I}|) \cdot |\mathcal{I}|^c)$ for some constant c. In the rest of the paper, we assume that $\mathcal{I}$ is a multi-set on $\mathcal{T}_U$, $|\mathcal{I}| = m \geq 3$, $|U| = n$, and TM is an arbitrary tournament majority graph of $\mathcal{I}$.

3   Parameterized Algorithms

3.1   The Parameter k_t

In this section, we show how to improve the $O^*(1.53^{k_t})$ running time by Betzler et al. [3] to $O^*(1.403^{k_t})$. We base our analysis on the following lemma; for simplicity, we use d to denote $|E(TM) - unanimity(\mathcal{I})|$ and use k to denote $|E(TM) - \sigma|$ for an arbitrary $\sigma \in OPT(\mathcal{I})$.

Lemma 2. *For any $\pi \in \mathcal{T}_U$,*

$$(d - |E(TM) - \pi|) + |E(TM) - \pi| \cdot \lceil m/2 \rceil \leq \tau(\pi, \mathcal{I}).$$

Proof. Each of the pairs $(a, b) \in E(TM) - \pi$ indicates that π opposes the ordering of $\{a, b\}$ suggested by the majority. Also, by the definition of dirty pairs, for each of the dirty pairs not in $E(TM) - \pi$, there exists a total order in $\mathcal{I}$ that disagrees with the pair's ordering in π. The number of such pairs is at least $d - |E(TM) - \pi|$. Therefore, the number of disagreements of π with total orders in $\mathcal{I}$ is at least $(d - |E(TM) - \pi|) + |E(TM) - \pi| \cdot \lceil m/2 \rceil$. $\square$

Since $m \geq 3$, Lemma 2 proves the following relationship between d and k:

Corollary 1. $d + k \leq k_t$.

Algorithm 3. OPTAGGREGATION1

  Require : $\mathcal{I}$

1 $TM \leftarrow$ a tournament majority graph of $\mathcal{I}$;

2 $O \leftarrow$ BOUNDEDSEARCHTREE2($TM, unanimity(\mathcal{I})\}$);

3 **return** $\sigma \in O$ that minimizes $\tau(\sigma, \mathcal{I})$;

The idea is to use MINFAS($TM, k_t - d$) for large values of d, and develop an algorithm that runs fast for small values of d. In the following, we give a search tree algorithm, shown in Algorithm 3, that finds an optimal aggregation in time $O^*(3^{(d/2)})$.

The algorithm gradually decides on the orderings of dirty pairs and uses a set L to keep track of the pairs of vertices ordered so far. Each branch is stopped when either all dirty pairs are ordered in L or the computed L does not correspond to any total order.

Compared to Betzler et al.'s algorithm [3], Algorithm 3 incorporates a tournament majority graph into the search tree, and branches on triples of dirty pairs that form a C_3 in TM, instead of all triples of dirty pairs. Using the ideas in MINFAS [12], we go one step further, and consider C_4's whenever possible. Since we will use this search tree algorithm for small values of d, we modify BOUND-EDSEARCHTREE1 to optimize the running time for small d's. More precisely, in places that BOUNDEDSEARCHTREE1 branches on minimal feedback arc sets of a cycle, BOUNDEDSEARCHTREE2, shown in Algorithm 4, branches on all feedback arc sets of the cycle (lines 4 and 4 of BOUNDEDSEARCHTREE2).

Algorithm 4. BOUNDEDSEARCHTREE2

  Require : G, L

1 **if** G does not have a C_3 **then**   /* no cycles remain */

2 | **return** $\{E(G)\}$;

3 **else if** G has a $C_3 = (V_c, E_c)$ with $E_c \cap L \neq \emptyset$ **then**

4 | **if** $E_c \subseteq L$ **then return** $\emptyset$;   /* L has a cycle */

5 | **else** $P \leftarrow \{\pi \in \mathcal{T}_{V_c} : \pi$ is consistent with $L\}$;

6 **else if** G has a $C_4 = (V_c, E_c)$ **then**

7 | $P \leftarrow \{\pi \in \mathcal{T}_{V_c} : \pi$ is consistent with $L\}$;

8 **else**   /* C_3's in G do not have common edges */

9 | let (V_c, E_c) be a C_3 in G;

10 | let e be a minimum-weight edge in E_c;

11 | $P \leftarrow \{E_c - \{e\} \cup rev(\{e\})\}$;

12 **return** $\bigcup_{\pi \in P}$ BOUNDEDSEARCHTREE2($(G - rev(\pi)) + \pi, L \cup \pi$);

Theorem 1. OPTAGGREGATION1($\mathcal{I}$) *returns an optimal aggregation of* $\mathcal{I}$ *in time* $O^*((\sqrt{3})^d)$.

Due to space limitations, we do not give a proof for Theorem 1.

Theorem 2. *An optimal aggregation can be found in $O^*(1.403^{k_t})$ time.*

Proof. If $d \geq \frac{2\log_2(1+\sqrt{2})k_t}{\log_2(3)+2\log_2(1+\sqrt{2})}$, then by Lemma 1 we can run MINFAS(TM, $k_t - d$) to obtain an optimal aggregation in time

$$O^*((1+\sqrt{2})^{k_t-d}) \leq O^*((1+\sqrt{2})^{(1-\frac{2\log_2(1+\sqrt{2})}{\log_2(3)+2\log_2(1+\sqrt{2})})k_t}) < O^*(1.403^{k_t}).$$

Otherwise, if $d < \frac{2\log_2(1+\sqrt{2})k_t}{\log_2(3)+2\log_2(1+\sqrt{2})}$, then by Theorem 1 we can run OPTAG-GREGATION1($\mathcal{I}$) to find the optimal aggregation in time

$$O^*((\sqrt{3})^d) \leq O^*((\sqrt{3})^{\frac{2\log_2(1+\sqrt{2})k_t}{\log_2(3)+2\log_2(1+\sqrt{2})}}) < O^*(1.403^{k_t}). \qquad \square$$

In the next section, we study parameters other than k_t.

3.2   The Parameters k_t/m, k_{avg}, and k_{max}

Since the value of k_t generally increases when m is increased, it is more reasonable to study the problem with respect to k_t/m or other parameters that do not depend on m. In this section, we consider the parameters k_t/m, k_{avg}, and k_{max}.

Again, Lemma 2 plays an essential role:

Corollary 2. *For any $\pi \in \mathcal{T}_U$, $|E(TM) - \pi| \leq \tau(\pi, \mathcal{I})/\lceil m/2 \rceil$.*

Corollary 3. *$k \leq k_t/\lceil m/2 \rceil$.*

Corollary 3 and Lemma 1 prove that we can use MINFAS($TM, k_t/\lceil m/2 \rceil$) to compute an optimal aggregation in $O^*(2.415^{k_t/\lceil m/2 \rceil})$ time.

Theorem 3. *An optimal aggregation can be found in time $O^*(2.415^{k_t/\lceil m/2 \rceil}) \leq O^*(2.415^{((m-1)/\lceil m/2 \rceil)k_{avg}}) \leq O^*(2.415^{((m-1)/\lceil m/2 \rceil)k_{max}})$.*

Note that for $m \geq 5$ the bound $O^*(2.415^{k_t/\lceil m/2 \rceil})$ of Theorem 3 is better than the bound $O^*(1.403^{k_t})$ in Theorem 2, with respect to k_t.

In the remainder of this section, we focus on the parameter k_{max} and show how to improve the running time $O^*(2.415^{((m-1)/\lceil m/2 \rceil)k_{max}}) \approx O^*(5.833^{k_{max}})$ to $O^*((4.829)^{k_{max}})$. The idea is to work with a total order in $\mathcal{I}$ that is close to some $\sigma \in OPT(\mathcal{I})$, and agrees with the majority of $\mathcal{I}$ in most pair orderings. The precise algorithm, called OptAggregation2, is shown in Algorithm 5.

The algorithm goes through every $\pi \in \mathcal{I}$. For every π, the algorithm assumes that π is close to some optimal aggregation, and starts the search by deciding on the ordering of the pairs in $E(TM) - \pi$, using the assumption to confine the search space.

Theorem 4. OPTAGGREGATION2$(\mathcal{I})$ *returns an optimal aggregation in time $O^*(4.829^{k_{max}})$.*

Algorithm 5. OPTAGGREGATION2

   Require : $\mathcal{I}$
1 $TM \leftarrow$ a tournament majority graph of $\mathcal{I}$;
2 $k_{max} \leftarrow \max\{\tau(\pi_1, \pi_2) : \pi_1, \pi_2 \in \mathcal{I}\}$;
3 Initialize Q to $E(TM)$;   /* max-weight subset of edges */
4 **foreach** $\pi \in \mathcal{I}$ **do**
5   **foreach** $S \subseteq (E(TM) - \pi)$ with $|S| \leq k_{max}$ **do**
6          $P_1 \leftarrow (E(TM) - \pi) - S$;
7          $P_2 \leftarrow$ MINFAS$(TM - P_1 + rev(P_1), k_{max} - |S|)$;
8   **if** weight$(P_1 \cup P_2) <$ weight(Q) **then** $Q \leftarrow P_1 \cup P_2$;
9   **end**
10 **end**
11 **return** $(E(TM) - Q) \cup rev(Q)$;

Proof. Since any computed P_2 in line 5 is a feedback arc set for $TM - P_1 + rev(P_1)$, $P_1 \cup P_2$ is always a feedback arc set for TM. By Observation 2, the algorithm is proved to be sound once we show that $P_1 \cup P_2$ is set to a minimum-weight feedback arc set for TM at some point. We suppose that σ is an optimal aggregation. There exists some π in line 5 such that $\tau(\sigma, \pi) \leq k_{max}$; since otherwise, $mk_{max} < \tau(\sigma, \mathcal{I})$, proving that $mk_{max} < \tau(\omega, \mathcal{I})$ for every $\omega \in \mathcal{I}$, which violates the definition of k_{max}.

The set $(E(TM) - \pi) - (E(TM) - \sigma)$ is among the enumerated S's in line 5, since it is a subset of $\sigma - \pi$ and therefore its size is at most k_{max}. The P_1 corresponding to this S will be $(E(TM) - \pi) \cap (E(TM) - \sigma)$.

We claim that weight(P_2) = weight$((E(TM) - \sigma) - P_1)$. The weight of P_2 is not larger than weight$((E(TM) - \sigma) - P_1)$, since $(E(TM) - \sigma) - P_1$ is a feedback arc set for $TM - P_1 + rev(P_1)$ that has no more than $k_{max} - |S|$ edges: the two sets $S = (E(TM) - \pi) - (E(TM) - \sigma)$ and $((E(TM) - \sigma) - P_1) = ((E(TM) - \sigma) - (E(TM) - \pi))$ are disjoint subsets of $E(TM)$, and in both sets, each edge connects a pair of vertices that are ordered differently by σ and π. Therefore, $|(E(TM) - \sigma) - P_1| + |S|$ is at most $\tau(\sigma, \pi)$, which is no more than k_{max}, due to the choice of π.

Consequently, the weight of $P_1 \cup P_2$ is at most weight(P_1)+weight$((E(TM) - \sigma) - P_1)$. Since $P_1 \subseteq (E(TM) - \sigma)$, this weight is weight$(E(TM) - \sigma)$. Therefore, $P_1 \cup P_2$ is a minimum-weight feedback arc set in some iteration. This proves that the algorithm is sound.

The OptAggregation2 algorithm can construct the majority graph in $O(mn^2)$ time. Also, computing the value of k_{max} takes at most $O(m^2n^2)$ time. The cost of the loop is dominated by the cost of lines 5 and 5. The first branching step, line 5, takes $O\left(\binom{|E(TM)-\pi|}{i}\right)$ time, for any $1 \leq i \leq k_{max}$. By Lemma 1, the second branching step, line 5, takes $O^*((2.415)^{k_{max}-i})$ time. Overall, the algorithm runs in time

$$O^*\left(m^2n^2 + m \cdot \sum_{1 \leq i \leq k_{max}} \binom{|E(TM)-\pi|}{i} \cdot 2.415^{(k_{max}-i)}\right) =$$
$$O^*\left(\sum_{1 \leq i \leq k_{max}} \binom{|E(TM)-\pi|}{i} \cdot 2.415^{(k_{max}-i)}\right)$$

By the definition of k_{max}, $\tau(\pi, \mathcal{I}) \leq (m-1)k_{max}$. Hence, Corollary 2 proves that $|E(TM) - \pi| < 2k_{max}$, and the running time is bounded by

$$O^*(\textstyle\sum_{1 \leq i \leq k_{max}} \binom{2k_{max}}{i} \cdot 2.415^{(k_{max}-i)}) =$$
$$O^*(2.415^{(-k_{max})} \textstyle\sum_{1 \leq i \leq k_{max}} \binom{2k_{max}}{i} \cdot 2.415^{(2k_{max}-i)}) =$$
$$O^*(2.415^{(-k_{max})} \textstyle\sum_{1 \leq i \leq 2k_{max}} \binom{2k_{max}}{i} \cdot 2.415^{(2k_{max}-i)}) =$$
$$O^*(2.415^{(-k_{max})}(1 + 2.415)^{2k_{max}}) < O^*(4.829^{k_{max}}) . \qquad \square$$

Theorem 4 improves the best previous running time of $O^*(16^{k_{max}})$ by Betzler et al. [4].

3.3   Enumerating Optimal Aggregations

In this section, we give an algorithm, shown in Algorithm 6, to enumerate $OPT(\mathcal{I})$. The key point is to focus on candidates that are ordered consecutively in an optimal aggregation. To this end, we define $seq(\pi)$ for a total order $\pi \in \mathcal{T}_U$ as $\{(a, b) :$ for no $c \in U$, $a <_\pi c <_\pi b\}$. Our algorithm uses the fact that elements of $seq(\pi)$ are edges of $M(\mathcal{I})$, for any optimal aggregation π. We also define $interval_\pi((u, v))$ for any $(v, u) \in \pi$ as $\{x : x \in U, v <_\pi x <_\pi u\}$. The order of u and v may appear strange in this definition; however, the current order makes more sense since we want to use the interval notation for edges whose reversals are in π. For a set $S = \{(u_1, v_1), (u_2, v_2), \ldots, (u_{|S|}, v_{|S|})\}$, we use $head(S)$ and $tail(S)$ to denote the sets $\{v_1, v_2, \ldots, v_{|S|}\}$ and $\{u_1, u_2, \ldots, u_{|S|}\}$.

The algorithm uses a total order σ to enumerate the optimal aggregations. Although any arbitrary σ is good for the enumeration, σ is chosen to be an optimal aggregation in order to have the desired running time.

For any total order π, $seq(\pi)$ has a set S of edges in common with $E(TM) - \sigma$, and for every edge $(u, v) \in S$, π orders zero or more candidates in U before u and v, and orders the others after u and v. As a result, any total order is mapped to a *partition representation*, which consists of a partitioning of $E(TM) - \sigma$ into $S = \{(u_1, v_1), (u_2, v_2), \ldots, (u_{|S|}, v_{|S|})\}$ and $E(TM) - \sigma - S$, and a sequence of partitions of the candidates in $interval_\sigma((u_i, v_i)) - head(S)$ into the candidates ordered before u_i, v_i and the candidates ordered after u_i, v_i, in π, for every $1 \leq i \leq |S|$. For example, assuming that $1 <_\sigma 2 <_\sigma 3 <_\sigma 4 <_\sigma 5 <_\sigma 6 <_\sigma 7$ and $E(TM) - \sigma = \{(3, 1), (5, 2), (7, 3)\}$, the total order $1 <_\pi 6 <_\pi 7 <_\pi 3 <_\pi 4 <_\pi 5 <_\pi 2$ is mapped to $S = \{(5, 2), (7, 3)\}$, and the presence of $(4, 5)$ in π indicates that $interval_\sigma((5, 2)) - head(S)$ is partitioned into $\{4\}$ and $\emptyset$, the presence of $(6, 7), (7, 4)$, and $(7, 5)$ in π indicates that $interval_\sigma((7, 3)) - head(S)$ is partitioned into $\{6\}$ and $\{4, 5\}$.

We will prove that π can be reconstructed from its partition representation if $seq(\pi) \subseteq E(TM)$. Consequently, the set of all optimal aggregations can be computed by going through all possible partition representations.

In Algorithm 6, the sets S and R specify a partition representation, and P holds potential orderings of the candidates in $interval_\sigma((u, v)) - head(S)$ before u, for any $(u, v) \in S$. The set Q is the total order reconstructed from the partition representation.

Algorithm 6. ENUMAGGREGATIONS

Require : $\mathcal{I}$

1   $OPT \leftarrow \emptyset$;

2   $\sigma \leftarrow$ OPTAGGREGATION1($\mathcal{I}$);

3   **foreach** tournament majority graph TM of $\mathcal{I}$ **do**

4   **foreach** $S \subseteq (E(TM) - \sigma)$ **do**

5   **if** $|head(S)| = |tail(S)| = |S|$ **then**

6            $P \leftarrow \{(x, u) : (u, v) \in S, x \in (interval_\sigma((u, v)) - head(S))\}$;

7   **foreach** $R \subseteq P$ **do**

8               $R \leftarrow R \cup rev(P - R) \cup S$;

9               $R \leftarrow R \cup \{(x, u) : (u, v) \in S, (x, v) \in R^+\}$;

10               $R \leftarrow R \cup \{(v, y) : (u, v) \in S, (u, y) \in R^+\}$;

11               $Q \leftarrow (\sigma - rev(R^+)) \cup R^+$;

12   **if** Q is transitive and $\tau(Q, \mathcal{I}) = \tau(\sigma, \mathcal{I})$ **then**
                 $OPT \leftarrow OPT \cup \{Q\}$;

13   **end**

14   **end**

15   **end**

16 **end**

17 **return** OPT;

Due to space limitations, the proofs of the following two lemmas are eliminated.

Lemma 3. *In Algorithm 6, if there exists $\pi \in \mathcal{T}_U$ that is consistent with R, $seq(\pi) \subseteq E(TM)$ and $seq(\pi) \cap (E(TM) - \sigma) = S$, then $Q = \pi$.*

Lemma 4. *In Algorithm 6, for any fixed TM, $|P| \leq 2k_t/m - |S|$.*

Theorem 5. ENUMAGGREGATIONS($\mathcal{I}$) *returns $OPT(\mathcal{I})$ in time $O^*(36^{k_t/m})$.*

Proof. ENUMAGGREGATIONS($\mathcal{I}$) iterates through all possible orderings of pairs $\{\{a, b\} : num_{(a,b)}(\mathcal{I}) = num_{(b,a)}(\mathcal{I})\}$. For any fixed ordering L of these pairs, the algorithm searches for the subset $OPT_L(\mathcal{I}) = \{\pi \in OPT(\mathcal{I}) : \pi$ is consistent with $L\}$. It divides $OPT(\mathcal{I})$ further in line 6, and looks for the subset $OPT_{L,S}(\mathcal{I})$ of $OPT_L(\mathcal{I})$ defined as $\{\pi \in OPT_L(\mathcal{I}) : seq(\sigma) \cap (E(TM) - \sigma) = S\}$. All potential S's are produced in line 6. Line 6 removes those S's that contain two edges with the same head or the same tail, since the seq of a total order cannot contain such edges. Finally, a set P of pairs is computed such that any decision on the orderings of the pairs in P narrows $OPT_{L,S}(\mathcal{I})$ down dramatically. Indeed, Lemma 3 proves that for any chosen R there is either one or zero $\pi \in OPT_{L,S}(\mathcal{I})$ that is consistent with R^+. Furthermore, in case there exists one such π, it is consistent with the transitive relation Q (in line 6). Consequently, we can produce all total orders in $OPT(\mathcal{I})$ by going through all possible R's and see if Q becomes transitive and it is indeed an optimal aggregation.

For any chosen TM, the number of iterations with $|S| = i$, $0 \leq i \leq |E(TM) - \sigma|$, will be $\binom{|E(TM)-\sigma|}{i} \times 2^{|P|}$. Due to Lemma 4, $|P| \leq 2k_t/m - |S|$. Therefore,

the number of iterations for each TM is at most $\sum_{0 \le i \le |E(TM)-\sigma|} \binom{|E(TM)-\sigma|}{i} \times$
$2^{2k_t/m-i}$. By Corollary 2, $|E(TM)-\sigma| \le 2k_t/m$. Therefore, this value is bounded
by

$$\sum_{0 \le i \le 2k_t/m} \binom{2k_t/m}{i} \times 2^{2k_t/m-i} = 2^{2k_t/m} \times (1+1/2)^{2k_t/m} = 9^{k_t/m}$$

Any edge $(a,b) \in L$ indicates that σ opposes the preference of exactly $m/2$ total
orders in $\mathcal{I}$. Therefore, $|L| \le 2k_t/m$. Since there are $2^{|L|}$ possible TM's, the total
number of iterations is bounded by $36^{k_t/m}$. $\qquad\square$

4   Concluding Remarks

In Sections 3.1 and 3.2, a parameterized algorithm for the feedback arc set
problem for tournaments [12] is used as a core algorithm. Any improvement in
the running time of this algorithm will improve our bounds.

In Section 3.3, the current bounds for odd and even m's have a large gap.
The question is whether we can make the $O^*(36^{k_t/m})$ upper bound for even m's
closer to the $O^*(9^{k_t/m})$ bound for odd m's. Furthermore, there are at most $4^{k_t/m}$
Hamiltonian paths in the majority graph if $|head(E(TM)-\sigma)| = |tail(E(TM)-\sigma)| = |E(TM) - \sigma|$, for some $\sigma \in OPT(\mathcal{I})$. It seems to us that the same bound,
instead of the current $9^{k_t/m}$ bound, should hold for the case this restriction is
released.

Acknowledgements. The author is grateful to Naomi Nishimura, Jonathan
Buss, and anonymous reviewers for helpful comments, and to Timothy Chan for
his suggestion to consider the parameter k_t/m.

References

1. Ailon, N., Charikar, M., Newman, A.: Aggregating inconsistent information: Ranking and clustering. Journal of the ACM 55(5), 1–27 (2008)
2. Bartholdi, J.J., Tovey, C.A., Trick, M.A.: Voting schemes for which it can be difficult to tell who won the election. Social Choice and Welfare 6(2), 157–165 (1989)
3. Betzler, N., Fellows, M.R., Guo, J., Niedermeier, R., Rosamond, F.A.: Fixed-parameter algorithms for Kemeny scores. In: Fleischer, R., Xu, J. (eds.) AAIM 2008. LNCS, vol. 5034, pp. 60–71. Springer, Heidelberg (2008)
4. Betzler, N., Fellows, M.R., Guo, J., Niedermeier, R., Rosamond, F.A.: How similarity helps to efficiently compute Kemeny rankings. In: AAMAS 2009: Proc. of the 8th Int. Conf. on Autonomous Agents and Multiagent Systems, pp. 657–664 (2009)
5. Biedl, T., Brandenburg, F.J., Deng, X.: Crossings and permutations. In: Healy, P., Nikolov, N.S. (eds.) GD 2005. LNCS, vol. 3843, pp. 1–12. Springer, Heidelberg (2006)
6. Charon, I., Hudry, O.: A survey on the linear ordering problem for weighted or unweighted tournaments. 4OR 5(1), 5–60 (2007)

7. Davenport, A., Kalagnanam, J.: A computational study of the Kemeny rule for preference aggregation. In: AAAI 2004: Proc. of the 19th National Conf. on Artificial Intelligence, pp. 697–702 (2004)
8. Dwork, C., Kumar, R., Naor, M., Sivakumar, D.: Rank aggregation methods for the web. In: WWW 2001: Proc. of the 10th Int. Conf. on World Wide Web, pp. 613–622 (2001)
9. Kemeny, J.G.: Mathematics without numbers. Daedalus 88, 575–591 (1959)
10. Kenyon-Mathieu, C., Schudy, W.: How to rank with few errors. In: STOC 2007: Proc. of the 39th Annual ACM Symp. on Theory of Computing, pp. 95–103 (2007)
11. Monjardet, B.: Tournois et ordres médians pour une opinion. Mathématiques et Sciences humaines, 55–73 (1973)
12. Raman, V., Saurabh, S.: Parameterized algorithms for feedback set problems and their duals in tournaments. Theoretical Computer Science 351(3), 446–458 (2006)
13. Williamson, D.P., van Zuylen, A.: Deterministic algorithms for rank aggregation and other ranking and clustering problems. In: Kaklamanis, C., Skutella, M. (eds.) WAOA 2007. LNCS, vol. 4927, pp. 260–273. Springer, Heidelberg (2008)

Computing Pathwidth Faster Than $2^{n\star}$

Karol Suchan[1,2] and Yngve Villanger[3]

[1] Facultad de Ingeniería y Ciencias, Universidad Adolfo Ibáñez, Santiago, Chile
[2] WMS, AGH University of Science and Technology, Cracow, Poland
karol@suchan.info
[3] University of Bergen, N-5020 Bergen, Norway
yngve.villanger@uib.no

Abstract. Computing the PATHWIDTH of a graph is the problem of finding a tree decomposition of minimum width, where the decomposition tree is a path. It can be easily computed in $\mathcal{O}^*(2^n)$ time by using dynamic programming over all vertex subsets. For some time now there has been an open problem if there exists an algorithm computing PATHWIDTH with running time $\mathcal{O}^*(c^n)$ for $c < 2^1$. In this paper we show that such an algorithm with $c = 1.9657$ exists, and that there also exists an approximation algorithm and a constant τ such that an $opt + \tau$ approximation can be obtained in $\mathcal{O}^*(1.89^n)$ time.

1   Introduction

PATHWIDTH is a graph parameter defined in the same way as TREEWIDTH with the exception that the decomposition tree is requested to be a path. Both these parameters where introduced by Robertson and Seymour [17] in their graph minor project, and both parameters have algorithmic applications. Examples of these are algorithms that use dynamic programming over a decomposition. Such algorithms will have running times consisting of a polynomial part, and an exponential part that only depends on the width of the decomposition.

The TREEWIDTH and PATHWIDTH problems have been substantially studied for two decades, and this has resulted in massive literature. For an introduction see [2]. Both problems are NP-hard already in cocomparability graphs [11], and PATHWIDTH is NP-hard even in some restricted subclasses of chordal graphs and in weighted trees [16]. Some examples of efficient algorithms include polynomial time algorithms for CIRCLE and CIRCULAR-ARC graphs [15,18], and PERMUTATION graphs [6], and fixed parameter tractable algorithms for both problems [3]. There also exist approximation algorithms for both problems, where Feige et al. [8] give the most recent algorithm for treewidth.

When it comes to exponential time algorithms the picture changes. For TREEWIDTH there are several results: Arnborg et al. give an $\mathcal{O}(n^{tw+2})$ time algorithm for treewidth, where tw is the treewidth of the graph[1]; Fomin et al.

$\star$ This work is supported by the Research Council of Norway and by the Basal-CMM program of CONICYT, Chile.
[1] $f(n) = \mathcal{O}^*(g(n))$ if there is a polynomial function $p(n)$ s.t. $f(n) \leq p(n)g(n)$.

J. Chen and F.V. Fomin (Eds.): IWPEC 2009, LNCS 5917, pp. 324–335, 2009.
© Springer-Verlag Berlin Heidelberg 2009

give $\mathcal{O}(c^n)$ time algorithms with $c < 2$[9,10]. These results are based on the property that the treewidth problem decomposes into independent subproblems if one bag or separator of the tree decomposition under construction is known. To be more precise, each of the connected components of the graph remaining when the vertices of the bag are removed defines one independent subproblem. Adapting this approach to the pathwidth problem faces a major difficulty, as the remaining connected components cannot be treated independently, but have to be divided into a left and right set, with the components in a same set being merged together. This reflects the fact that in a tree decomposition there can be an arbitrary number of independent branches whereas in a path decomposition there may be only two of them. Thus, for a star on $n + 1$ vertices there will be 2^n possible partitions.

As a result, PATHWIDTH is often mentioned in the same breath as CUTWIDTH, DIRECTED OPTIMAL LINEAR ARRANGEMENT, DIRECTED FEEDBACK ARC SET (for a longer list see [4]), since all three can be expressed as vertex ordering problems, and for each of them the best known exact algorithm solves the problem in $\mathcal{O}^*(2^n)$ time. Furthermore, the bound can be reached for all three problems by dynamic programming over the 2^n different vertex subsets. This technique is often refereed to as the Held and Karp algorithm [13]. In the end of [4] the following citation can be found *"On the more theoretical side, it is interesting to try to improve the time bounds. Some problems appear to be hard, e.g., to improve upon the $\mathcal{O}^*(2^n)$ time for Pathwidth."*

In this paper we show that the PATHWIDTH problem can be solved in (asymptotic) time less than 2^n by giving an algorithm that runs over $\mathcal{O}^*(c^n)$ vertex partitions, with $c = 1.9657$. In addition to this, we show that there exits a constant τ such that an *opt* $+ \tau$ approximation can be obtained in $\mathcal{O}^*(1.89^n)$ time.

2   Preliminaries

All considered graphs are simple and undirected. For a graph $G = (V, E)$ we denote by $n = |V|$ the number of vertices and by $m = |E|$ the number of edges. The neighborhood of a vertex v is defined as $N(v) = \{u \in V : \{u, v\} \in E\}$, and the closed neighborhood is defined as $N[v] = N(v) \cup \{v\}$. The cardinality $|N(v)|$ is called the degree of v. For a vertex set W, we define its neighborhood as $N(W) = \bigcup_{v \in W} N(v) \setminus W$, and its closed neighborhood as $N[W] = N(W) \cup W$. A graph $G' = (V', E')$ with $V' \subseteq V$ and $E' \subseteq E$ is called a subgraph of G. For a vertex set $W \subseteq V$, the subgraph of G induced by W is denoted $G[W] = (W, E_W)$, where E_W is the restriction of E to edges having both incident vertices in W. A graph G is not connected if it is possible to partition its vertex set into two non-empty subsets, such that no edge of G is incident to vertices in both parts; otherwise, G is connected. A clique is a graph containing an edge for each pair of distinct vertices. A cycle is a connected graph in which all vertices are of degree 2. A path is a connected graph in which all vertices are of degree at most 2 and that contains no cycle as a subgraph. In an intuitive way, we sometimes consider a path to be a permutation of its vertex set. A connected component is an inclusion maximal subset of vertices that induces a connected subgraph.

For two vertices $a, b \in V$, a set $S \subseteq V$ is an a, b-separator if a and b belong to different connected components of $G[V \setminus S]$. S is a minimal a, b-separator if no proper subset of S is an a, b-separator. In general, S is a minimal separator in G if there exist $a, b \in V$, such that S is a minimal a, b-separator.

The pathwidth of a graph G is defined through *path decompositions*. A *path decomposition* of a graph $G = (V, E)$ is a pair (χ, P) in which $P = (V_P, E_P)$ is a path and $\chi = \{\chi_i \mid i \in V_P\}$ is a family of subsets of V, called *bags*, such that

(i) $\bigcup_{i \in V_P} \chi_i = V$;
(ii) for each edge $\{u, v\} \in E$ there exists an $i \in V_P$ such that both u and v belong to χ_i; and
(iii) for all $v \in V$, the set of nodes $\{i \in V_P \mid v \in \chi_i\}$ induces a connected subgraph of P.

The maximum of $|\chi_i| - 1$, $i \in V_P$, is called the *width* of the path decomposition. The *pathwidth* of a graph G, denoted by $pw(G)$, is the minimum width taken over all path decompositions of G. To help distinguish between the vertices of a graph G and the vertices of its decomposition path P we use the term *node* for the latter.

Treewidth is defined analogously to pathwidth, but is based on *tree decompositions* in place of *path decompositions*. A *tree decomposition* of a graph $G = (V, E)$, is a pair (χ, T) in which $T = (V_T, E_T)$ is a tree and $\chi = \{\chi_i \mid i \in V_T\}$ is a family of subsets of V, called *bags*, such that conditions $(i), (ii), (iii)$ of the path decomposition, with T put in place of P, hold. The treewidth is the minimum width over all possible tree decompositions.

3   Computation of Pathwidth

A naive and well known way to compute the pathwidth of a graph, is to enumerate all $n!$ vertex orderings. Fix a permutation α, called a vertex ordering, of the vertex set V; $\alpha(i)$ gives the ith vertex in α and $\alpha^{-1}(v)$ gives the position of v in α. Given a graph $G = (V, E)$ and a vertex ordering α, let $V_j = \bigcup_{i=1}^{j} \alpha(i)$. A path decomposition $(\chi, P)_\alpha$ of G can now be obtained from α as follows. For $i \in \{1, \ldots, n\}$ let χ_i be a bag containing $N[V_i] \setminus V_{i-1}$. In the opposite direction, we can consider a provided path decomposition (χ, P), where $\chi_1, \chi_2, \ldots, \chi_r$ are the bags in the order defined by P; then pick a vertex ordering α_P, such that for all pairs $u, v \in V$ for which all bags containing u have smaller index than those containing v ($u \in \chi_i$ and $v \in \chi_j$ implies that $i < j$) there is $\alpha_P^{-1}(u) < \alpha_P^{-1}(v)$. One simple way to obtain such an ordering is to number vertices in the order they disappear in $\bigcup_{j=i}^{r} \chi_j$ for increasing values of i. Notice that α_P defines a path decomposition of the same width as (χ, P), since by properties of a path decomposition χ_j contains $N(W)$ where $W = \bigcup_{i=1}^{j} \chi_i \setminus \chi_j$. Thus, pathwidth can be computed by enumerating all vertex orderings and returning the minimum obtained width.

This is not an efficient way to compute the pathwidth since $n!$ orderings have to be checked, but it provides the basic idea we will use to compute the pathwidth in $\mathcal{O}^*(c^n)$ time, for $c < 2$. With some simple arguments this approach can be adapted, so that only 2^n different vertex sets need to be considered. Let us introduce a scheme of dynamic programming of vertex sets of increasing cardinality that achieves this goal.

Let $pw(U, S)$ be defined as the minimum pathwidth of $G[U \cup S]$ where S is contained in the last bag. Recall that we consider a path as a permutation of nodes, hence the ordering. Let $\tilde{U} = V \setminus N[U]$ for a vertex set U. Consider now a vertex ordering α providing a path decomposition of minimum width, and let $V_j = \bigcup_{i=1}^{j} \alpha(i)$. Then $pw(V_1, N(V_1)) = |N[V_1]| - 1$ and $pw(V_i, N(V_i)) \leq max(pw(V_{i-1}, N(V_{i-1})), |(N(\tilde{V}_{i-1}) \cup N[v_i]) \setminus V_{i-1}| - 1)$. Finally, the pathwidth of G is $pw(V_n, \emptyset)$ - which gives us the minimum pathwidth where there are no restrictions on the last bag.

In the general case we do not know the optimal ordering, so for a vertex set U we simply set $pw(U, N(U))$ to be the minimum value obtained by testing every vertex of U as the last vertex added to the set. This is computable since the values for all smaller vertex sets are computed beforehand. One consequence of this is that the value for each of the 2^n vertex sets has to be computed. More formally, for every vertex set $U \subseteq V$ of increasing size we get that

$$pw(U, N(U)) = \min_{u \in U}(max(pw(U \setminus \{u\}, N(U \setminus \{u\})), |N[U] \setminus (U \setminus \{u\})| - 1)). \quad (1)$$

To break the 2^n barrier we need some more insight into the structure of path decompositions. This comes from the study of minimal interval completions - let us briefly introduce these.

A graph $H = (V, F)$ is an *interval graph* if each vertex of H can be assigned an interval on the real line, such that $\{u, v\} \in F$ if and only if the intervals of u and v intersect. An interval graph H is said to be an *interval completion* of $G = (V, E)$ if H is an interval graph and $E \subseteq F$. In the case where $H' = (V, F')$ is not an interval graph for any edge set F', $E \subseteq F' \subset F$, we say that H is a *minimal interval completion* of G. Any interval graph H can be represented by a structure called a *clique path* (χ, P), where χ is the set of maximal cliques of H, each of which is associated to a node of the path P and, for each vertex $u \in V$, the set of nodes associated to maximal cliques of H containing u induces a subpath of P. Notice that this is also a path decomposition of G. If H is a minimal interval completion of G, we call the path decomposition (χ, P) of G defined by the clique path of H a *minimal path decomposition* of G. Notice that each bag of a minimal path decomposition of G is a maximal clique of H.

As it is enough to consider the set of its *minimal triangulations* in order to compute the *treewidth* of a graph, it is also enough to consider the *minimal interval completions* if we want to compute the *pathwidth* of a graph. Let (χ, P) be a path decomposition of minimum width. By adding edges necessary to make each bag χ_i induce a clique, for each $\chi_i \in \chi$, we obtain an interval completion H of G. Now remove edges from H until a minimal interval completion H' is obtained. Clearly, the minimal path decomposition defined by a clique path of

H' will have the same width as (χ, P). Thus, it is enough to consider minimal path decompositions, and not all possible path decompositions.

4   Faster Computation of Pathwidth

Minimal path decompositions, as path cliques of minimal interval completions, have some useful properties that we state in the following proposition.

Proposition 1. *Let (χ, P) be a minimal path decomposition of G. Take the bags $\chi_1, \chi_2, \ldots, \chi_r$ in the order defined by P, and let H be the minimal interval completion defined by (χ, P). Then each minimal separator S of H is given by $\chi_j \cap \chi_{j+1}$, for some j (cf. [7,14]). Moreover, if $S = \chi_j \cap \chi_{j+1}$, then $S = N(\bigcup_{i=1}^{j} \chi_i \setminus S)$ and $S = N(\bigcup_{i=j+1}^{r} \chi_i \setminus S)$ (cf. [12]).*

For a vertex set $U \subset V$, we define $U^* = N[U] \setminus N(\tilde{U})$ as the *full set of U*, that is the set of vertices in $N[U]$ that do not have a neighbor in $V \setminus N[U]$. If $U = U^*$ we will refer to the set U as a *full set*.

Since it is sufficient to consider minimal path decompositions to compute pathwidth, and each prefix of a minimal path decomposition minus the adjacent separator corresponds to a full set $\bigcup_{i=1}^{j} \chi_i \setminus S$, it is sufficient to consider only vertex sets U, where $U = U^*$. But we need to be able to compute each full set, by means of polynomial time computation, directly from full sets of smaller cardinality - to avoid the bottleneck of checking $\Omega(2^n)$ subsets of V. One may notice the resemblance of this approach to the use of certain structures related to minimal triangulations to compute treewidth (cf. [9]).

Consider again the optimal vertex ordering α, and like before let $V_j = \bigcup_{i=1}^{j} \alpha(i)$, and in difference to before, let $U_j = V_j^*$. We now want to show that U_i can be obtained from U_{i-1} and $v_i = \alpha(i)$. Observe first by definition that $V_i \subseteq U_i$ and that $\tilde{V}_i = \tilde{U}_i$. If $v_i \notin N[\tilde{U}_{i-1}]$, then $v_i \in U_{i-1}$ and as a consequence $U_{i-1} = U_i$. Consider now the remaining case when $v_i \notin U_{i-1}$. Let $U = U_{i-1} \cup \{v_i\}$ and out goal is to show that $U^* = U_i$. First of all, we have that $V_i \subseteq U$, which means that $N[V_i] \subseteq N[U]$, and since every vertex in $U_{i-1} \setminus V_{i-1}$ only has neighbors in $N[V_i]$, then we get that $N[V_i] = N[U]$. Thus $V_i^* = U^*$. This means that the algorithm is exactly as before, but only runs over full sets and not over all 2^n vertex subsets.

For a full set U we say that a minimal interval completion H is an *extension* of U if $N(U)$ is a minimal separator, separating U from $\tilde{U}$ in H. In an analogous way, a minimal path decomposition (χ, P) of G is an extension of the full set U if there an integer j such that $N(U) = \chi_j \cap \chi_{j+1}$ and $U = \bigcup_{i=1}^{j} \chi_i \setminus N(U)$.

Observation 1. *For a full set U we have that $pw(U, N(U)) \geq |N(U)|$.*

Proof. Let (χ, P) be a minimal path decomposition of minimum width which is also an extension of the full set U. Let j be the integer such that $N(U) = \chi_j \cap \chi_{j+1}$ and $U = \bigcup_{i=1}^{j} \chi_i \setminus N(U)$. Since (χ, P) is a minimal path decomposition, it follows that $\chi_j \not\subseteq \chi_{j+1}$. Therefore, $|\chi_j| > |\chi_j \cap \chi_{j+1}| = |N(U)|$, and the claim statement follows. $\square$

Observation 2. *For a full set U, $\tilde{U}$ is also a full set.*

4.1   Bounding the Number of Base Sets

We will now give an algorithm that verifies if the pathwidth of a given graph G is at most k in $\mathcal{O}^*(c^n)$ time for $c < 2$. Part one of this algorithm consist of generating a set of vertex partitions that the algorithm can use as intermediate steps to verify the existence of a path decomposition of width k. We will now define a base $\mathcal{B}$ containing 2 types of pairs of vertex sets:

1. (U, S) where $|U| \leq 0.404n$, and $S = N(U)$,
2. (U, S) where $|S| \leq 0.192n$, $U \subset V \setminus S$, $N(U) \subseteq S$, and each connected component of $G[\tilde{U}]$ contains at least three vertices.

It is important to notice that we only want to take advantage of a pair (U, S) if S turns out to be a minimal separator in a minimal interval completion of minimum width. This makes it safe to contain S in one bag, and by Observation 1 it follows that the pathwidth is at least $|S|$, that is the largest bag contains at least $|S| + 1$ vertices.

Lemma 1. *The number of pairs of vertex sets in $\mathcal{B}$ is $\mathcal{O}(n \cdot 1.9657^n)$.*

Proof. The number of pairs of vertex sets of type 1 is at most the number of vertex sets of size $0.404n$, thus

$$\sum_{i=1}^{0.404n} \binom{n}{i} \leq n \cdot \binom{n}{0.404n} = \mathcal{O}(n \cdot 1.9633^n).$$

For type 2, the number is at most

$$\sum_{i=1}^{0.192n} \binom{n}{i} \cdot 2^{(n-i)/3} \leq n \cdot \binom{n}{0.192n} \cdot 2^{(n-0.192n)/3} = \mathcal{O}(n \cdot 1.9657^n). \qquad \square$$

Our goal now will be to compute the pathwidth of G in $\mathcal{O}^*(|\mathcal{B}|)$ time.

4.2   Extending from the Base Sets

The set of pairs contained in $\mathcal{B}$ at this point will be referred to as *base pairs*, and these will be used to generate at most $n^2|\mathcal{B}|$ new pairs that will be added to the set $\mathcal{F}$. For each pair $(U, S) \in \mathcal{B}$ of type 1, we can compute $pw(U, S)$ by dynamic programming on increasing size over the set of pairs of type 1. The procedure for this is given in Equation 1. Put these pairs as the initial content of the set $\mathcal{F}$. For all of these pairs, U is a full set. Notice that initially we cannot compute the pathwidth of a type 2 pair (U, S). The function $pw(U, S)$ will be computed if pair (U, S) is used as a base for a pair added to the set $\mathcal{F}$.

All remaining pairs added to $\mathcal{F}$ will be created from a pair already in $\mathcal{F}$, by one of three rules. Before describing these rules, the vertices are numbered in any order from 1 to n, where $\ell(v)$ refers to the number assigned to vertex v. The purpose of the numbering is to ensure that a rule creates a unique output from a given pair. Some of the pairs in $\mathcal{F}$ will also be assigned a mark.

Rule 1. *(Monotone Push Rule) Given a pair $(U, S) \in \mathcal{F}$, and a vertex $u \in S$ such that $|N(u) \cap \tilde{U}| = 1$ and $|N(v) \cap \tilde{U}| > 1$ for any vertex $v \in S$ where $\ell(v) < \ell(u)$. Let $W = (U \cup \{u\})^*$ and add the pair $(W, N(W))$ to the set $\mathcal{F}$, and set $pw(W, N(W)) = pw(U, S)$. Mark $(W, N(W))$ if pair (U, S) is marked, and $(W, N(W))$ is not a pair in $\mathcal{B}$.*

Lemma 2. *The Monotone Push Rule is safe for pathwidth.*

Proof. Let us assume that there exists an interval completion H and a minimal path decomposition $\chi_1, \chi_2, \ldots, \chi_r$ of width k, where $\bigcup_{i=1}^{j} \chi_i \setminus S = U$ and $\bigcup_{i=1}^{j} \chi_i \setminus U = S$. Since S is a minimal separator in H, then $|S| \leq k$. Let w be the unique neighbor of u in $\tilde{U}$, and let p be the smallest number such that $w \in \chi_p$. Vertex u is contained in S and thus also in χ_j, and in χ_t for $j \leq t \leq q$ where q is the highest number such that $u \in \chi_q$. Because of the edge uw we have that $p \leq q$. Construct a new path decomposition as follows: Remove u from any bag χ_t where $j < t$, and add w to any χ_t where $j < t < p$, and insert a bag between χ_j and χ_{j+1} containing $S \cup \{w\}$. Clearly this is a legal path decomposition of the same width. $\square$

Rule 2. *(Component Push Rule) Given a pair $(U, S) \in \mathcal{F}$ where Rule 1 can not be applied, and a vertex $u \in \tilde{U}$ such that $|S| + |C| \leq k + 1$ where C is the connected component of $G[V \setminus (U \cup S)]$ containing u, and for any vertex $v \in \tilde{U}$ where $\ell(v) < \ell(u)$ the connected component C_v of $G[V \setminus (U \cup S)]$ containing v we have that $|S| + |C_v| \geq k + 1$. Let $W = (U \cup C)^*$, and add the pair $(W, N(W))$ to $\mathcal{F}$ and set $pw(W, N(W)) = max(pw(U, S), |S| + |C| - 1)$. Mark $(W, N(W))$ if pair (U, S) is marked, and $(W, N(W))$ is not a pair in $\mathcal{B}$.*

Lemma 3. *The Component Push Rule is safe for pathwidth.*

Proof. The proof is similar as the one for Rule 1. Let us assume that there exists an interval completion H and a minimal path decomposition $\chi_1, \chi_2, \ldots, \chi_r$ of width k, where $\bigcup_{i=1}^{j} \chi_i \setminus S = U$ and $\bigcup_{i=1}^{j} \chi_i \setminus U = S$. Insert a bag between χ_j and χ_{j+1} containing $S \cup C$, and remove vertices of C from χ_t for each $t > j$. Clearly this is a legal path decomposition of width at most k. $\square$

For each pair in $\mathcal{F}$, at each step of its construction, at most one of Rules 1, 2 can be applied in only one possible way. Therefore these two rules can be applied recursively without producing more than $\mathcal{O}(n)$ new pairs. The final rule is a brute force search rule and applying it recursively would yield an exponential growth in the number of new pairs. Thus, we use the marks to apply it at most once in any sequence of rule applications starting from a base pair.

Rule 3. *(Push Rule) Given an unmarked pair $(U, S) \in \mathcal{F}$, where neither of Rules 1,2 applies, and a vertex $u \in S$ such that $max(pw(U, S), |N[U \cup \{u\}] \setminus U| - 1) \leq k$. Let $W = (U \cup \{u\})^*$ and add the marked pair $(W, N(W))$ to the set $\mathcal{F}$, and set $pw(W, N(W)) = max(pw(U, S), |N[U \cup \{u\}] \setminus U| - 1)$.*

Rule 3 is safe by Equation 1.

4.3   The Algorithm

It is now time to populate the set $\mathcal{F}$, and verify if the pathwidth is at most k. First step is to add the seeds, that is type 1 pairs from $\mathcal{B}$. Notice that $pw(U,S) = pw(U^*, N(U^*))$, so for every type 1 pair in $\mathcal{B}$, where $pw(U,S) \leq k$, add an unmarked full set $(U^*, N(U^*))$ to $\mathcal{F}$ and set $pw(U^*, N(U^*)) = pw(U,S)$. From this set of seeds, rules 1,2,3 are applied if possible. Eventually, when no rule can be further applied, the set $\mathcal{F}$ satisfies:

Lemma 4. $|\mathcal{F}| \leq n^2 |\mathcal{B}|$.

Proof. Let us assume that $\mathcal{B} \subseteq \mathcal{F}$. All elements of $\mathcal{F} \setminus \mathcal{B}$ are obtained by applying Rule 1, Rule 2, or Rule 3. Thus we can say that a pair $(U, S) \in \mathcal{F} \setminus \mathcal{B}$ is obtained from a base case (W, R) contained in $\mathcal{B}$. Let $\mathcal{F}(W, R)$ be the set of elements in $\mathcal{F} \setminus \mathcal{B}$ originating from the base case (W, R), and where none of the intermediate pairs is a base pair. This gives a partitioning of $\mathcal{F} \setminus \mathcal{B}$ into $|\mathcal{B}|$ sets. Let $\mathcal{F}(W, R)$ be a set of maximum cardinality, and notice that $|\mathcal{F}| \leq |\mathcal{F}(W, R)| \cdot |\mathcal{B}|$. We want to prove that $|\mathcal{F}(W, R)| \leq n^2$. First observation is that Rule 1 and Rule 2 are applied recursively from (W, R) until a pair (W', R') is obtained, where neither of the two rules applies. Since Rule 2 is only applied if Rule 1 does not apply, and the output for both rules is unique due to the labeling, there is a directed path from (W, R) to (W', R'), with no extra leaves. Since the size of the full set increases by at least one at each time a rule is applied, the length of the path is bounded by n. Next, by Rule 3, there might be one marked pair created for each vertex $x \in V \setminus W'$. Thus we obtain at most n marked pairs, each of which might be further developed by Rule 1 and Rule 2 into a path of at most n pairs. But notice that all these pairs are marked, so when Rules 1, 2 cannot be further applied, the development stops. By this construction we obtain a tree, where (W, R) is the root, and with at most $|V \setminus W'|$ leaves and where the distance from the leaves to the root is at most n. Thus, the number of vertices, and thus also pairs in $\mathcal{F}(W, R)$ is at most n^2. $\qquad\square$

Lemma 5. *Consider the set $\mathcal{F}$ computed for a value k, where none of the rules can be applied. Then the pathwidth is at most k if and only if there exists a full set U such that*

- $(U, N(U)) \in \mathcal{F}$,
- $(\tilde{U}, N(U)) \in \mathcal{F}$,
- $pw(U, N(U)) \leq k$, *and*
- $pw(\tilde{U}, N(U)) \leq k$.

Proof. Notice that $pw(U, N(U)) \leq k$ implies that there exists a path decomposition of $G[N[U]]$ of width at most k, where $N(U)$ is contained in the last bag. In the case where $pw(U, N(U)) \leq k$ and $pw(\tilde{U}, N(U)) \leq k$, it is possible to put the two path decompositions of respectively $G[N[U]]$ and $G[V \setminus U]$ next to each other and obtain a path decomposition for G of width k. We can see it as the one extends the other.

For the opposite direction, assume on the contrary that the pathwidth of G is k. Let α be a vertex ordering providing a path decomposition of width at most k. Let $V_j = \bigcup_{i=1}^{j} \alpha(i)$, and let $U_j = V_j^*$. Define $b(\alpha)$ to be the largest integer j such that $(U_i, N(U_i)) \in \mathcal{F}$ and $pw(U_i, N(U_i)) \leq k$ for every $i \in \{1, \dots, j\}$. Let now α be the vertex ordering with largest $b(\alpha)$ among all orderings providing a path decomposition of width at most k.

Let $j = b(\alpha)$, and consider the pair (U_j, S_j). If $|S_j| < k$, then $G[\tilde{U}_j]$ contains no connected component C where $|C| \leq 2$, since Rule 2 can be applied to these. By the safeness of Rule 2 it follows that there exists an ordering α' which provides a path decomposition of width at most k, where $b(\alpha) < b(\alpha')$. Notice that Rule 2 is applied if possible.

Since Rule 2 cannot be applied, then every connected component of $G[\tilde{U}_j]$ contains at least three vertices. Furthermore, (U_j, S_j) is not of type 1 in $\mathcal{B}$ since this would imply that (U_j, S_j) is unmarked and (U_{i+1}, S_{i+1}) would have been obtained by Rule 3.

As a result $0.404n \leq |U_i|$. By the conditions of the lemma $(\tilde{U}_j, S_j) \notin \mathcal{F}$, and since all pairs in $\mathcal{B}$ of type 1 are added to $\mathcal{F}$ we get that $0.404n \leq |\tilde{U}_j|$ and thus $|S_j| < 0.192n$. Now (U_j, S_j) is of type 2 in $\mathcal{B}$, since $|S_j| < 0.192n$ and $G[\tilde{U}_j]$ only contains connected components of size at least three, which is also a contradiction since (U_j, S_j) would be unmarked and (U_{j+1}, S_{j+1}) would be obtained by Rule 3.

So we know that $|S_j| = k$. Let x be the vertex such that $U_{j+1} = (U_j \cup \{x\})^*$. Observe that there are no connected components in $G(\tilde{U}_j)$ of size 1, since $|S_j| = k$ and Rule 2 can then be applied to these. This implies that x has at least one neighbor in $\tilde{U}_j$. By fixing the next vertex to be x we have also fixed the next bag to be $S_j \cup (N[x] \setminus U_j)$. This implies that $|N(x) \cap \tilde{U}_j| = 1$ and that $x \in S_j$, since otherwise $k + 1 < |S_j \cup (N[x] \setminus U_j)|$. Thus, this is a monotone push, and Rule 1 can be applied, something which is a contradiction to the assumption that α is the vertex ordering with the maximum value of $b(\alpha)$.   □

Theorem 3. *There exists an algorithm that verifies if the pathwidth of a graph G on n vertices is at most k in $\mathcal{O}^*(1.9657^n)$ time.*

Proof. By Lemma 1 the set $\mathcal{B}$ contains at most $\mathcal{O}^*(1.9657^n)$ pairs, and by Lemma 4, the same bound holds for $\mathcal{F}$. Finally by Lemma 5, $\mathcal{F}$ contains a pair (U, S) such that

- $(U, N(U)) \in \mathcal{F}$,
- $(\tilde{U}, N(U)) \in \mathcal{F}$,
- $pw(U, N(U)) \leq k$, and
- $pw(\tilde{U}, N(U)) \leq k$

if and only if the pathwidth of G is at most k.   □

5   Approximation of the Pathwidth

This section gives an algorithm that approximates the pathwidth up to $opt + \tau$ for some constant τ, in $\mathcal{O}^*(\binom{n}{tn})$ time, where $0 < t < 1$ is a constant such that

$\binom{n}{(1-2t)n} \cdot 2^{2tn/\tau} \leq \binom{n}{tn}$. For instance, setting τ to $2, 20, 200$, and 10.000 gives algorithms with running times $\mathcal{O}^*(1.9427^n), \mathcal{O}^*(1.9047^n), \mathcal{O}^*(1.8921^n), \mathcal{O}^*(1.88997^n)$, respectively. The approximation algorithm can be obtained by a slight adaptation of the exact algorithm for pathwidth. The base set $\mathcal{B}$ is obtained with two types of pairs as before.

1. (U, S) where $|U| \leq tn$, and $S = N(U)$,
2. (U, S) where $|S| \leq (1 - 2t)n$, $U \subset V \setminus S$, $N(U) \subseteq S$, and each connected component of $G[\tilde{U}]$ contains at least τ vertices.

The second modification is to replace Rule 2 by the rule given below.

Rule 4. *(Component push Rule) Given a pair $(U, S) \in \mathcal{F}$ where Rule 1 cannot be applied, and a vertex $u \in \tilde{U}$ such that $|S| + |C| \leq k + 1 + \tau$, where C is the connected component of $G[V \setminus (U \cup S)]$ containing u, and for any vertex $v \in \tilde{U}$, where $\ell(v) < \ell(u)$, the connected component C_v of $G[V \setminus (U \cup S)]$ containing v we have that $|S| + |C_v| \geq k + 1 + \tau$. Then let $W = U \cup C$, and add the pair (W, S) to $\mathcal{F}$ and set $pw(W, S) = max(pw(U, S), |S| + |C| - 1 - \tau)$. Mark (W, S) if pair (U, S) is marked, and $(W, N(W))$ is not a pair in $\mathcal{B}$.*

In an analogous way as before, we create a set $\mathcal{F}$ using type 1 in $\mathcal{B}$ as the seeds, and then apply Rules 1,4, and 3 if possible. Rule 4 behaves exactly as Rule 2 when it comes to generating new pairs for the set $\mathcal{F}$, thus by Lemma 4 $|\mathcal{F}| \leq n^2|\mathcal{B}|$.

Corollary 1. *Consider the set $\mathcal{F}$ computed for a value k, where no new information can be obtained by applying a rule. Then the pathwidth is at most $k + \tau$ if and only if there exists a full set U such that*

- $(U, N(U)) \in \mathcal{F}$,
- $(\tilde{U}, N(U)) \in \mathcal{F}$,
- $pw(U, N(U)) \leq k$, and
- $pw(\tilde{U}, N(U)) \leq k$.

Theorem 4. *There exists an approximation algorithm that verifies if the pathwidth of a graph G on n vertices is at most $k + \tau$ in $\mathcal{O}^*((\binom{n}{(1-2t)n}) \cdot 2^{2tn/\tau} + \binom{n}{tn}))$ time for some constant $0 \leq t \leq 1$.*

Proof. By Lemma 1 the set $\mathcal{B}$ contains at most $\mathcal{O}^*((\binom{n}{(1-2t)n}) \cdot 2^{2tn/\tau} + \binom{n}{tn}))$ pairs, and by Lemma 4 the same bound holds for $\mathcal{F}$. Finally by Corollary 1 $\mathcal{F}$ contains a pair (U, S) such that

- $(U, N(U)) \in \mathcal{F}$,
- $(\tilde{U}, N(U)) \in \mathcal{F}$,
- $pw(U, N(U)) \leq k$, and
- $pw(\tilde{U}, N(U)) \leq k$

if and only if the pathwidth of G is at most $k + \tau$. $\qquad\square$

6   Conclusion and Open Questions

It is well known that Held and Karp's dynamic programming algorithm [13] can be adapted to solve a long sequence of problems which are definable as a vertex ordering problem. Their approach gives running time $\mathcal{O}^*(2^n)$, and is based on doing dynamic programming over all the 2^n vertex subsets of the graph. Some examples of such problems are TREEWIDTH, PATHWIDTH, CUTWIDTH, DIRECTED OPTIMAL LINEAR ARRANGEMENT, and DIRECTED FEEDBACK ARC SET. Among these problems, treewidth is the only problem prior to this work where an $\mathcal{O}^*(c^n)$ algorithm for $c < 2$ was known. It is an open question if other problems that are expressible as a vertex ordering problems and can be solved by Held and Karp's approach, have an algorithm of running time $\mathcal{O}^*(c^n)$ for $c < 2$.

One of the techniques used to compute the pathwidth is to guess a balanced separator in the cases where the rules do not apply. In [5] the approach of guessing the middle cut or bag was used to give an $\mathcal{O}^*(2.9512^n)$ polynomial space algorithm for computing the treewidth of a graph. It is not clear how or if it is possible to extend this approach to other problems in the list above. Guessing a balanced separator has a natural limitation in $\binom{n}{n/3}$, which is the lower bound obtained for the additive approximation algorithm. Finally the possible improvement of the running time for computing pathwidth exact is left as an open question, and especially if the bound $\binom{n}{n/3}$ can be obtained. Improvement on the running time would probably require some new ideas, since both the Rule's 1 and 2 rely on arriving at a very similar situation. For instance if Rule 1 was extended to have two neighbors instead of one, then the order of applying the rule would make a difference.

References

1. Arnborg, S., Corneil, D.G., Proskurowski, A.: Complexity of finding embeddings in a k-tree. SIAM J. Algebraic Discrete Methods 8, 277–284 (1987)
2. Bodlaender, H.L.: A tourist guide through treewidth. Acta Cybern. 11, 1–22 (1993)
3. Bodlaender, H.L.: A linear-time algorithm for finding tree-decompositions of small treewidth. SIAM J. Comput. 25, 1305–1317 (1996)
4. Bodlaender, H.L., Fomin, F.V., Koster, A.M.C.A., Kratsch, D., Thilikos, D.M.: On exact algorithms for treewidth, Tech. Rep. UU-CS-2006-032, Department of Information and Computing Sciences, Utrecht University (2006)
5. Bodlaender, H.L., Fomin, F.V., Koster, A.M.C.A., Kratsch, D., Thilikos, D.M.: On exact algorithms for treewidth. In: Azar, Y., Erlebach, T. (eds.) ESA 2006. LNCS, vol. 4168, pp. 672–683. Springer, Heidelberg (2006)
6. Bodlaender, H.L., Kloks, T., Kratsch, D.: Treewidth and pathwidth of permutation graphs. SIAM J. Discrete Math. 8, 606–616 (1995)
7. Buneman, P.: A characterization of rigid circuit graphs. Discrete Math. 9, 205–212 (1974)
8. Feige, U., Hajiaghayi, M., Lee, J.R.: Improved approximation algorithms for minimum weight vertex separators. SIAM J. Comput. 38, 629–657 (2008)

9. Fomin, F.V., Kratsch, D., Todinca, I., Villanger, Y.: Exact algorithms for treewidth and minimum fill-in. SIAM J. Comput. 38, 1058–1079 (2008)
10. Fomin, F.V., Villanger, Y.: Treewidth computation and extremal combinatorics. In: Aceto, L., Damgård, I., Goldberg, L.A., Halldórsson, M.M., Ingólfsdóttir, A., Walukiewicz, I. (eds.) ICALP 2008, Part I. LNCS, vol. 5125, pp. 210–221. Springer, Heidelberg (2008)
11. Habib, M., Mhring, R.H.: Treewidth of cocomparability graphs and a new order-theoretic parameter. Order 11, 47–60 (1994)
12. Heggernes, P., Suchan, K., Todinca, I., Villanger, Y.: Characterizing minimal interval completions. In: Thomas, W., Weil, P. (eds.) STACS 2007. LNCS, vol. 4393, pp. 236–247. Springer, Heidelberg (2007)
13. Held, M., Karp, R.M.: A dynamic programming approach to sequencing problems. In: Proceedings of the 1961, 16th ACM national meeting, pp. 71.201–71.204. ACM, New York (1961)
14. Ho, C.W., Lee, R.C.T.: Counting clique trees and computing perfect elimination schemes in parallel. Information Processing Letters 31, 61–68 (1989)
15. Kloks, T.: Treewidth of circle graphs. Int. J. Found. Comput. Sci. 7, 111–120 (1996)
16. Mihai, R., Todinca, I.: Pathwidth is np-hard for weighted trees. In: Faw, X., Deng, J.E. (eds.). LNCS, vol. 5598, pp. 181–195. Springer, Heidelberg (2009)
17. Robertson, N., Seymour, P.D.: Graph minors. II. Algorithmic aspects of tree-width. Journal of Algorithms 7, 309–322 (1986)
18. Suchan, K., Todinca, I.: Pathwidth of circular-arc graphs. In: Brandstädt, A., Kratsch, D., Müller, H. (eds.) WG 2007. LNCS, vol. 4769, pp. 258–269. Springer, Heidelberg (2007)

Author Index